# Mechanisms and Machine Science 135

This book series establishes a well-defined forum for monographs, edited Books, and proceedings on mechanical engineering with particular emphasis on MMS (Mechanism and Machine Science). The final goal is the publication of research that shows the development of mechanical engineering and particularly MMS in all technical aspects, even in very recent assessments. Published works share an approach by which technical details and formulation are discussed, and discuss modern formalisms with the aim to circulate research and technical achievements for use in professional, research, academic, and teaching activities.

This technical approach is an essential characteristic of the series. By discussing technical details and formulations in terms of modern formalisms, the possibility is created not only to show technical developments but also to explain achievements for technical teaching and research activity today and for the future.

The book series is intended to collect technical views on developments of the broad field of MMS in a unique frame that can be seen in its totality as an Encyclopaedia of MMS but with the additional purpose of archiving and teaching MMS achievements. Therefore, the book series will be of use not only for researchers and teachers in Mechanical Engineering but also for professionals and students for their formation and future work.

The series is promoted under the auspices of International Federation for the Promotion of Mechanism and Machine Science (IFToMM).

Prospective authors and editors can contact Mr. Pierpaolo Riva (publishing editor, Springer) at: pierpaolo.riva@springer.com

Indexed by SCOPUS and Google Scholar.

conference. Their invaluable contributions ensured the success of this event, and we deeply grateful for their dedication and commitment.

Finally, we would like to express our gratitude to all the participants who attended conference and contributed to the vibrant and stimulating atmosphere. Your presence l engagement enriched the conference experience, and we hope that this book will ve as a lasting testament to the quality and impact of the research presented at RAAD 23.

ıe 2023

Tadej Petrič
Leon Žlajpah
Aleš Ude

Tadej Petrič · Aleš Ude · Leon Žlajpah
Editors

# Advances in Service and Industrial Robotics

## RAAD 2023

Springer

*Editors*
Tadej Petrič
Department of Automatics, Biocybernetics, and Robotics
Jožef Stefan Institute
Ljubljana, Slovenia

Aleš Ude
Department of Automatics, Biocybernetics, and Robotics
Jožef Stefan Institute
Ljubljana, Slovenia

Leon Žlajpah
Department of Automatics, Biocybernetics, and Robotics
Jožef Stefan Institute
Ljubljana, Slovenia

ISSN 2211-0984 ISSN 2211-0992 (electronic)
Mechanisms and Machine Science
ISBN 978-3-031-32605-9 ISBN 978-3-031-32606-6 (eBook)
https://doi.org/10.1007/978-3-031-32606-6



This Springer imprint is published by the registered company Springer Nature Switzerland AG
The registered company address is: Gewerbestrasse 11, 6330 Cham, Switzerland

# Preface

This book is a collection of 57 papers selected from the 32nd Interna on Robotics in Alpe-Adria-Danube Region (RAAD 2023), which wa Balance Hotel in Bled, Slovenia, on June 14–16, 2023. The confe platform for robotics researchers from 16 countries, including thos Adria-Danube Region and beyond, to exchange ideas and present th findings. The papers were chosen after a rigorous review process and c of topics and ongoing trends in robotics research. The book is orga tracks, corresponding to the various chapters of the proceedings:

- Collaborative robotics
- Innovative robots design
- Kinematics and dynamics
- Medical robotics
- Mobile robotics and localization
- Motion planning and control
- Perception and learning
- Service robotics.

The RAAD conference is known for its open and collegial atmosphe discussion of current trends and exchange of recent results in robotics r tion to the technical sessions, plenary talks by renowned experts are a vi the conference. We are honored and grateful to have had the following share their insights on current topics:

- Prof. Dongheui Lee, Faculty of Electrical Engineering and Informat TU Wien
- Dr. Sylvain Calinon, Idiap Research Institute
- Prof. Marko Munih

Organizing successful scientific events requires the dedication of aut ers, as well as the generous support of our sponsors. We extend our he to the International Federation for the Promotion of Mechanism and M (IFToMM) for granting us the opportunity to organize this conference u tutional umbrella. We are especially appreciative of the financial supp the following companies:

- OPL industrijska avtomatizacija d.o.o.
- ROBOTEH d.o.o.
- Yaskawa Slovenia d.o.o.
- RLS Merilna tehnika d.o.o.

We would like to express our special thanks and appreciation to the tional Scientific Committee for their unwavering support and tireless effo



the
are

the
an
se
20

Ju

# Organization

## General Chair

| | |
|---|---|
| Petrič, Tadej | Jožef Stefan Institute, Slovenia |

## Co-chairs

| | |
|---|---|
| Ude, Aleš | Jožef Stefan Institute, Slovenia |
| Žlajpah, Leon | Jožef Stefan Institute, Slovenia |

## International Scientific Committee (RAAD ISC)

| | |
|---|---|
| Aspragathos, Nikos A. | University of Patras, Greece |
| Berns, Karsten | University of Kaiserslautern, Germany |
| Borangiu, Theodor | Polytech. University of Bucharest, Romania |
| Brandstötter, Mathias | Joanneum Research Robotics, Austria |
| Budinská, Ivana | Slovak Academy of Sciences, Slovakia |
| Carbone, Giuseppe | University of Calabria, Italy |
| Dobrovodský, Karol | Slovak Academy of Sciences, Slovakia |
| Ferraresi, Carlo | Polytechnic of Turin, Italy |
| Gasparetto, Alessandro | University of Udine, Italy |
| Ivanescu, Nick Andrei | Polytech. University of Bucharest, Romania |
| Jovanović, Kosta | University of Belgrade, Serbia |
| Kamnik, Roman | University of Ljubljana, Slovenia |
| Kronreif, Gernot | ACMIT, Austria |
| Laribi, Med Amine | University of Poitiers, France |
| Mueller, Andreas | Johannes Kepler University, Austria |
| Petrič, Tadej | Jožef Stefan Institute, Slovenia |
| Petrović, Ivan | University of Zagreb, Croatia |
| Pîsla, Doina | Technical University of Cluj-Napoca, Romania |
| Rodić, Aleksandar | Mihajlo Pupin Institute, Serbia |
| Tar, József K. | Óbuda University, Hungary |
| Zeghloul, Said | University of Poitiers, France |
| Žlajpah, Leon | Jožef Stefan Institute, Slovenia |

## Local Arrangement Chairs

| | |
|---|---|
| Brecelj, Tilen | Jožef Stefan Institute, Slovenia |
| Deniša, Miha | Jožef Stefan Institute, Slovenia |
| Gams, Andrej | Jožef Stefan Institute, Slovenia |
| Kovač, Igor | Jožef Stefan Institute, Slovenia |
| Nemec, Bojan | Jožef Stefan Institute, Slovenia |

# Contents

## Mobile Robotics and Localisation



## Medical Robotics

# Perception and Learning

# A Lightweight Method for Detecting Dynamic Target Occlusions by the Robot Body

Savvas Sampaziotis, Sotiris Antonakoudis, Marios Kiatos, Fotios Dimeas(✉), and Zoe Dougleri

Aristotle University of Thessaloniki, Thessaloniki, Greece
{sampazio,antonakos,dimeasf,doulgeri}@ece.auth.gr

**Abstract.** Autonomous robots face significant challenges in accurately perceiving their environment due to occlusions. The robot itself may hide targets of interest from the camera while moving within the field of view, which can lead to safety hazards or failures in task execution. For example, if a target of interest is partially occluded by the robot, detecting and grasping it correctly, becomes very challenging. To solve this problem, we propose a novel computationally lightweight method to determine the areas that the robot occludes. Our approach uses the Unified Robot Description Format (URDF) to generate a virtual depth image of the 3D robot model. With the virtual depth image we can effectively determine the partially occluded areas to improve the robustness of the information given by the perception system. Due to the real-time capabilities of the method, it can successfully detect occlusions of moving targets by the moving robot. We validate the effectiveness of the method in an experimental setup using a 6-DoF robot arm and an RGB-D camera by detecting and handling occlusions for two tasks: Pose estimation of a moving object for pickup and human tracking for robot handover. The code is available in https://github.com/auth-arl/virtual_depth_image.

**Keywords:** occlusion detection · URDF · virtual depth image

## 1 Introduction

Robotic systems that need to operate in unstructured environments and to present versatility in their operation, such as for object pickup, grasping in cluttered environments and avoiding obstacles, require machine vision for enhanced perception capabilities. This is achieved with cameras that cover the robot's workspace and provide the necessary information that is critical to the robot's operation. However, occlusions can become a serious problem. These are blind spots of targets of interest in the camera view. A target may be an object that the robot needs to manipulate (e.g., for pick & place) or a human in the workspace that the robot needs to interact with. Such blind spots can have a negative impact on the performance of the perception methods. An error in perception


T. Petrič et al. (Eds.): RAAD 2023, MMS 135, pp. 3–11, 2023.
https://doi.org/10.1007/978-3-031-32606-6_1

can have a serious impact on the robot's performance, preventing it from completing its task successfully and risking damage from a possible misbehavior.

Perception systems use cameras in various configurations to reduce occlusions, depending on the application. For example, most object detection systems for pickup use overhead cameras [12]. Human tracking systems use cameras mounted at an angle rather than overhead to keep the human body in front view and cover a larger field of view towards the collaborative workspace [8]. One way to reduce occlusions is to use multiple cameras with different viewing angles [9]. However, depth-sensing cameras with structured light or lidar technologies can be quite expensive and interfere with each other. In addition, image processing of additional cameras multiplies computational costs. A common method to prevent occlusion is a camera-in-hand configuration. However, this is not always practical, for example, when the entire working area must be covered or the measurement distance is less than the minimum allowable distance of the depth sensor. Moreover, the moving camera causes motion blur.

Occlusion detection has been studied in the literature to improve the performance of perception systems for object tracking. In [12], a simple approach is proposed to detect occluded areas in the sensing data (e.g., point cloud) using a predefined height threshold. When the robot arm is above the target objects that are laid on a table, any point above the threshold is classified as occluded. This approach has significant limitations in the operation of the robot and in the flexibility of the setup, since the classification is based on a division surface. The authors in [7] address occlusions of target objects from the robot by training a deep neural network with a dataset containing images of occluded objects by the robot arm. However, this method requires a large amount of data to be trained and the inference time can be considerably high.

Detecting occlusion specifically for human tracking aims to improve the safety aspect of human-robot collaboration in shared workspaces. The authors in [5] propose an occlusion handling method for a mobile robot that follows humans using color and disparity information from a stereo camera. However, it depends heavily on the discrete depth information capability to detect occlusions, which may affect the accuracy in close proximity to the target from the occlusion. Shu et al. [10] proposed a method for detecting occlusions of human body parts in skeleton tracking. Although this method can obtain regional features, it scans the scene where the features are extracted, which has high computational cost.

Information about the structure of the robot, such as the URDF model, has been used in the literature to improve perception. URDF can be used to model robot arms using a hierarchical structure with XML trees and is well supported in Robotics Operating System (ROS). It can contain information about the kinematics, link inertia values, visual representation, and even the collision model of a robot, which is usually a slightly enlarged simplified body of each link. URDF filtering [3] was used in [8] along with an RGB-D camera to avoid ambiguities in tracking the human skeleton when the hands get very close to the robot. However, URDF filtering does not distinguish whether the targets are actually occluded or whether they are between the line of sight from

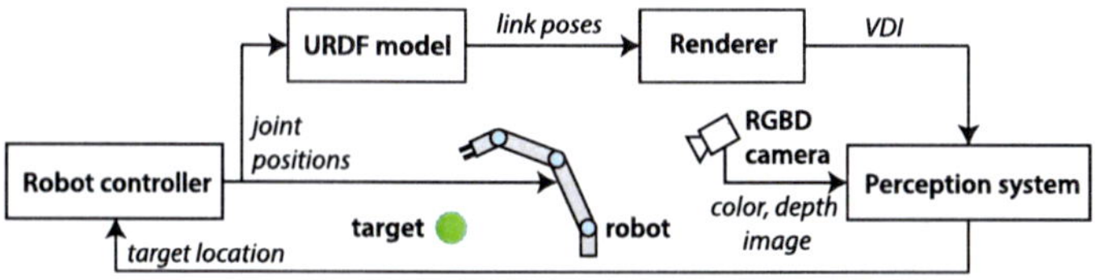


**Fig. 1.** Block diagram of the architecture that utilizes the virtual depth image (VDI).

the camera to the robot, because the area of the robot model is subtracted from the depth image, causing loss of useful depth information.

This work proposes a computationally lightweight method for determining the areas that the robot occludes from the camera. The contribution compared to methods that subtract the robot model from the depth image, is in the generation of a virtual depth image from the URDF model with minimal computational resources and the fusion of this virtual depth image with the depth image captured by the camera sensor to determine the area that the robot actually occludes. Furthermore, our method runs in real-time with a single RGB-D camera and is independent of the application or perception system. In the experiments, we demonstrate the proposed occlusion detection and how occlusions can be handled in two different use cases.

## 2 Methodology

Consider a robot receiving input from a perception system that detects and tracks targets of interest related to its task. The perception system uses an RGB-D camera with the robot being either fully or partially in its field of view. During task execution the robot may partially or fully occlude the targets, which may affect the target detection accuracy of the perception system.

Having the URDF model and the transformation of the camera's coordinate frame from the robot base, we use the joint position measurements of the robot to render the virtual depth image (VDI). The VDI is a noise-free depth image of the robot from the camera's point of view, from which a perception system can infer whether the target is partially or fully occluded, improving the robustness of the perception system. The block diagram in Fig. 1 shows a general architecture that uses the VDI, indicating that it is independent of the perception system.

### 2.1 Rendering the Virtual Depth Image

The first step of the method is to render depth images of the robot from the camera's point of view. We use a virtual camera with a pinhole camera model, where the intrinsic and extrinsic parameters correspond to the known parameters of the actual camera. The image distortion of the actual camera is compensated to match the pinhole camera model. To render the image, the 3D model is taken from the URDF and the joint positions of the actual robot. The 3D model

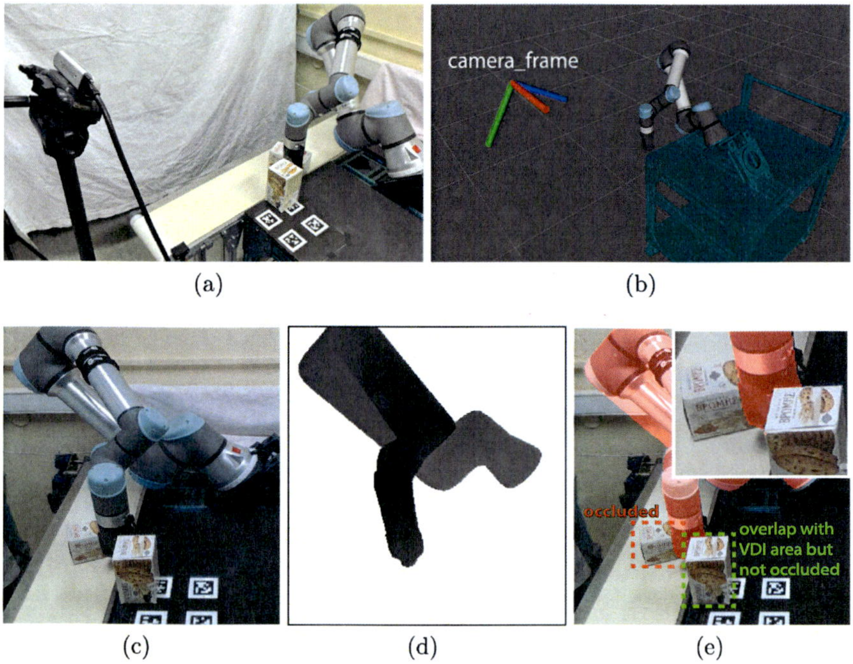


**Fig. 2.** (a) Setup with a robot, an RGB-D camera and two objects (one in front and one behind the robot), (b) the digital visualization, (c) the color view from the camera, (d) the rendered VDI including the end-effector, (e) the augmented view highlighting the areas occluded by the robot.

need not be an exact representation of the robot, as fine details will have a negligible effect on the final image for occlusions. For this purpose, we use the collision model, which is a convex approximation of each robot link in Standard Triangle Language (STL). The geometry of each link is represented by a mesh of triangles defined by a set of vertices. The use of the collision model is a conservative approach for detecting occlusions, because it increases the area marked as occluded in the VDI, but on the other hand it can compensate for small calibration errors of the camera.

For the implementation of the renderer we used the OpenGL library. This library allows us to build a simple graphics pipeline that takes advantage of a computer's graphics processor and achieve a high frame rate while using minimal CPU resources. The frame rate of the virtual camera must match that of the real camera to produce a matching image pair. The renderer first obtains the joint positions of the robot, then transforms the pose of each robot link from the world frame to the camera frame and finally renders all 3D models to generate each image. The output is a noise-free virtual depth image of the robot and any other geometry in the URDF we choose to render, as shown in Fig. 2.

The implemented graphics pipeline is described by the MVP (Model-View-Projection) homogeneous transformation matrix $\mathbf{g}_{MVP} \in \mathrm{R}^{4\times4}$, which is applied on all vertices of each $i^{th}$ robot link to recreate it from the camera view:

$$\mathbf{g}_{MVP} = \mathbf{g}_P \mathbf{g}_V \mathbf{g}_M^i. \tag{1}$$

The model matrix $\mathbf{g}_M^i$ is unique for each 3D object rendered in the scene. It is defined by each robot link's pose in respect to world frame and is derived from the robot's forward kinematic model and joint positions. The view matrix $\mathbf{g}_V$ is the homogeneous transformation from the world frame to the camera frame, which can be obtained from a camera registration procedure. The camera projection matrix $\mathbf{g}_P$ transforms the vertices into the camera space. The matrix parameters are defined by the actual RGB-D camera intrinsics:

$$\mathbf{g}_P = \begin{bmatrix} \frac{f_x}{c_x} & 0 & 0 & 0 \\ 0 & \frac{f_y}{c_y} & 0 & 0 \\ 0 & 0 & -\frac{f+n}{f-n} & -\frac{2fn}{f-n} \\ 0 & 0 & -1 & 0 \end{bmatrix}, \tag{2}$$

where $f_x, f_y$ and $c_x, c_y$ are the focal lengths and principal point of the pinhole model of the actual depth camera. The $f$ and $n$ parameters are called the "far" and "near" plane and refer to the maximum and minimum depth that will be rendered. Their values are chosen to match the actual depth camera range.

### 2.2 Occlusion Detection and Handling

Let $p_{ij}$ be a pixel at an area of interest on the color image of the RGB-D camera with coordinates $i, j$. By comparing the actual and virtual depth measurements $d_{ij}$ and $d_{ij}^V$ respectively, we can deduce whether this pixel is occluded by the robot and whether it can be safely de-projected to 3D space using the depth measurements, as shown in Fig. 3. Specifically, if $d_{ij} < d_{ij}^V - \epsilon$ then the point of interest is between the camera and the robot and no occlusion occurs. $\epsilon$ is a small positive value that compensates for the depth measurement noise and calibration errors. On the contrary, if $d_{ij} \geq d_{ij}^V - \epsilon$ then the depth measurement is derived from the robot's body and the point of interest is occluded.

The way to handle a detected occlusion depends on the method of the perception system. Different methods for target detection have different robustness to the occlusion percentage of the target [6]. In pose estimation of a moving object, for example, when the occlusion percentage of the object is higher than the acceptable occlusion threshold, future poses of the object can be predicted based on the current non-occluded estimates of pose and velocity [1].

## 3 Experimental Evaluation

To evaluate the proposed method, we conducted two experiments using a UR5e 6-DOF robot arm and an Intel Realsense D415 RGB-D camera. In the first

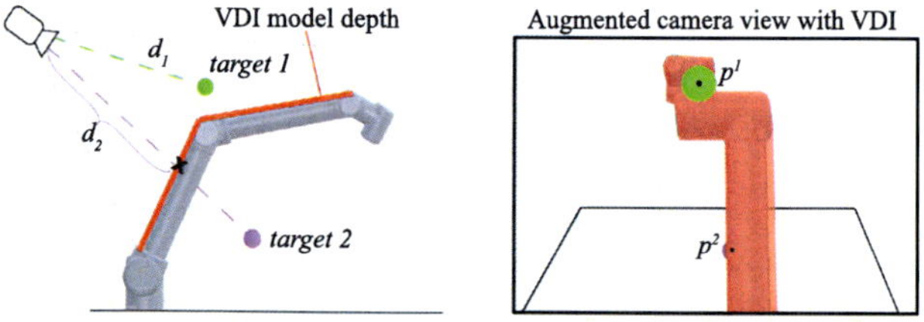


**Fig. 3.** Two targets that lie within the VDI area. Without the occlusion mask, de-projecting the partially occluded target 2, results in wrong distance $d_2$. Target 1 is correctly detected and measured despite being within the VDI area.

experiment, we demonstrate how we can handle occlusions that affect the pose estimation of a moving object for robot pickup. In the second one, we show a successful handover task in which the detected human is occluded by the robot during task execution. Our method is implemented with ROS2 on a Ryzen 3700x CPU and a GTX1060 GPU. The virtual depth image of the robot arm is rendered in real-time at 30 fps rate, equal to the camera frame rate. The average calculation time of each VDI frame is less than 10 ms. The values of the parameters in the experiments are $n = 0.5$ m, $f = 3$ m, $\epsilon = 0.01$ m.

### 3.1 Pose Estimation of a Moving Target Object for Pickup

Robot pickup of objects moving on a conveyor belt requires a perception system that estimates the pose of the target object. In this experiment, we use a perception system based on Mask-RCNN network [4], fine-tuned on synthetic images, to detect the mask of the target object and a subsequent feature matching method to estimate the 6-DOF pose of the target based on SIFT features [11]. This two-stage approach is very common for estimating the pose of an object [13]. However, during task execution, the robot may partially occlude the moving object, leading to wrong estimates of its pose.

To deal with this problem, the proposed VDI is used with an object tracking method that predicts the pose of the object when the occlusion threshold is higher than 5%, based on the last estimates of the pose and the velocity of the object before it got occluded. When the percentage drops below that threshold, the tracker uses again the estimates from the pose detection method as described above. With this approach, the object can be accurately tracked despite the wrong pose estimates during partial occlusions, as shown in of Fig. 4.

### 3.2 Human Pose Estimation for Robot Handover

The proposed method is also applied to a handover task. In particular, the robot should give an item to the left hand of a human as shown in Fig. 5. To achieve that a perception system detects the wrist of the human as the target location for the handover. The AlphaPose algorithm [2] is used to detect the human skeleton 2D keypoints on the RGB image. Using the depth image, we de-project the

Tadej Petrič · Aleš Ude · Leon Žlajpah
Editors

# Advances in Service and Industrial Robotics

RAAD 2023

Springer

*Editors*
Tadej Petrič
Department of Automatics, Biocybernetics, and Robotics
Jožef Stefan Institute
Ljubljana, Slovenia

Aleš Ude
Department of Automatics, Biocybernetics, and Robotics
Jožef Stefan Institute
Ljubljana, Slovenia

Leon Žlajpah
Department of Automatics, Biocybernetics, and Robotics
Jožef Stefan Institute
Ljubljana, Slovenia

ISSN 2211-0984 ISSN 2211-0992 (electronic)
Mechanisms and Machine Science
ISBN 978-3-031-32605-9 ISBN 978-3-031-32606-6 (eBook)
https://doi.org/10.1007/978-3-031-32606-6



This Springer imprint is published by the registered company Springer Nature Switzerland AG
The registered company address is: Gewerbestrasse 11, 6330 Cham, Switzerland

# Preface

This book is a collection of 57 papers selected from the 32nd International Conference on Robotics in Alpe-Adria-Danube Region (RAAD 2023), which was held at the Rikli Balance Hotel in Bled, Slovenia, on June 14–16, 2023. The conference served as a platform for robotics researchers from 16 countries, including those from the Alpe-Adria-Danube Region and beyond, to exchange ideas and present their latest research findings. The papers were chosen after a rigorous review process and cover a broad range of topics and ongoing trends in robotics research. The book is organized into several tracks, corresponding to the various chapters of the proceedings:

- Collaborative robotics
- Innovative robots design
- Kinematics and dynamics
- Medical robotics
- Mobile robotics and localization
- Motion planning and control
- Perception and learning
- Service robotics.

The RAAD conference is known for its open and collegial atmosphere, which fosters discussion of current trends and exchange of recent results in robotics research. In addition to the technical sessions, plenary talks by renowned experts are a vital component of the conference. We are honored and grateful to have had the following plenary speakers share their insights on current topics:

- Prof. Dongheui Lee, Faculty of Electrical Engineering and Information Technology, TU Wien
- Dr. Sylvain Calinon, Idiap Research Institute
- Prof. Marko Munih

Organizing successful scientific events requires the dedication of authors and reviewers, as well as the generous support of our sponsors. We extend our heartfelt gratitude to the International Federation for the Promotion of Mechanism and Machine Science (IFToMM) for granting us the opportunity to organize this conference under their institutional umbrella. We are especially appreciative of the financial support provided by the following companies:

- OPL industrijska avtomatizacija d.o.o.
- ROBOTEH d.o.o.
- Yaskawa Slovenia d.o.o.
- RLS Merilna tehnika d.o.o.

We would like to express our special thanks and appreciation to the RAAD International Scientific Committee for their unwavering support and tireless efforts in promoting the conference. Their invaluable contributions ensured the success of this event, and we are deeply grateful for their dedication and commitment.

Finally, we would like to express our gratitude to all the participants who attended the conference and contributed to the vibrant and stimulating atmosphere. Your presence and engagement enriched the conference experience, and we hope that this book will serve as a lasting testament to the quality and impact of the research presented at RAAD 2023.

June 2023

Tadej Petrič
Leon Žlajpah
Aleš Ude

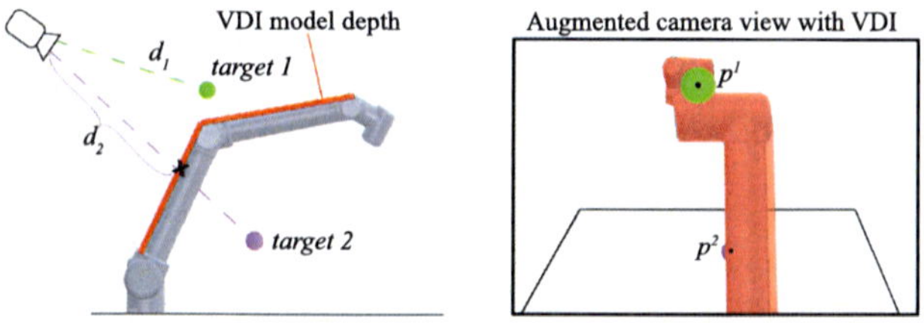


**Fig. 3.** Two targets that lie within the VDI area. Without the occlusion mask, de-projecting the partially occluded target 2, results in wrong distance $d_2$. Target 1 is correctly detected and measured despite being within the VDI area.

experiment, we demonstrate how we can handle occlusions that affect the pose estimation of a moving object for robot pickup. In the second one, we show a successful handover task in which the detected human is occluded by the robot during task execution. Our method is implemented with ROS2 on a Ryzen 3700x CPU and a GTX1060 GPU. The virtual depth image of the robot arm is rendered in real-time at 30 fps rate, equal to the camera frame rate. The average calculation time of each VDI frame is less than 10 ms. The values of the parameters in the experiments are $n = 0.5$ m, $f = 3$ m, $\epsilon = 0.01$ m.

### 3.1 Pose Estimation of a Moving Target Object for Pickup

Robot pickup of objects moving on a conveyor belt requires a perception system that estimates the pose of the target object. In this experiment, we use a perception system based on Mask-RCNN network [4], fine-tuned on synthetic images, to detect the mask of the target object and a subsequent feature matching method to estimate the 6-DOF pose of the target based on SIFT features [11]. This two-stage approach is very common for estimating the pose of an object [13]. However, during task execution, the robot may partially occlude the moving object, leading to wrong estimates of its pose.

To deal with this problem, the proposed VDI is used with an object tracking method that predicts the pose of the object when the occlusion threshold is higher than 5%, based on the last estimates of the pose and the velocity of the object before it got occluded. When the percentage drops below that threshold, the tracker uses again the estimates from the pose detection method as described above. With this approach, the object can be accurately tracked despite the wrong pose estimates during partial occlusions, as shown in of Fig. 4.

### 3.2 Human Pose Estimation for Robot Handover

The proposed method is also applied to a handover task. In particular, the robot should give an item to the left hand of a human as shown in Fig. 5. To achieve that a perception system detects the wrist of the human as the target location for the handover. The AlphaPose algorithm [2] is used to detect the human skeleton 2D keypoints on the RGB image. Using the depth image, we de-project the

The implemented graphics pipeline is described by the MVP (Model-View-Projection) homogeneous transformation matrix $\mathbf{g}_{MVP} \in \mathrm{R}^{4\times4}$, which is applied on all vertices of each $i^{th}$ robot link to recreate it from the camera view:

$$\mathbf{g}_{MVP} = \mathbf{g}_P \mathbf{g}_V \mathbf{g}_M^i. \tag{1}$$

The model matrix $\mathbf{g}_M^i$ is unique for each 3D object rendered in the scene. It is defined by each robot link's pose in respect to world frame and is derived from the robot's forward kinematic model and joint positions. The view matrix $\mathbf{g}_V$ is the homogeneous transformation from the world frame to the camera frame, which can be obtained from a camera registration procedure. The camera projection matrix $\mathbf{g}_P$ transforms the vertices into the camera space. The matrix parameters are defined by the actual RGB-D camera intrinsics:

$$\mathbf{g}_P = \begin{bmatrix} \frac{f_x}{c_x} & 0 & 0 & 0 \\ 0 & \frac{f_y}{c_y} & 0 & 0 \\ 0 & 0 & -\frac{f+n}{f-n} & -\frac{2fn}{f-n} \\ 0 & 0 & -1 & 0 \end{bmatrix}, \tag{2}$$

where $f_x, f_y$ and $c_x, c_y$ are the focal lengths and principal point of the pinhole model of the actual depth camera. The $f$ and $n$ parameters are called the "far" and "near" plane and refer to the maximum and minimum depth that will be rendered. Their values are chosen to match the actual depth camera range.

### 2.2 Occlusion Detection and Handling

Let $p_{ij}$ be a pixel at an area of interest on the color image of the RGB-D camera with coordinates $i, j$. By comparing the actual and virtual depth measurements $d_{ij}$ and $d_{ij}^V$ respectively, we can deduce whether this pixel is occluded by the robot and whether it can be safely de-projected to 3D space using the depth measurements, as shown in Fig. 3. Specifically, if $d_{ij} < d_{ij}^V - \epsilon$ then the point of interest is between the camera and the robot and no occlusion occurs. $\epsilon$ is a small positive value that compensates for the depth measurement noise and calibration errors. On the contrary, if $d_{ij} \geq d_{ij}^V - \epsilon$ then the depth measurement is derived from the robot's body and the point of interest is occluded.

The way to handle a detected occlusion depends on the method of the perception system. Different methods for target detection have different robustness to the occlusion percentage of the target [6]. In pose estimation of a moving object, for example, when the occlusion percentage of the object is higher than the acceptable occlusion threshold, future poses of the object can be predicted based on the current non-occluded estimates of pose and velocity [1].

## 3 Experimental Evaluation

To evaluate the proposed method, we conducted two experiments using a UR5e 6-DOF robot arm and an Intel Realsense D415 RGB-D camera. In the first

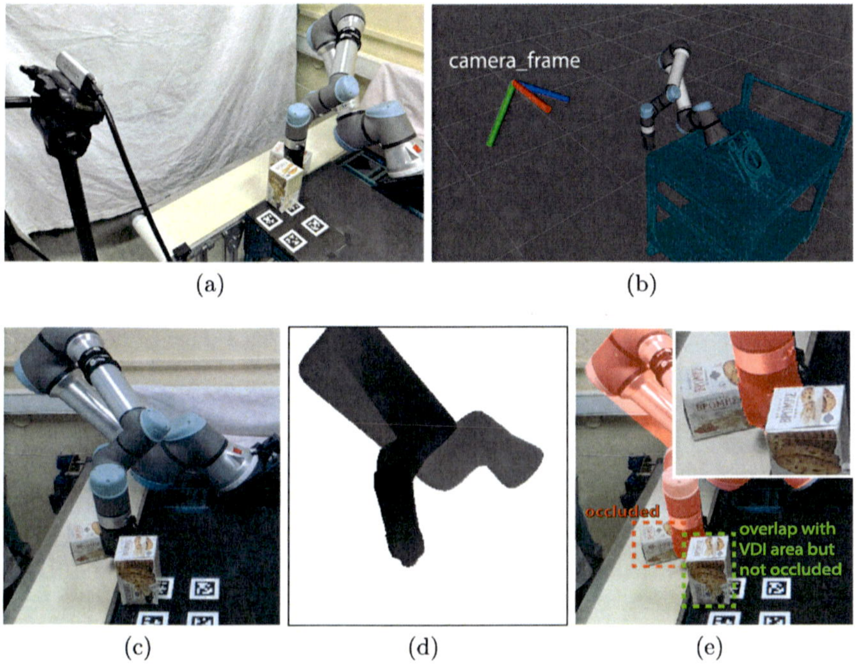


**Fig. 2.** (a) Setup with a robot, an RGB-D camera and two objects (one in front and one behind the robot), (b) the digital visualization, (c) the color view from the camera, (d) the rendered VDI including the end-effector, (e) the augmented view highlighting the areas occluded by the robot.

need not be an exact representation of the robot, as fine details will have a negligible effect on the final image for occlusions. For this purpose, we use the collision model, which is a convex approximation of each robot link in Standard Triangle Language (STL). The geometry of each link is represented by a mesh of triangles defined by a set of vertices. The use of the collision model is a conservative approach for detecting occlusions, because it increases the area marked as occluded in the VDI, but on the other hand it can compensate for small calibration errors of the camera.

For the implementation of the renderer we used the OpenGL library. This library allows us to build a simple graphics pipeline that takes advantage of a computer's graphics processor and achieve a high frame rate while using minimal CPU resources. The frame rate of the virtual camera must match that of the real camera to produce a matching image pair. The renderer first obtains the joint positions of the robot, then transforms the pose of each robot link from the world frame to the camera frame and finally renders all 3D models to generate each image. The output is a noise-free virtual depth image of the robot and any other geometry in the URDF we choose to render, as shown in Fig. 2.

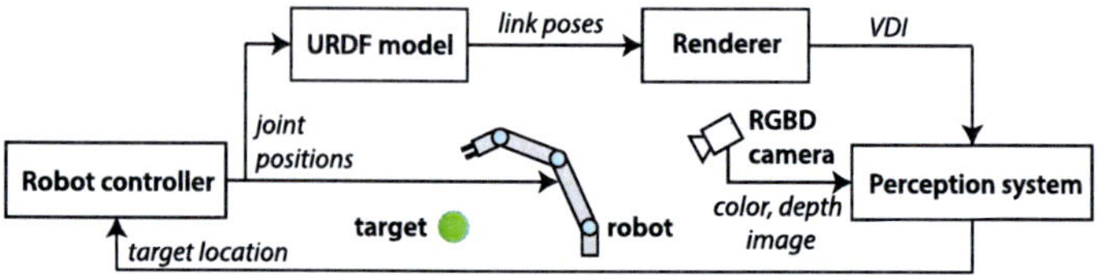


**Fig. 1.** Block diagram of the architecture that utilizes the virtual depth image (VDI).

the camera to the robot, because the area of the robot model is subtracted from the depth image, causing loss of useful depth information.

This work proposes a computationally lightweight method for determining the areas that the robot occludes from the camera. The contribution compared to methods that subtract the robot model from the depth image, is in the generation of a virtual depth image from the URDF model with minimal computational resources and the fusion of this virtual depth image with the depth image captured by the camera sensor to determine the area that the robot actually occludes. Furthermore, our method runs in real-time with a single RGB-D camera and is independent of the application or perception system. In the experiments, we demonstrate the proposed occlusion detection and how occlusions can be handled in two different use cases.

## 2 Methodology

Consider a robot receiving input from a perception system that detects and tracks targets of interest related to its task. The perception system uses an RGB-D camera with the robot being either fully or partially in its field of view. During task execution the robot may partially or fully occlude the targets, which may affect the target detection accuracy of the perception system.

Having the URDF model and the transformation of the camera's coordinate frame from the robot base, we use the joint position measurements of the robot to render the virtual depth image (VDI). The VDI is a noise-free depth image of the robot from the camera's point of view, from which a perception system can infer whether the target is partially or fully occluded, improving the robustness of the perception system. The block diagram in Fig. 1 shows a general architecture that uses the VDI, indicating that it is independent of the perception system.

### 2.1 Rendering the Virtual Depth Image

The first step of the method is to render depth images of the robot from the camera's point of view. We use a virtual camera with a pinhole camera model, where the intrinsic and extrinsic parameters correspond to the known parameters of the actual camera. The image distortion of the actual camera is compensated to match the pinhole camera model. To render the image, the 3D model is taken from the URDF and the joint positions of the actual robot. The 3D model can have a serious impact on the robot's performance, preventing it from completing its task successfully and risking damage from a possible misbehavior.

Perception systems use cameras in various configurations to reduce occlusions, depending on the application. For example, most object detection systems for pickup use overhead cameras [12]. Human tracking systems use cameras mounted at an angle rather than overhead to keep the human body in front view and cover a larger field of view towards the collaborative workspace [8]. One way to reduce occlusions is to use multiple cameras with different viewing angles [9]. However, depth-sensing cameras with structured light or lidar technologies can be quite expensive and interfere with each other. In addition, image processing of additional cameras multiplies computational costs. A common method to prevent occlusion is a camera-in-hand configuration. However, this is not always practical, for example, when the entire working area must be covered or the measurement distance is less than the minimum allowable distance of the depth sensor. Moreover, the moving camera causes motion blur.

Occlusion detection has been studied in the literature to improve the performance of perception systems for object tracking. In [12], a simple approach is proposed to detect occluded areas in the sensing data (e.g., point cloud) using a predefined height threshold. When the robot arm is above the target objects that are laid on a table, any point above the threshold is classified as occluded. This approach has significant limitations in the operation of the robot and in the flexibility of the setup, since the classification is based on a division surface. The authors in [7] address occlusions of target objects from the robot by training a deep neural network with a dataset containing images of occluded objects by the robot arm. However, this method requires a large amount of data to be trained and the inference time can be considerably high.

Detecting occlusion specifically for human tracking aims to improve the safety aspect of human-robot collaboration in shared workspaces. The authors in [5] propose an occlusion handling method for a mobile robot that follows humans using color and disparity information from a stereo camera. However, it depends heavily on the discrete depth information capability to detect occlusions, which may affect the accuracy in close proximity to the target from the occlusion. Shu et al. [10] proposed a method for detecting occlusions of human body parts in skeleton tracking. Although this method can obtain regional features, it scans the scene where the features are extracted, which has high computational cost.

Information about the structure of the robot, such as the URDF model, has been used in the literature to improve perception. URDF can be used to model robot arms using a hierarchical structure with XML trees and is well supported in Robotics Operating System (ROS). It can contain information about the kinematics, link inertia values, visual representation, and even the collision model of a robot, which is usually a slightly enlarged simplified body of each link. URDF filtering [3] was used in [8] along with an RGB-D camera to avoid ambiguities in tracking the human skeleton when the hands get very close to the robot. However, URDF filtering does not distinguish whether the targets are actually occluded or whether they are between the line of sight from

# A Lightweight Method for Detecting Dynamic Target Occlusions by the Robot Body

Savvas Sampaziotis, Sotiris Antonakoudis, Marios Kiatos, Fotios Dimeas(✉), and Zoe Dougleri

Aristotle University of Thessaloniki, Thessaloniki, Greece
{sampazio,antonakos,dimeasf,doulgeri}@ece.auth.gr

**Abstract.** Autonomous robots face significant challenges in accurately perceiving their environment due to occlusions. The robot itself may hide targets of interest from the camera while moving within the field of view, which can lead to safety hazards or failures in task execution. For example, if a target of interest is partially occluded by the robot, detecting and grasping it correctly, becomes very challenging. To solve this problem, we propose a novel computationally lightweight method to determine the areas that the robot occludes. Our approach uses the Unified Robot Description Format (URDF) to generate a virtual depth image of the 3D robot model. With the virtual depth image we can effectively determine the partially occluded areas to improve the robustness of the information given by the perception system. Due to the real-time capabilities of the method, it can successfully detect occlusions of moving targets by the moving robot. We validate the effectiveness of the method in an experimental setup using a 6-DoF robot arm and an RGB-D camera by detecting and handling occlusions for two tasks: Pose estimation of a moving object for pickup and human tracking for robot handover. The code is available in https://github.com/auth-arl/virtual_depth_image.

**Keywords:** occlusion detection · URDF · virtual depth image

## 1 Introduction

Robotic systems that need to operate in unstructured environments and to present versatility in their operation, such as for object pickup, grasping in cluttered environments and avoiding obstacles, require machine vision for enhanced perception capabilities. This is achieved with cameras that cover the robot's workspace and provide the necessary information that is critical to the robot's operation. However, occlusions can become a serious problem. These are blind spots of targets of interest in the camera view. A target may be an object that the robot needs to manipulate (e.g., for pick & place) or a human in the workspace that the robot needs to interact with. Such blind spots can have a negative impact on the performance of the perception methods. An error in perception


T. Petrič et al. (Eds.): RAAD 2023, MMS 135, pp. 3–11, 2023.
https://doi.org/10.1007/978-3-031-32606-6_1

# Perception and Learning

## Mobile Robotics and Localisation



## Medical Robotics

# Contents

## Local Arrangement Chairs

| | |
|---|---|
| Brecelj, Tilen | Jožef Stefan Institute, Slovenia |
| Deniša, Miha | Jožef Stefan Institute, Slovenia |
| Gams, Andrej | Jožef Stefan Institute, Slovenia |
| Kovač, Igor | Jožef Stefan Institute, Slovenia |
| Nemec, Bojan | Jožef Stefan Institute, Slovenia |

# Organization

## General Chair

| | |
|---|---|
| Petrič, Tadej | Jožef Stefan Institute, Slovenia |

## Co-chairs

| | |
|---|---|
| Ude, Aleš | Jožef Stefan Institute, Slovenia |
| Žlajpah, Leon | Jožef Stefan Institute, Slovenia |

## International Scientific Committee (RAAD ISC)

| | |
|---|---|
| Aspragathos, Nikos A. | University of Patras, Greece |
| Berns, Karsten | University of Kaiserslautern, Germany |
| Borangiu, Theodor | Polytech. University of Bucharest, Romania |
| Brandstötter, Mathias | Joanneum Research Robotics, Austria |
| Budinská, Ivana | Slovak Academy of Sciences, Slovakia |
| Carbone, Giuseppe | University of Calabria, Italy |
| Dobrovodský, Karol | Slovak Academy of Sciences, Slovakia |
| Ferraresi, Carlo | Polytechnic of Turin, Italy |
| Gasparetto, Alessandro | University of Udine, Italy |
| Ivanescu, Nick Andrei | Polytech. University of Bucharest, Romania |
| Jovanović, Kosta | University of Belgrade, Serbia |
| Kamnik, Roman | University of Ljubljana, Slovenia |
| Kronreif, Gernot | ACMIT, Austria |
| Laribi, Med Amine | University of Poitiers, France |
| Mueller, Andreas | Johannes Kepler University, Austria |
| Petrič, Tadej | Jožef Stefan Institute, Slovenia |
| Petrović, Ivan | University of Zagreb, Croatia |
| Pîsla, Doina | Technical University of Cluj-Napoca, Romania |
| Rodić, Aleksandar | Mihajlo Pupin Institute, Serbia |
| Tar, József K. | Óbuda University, Hungary |
| Zeghloul, Said | University of Poitiers, France |
| Žlajpah, Leon | Jožef Stefan Institute, Slovenia |

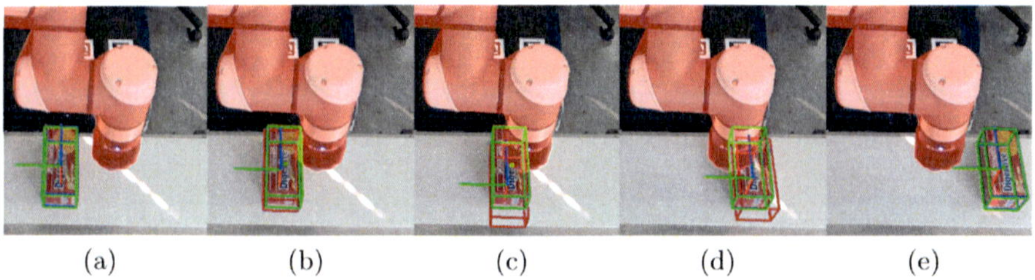


**Fig. 4.** Pose estimation of a moving object on a conveyor. When the object is partially occluded by the robot, the object detection system provides inaccurate estimates of the object's pose (red bounding boxes). Utilizing the VDI, we can accurately track the object (green bounding boxes) based on its position and velocity. (Color figure online)

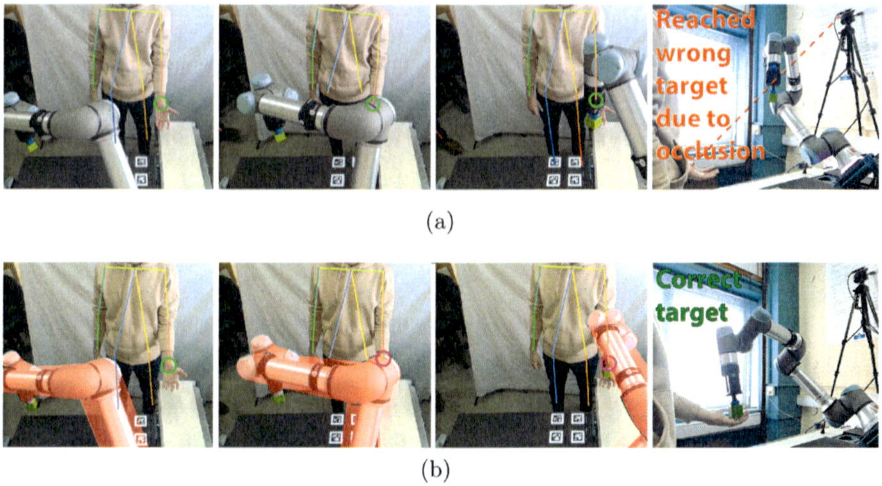


**Fig. 5.** Handover experiment where the robot reaches the detected left hand of the user. (a) With simple de-projection the robot reaches the wrong target because of occlusion. (b) With the VDI the occlusions are handled and the robot reaches the correct target.

keypoint that corresponds to the left wrist to acquire their 3D position. During the handover, the target position is continuously updated in case the human moves their hand. However, during the motion of the robot, occlusions occur as depicted in Fig. 5. For example, the robot's elbow occludes the human hand while it is approaching the human and the robot reaches the wrong target position for the handover due to incorrect de-projection, as shown in Fig. 5a. This happens because the human pose estimation algorithm predicts correctly the left wrist keypoint of the human, even though the human is partially occluded. However, the de-projection of that 2D keypoint provides a 3D position that lies on the robot's elbow instead of the human hand. Using the VDI we can detect that this point is occluded and then handle the occlusion by using the last non-occluded target position, as shown in Fig. 5b. In this case, the robot reaches the correct target for handover, right over the hand of the user. Compared to relying on noisy depth measurements to handle occlusions, the VDI provides a mechanism to make perception capabilities more robust for robotic integration.

## 4 Conclusions

In this work we presented a method to detect occluded areas in the view of a camera by rendering the URDF. The generated Virtual Depth Image is noise-free, is calculated in real-time and can be used by a perception system to detect if a target is occluded by the robot or not, even if it overlaps with the VDI area. We demonstrated the effectiveness of the method in two use cases that utilized a perception system for object tracking and another for human-tracking. The results suggest that this can be an effective way to detect occlusions and treat them appropriately, regardless of the perception method that is used.

**Acknowledgements.** Co-financed by the European Regional Development Fund of the European Union and Greek national funds through the Operational Program Competitiveness, Entrepreneurship and Innovation, under the call RESEARCH - CREATE - INNOVATE (project code: T2EDK-02358).

## References

1. Chandel, H., Vatta, S.: Occlusion detection and handling: a review. Int. J. Comput. Appl. **120**(10), 33–38 (2015)
2. Fang, H.S., Xie, S., Tai, Y.W., Lu, C.: RMPE: regional multi-person pose estimation. In: ICCV (2017)
3. Github: Realtime urdf filter. https://github.com/blodow/realtime_urdf_filter
4. He, K., Gkioxari, G., Dollár, P., Girshick, R.: Mask R-CNN. In: Proceedings of the IEEE International Conference on Computer Vision, pp. 2961–2969 (2017)
5. Isobe, Y., Masuyama, G., Umeda, K.: Occlusion handling for a target-tracking robot with a stereo camera. Robomech J. **5**(1), 1–13 (2018)
6. Kuipers, T.P., Arya, D., Gupta, D.K.: Hard occlusions in visual object tracking. In: Bartoli, A., Fusiello, A. (eds.) ECCV 2020. LNCS, vol. 12539, pp. 299–314. Springer, Cham (2020). https://doi.org/10.1007/978-3-030-68238-5_22
7. Luo, Y., et al.: Calibration-free monocular vision-based robot manipulations with occlusion awareness. IEEE Access **9**, 85265–85276 (2021)
8. Magrini, E., Luca, A.D.: Estimation of contact forces using a virtual force sensor. In: IEEE/RSJ International Conference on Intelligent Robots and Systems (2014)
9. Melchiorre, M., Scimmi, L.S., Pastorelli, S.P., Mauro, S.: Collison avoidance using point cloud data fusion from multiple depth sensors: a practical approach. In: 2019 23rd International Conference on Mechatronics Technology (ICMT). IEEE (2019)
10. Shu, G., Dehghan, A., Oreifej, O., Hand, E., Shah, M.: Part-based multiple-person tracking with partial occlusion handling. In: 2012 IEEE Conference on Computer Vision and Pattern Recognition, pp. 1815–1821. IEEE (2012)
11. Tang, J., Miller, S., Singh, A., Abbeel, P.: A textured object recognition pipeline for color and depth image data. In: 2012 IEEE International Conference on Robotics and Automation, pp. 3467–3474. IEEE (2012)

12. Yoshioka, K., Okuni, H., Ta, T.T., Sai, A.: Through the looking glass: diminishing occlusions in robot vision systems with mirror reflections. In: 2021 IEEE/RSJ International Conference on Intelligent Robots and Systems (IROS). IEEE (2021)
13. Zaidi, S.S.A., Ansari, M.S., Aslam, A., Kanwal, N., Asghar, M., Lee, B.: A survey of modern deep learning based object detection models. Digit. Signal Process. **126**, 103514 (2022)

Check for updates

# Probabilistic Online Robot Learning via Teleoperated Demonstrations for Remote Elderly Care

Floris Meccanici[1,2], Dimitrios Karageorgos[2], Cock J. M. Heemskerk[2], David A. Abbink[1], and Luka Peternel[1](✉)

[1] Cognitive Robotics, Delft University of Technology, Delft, The Netherlands
l.peternel@tudelft.nl

[2] Heemskerk Innovative Technology B.V., Delft, The Netherlands

**Abstract.** Daily household tasks involve manipulation in cluttered and unpredictable environments and service robots require complex skills and adaptability to perform such tasks. To this end, we developed a teleoperated online learning approach with a novel skill refinement method, where the operator can make refinements to the initially trained skill by a haptic device. After a refined trajectory is formed, it is used to update a probabilistic trajectory model conditioned to the environment state. Therefore, the initial model can be adapted when unknown variations occur and the method is able to deal with different object positions and initial robot poses. This enables human operators to remotely correct or teach complex robotic manipulation skills. Such an approach can help to alleviate shortages of caretakers in elderly care and reduce travel time between homes of different elderly to reprogram the service robots whenever they get stuck. We performed a human factors experiment on 18 participants teaching a service robot how to empty a dishwasher, which is a common daily household task performed by caregivers. We compared the developed method against three other methods. The results show that the proposed method performs better in terms of how much time it takes to successfully adapt a model and in terms of the perceived workload.

**Keywords:** Learning from Demonstration · Teleoperation · Online Learning

## 1 Introduction

Due to an ageing society and shortages of the workforce, elderly care at homes and in nursing homes is becoming one of the key societal challenges [7]. To mitigate the lack of caregivers with respect to the increasing numbers of elderly, we see service robots and teleoperation as one of the most promising solutions. We envision deploying multiple robots to various locations, while caregivers can teach and correct them remotely when they get stuck, thus eliminating travel time.

Most of the existing Learning from Demonstration (LfD) is done by humans kinaesthetically guiding the robot on how to perform manipulation tasks [2,8,14], which requires physical presence of the teacher and thus does not fit with our vision. On the other hand, robot teaching can also be done through teleoperation, where the teacher performs demonstration remotely through various interfaces [12,15]. Skills for complex


T. Petrič et al. (Eds.): RAAD 2023, MMS 135, pp. 12–19, 2023.
https://doi.org/10.1007/978-3-031-32606-6_2

manipulation tasks are typically encoded by trajectories such as Dynamic Movement Primitives (DMPs) [13]. However, the deterministic nature of DMPs offers limited generalisation with respect to unpredictable situations. On the other hand, probabilistic trajectories such as Probabilistic Movement Primitives (ProMPs) [9] use a probability distribution inferred from multiple demonstrations for improved generalisation.

A general LfD approach is to train a skill model offline and expect it to perform well afterwards. However, problems arise when the trained model is incorrect, either due to insufficient training or unknown variations that have occurred in the environment. This demands methods that are able to refine the initially trained model, of which the state-of-the-art can be divided into online [1,4,11,14] and active learning [3,8]. In online learning, the operator is able to refine the motion during the execution time, while active learning uses an uncertainty measurement to query the operator for a new demonstration. In care scenarios, we want to adapt the model towards unknown changes in the environment, and since active learning needs to encode the environment to determine the uncertainty in the prediction, these methods are not well suited. On the other hand, in online learning, the operator has the role to intervene when the model needs to be adapted, and is, therefore, better suited in an unknown environment. State-of-the-art online learning methods mainly use kinaesthetic teaching to refine the motion [4,14], which are not suitable for the remote elderly care service.

Teleoperated online teaching is preferred in the home-care scenario, because it does not require an operator to be constantly physically present at the robot and thus enables easy switching between teaching multiple robots at different locations. The methods in [10,15] enabled teleoperated teaching, however, the learning process was done offline. The method in [12] enabled online teaching by encoding the variance indirectly through a separate deterministic stiffness trajectory. However, encoding variance indirectly through deterministic methods makes the skill model less rich and less generalisable. The method examined in [6] used probabilistic encoding in combination with teleoperated teaching, however, they offered no online refinement mode for specific corrections/updates of the trained trajectory. Some online teleoperated teaching methods used a shared control approach to arbitrate the level of autonomy between the robot skill and the human teacher [1,11]. Since we want humans to be readily in control when needed and to quickly adapt the motion in order to deal with unknown/changing parts of the environment, these types of methods are also not well suited for our problem. Therefore, the existing literature is missing a teleoperated online learning method that allows for quick refinements of the executed predicted motion of a probabilistic skill model to deal with changing environment.

To close this gap we develop a teleoperated online learning approach with a novel refinement method that can quickly adapt the executed predicted motion using a haptic device. The method is based on ProMP, where the detected object is an input and the probabilistic trajectory is an output. To analyse the usability of the proposed method, we conduct an extensive human factors experiment.

## 2 Method Design

To achieve the desired teleoperated online learning, we built an LfD framework by using ProMPs (see Fig. 1). Inspired by work in kinaesthetic teaching [4], we conditioned

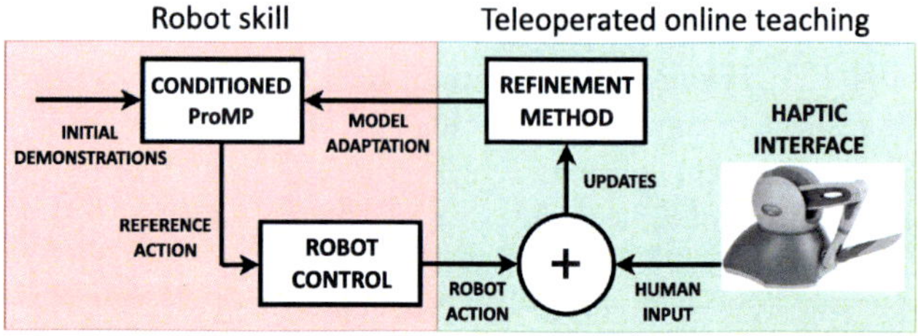


**Fig. 1.** Overview of the proposed framework. Initial demonstrations are used to train a skill model offline using conditioned-ProMPs. This model generates an action, which is adapted online using the proposed novel teleoperated online refinement method, and is used to updated the model.

ProMPs on an external state variable $\boldsymbol{s}$ (in our case different object positions and initial robot poses). After using the conditioned-ProMPs to train an initial skill model offline, the predictions were refined in an online manner by the proposed novel teleoperated online refinement method. The refinements were then used to update the skill model.

To enable the operator to effectively adapt the executed trajectory online via teleoperation, we developed a novel online refinement method, where both visual and haptic feedback was provided to the human operator. The online learning started with an initial skill model from ProMPs that produces a desired reference trajectory $\boldsymbol{\tau}_d$, which was used by the robot controller to produce the actually executed trajectory $\boldsymbol{\tau}_a$ (in the case of ideal controller $\boldsymbol{\tau}_a = \boldsymbol{\tau}_d$). Then the operator was able to adapt this trajectory in a shared control manner (Fig. 1), resulting in human refinement trajectory $\boldsymbol{\tau}_{hr}$. To do this, the operator moved the master (haptic) device in the desired adaptation direction and, while doing so, he/she felt a force proportional to the magnitude of the refinement.

Rather than adapting the position at the current time step, we adapted the position of the executed trajectory at the next time step by using the current master position (Fig. 2). To achieve this, we calculated the position vector in the next step $i+1$ relative to the current step $i$ (called the $i+1$ and $i$ frame, respectively) as ${}^{i}\boldsymbol{p}_{i+1} ={}^{i} R_b({}^{b}\boldsymbol{p}_{i+1} -{}^{b} \boldsymbol{p}_i)$, where position vector is $\boldsymbol{p} = [x,\ y,\ z]^T$ and ${}^{i}R_b$ is a rotation matrix between the base frame and the current frame. The master position was normalised so that the zero position was approximately in the middle of the workspace, which can be seen in Fig. 2. To get the position of the human intervention (green dot in Fig. 2) combined with the reference position, we added this normalised master position ${}^{i}\boldsymbol{p}_m$ to ${}^{i}\boldsymbol{p}_{i+1}$ as ${}^{i}\boldsymbol{p}_{i+1,new} ={}^{i} \boldsymbol{p}_{i+1} +{}^{i} \boldsymbol{p}_m$. The end-effector position expressed in the base frame, which was used for the robot control, was obtained by ${}^{b}\boldsymbol{p}_{i+1,new} ={}^{b} \boldsymbol{p}_i +{}^{b} R_i({}^{i}\boldsymbol{p}_{i+1,new})$.

After the operator created $\boldsymbol{\tau}_{hr}$, which is ${}^{b}\boldsymbol{p}_{i+1,new}$ at every time step, the reference $\boldsymbol{\tau}_d$ was updated using $\boldsymbol{\tau}_d^{\text{new}} = \boldsymbol{\tau}_d^{\text{old}} + \alpha(\boldsymbol{\tau}_{hr} - \boldsymbol{\tau}_a)$, where $\alpha \in [0, 1]$ defined how much the difference between $\boldsymbol{\tau}_{hr}$ and $\boldsymbol{\tau}_a$ changes $\boldsymbol{\tau}_d^{\text{old}}$. This value can be adapted based on the confidence of the operator in its refinement, but in this application, $\alpha$ is set to 1, as in [4]. This loop continued until the prediction was successful. Visual feedback to the operator was provided through the wrist and head camera views, and additionally, the point cloud of the environment was shown, enabling visualisation of the execution ($\boldsymbol{\tau}_a$) and refined trajectory ($\boldsymbol{\tau}_d^{\text{new}}$) with respect to the environment.

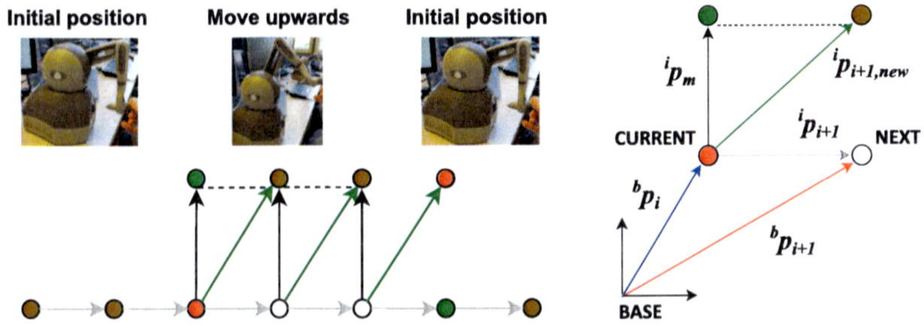


**Fig. 2.** Illustration of corrections to the executed trajectory. The green and red dots represent the master and slave (remote) robot positions, respectively. When the master and slave are in the same position, the dot has a brown colour. The white dots represent the reference trajectory, where the executed trajectory converges if the operator does not apply any input. (Color figure online)

After refining the trajectory, $\boldsymbol{\tau}_d^{\text{new}}$ was used to update the conditioned-ProMPs. First, new ProMP weights $\boldsymbol{w}$ were calculated from $\boldsymbol{\tau}_d^{\text{new}}$, after which a vector $\boldsymbol{x}$ was created by appending condition $\boldsymbol{s}$ (i.e., different object positions and initial robot poses) to ProMP weights $\boldsymbol{w}$ as $\boldsymbol{x} = [\boldsymbol{w}^T, \boldsymbol{s}^T]$. $\boldsymbol{x}$ was then used with Welford's method for updating the mean and covariance incrementally, which means that one data sample was used [4]. This method was able to quickly update the mean and covariance, without having to store all previous data. To align demonstrated trajectories with different time durations, we employed Dynamic Time Warping.

## 3 Human Factors Experiment

Eighteen participants volunteered for this experiment and signed an informed consent form prior to participation. The participants were asked about their gaming and teleoperation experience since this information was relevant to the analysis. The research was approved by the Human Research Ethics Committee of Delft University of Technology.

### 3.1 Experiment Setup

The experiment setup is shown in Fig. 3. The goal of the human factors experiment was to compare the proposed method against three other methods for adapting a trained skill model. These methods were obtained by a combination of two learning mechanisms (online or offline learning) and two teaching devices (haptic interface or teach pendant), leading to four experimental conditions. In online learning, the refinement was done during the trajectory execution, while in offline learning the trajectory had to be recreated as a whole prior to execution. The developed online learning combined with a haptic stylus interface (Geomagic Touch) is the proposed method (*OnStyl*). The teach pendant is still the most common device used in the industry and was emulated with a generic PC keyboard and employed either in offline *OffKey*) or online (*OnKey*) learning. The teleoperated offline LfD was performed using a haptic stylus interface (*OffStyl*), where demonstrated trajectories were recorded and then ProMP weights were recalculated offline. Before starting each experiment condition (method), the participants went through familiarisation trials to minimise the human learning effect on the results. We counterbalanced the effects of the order in which the participants are presented with different experiment conditions by using a balanced Latin Square experimental design.

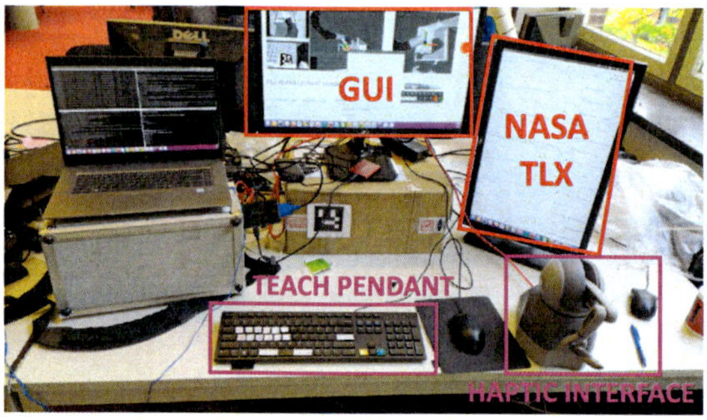


**Fig. 3.** Experiment setup with the teaching devices (purple), as well as graphical user interface (GUI) for visual feedback and a monitor to fill in the subjective questionnaire (NASA TLX).

We hypothesised that *the proposed method, which is a combination of online learning and haptic stylus, has the lowest refinement time and workload.* To test the hypothesis, we used an experiment task of moving dishes out of a dishwasher, which is one of the common daily tasks performed by caregivers. The complexity of this task comes from a cluttered and unpredictable environment with many environmental constraints. The participants had to operate a service humanoid robot to teach and refine the skill model for the given task in a simulated environment. Each participant had to adapt the model for one object, where its positions differed among the participants. Three initial models (i.e., reaching trajectories) had to be adapted with each of the four methods per participant. For each model, the participant had eight attempts to update the model, and per update ten refinement attempts. A refinement was defined as the creation of an adaptation of the trajectory (or previous refinement). If the participant failed to adapt the model within these attempts, we classified it as a failure and excluded this specific participant data from the refinement time hypotheses testing.

We analysed the performance in terms of how much time it takes the human users to successfully adapt a model (refinement time). Additionally, we analysed the perceived workload by the participants, which was evaluated using NASA TLX [5]. Finally, after the experiment, each participant was asked which method they liked the most and why. We analysed the significance of the results with a statistical test. Because both the refinement time and the workload data were not normally distributed, we used Wilcoxon signed rank test instead of a one-sided t-test. The level of significance was set to 0.05.

### 3.2 Results

An example of reaching for four different object positions is shown in Fig. 4. Refinements were only necessary for the first two object positions, after which the model was able to generalise for the other two object positions without further refinements.

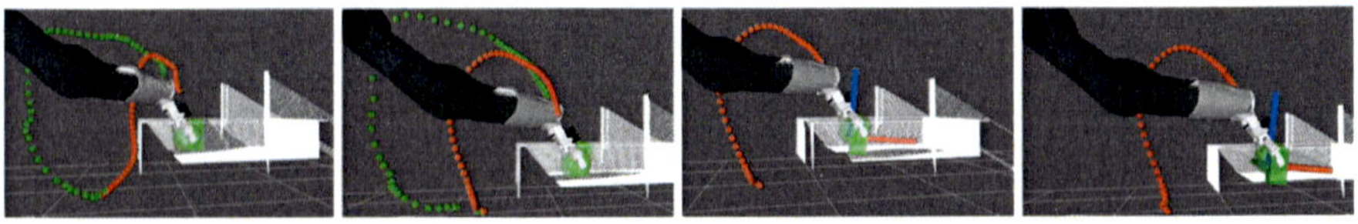

**Fig. 4.** Reaching for four different object positions inside the dishwasher. The executed (red) and refined (green) trajectories are indicated with point clouds. (Color figure online)

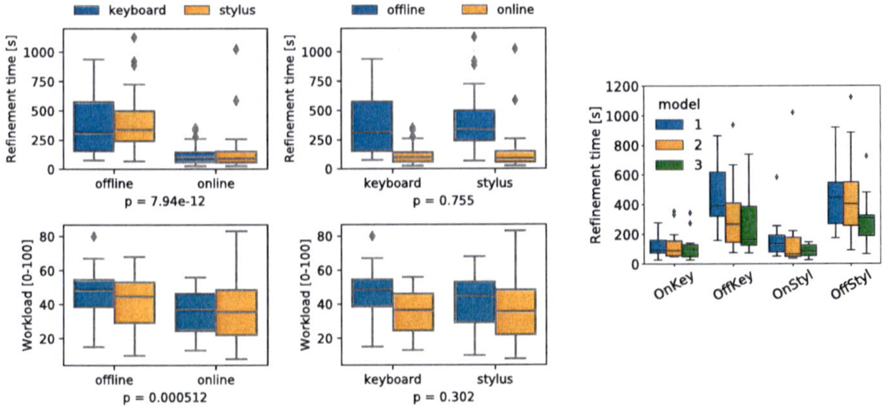


**Fig. 5.** Left four graph: main results. Right graph: refinement time per model. Blue, orange and green colours represent models 1, 2 and 3, respectively. The boxplots report the median (M), first quartile (25) and third quartile (75) of the refinement time and workload. The statistical significance of the difference between conditions is indicated by p-values below the graphs. (Color figure online)

Out of 18 participants, six failed either in the training or to adapt a model in one of the methods. This means that in total 12 participants were used for the data analysis of the refinement time hypothesis, thus 36 models were evaluated per each method (experiment condition). Figure 5 shows the experiment results, where the distribution of this data and the refinement time per model and per method are depicted. The online methods performed much better compared to the offline methods in both refinement time and workload (i.e., lower values indicate better in both cases). The difference in refinement time was statistically significant both online ($p < 0.001$) and offline ($p < 0.001$). However, there was no statistically significant difference in refinement time ($p = 0.755$) and workload ($p = 0.302$) between the keyboard and stylus.

No Wilcoxon signed rank test can be performed on any background parameter except the gaming experience since the sample sizes are not equal. The participants with high gaming experience perceived a lower workload for all methods compared to having low gaming experience. The difference was statistically significant ($p = 0.048$). On the other hand, there was no statistically significant difference between having high or low gaming experience in terms of the refinement time ($p = 0.189$).

The results of the questionnaire regarding the method preference showed a strong preference for the proposed method (*OnStyl*), where 8 out of 18 participants voted for it.

The second most popular was the online learning method using teach pendant (*OnKey*) with 4 votes. The least popular were offline methods (*OffKey* and *OffStyl*), both receiving 3 votes each.

## 4 Discussion

Results show that, by using the developed method, human operators are able to adapt an initially trained model to account for unknown variations: goal deviation and an unforeseen obstacle. The method was able to update the model for different object positions. It should be noted that the amount of conditions to refine is dependent on the number of initial trajectories used to train the initial model, and how complex the motion is.

The significant difference in refinement time and workload in favour of online methods can most likely be attributed to the operator only performing small adjustments to the executed trajectory in online learning, instead of completely teaching a new one as in offline learning. This means that online methods are inherently faster in creating a single refinement and a lower amount of input from the operator is needed, possibly resulting in a lower workload as well. Interestingly, there is also a lower variability in refinement time in the online compared to offline methods (Fig. 5).

Another influence on the refinement time is the task execution strategy, which tended to change within and between the methods. The within change can be explained by the right graph of Fig. 5, where the refinement time decreases as a function of the model number. This gives an indication that either the operator has a constant strategy in its mind and is figuring out how to translate this strategy to a demonstration using the specific method, or the strategy changes and the operator is figuring out what strategy works best. Since the offline methods have a higher slope in the medians of the refinement time compared to online methods and the method exposure was counterbalanced, it seems more reasonable that the participants had a more or less constant strategy, but had more trouble using the offline methods to convey this strategy. This suggests that more operator training is needed to use offline methods effectively.

No significant difference in the type of teaching device (stylus or keyboard) in both refinement time and workload could be because the intuitiveness of the interface is person dependent. Some participants reported *OnKey* to be easy and less sensitive than *OnStyl* and preferred the limited degrees of freedom (DoF). On the other hand, others felt that they could easily press the wrong buttons on the teach pendant and found it hard to figure out the axes. Another explanation could lie in the complexity of the corrections, wherein in the examined cases the corrections were mostly in one DoF, i.e., movement along the vertical axis to avoid the dishwasher basket. Increasing the complexity of correction might favour *OnStyl* but additional study is required to provide that insight.

Another explanation for finding no significant difference between using the stylus and keyboard is that there is a difference in the implementation of the offline methods. In *OffKey* the participants could take their time, as it did not matter what was done between the waypoints as long as the waypoints are correctly specified. *OffStyl*, on the other hand, continuously tracked the demonstrated motion, and when the operator made a small mistake this could translate into an unsuccessful demonstration more easily. Therefore, *OffStyl* could be implemented similarly as the teach pendant, where the operator can specify waypoints and interpolate between them. This is expected to show statistical results in favour of the stylus since the definition of offline will be more similar between the stylus and the keyboard implementation.

The proposed method was tested on one representative task in elderly care, i.e., emptying a dishwasher. In future, other common daily tasks found in elderly care should be examined, such as setting the table for breakfast and cleaning it afterwards. The method itself could be improved by enabling robots to also learn cognitive skills.

## References


1. Abi-Farraj, F., Osa, T., Peters, N.P.J., Neumann, G., Giordano, P.R.: A learning-based shared control architecture for interactive task execution. In: 2017 IEEE International Conference on Robotics and Automation (ICRA), pp. 329–335. IEEE (2017)
2. Calinon, S., Sardellitti, I., Caldwell, D.G.: Learning-based control strategy for safe human-robot interaction exploiting task and robot redundancies. In: 2010 IEEE/RSJ International Conference on Intelligent Robots and Systems, pp. 249–254. Citeseer (2010)
3. Conkey, A., Hermans, T.: Active learning of probabilistic movement primitives. In: 2019 IEEE-RAS 19th International Conference on Humanoid Robots (Humanoids), pp. 1–8. IEEE (2019)
4. Ewerton, M., Maeda, G., Kollegger, G., Wiemeyer, J., Peters, J.: Incremental imitation learning of context-dependent motor skills. In: 2016 IEEE-RAS 16th International Conference on Humanoid Robots (Humanoids), pp. 351–358. IEEE (2016)
5. Hart, S.G., Staveland, L.E.: Development of NASA-TLX (task load index): results of empirical and theoretical research. In: Advances in Psychology, vol. 52, pp. 139–183. Elsevier (1988)
6. Havoutis, I., Calinon, S.: Learning from demonstration for semi-autonomous teleoperation. Auton. Robot. **43**(3), 713–726 (2019)
7. Kovner, C.T., Mezey, M., Harrington, C.: Who cares for older adults? Workforce implications of an aging society. Health Aff. **21**(5), 78–89 (2002)
8. Maeda, G., Ewerton, M., Osa, T., Busch, B., Peters, J.: Active incremental learning of robot movement primitives. In: Conference on Robot Learning, pp. 37–46. PMLR (2017)
9. Paraschos, A., Daniel, C., Peters, J.R., Neumann, G.: Probabilistic movement primitives. In: Advances in Neural Information Processing Systems, pp. 2616–2624 (2013)
10. Pervez, A., Latifee, H., Ryu, J.H., Lee, D.: Motion encoding with asynchronous trajectories of repetitive teleoperation tasks and its extension to human-agent shared teleoperation. Auton. Robot. **43**(8), 2055–2069 (2019)
11. Peternel, L., Oztop, E., Babič, J.: A shared control method for online human-in-the-loop robot learning based on locally weighted regression. In: 2016 IEEE/RSJ International Conference on Intelligent Robots and Systems (IROS), pp. 3900–3906. IEEE (2016)
12. Peternel, L., Petrič, T., Oztop, E., Babič, J.: Teaching robots to cooperate with humans in dynamic manipulation tasks based on multi-modal human-in-the-loop approach. Auton. Robot. **36**(1–2), 123–136 (2014)
13. Saveriano, M., Abu-Dakka, F.J., Kramberger, A., Peternel, L.: Dynamic movement primitives in robotics: a tutorial survey (2021). arXiv preprint arXiv:2102.03861
14. Saveriano, M., An, S.i., Lee, D.: Incremental kinesthetic teaching of end-effector and null-space motion primitives. In: 2015 IEEE International Conference on Robotics and Automation (ICRA), pp. 3570–3575. IEEE (2015)
15. Schol, J., Hofland, J., Heemskerk, C.J., Abbink, D.A., Peternel, L.: Design and evaluation of haptic interface wiggling method for remote commanding of variable stiffness profiles. In: 2021 20th International Conference on Advanced Robotics (ICAR), pp. 172–179. IEEE (2021)

# Exploring the Potential and Challenges of Reinforcement Learning for Object Pushing with Robots

Domen Težak(✉), Marko Munih, Matjaž Mihelj, and Janez Podobnik

Faculty of Electrical Engineering, University of Ljubljana, Ljubljana, Slovenia
dt4907@student.uni-lj.si
https://fe.uni-lj.si

**Abstract.** This paper describes a 2D simulation environment for reinforcement learning of real robots that incorporates physical simulation using the PyBox2D framework. Using reinforcement learning, we aim to explore different strategies for playing air hockey and non-prehensile object manipulation. To achieve this, we first trained our models in simulation and then transferred the successful ones to a Franka Emika Panda robot for testing on a real system. Our research primarily focused on two main tasks: playing air hockey and non-prehensile object manipulation, with an emphasis on pushing. We present the results obtained from these experiments and discuss the potential future work that can be done to improve the models' performance. The main contribution of this work is to demonstrate the effectiveness of using reinforcement learning in combination with simulation to train models for complex manipulation tasks. The successful transfer of the models from simulation to the real world highlights the potential of this approach in real-world applications.

**Keywords:** robot · reinforcement learning · pushing tasks · PyBox2D · simulation

## 1 Introduction

In recent years, we have seen rapid development in the field of artificial intelligence and machine learning. Reinforcement learning is one of the fields of machine learning. In this context, it is not surprising that much of the research in robotics is focused on this topic. Key research areas in robotics include trajectory generation, navigation in mobile robotics and object manipulation. Some of the leading research in recent years include the use of reinforcement learning for the control of bipedal robots [1], a hybrid model of reinforcement learning for visual-linguistic navigation [2], and decentralised collision avoidance of multiple robots using reinforcement learning [3].

The authors acknowledge the financial support from the Slovenian Research Agency (research core funding No. P2-0228).


T. Petrič et al. (Eds.): RAAD 2023, MMS 135, pp. 20–28, 2023.
https://doi.org/10.1007/978-3-031-32606-6_3

This paper first focuses on learning a strategy for playing air hockey in simulation using reinforcement learning algorithms, which are later transferred to the Panda robot. A related study [4] used probabilistic theory instead of reinforcement learning to control a robot playing air hockey. In the second part of the paper, we focus on the manipulation of objects without grasping (non-prehensile manipulation), in particular, pushing. The related work in this area involves transferring non-prehensile manipulation techniques from simulation to a physical system [5], multiple agent collaboration in pushing objects tasks, and object relocation using deep learning [6]. The aim of this work is to learn different strategies in a simulation using reinforcement learning, transfer them to a robot system and evaluate such an approach to control the robot.

## 2 Methodology

The air hockey simulator project [7] was used as a basis, which already contained the complete framework for learning the strategy but lacked the physical simulation of the environment. The first step was, therefore, to build on the initial project by implementing the physical simulation.

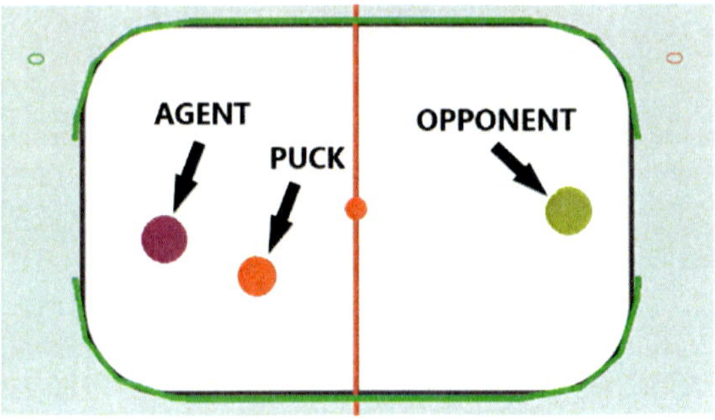


**Fig. 1.** PyBox2D air hockey environment.

### 2.1 Simulation Environment

The physical simulation was implemented using the PyBox2D Python library for physical simulation of 2D objects, which was sufficient for our needs as all agent and object movements were performed in the plane (See Fig. 1). The library uses SI units and allows the creation of dynamic and static objects. Dynamic objects have finite mass, forces and torques can be applied, causing them to move. Static objects are considered to have infinite mass and are therefore immobile. In our case, the agent, puck and opponent are dynamic objects, defined by their shape, mass, elasticity, velocity constraints and interaction with the surface (friction), while the edges of the playing surface are static objects and, therefore stationary.

The simulation is running with a time step of 1/60 s, with two position corrections and six velocity corrections per step, as recommended in documentation, to optimise the simulation performance when resolving collisions and updating the velocities and positions of all dynamic objects.

### 2.2 Reinforcement Learning Algorithms

Stable Baselines3 is a collection of reliable implementations of machine learning algorithms in Pytorch. This repository aims to facilitate the use of these algorithms and also serves as a good basis for building new projects and comparing them with existing solutions in reinforcement learning.

**DQN (Deep Q Network)** is a neural network architecture used in reinforcement learning. It consists of two neural networks and an experience replay cache. The first network (Q network) predicts the values of Q for each state, while the second network (target network) predicts the target Q value. Q value is a measure of the overall expected reward assuming the agent is in known state and performs chosen action, and then continues operating until the end of the episode following some policy. To train the DQN, samples are collected by interacting with the environment. The samples are stored in the experience replay cache and used as training data. The neural network is trained by inputting batches of data from the experience replay cache. The loss function used is known as the Mean Squared Error (MSE) loss. The MSE loss is calculated by taking the difference between the target Q-value and the predicted Q-value for a given action in a given state, squaring this difference, and then taking the average over all actions and states. Gradient descent is used to update the weights of the Q network to minimize the loss function. The weights of the target network are only updated every T time steps to ensure stability and efficiency of learning. The DQN algorithm uses the $\epsilon$-greedy method to select an action in each state. The action with the maximum Q value is chosen as the optimal action. The reward obtained from the transition is added to the target Q value to calculate the loss function. The DQN algorithm is a powerful tool for solving complex reinforcement learning problems by training neural networks to predict Q values and optimize actions [9].

**TQC (Truncated Quantile Critics)** falls under actor-critic algorithms, where we have two separate neural networks. The actor performs actions in the environment and learns through the critic's estimate of the value of the action at a given time, while the critic learns to estimate the values of all possible actions in each state, thus indirectly changing the agent's strategy. The TQC algorithm replaces the value of the Q function by its probability distribution on a given interval, uses multiple critics (neural networks), then combines the predictions of all critics and removes a certain number of maximum predictions, which eliminates the problem of overestimating the value of the Q function and increases the efficiency of the agent. A more detailed description and the study of the above algorithm can be found in Kuznetsov et al. [10].

**OpenAI Gym** [11] is a collection of various environments for teaching agents with reinforcement learning, which are also compatible with the Stable Baselines3 algorithms. More importantly, it also enables the integration of the custom environment we develop ourselves. This simplifies workflow on our project and makes it possible to use the algorithms for reinforcement learning on our agent.

### 2.3 Hardware

Robotics system consists of Franka Emika Panda robot and touch screen, which acts as a display for augmented reality visualization for task and a sensor for detecting position of objects on the screen (see Fig. 2 left).

**Fig. 2.** System configuration and data exchange between simulation and robot.

Collaborative robot **Panda robot** has seven degrees of freedom and a load capacity of three kilograms. A 3D-printed tool was inserted into the gripper, representing the puck and our agent. The tool was placed over a screen displaying the environment created in Unity.

**Infrared touch screen** operates based on light-beam interruption, commonly referred to as beam break, to determine the location of multiple touch events. Infrared uses an array of X-Y infrared LED and photodetector pairs around the edges of the screen to detect a disruption in the pattern of LED beams. These LED beams cross each other in vertical and horizontal patterns. As an object touches the screen, it interrupts the light beam which is detected at the sensor side. A major benefit of such a system is that it can detect multiple inputs, which we used to determine the object's orientation on the screen.

### 2.4 Robot Control

The joint torques of the robot were calculated based on deviations from the desired robot end effector pose (in the external coordinates) using a gravity-compensated proportional-differential (PD) controller and the contribution of Coriolis and centripetal forces. The required torques $\tau$ can be written as

$$\tau = J_A^T(q)(K_p\widetilde{x} - K_d J_A(q)\dot{q}) + \widehat{C}(q,\dot{q})\dot{q} + \widehat{g}(q), \tag{1}$$

where $\widetilde{x}$ represents the deviation between the desired pose $x_r$ and the actual pose of the end effector x

$$\widetilde{x} = x_r - x \tag{2}$$

$J_A(q)$ represents the analytical Jacobi matrix, $\widehat{C}(q, \dot{q})$ represents the model of the contributions of the Coriolis and centripetal forces, $\widehat{g}(q)$ is the model of the contributions of the gravitational forces. The matrices $K_p$ and $K_d$ represent the diagonal matrices of the proportional and differential gains.

Figure 2 right shows the overall architecture of the system and the data exchange. In addition to sending the reference pose $x_r$ to the robot controller, we received the puck's position, which was detected on the touch screen and then sent from Unity to the computer where the simulation was running. This resulted in a working system that tracked changes in the puck's positions and the agent's positions.

### 2.5 Agent Learning and Testing on the Robot

All models were trained on the computer with 11th Gen Intel i7-11700 CPU and NVIDIA GeForce RTX 3060 GPU. First, we defined all the necessary object parameters for the task. Then we chose either the DQN or the TQC algorithm, set the learning parameters and ran the model learning script. We observed the results of the average rewards per learning episode via TensorBoard running in the browser. When we noticed that the model was no longer improving, we stopped learning and saved the best model with the highest average reward. For example to learn agent how to play air hockey it took about 2 million episodes, meanwhile learning agent both pushing tasks resulted close to 3 million episode before models were not improving anymore. This translated to couple of hours of training for each model on our GPU. We then tested this model on the robot by sending the agent's position in the simulation environment as a reference position to the robot controller. The puck's position on the screen was also detected and, in this case, used as an observation of the environment.

## 3 Results

In this section we present results of the simulations and testing on the robot for various tasks using reinforcement learning.

### 3.1 Playing Air Hockey in Simulation

State space consisted of positions and velocities of our agent, opponent and puck. Action space using TQC was 2D vector on interval between $-1$ and 1, which gave us direction in which our agent should move. We approached teaching the agent using a sparse reward. The agent got a positive reward if he scored a goal and a negative reward if the opponent scored. After the learning process, our agent was as good as the opponent programmed heuristically with predetermined rules since when playing one against the other, neither consistently beat the other. Rather than trying to improve the agent, we focused on the pushing task.

### 3.2 Pushing Objects

We focused on the non-prehensile object manipulation, which is also an interesting area of research in the field of reinforcement learning. State space consisted of position and velocity of our agent and puck, we also added body and target orientation in the target pose task. Action space was the same as in the air hockey task. Reward function included positive rewards if agent came closer to the puck and if puck was pushed closer to the target position. Later we also included positive reward if the orientation error between body and target orientation was reduced. Negative rewards were given if opposite of mentioned things happened.

**Pushing the Object to a Target Point**
First, we wanted the agent to be able to push the puck to any point on the screen. For all further model learning, we used the TQC algorithm, which proved more reliable than the DQN algorithm. Threshold for successful episode was set at 5 mm deviation from target point. In the simulation, the agent successfully pushed puck to target position on average in 7 out of 10 tries. The main reason for unsuccessful episodes was puck getting stuck on the wall of the environment and agent being unable to circle around the puck to push it from the wall. When tested on a real system, the robot was able, in most cases, to mimic the simulation (see Fig. 3). However, there were also problems detecting the correct position of the puck because the screen did not correctly detect centre of the puck, and this position is then compared to the centre of the puck in the simulation. Also, the centre of our tool is not always aligned with the robot's end effector. This is due to the tool being slightly displaced from its starting position when pushing objects, as for safety reasons, it is attached via a stiff spring in the gripper.

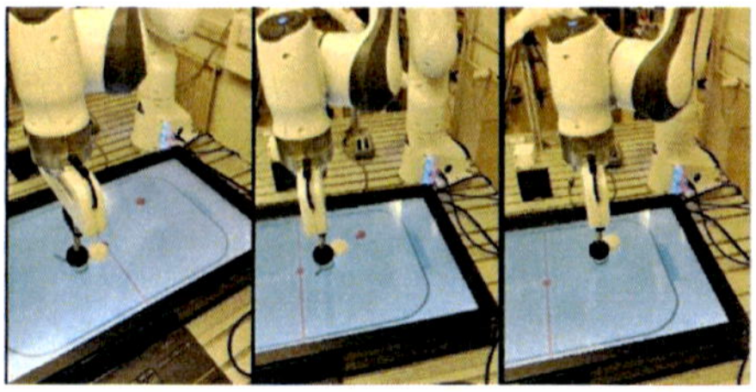

**Fig. 3.** Episode progress of pushing object to target point on a robot.

**Pushing the Object to a Target Position and Orientation**
We upgraded the previous task by introducing the target orientation of the object at the target point (see Fig. 4). In this task, we used a 3D-printed triangular object. The triangle stands on three legs located in the corners of the triangle (see Fig. 5).

We can read three touches on the screen to calculate the triangle's orientation. In the case of three separate touch recognitions $(P_1, P_2, P_3)$, we must

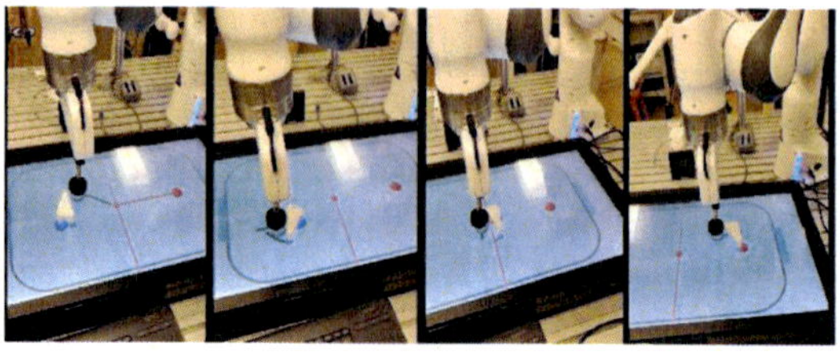

**Fig. 4.** Episode progress of pushing object to target pose on a robot.

determine which touch corresponds to which triangle point $(A, B, C)$. Finally, we can calculate triangle orientation from obtained triangle points.

Thresholds for completed episode were set at 5 mm for position and 5° for orientation. In the simulation, the agent aligns the triangle to its target position on average in 6 out of 10 episodes. In a few specific cases, it achieves the correct orientation and approaches the target point but fails to reach it in the required number of steps. Testing on the robot shows that it manages to orient the triangle correctly and starts pushing it to its target position. Problems arise around the target position when the robot tries to make small corrections. Due to the differences between the positions in the simulation and on the screen, the robot fails to properly move around the object, increasing the orientation error and making the final result worse. Figure 5 shows an episode in which the robot correctly oriented a triangle and pushed it close to the target point.

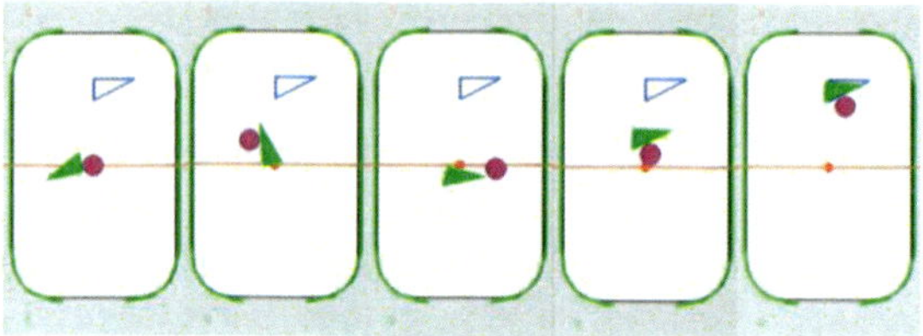

**Fig. 5.** Episode progress of pushing object to target pose in simulation.

## 4 Discussion

As part of this work, we have been able to build a custom 2D environment with a physics simulation in a PyBox2D environment and make the project compatible with the OpenAI Gym application programming interface, which allowed us to use reinforcement learning algorithms to learn different tasks in our environment. The agent successfully learned to play air hockey in the simulation using DQN and TQC algorithms. We were then able to transfer the model to the robot and successfully demonstrate the working principle of our system. In the second part, we explored the potential of our approach for object pushing tasks. We successfully taught the agent to push objects of different shapes in simulation, first to a target point and then to the correct pose, which included correct orientation. When tested on a robot, our approach showed potential, as the robot was able to correctly orient the object and push it to the target or at least close to the target. However, there were also problems when comparing the real position and the position in the simulation. Inaccurate detection of the centre of the puck on the screen and offset of the tool prevented the successful alignment of the object to a final position near the target point. These issues should be addressed in future work to approach the performance achieved by the same models in simulation.

## 5 Conclusion and Future Work

The reinforcement learning tasks demonstrated in this work represent only a fraction of the potential research opportunities in this field. In addition to upgrading the system, as previously mentioned, there are numerous topics related to pushing that could be explored further. For example, future work could focus on pushing objects amidst environmental obstacles or sequential pushing of multiple objects, which could translate to an in-plane assembly task. These research avenues have the potential to broaden the scope of knowledge in the field of manipulation and enhance the effectiveness of reinforcement learning techniques for real-world applications.

## References

1. Li, Z., Zhang, W., Chen, Y., Yu, H., Li, D.: Reinforcement learning for robust parameterized locomotion control of bipedal robots. In: 2021 IEEE International Conference on Robotics and Automation (ICRA) (2021)
2. Wang, X., Xiong, W., Wang, H., Wang, W.Y.: Look before you leap: bridging model-free and model-based reinforcement learning for planned-ahead vision-and-language navigation. In: Ferrari, V., Hebert, M., Sminchisescu, C., Weiss, Y. (eds.) ECCV 2018. LNCS, vol. 11220, pp. 38–55. Springer, Cham (2018). https://doi.org/10.1007/978-3-030-01270-0_3
3. Long, P., Fan, T., Liao, X., Liu, W., Zhang, H., Pan, J.: Towards optimally decentralized multi-robot collision avoidance via deep reinforcement learning. In: 2018 IEEE International Conference on Robotics and Automation (ICRA) (2018)
4. AlAttar, A., Rouillard, L., Kormushev, P.: Autonomous air-hockey playing cobot using optimal control and vision-based Bayesian tracking. In: Althoefer, K., Konstantinova, J., Zhang, K. (eds.) TAROS 2019. LNCS (LNAI), vol. 11650, pp. 358–369. Springer, Cham (2019). https://doi.org/10.1007/978-3-030-25332-5_31
5. Lowrey, K., Kolev, S., Dao, J., Rajeswaran, A., Todorov, E.: Reinforcement learning for non-prehensile manipulation: transfer from simulation to physical system. In: 2018 IEEE International Conference on Simulation, Modeling, and Programming for Autonomous Robots (SIMPAR) (2018)

6. Yuan, W., Hang, K., Kragic, D., Wang, M.Y., Stork, J.A.: End-to-end nonprehensile rearrangement with deep reinforcement learning and simulation-to-reality transfer. Robot. Auton. Syst. **119**, 119–134 (2019)
7. Rakhmati, A.: Air hockey simulator. https://github.com/arakhmati/gym-air-hockey
8. Sutton, R.S., Barto, A.G.: Reinforcement Learning: An Introduction. MIT Press, Cambridge (2018)
9. Ana, U.: Globoko spodbujevalno učenje robotskih strategij. Univerza v Ljubljani, Magistrsko delo (2022)
10. Kuznetsov, A., Shvechikov, P., Grishin, A., Vetrov, D.: Controlling overestimation bias with truncated mixture of continuous distributional quantile critics (2020)
11. OpenAI. "Gym". https://www.gym.ai

# In-Hand Pose Refinement Based on Contact Point Information

Iñigo Iturrate(✉), Yitaek Kim, Aljaz Kramberger, and Christoffer Sloth

The Maersk Mc-Kinney Moller Institute, University of Southern Denmark, Odense, Denmark
{inju,yik,alk,chsl}@mmmi.sdu.dk

**Abstract.** This paper presents a method for in-hand pose refinement using point pair features and robust estimation. The algorithm assumes that an initial pose distribution is available, and uses this for an initial guess of the correspondences between source and target data. The data needed for the algorithm is points and surface normal vectors that can be estimated from force-torque data; thus, tactile sensor arrays are not needed. The method is demonstrated on the in-hand pose estimation of an object in an articulated five finger hand.

**Keywords:** tactile manipulation · in-hand pose estimation · robust optimization

## 1 Introduction

Pose estimation has traditionally been approached as a computer-vision problem, where the pose of an object must be estimated given either 2D images [10] or 3D pointclouds [8] as input. However, once a robot manipulates an object, the view could become occluded. Although some vision solutions address in-hand pose estimation even for fine manipulation tasks [9], further handling of the object after its pose has been estimated assumes a stable grasp. In recent years, with a push towards more flexible robots with increasingly human-like capabilities, there has been a move towards more dexterous articulated grippers. Due to this trend, tactile pose estimation has become an active topic of research.

A wide range of methods for in-hand pose estimation rely on high-resolution tactile sensor arrays for tactile pose estimation [1,2,4,16]. Although the sensing modality of specific tactile sensor arrays may vary, they have in common that the output is similar to a pressure map or imprint on the sensor surface, often with multiple sensing modalities available [15]. For this reason, 2D computer vision methods can also be used on this data. A main limitation of the above approach is, however, that tactile sensor arrays are typically either expensive, fragile, or both, and as such may not be suited for industrial applications.

Instead of tactile sensor arrays, some approaches utilize force-torque sensor data [6,12,13,17], which is more readily available in many robot systems. A common approach is to use a particle filter [6,13] to iteratively refine the


T. Petrič et al. (Eds.): RAAD 2023, MMS 135, pp. 29–36, 2023.
https://doi.org/10.1007/978-3-031-32606-6_4

pose estimate based on contact between the target object and the environment. Alternatively, in our previous work [12], inspired by 3D computer vision methods based on point clouds, we proposed a method for contact-based pose estimation using robust optimization that relies on sliding the robot over the object's surface in order to generate pairs of points and associated surface normal vectors, from which correspondences are found using point pair feature and used as input to a robust pose estimation algorithm. The surface normal estimation can be accomplished using only force-torque sensor information [14]. As the algorithm has a combinatorial element, the execution time is high, and it is not suitable for online execution. In this paper, we propose to extend our previous framework by considering an initial guess for the pose distribution, reducing the search to points within a given confidence interval and thus also reducing the search time to allow for online execution.

## 2 Problem Formulation

We consider an in-hand 6D pose estimation problem using a multi-finger dexterous manipulator, where two (or more) fingers hold the object in hand, while another tactilely explores the object surface to uniquely determine its pose. We assume that a pose distribution is known a priori. In the given case, the distribution could be derived from how the object is constrained by the grasp, but in a more general sense it could, for example, come from a noisy computer-vision system, as considered in [5], or directly from a probabilistic pose distribution estimator [11]. While we assume that a CAD model of the object is available, we do not make any assumptions about the shape of the object itself. Figure 1 shows an overview of the proposed problem description and the considered object and manipulator.

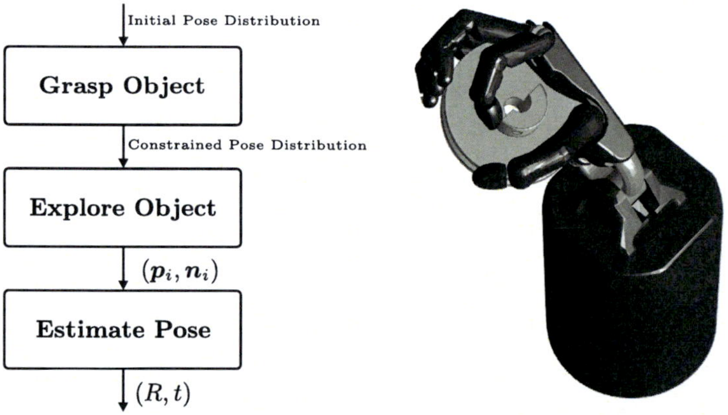


**Fig. 1.** Illustration of the proposed pose estimation from initial pose distribution to pose estimate given by $(R, t)$ based on positions and surface normals $(\boldsymbol{p}_i, \boldsymbol{n}_i)$. The considered object and robot hand is shown to the right.

## 3 In-Hand Pose Estimation

This section outlines in-hand pose estimation based on position measurements and surface normal vectors obtained from an articulated robot hand and is based on [12]. The estimation has the following steps:

1. **Data Collection:** We assume that position and surface normals are obtained from manipulation of an object $(\boldsymbol{p}_i, \boldsymbol{n}_i)$, where $\boldsymbol{p}_i \in \mathbb{R}^3$ is the measured position, and $\boldsymbol{n}_i \in \mathbb{R}^3$ is the estimated surface normal vector for $i = 1, \ldots, N$.
2. **Correspondence Generation:** Find correspondences between source data $(\boldsymbol{p}_i, \boldsymbol{n}_i)$ and target data $(\boldsymbol{q}_j, \boldsymbol{m}_j)$, i.e. a function $\gamma : \{1, \ldots, N\} \rightarrow \{1, \ldots, M\}$. The algorithm for finding correspondences is explained in Sect. 3.1.
3. **Pose Estimation:** Find a homogeneous transformation $(R, t) \in SE(3)$ from target to source data as explained in Sect. 3.2.

### 3.1 Correspondence Generation

The pose estimation presented in Sect. 3.2 computes a homogeneous transformation between a collection of pairs of points; hence, correspondences must be available to generate an initial parring of source data (measurements) and target data (CAD data). The correspondences are generated using the Point Pair Feature (PPF) method [7], where a PPF is defined as follows.

**Definition 1 (Point Pair Feature).** *A Point Pair Feature (PPF) is a tuple*

$$\mathbf{F}(\boldsymbol{p}_1, \boldsymbol{p}_2) = \big(v_d, \angle(\boldsymbol{n}_1, \boldsymbol{v}_d), \angle(\boldsymbol{n}_2, \boldsymbol{v}_d), \angle(\boldsymbol{n}_1, \boldsymbol{n}_2)\big) \tag{1}$$

*where $\boldsymbol{p}_1, \boldsymbol{p}_2 \in \mathbb{R}^3$ are points on a surface, $\boldsymbol{n}_1, \boldsymbol{n}_2 \in \mathbb{R}^3$ are surface normal vectors at each point, $v_d = ||\boldsymbol{p}_2 - \boldsymbol{p}_1||_2$ is the distance, and $\angle(\boldsymbol{a}, \boldsymbol{b})$ is the unsigned angle between vectors $\boldsymbol{a}$ and $\boldsymbol{b}$.*

Point pair features are computed between all pairs of points in most computer vision algorithms; consequently, the computational burden of generating correspondences is high. Also, cameras provide a dense point cloud whereas only a line of contact points is available in this application. Therefore, it is necessary to use a dedicated algorithm for finding the correspondences with the limited input data; we use a modification of the algorithm provided in [12]. This algorithm traverses the source data and identifies a best direction in a neighborhood. The algorithm has the following steps:

1. **Partition the Source Data into Segments:** Partition the source data $(\boldsymbol{p}_i, \boldsymbol{n}_i)$ into segments consisting of all points within a given radius $R$.
2. **Find Direction:** The best direction of the source data with respect to the CAD model is identified in each segment with respect to a given initial point in the target data.
3. **Connect Segments:** Ensure that neighboring segments are adjacent in the target data.

Remark that step 2 in the algorithm requires an initial point in the target data; thus, if no prior knowledge about this point exists then an exhaustive search through all target points is necessary; this is a main source of the high computational complexity. We assume that an initial distribution of poses is available; therefore, we modify the algorithm as follows:

- **Restrict Initial Points:** The initial point in the target data used in Step 2 of the algorithm is restricted according to the initial pose distribution.
- **Restrict Possible Directions:** The possible directions found in Step 2 are constrained according to the initial pose distribution.

### 3.2 Robust Pose Estimation

To obtain the transformation between the hand frame and the object frame, an optimization problem is formulated based on the correspondences given in Sect. 3.1. Since the correspondences are likely to be wrong, we use robust optimization that is less sensitive to outliers than other optimization algorithms. In particular, we consider the optimization problem

$$\min_{(R,t)\in SE(3)} \sum_{i=1}^{N} \rho\left(\left|\left|r(\boldsymbol{p}_i, \boldsymbol{q}_{\gamma(i)})\right|\right|_2\right) \tag{2}$$

where $\rho(\cdot)$ is a robust cost function and $r(\boldsymbol{p}_i, \boldsymbol{q}_{\gamma(i)})$ is the residual error between source data $\boldsymbol{p}_i$ and transformed target data $\boldsymbol{q}_{\gamma(i)}$. We use the method proposed in [18]. Specifically, the Black-Rangarajan duality is used for reformulating (2) using an outlier process $\Phi_\rho(W_i)$ that depends on weights $W_i$ and robust cost function $\rho(\cdot)$, and graduated non-convexity introduces an adaptive parameter $\mu$ that is gradually decreased to recover the robust cost functions $\rho(\cdot)$ from an initially convex surrogate cost function $\rho_\mu(\cdot)$. The algorithm used for solving optimization problem (2) is shown below.

**Algorithm 1 (Robust Estimation)**
***Input:*** *functions* $\rho_\mu$ *and* $\Phi_\rho$, *initial value* $\mu_0 \in \mathbb{R}$, *bounds* $m_{\max}$ *and* $\mu_{\min}$
***Output:*** $(R, t)$, $W_i$
***Procedure:***

*1:* $\mu \leftarrow \mu_0$
*2:* ***while*** $\mu < \mu_{\min}$ ***do***
*3:* $\quad m \leftarrow 0,\ W_i \leftarrow 1$
*4:* $\quad$ ***while*** $m < m_{\max}$ ***do***
*5:* $\quad\quad (R,t)^{(m)} = \underset{(R,t)\in SE(3)}{\arg\min} \sum_{i=1}^{N} W_i^{(m-1)} \left|\left|r(\boldsymbol{p}_i, \boldsymbol{q}_{\gamma(i)})\right|\right|_2^2$
*6:* $\quad\quad W_i^{(m)} = \underset{W_i\in[0,1]}{\arg\min} \sum_{i=1}^{N} \left[W_i \left|\left|r(\boldsymbol{p}_i, \boldsymbol{q}_{\gamma(i)})\right|\right|_2^2 + \Phi_{\rho_\mu}(W_i)\right]$
*7:* $\quad$ ***end while***
*8:* $\quad$ *decrease* $\mu$
*9:* ***end while***

The optimization problems in Algorithm 1 are formulated as Quadratically Constrained Quadratic Programs (QCQP), see [3] for details. We adopt the definition of residual $r$ from [12] that includes both pose data and surface normal vectors. Finally, it is seen that $W_i$ will go to zero if $\boldsymbol{p}_i$ is an outlier, i.e. if the correspondence associated to the $i$th point is found to be wrong; thus, these weights can be used for approximating the outlier ratio.

## 4 Evaluation

We evaluated our algorithm in simulation on part of an industrial pulley assembly, which presents no rotational symmetry. The part was grasped by the edges, as depicted in Fig. 1, while another finger explored other parts of the part's surface following the path in Fig. 2. The algorithm was initialized for all points within a radius of 3 mm of the mean value given by the prior pose distribution.

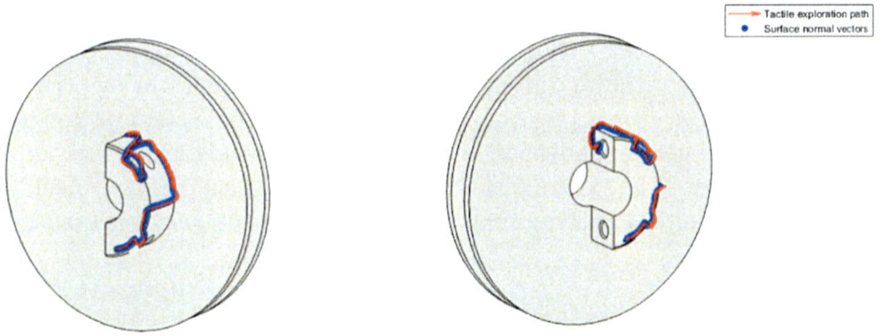


**Fig. 2.** Tactile exploration path (blue) alongside corresponding surface normal vectors (red).

Table 1 shows the resulting absolute error of the pose estimation for position and orientation, using Graduated Non-Convexity (GLS) with either Truncated Least Squares (TLS) or Geman-McClure (GM) as the robust cost function. This is compared with a standard Least-Squares cost function. Overall, TLS performed the best for this part, with a positional error of $5.46 \times 10^{-4}$ mm and an angular error of $9.12 \times 10^{-5}$ degrees, while GM slightly outperformed it for the orientation only, with an error of $1.55 \times 10^{-5}$ degrees. Meanwhile, the non-robust LS cost function was severely impacted by outliers in the point pair correspondences and showed much larger errors in both position and orientation.

We were interested in evaluating whether it was possible to determine a stop condition for the tactile exploration path based on the outlier ratio of the robust cost function. We approximated the outlier ratio as $1 - \mu_{\mathbf{W}}$, where $\mu_{\mathbf{W}}$ is the mean of the weights defined in Algorithm 1. It is expected that, as the optimization problem converges, the outlier ratio will become progressively lower as the number of true correspondences increases. Figure 3 compares the best

**Table 1.** Absolute pose estimation error for the position and orientation using Graduated Non-Convexity (GLS) with either Truncated Least Squares (TLS) or Geman-McClure (GM) robust cost functions, compared to standard Least-Squares (LS).

| | Absolute Error | |
|---|---|---|
| | Position [mm] | Orientation [°] |
| TLS (GNC) | $\mathbf{5.46 \times 10^{-4}}$ | $9.12 \times 10^{-5}$ |
| GM (GNC) | $6.94 \times 10^{-2}$ | $\mathbf{1.55 \times 10^{-5}}$ |
| LS | 3.69 | 2.068 |

performing run of all initial guesses based on the initial pose distribution with a second less well-performing run based on a sub-optimal start guess. Observe that, while the sub-optimal run shows an increasing tendency until the number of path correspondences reaches 348, the optimal start guess shows no such tendency, and in fact shows a decreasing tendency after 326 correspondences. Note also that, even for a shorter exploration path, the best-performing start guess exhibits a consistently lower outlier ratio. We suggest that this metric can be used as an early stop condition for the algorithm, forcing the tactile exploration to terminate once a desired outlier (or rather, inlier) ratio is achieved. Note that for paths with less than 305 correspondences, the pose estimation algorithm was unable to converge. Figure 4 shows the physical exploration paths on the object that correspond to 303 and 349 correspondences, respectively, on the best performing start guess.

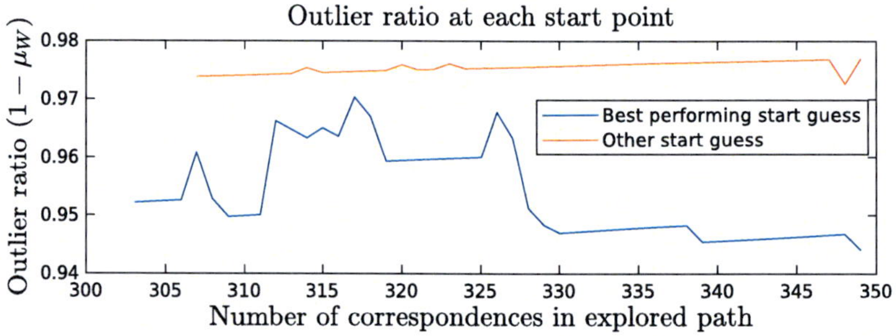


**Fig. 3.** Approximation of the outlier ratio of correspondences for a given length of a path (given as a number of correspondences for the total path length), compared for the best-performing start guess (blue) within the initial search area based on the prior pose distribution, and for a less well-performing start guess (red).

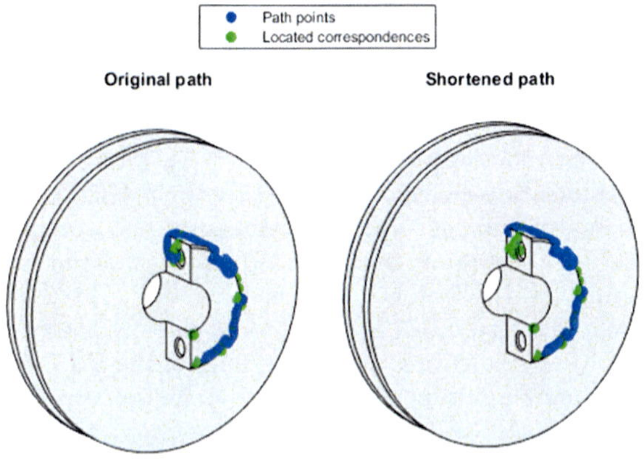


**Fig. 4.** Exploration paths (blue) and located point pair correspondences (green) for the full exploration path (left) with 349 correspondences, and a shortened path (right) with 303 correspondences. The path lengths in this graph should be seen in relation to the results for the outlier ratio shown in Fig. 3.

## 5 Conclusion

In this paper, we have proposed a framework for in-hand pose estimation using a dexterous hand. We assume the availability of a prior pose distribution, which could either be given by a noisy computer vision-based pose estimate or by the way the object is constrained in the hand. Based on this, we define an initial search region within a confidence interval of the mean of the pose estimate distribution. This is used to set the starting guess for a robust pose estimation algorithm based on point pair features and robust estimation. Our results show that robust pose estimation significantly reduces the estimation error compared to standard least-squares-based fitting. They also show a tendency in the outlier ratio of weight optimization step of the optimization algorithm that could be used as a stop condition for early termination of the tactile exploration phase, greatly speeding up the estimation process.

Our future work will consider the use of Bayesian Optimization to more efficiently choose the samples for the tactile exploration phase.

## References

1. Bauza, M., Bronars, A., Rodriguez, A.: Tac2Pose: Tactile object pose estimation from the first touch. arXiv preprint arXiv:2204.11701 (2022)
2. Bimbo, J., Luo, S., Althoefer, K., Liu, H.: In-hand object pose estimation using covariance-based tactile to geometry matching. IEEE Robot. Autom. Lett. **1**(1), 570–577 (2016)
3. Briales, J., Gonzalez-Jimenez, J.: Convex global 3D registration with Lagrangian duality. In: 2017 IEEE Conference on Computer Vision and Pattern Recognition (CVPR), pp. 5612–5621 (2017)

4. De Farias, C., Marturi, N., Stolkin, R., Bekiroglu, Y.: Simultaneous tactile exploration and grasp refinement for unknown objects. IEEE Robot. Autom. Lett. **6**(2), 3349–3356 (2021)
5. von Drigalski, F., et al.: Precise multi-modal in-hand pose estimation using low-precision sensors for robotic assembly. In: 2021 IEEE International Conference on Robotics and Automation (ICRA), pp. 968–974. IEEE (2021)
6. von Drigalski, F., Kasaura, K., Beltran-Hernandez, C.C., Hamaya, M., Tanaka, K., Matsubara, T.: Uncertainty-aware manipulation planning using gravity and environment geometry. IEEE Robot. Autom. Lett. **7**(4), 11942–11949 (2022)
7. Drost, B., Ulrich, M., Navab, N., Ilic, S.: Model globally, match locally: Efficient and robust 3D object recognition. In: Proceedings of the IEEE Computer Society Conference on Computer Vision and Pattern Recognition, pp. 998–1005 (2010)
8. Hagelskjær, F., Buch, A.G.: PointVoteNet: Accurate object detection and 6 DoF pose estimation in point clouds. In: 2020 IEEE International Conference on Image Processing (ICIP), pp. 2641–2645. IEEE (2020)
9. Hagelskjær, F., Kraft, D.: In-hand pose estimation and pin inspection for insertion of through-hole components. In: 2022 IEEE 18th International Conference on Automation Science and Engineering (CASE), pp. 382–389. IEEE (2022)
10. Hodaň, T., et al.: BOP: Benchmark for 6D object pose estimation. In: Ferrari, V., Hebert, M., Sminchisescu, C., Weiss, Y. (eds.) ECCV 2018. LNCS, vol. 11214, pp. 19–35. Springer, Cham (2018). https://doi.org/10.1007/978-3-030-01249-6_2
11. Iversen, T.M., Haugaard, R.L., Buch, A.G.: Ki-Pode: Keypoint-based implicit pose distribution estimation of rigid objects. arXiv preprint arXiv:2209.09659 (2022)
12. Kim, Y., Kramberger, A., Buch, A.G., Sloth, C.: Contact-based pose estimation of workpieces for robotic setups. In: Proceedings of the 2023 IEEE International Conference on Robotics and Automation (ICRA) (2023)
13. Sipos, A., Fazeli, N.: Simultaneous contact location and object pose estimation using proprioception and tactile feedback. In: 2022 IEEE/RSJ International Conference on Intelligent Robots and Systems (IROS), pp. 3233–3240. IEEE (2022)
14. Sloth, C., Kramberger, A., Diget, E.L., Iturrate, I.: Towards contact point and surface normal estimation for control of flexible tool. In: 2022 American Control Conference (ACC), pp. 500–505. IEEE (2022)
15. Taylor, I.H., Dong, S., Rodriguez, A.: Gelslim 3.0: High-resolution measurement of shape, force and slip in a compact tactile-sensing finger. In: 2022 International Conference on Robotics and Automation (ICRA), pp. 10781–10787. IEEE (2022)
16. Villalonga, M.B., Rodriguez, A., Lim, B., Valls, E., Sechopoulos, T.: Tactile object pose estimation from the first touch with geometric contact rendering. In: Conference on Robot Learning, pp. 1015–1029. PMLR (2021)
17. Von Drigalski, F., et al.: Contact-based in-hand pose estimation using Bayesian state estimation and particle filtering. In: 2020 IEEE International Conference on Robotics and Automation (ICRA), pp. 7294–7299. IEEE (2020)
18. Yang, H., Antonante, P., Tzoumas, V., Carlone, L.: Graduated non-convexity for robust spatial perception: From non-minimal solvers to global outlier rejection. IEEE Robot. Autom. Lett. **5**(2), 1127–1134 (2020)

# Explainable Object Detection in the Field of Search and Rescue Robotics

Peter Hönig[1,2](✉) and Wilfried Wöber[2,3]

[1] Automation and Control Institute, TU Wien, Vienna, Austria
hoenig@acin.tuwien.ac.at

[2] Department of Industrial Engineering, University of Applied Sciences Technikum Wien, Vienna, Austria
woeber@technikum-wien.at

[3] Institute for Integrative Nature Conservation Research, University of Natural Resources and Life Sciences Vienna, Vienna, Austria

**Abstract.** State-of-the-art object detection in search and rescue robotics relies on CNNs, which reach excellent scores but lack explainability due to their "black box" characteristic. In such a domain, mission failure or misdetections can have drastic consequences. Therefore research should strive to increase the explainability regarding the CNNs' classification strategies. In this paper, existing methods for object detection are applied and investigated in the context of search and rescue robotics in order to compose a fully explainable pipeline. Unlike existing object detection methods, the presented method is based on an exhaustive model investigation concerning post-hoc explainability. The method is applied to detecting handwheels of gate-valves, with a post-hoc analysis of the classification strategies learned by the object detector. In order to train and test the object detection model, a novel dataset is composed, including 2447 images, nine handwheel types, and 6696 annotations. Five CNN object detectors (R-CNN, Faster R-CNN, RetinaNet, YOLOv5s, and SSD) are compared based on $\mathrm{mAP}^{0.5}$ and $\mathrm{mAP}^{0.5:0.95}$. Two CNN object detectors' classification strategies were investigated with the SpRAy method. Comparing $\mathrm{mAP}^{0.5}$, $\mathrm{mAP}^{0.5:0.95}$, and inference times reveals that YOLOv5s is the superior model across all categories. SpRAy analysis of R-CNN and YOLOv5s did not reveal any abnormal classification strategies, which indicates a well-balanced dataset. YOLOv5s appears to have learned different classification strategies for different handwheel types. Our handwheels dataset is available at: https://www.kaggle.com/hoenigpet/handwheels-for-gatevalves.

**Keywords:** object detection · pose estimation · explainable artificial intelligence · convolutional neural networks

## 1 Introduction

Search and rescue robotics (SAR) is a subdomain of mobile robotics, where robots are deployed in disaster sites and aid first responders in various tasks,


T. Petrič et al. (Eds.): RAAD 2023, MMS 135, pp. 37–44, 2023.
https://doi.org/10.1007/978-3-031-32606-6_5

including mobile manipulation [25]. In order to perform manipulation, object detection and pose estimation are necessary [27]. These methods combine classic computer vision [34] and machine learning [19], or machine learning exclusively [33]. Machine learning, especially deep learning, offers higher flexibility and accuracy than classic computer vision approaches. The downside is needing more transparency and explainability, reducing trust in these decision-making systems [5,30]. Object detection and pose estimation methods based on machine learning are frequently used in SAR [9,18,19]. However, these machine learning applications still need to be thoroughly investigated regarding explainability. We claim that in SAR, where quality and completion of tasks are crucial, the explainability of decision-making systems contributes to more trust in such systems.

In order to overcome the "black box" characteristic of machine learning methods, explainable artificial intelligence (XAI) [1,5,28] methods, such as layerwise relevance propagation (LRP) [6] and spectral relevance analysis (SpRAy) [14], have been developed. These methods attempt to explain the models' classification strategies, get insight into which features a model has learned, and reveal how well the model generalizes [8]. Furthermore, a "Clever-Hans" effect [14], a significant flaw in the dataset preventing the model from learning representative features and resulting in corrupt classification strategies, can be identified. Therefore, XAI methods can help improve prediction models and datasets, increasing explainability and trust in object detection.

This paper tackles the XAI problem for search and rescue robots. A pipeline composed of existing object detection and XAI methods is presented and evaluated using a task from the ELROB challenge 2022 [24], namely detecting handwheels of gate-valves. Our presented pipeline is visualized in Fig. 1. The contri-

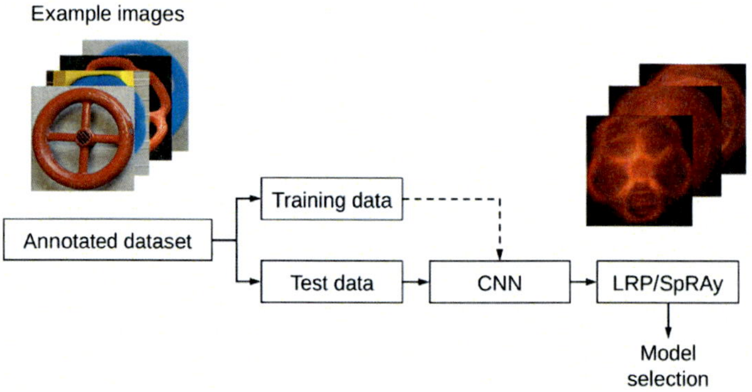


**Fig. 1.** Visualization of the proposed pipeline. In addition to the quantitative analysis, we rely on XAI methods to manually investigate the CNN's decision.

bution of this paper is the presentation of a novel handwheels dataset for SAR applications, the XAI-based investigation of the respective dataset and a quantitative and qualitative analysis of the results based on state-of-the-art object detectors.

This paper proceeds with an overview of the state of the art in Sect. 2. Selected methods are described in Sect. 3. Results are presented and discussed in Sect. 4. A final summary and outlook are given in Sect. 5.

## 2 State of the Art

Object detection in SAR is recently based on convolutional neural networks (CNNs) [11]. In [9], a framework was implemented to detect victims based on YOLOv3 [21] and RetinaNet [15]. A stair pose estimator for robots in disaster scenarios was proposed in [18], based on SSD [17]. Handwheels of gate-valves were previously detected in [19], using Faster R-CNN [23]. Lately, however, these deep learning-based models have been reported to cheat [14,31] under certain conditions, which is critical in safety-related domains. Furthermore, deep learning models and their applications were reported to have methodological shortcomings [20], limiting their usage, especially in SAR. In order to increase reliability and detect cheating behavior, explainability methods such as LRP [6] or GradCAM [26] were developed to investigate the decisions of neural networks post-hoc by applying them after model training. These methods investigate the reasons for a model's decision based on an input image or image database [2,4]. These approaches differ from traditional machine learning explainability definitions such as [7], which claim that reliable artificial intelligence-based systems must fundamentally understand their environment's explanatory factors. Hence, traditional model investigation methods such as *Eigenfaces* for face detection [29] investigate the parameters of the models and do not rely on sample prediction.

Previously, these traditional methods were applied for recent machine learning models [31,32], where the authors provide an explainable machine learning-based object classification pipeline. Nevertheless, the abovementioned pipeline is inferior to supervised CNNs regarding prediction accuracy. Furthermore, these models were not adapted for object detection. Therefore object detection applications are still driven by supervised CNNs, which must be investigated carefully using quantitative metrics such as detection accuracy and qualitative model investigation [14]. The number of available implementations of CNN-based object detector explanation pipelines limits this investigation. LRP-based model investigation software is available for R-CNN [3] and YOLOv5s [13].

Based on the literature above, we argue that a researcher's core task for computer vision-based SAR applications must be the careful and exhaustive investigation of deep learning models with existing explanation pipelines. Furthermore, we argue that model explanation must be preferred to model prediction accuracy if both models perform similarly.

## 3 Materials and Methods

In order to evaluate the explainability of the CNN models, a novel image dataset was composed. To our knowledge, no public handwheel dataset was available during our research. 2447 images were recorded in 13 scenes, using nine different handwheel types, resulting in 6696 handwheel annotations. Figure 2 shows a random selection of handwheel images.

**Fig. 2.** Randomly selected images of the nine handwheel types used in this study.

The 13 scenes were recorded using a smartphone camera with 30fps and $1920 \times 1080$ resolution, moving with approximately constant velocity. The camera was tilted randomly during recording, with a maximum of approximately $+/- 30°$ around each axis. Out of the recorded videos, every $15^{th}$ frame was saved as an image to the dataset. The dataset was annotated according to the COCO annotation format [16] and is publicly available at https://www.kaggle.com/hoenigpet/handwheels-for-gatevalves.

After dataset preparation, five CNN object detectors, namely R-CNN [10], Faster R-CNN [23], RetinaNet [15], YOLOv5s [12], and SSD [17], were implemented, trained, and tested by calculating mean average precision (mAP). 10-fold cross-validation was used [22] for training and testing the CNNs. All five CNNs were trained in order to evaluate the overall performance. Two CNN object detectors (R-CNN and YOLOv5s) have available XAI tools. Therefore, the R-CNN toolbox of [3], which analyzes the ResNet18 backbone of R-CNN, and the YOLOv5 toolbox of [13] were used to apply LRP and SpRAy. To this end, we aim to train all object detectors, evaluate their performance and investigate the post-hoc model explanation. Motivated by SAR and the black-box problem, we argue that explainable models are preferred if the model performance does not differ significantly.

For each test sample, LRP backpropagation results in a heatmap, which indicates areas that are highly influential for the classification decision. With SpRAy, an eigengap analysis of all heatmaps is conducted. Based on that eigengap analysis, the count of clusters is defined [14]. All heatmaps within each cluster are averaged to analyze each cluster's classification strategy. Out of all implemented CNN object detectors, the model with the best performance (mAP and inference time) and most reasonable classification strategies was chosen for object detection. An RTX 3060 GPU with 12GB RAM and a Ryzen 5 CPU with a 3.8 GHz base clock were used.

## 4 Results

The object detection performance is shown in Table 1. This investigation unveils that YOLOv5s has the highest $mAP^{0.5}$ and $mAP^{0.5:0.95}$ values across all ten dataset folds. Furthermore, YOLOv5s showed the fastest inference time with

0.013s. R-CNN has the least mAP scores and the slowest inference time out of all implemented CNN object detectors. Table 1 also reveals that SSD matches the $mAP^{0.5}$ score of YOLOv5s on the handwheels dataset while outperforming YOLOv5s regarding the $mAP^{0.5}$ score on the COCO/PASCAL benchmarking datasets. Based on the performance of SSD, an XAI toolbox should be developed to investigate if these scores are achieved while relying on representative features.

**Table 1.** $mAP^{0.5}$ on benchmark (BM; COCO and PASCAL) datasets for comparison, $mAP^{0.5}$, $mAP^{0.5:0.95}$ and inference per image over all folds; green cells indicate the best value within category, red cells indicate the worst value within category

| Model | $mAP^{0.5}$ BM | $mAP^{0.5}$ | $mAP^{0.5:0.95}$ | Inference [s] |
|---|---|---|---|---|
| R-CNN (ResNet18) | 0.533 | 0.588 | 0.372 | 7.7 |
| Faster R-CNN | 0.704 | 0.988 | 0.935 | 0.130 |
| RetinaNet | 0.591 | 0.989 | 0.930 | 0.117 |
| YOLOv5s | 0.568 | 0.994 | 0.980 | 0.013 |
| SSD | 0.743 | 0.994 | 0.871 | 0.053 |

Regarding the explainability of classification, the eigengap analysis of SpRAy neither showed significant eigengaps for R-CNN nor YOLOv5s. The number of eigengaps correlates with the number of clusters, as described in [14]. Since no significant eigengaps could be identified visually, the number of clusters was defined as nine. Therefore each handwheel type results in a classification cluster. Based on that assumption, it was expected that the average heatmaps of each cluster must show representative visual features according to the respective handwheel type, such as the count of bars and size of the centerpiece and outer ring.

Figure 3 visualizes the averaged heatmaps for each element in each of the nine clusters for ResNet18 (left) and YOLOv5s (right). The SpRAy analysis of ResNet18 shows that while in clusters four, five, and eight, attribution can be observed in the center of the heatmaps, resembling the centerpiece of a handwheel, the remaining clusters show much less attribution overall. Clusters three, six, and seven show increased attribution in the corners of the heatmaps. After analyzing the corresponding test samples of these clusters for a possible "Clever-Hans" effect, no dataset bias could be identified.

For the SpRAy analysis of YOLOv5s (Fig. 3, right), the averaged heatmaps of clusters one, three, and five clearly resemble the handwheel types four, six, and seven (refer to Fig. 2). For the remaining clusters, attribution is mostly located in the center of the heatmaps. This indicates strong reliance on the area of the inner piece of the handwheel. No attribution is located at the edges or in areas that would not be representative for a handwheel. YOLOv5s did apparently not learn different classification strategies for each handwheel type. However, different classification strategies have been learned for handwheel types four, six, and seven. In contrast, YOLOv5s concentrates on the areas of the bars, the outer ring, and the centerpiece for the remaining handwheel types.

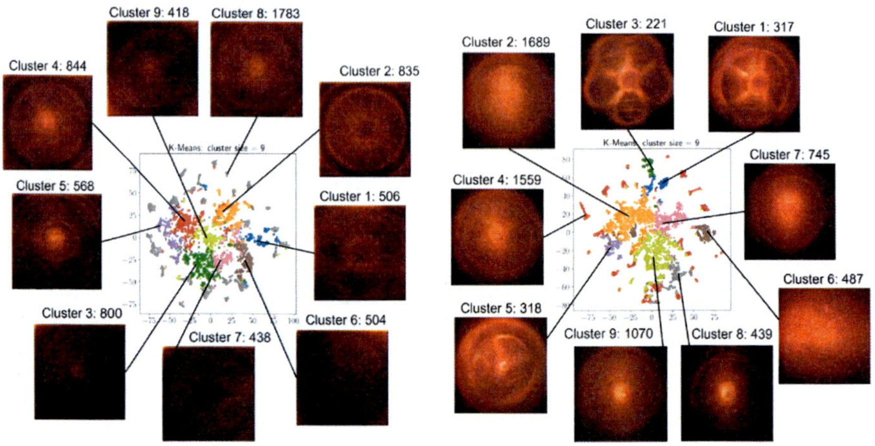


**Fig. 3.** Average heatmaps of nine k-means clusters based on ResNet18 (left) and YOLOv5s (right)

Since YOLOv5s was superior in performance and showed no abnormal classification strategies, it was chosen for the explainable object detection and pose estimation pipeline. The different classification strategies learned by ResNet18 and YOLOv5s, even though identical training and test data were used, could potentially result from different data augmentation techniques, loss functions, learning rates, and, most of all, model architecture.

## 5 Summary and Outlook

In this paper, an exhaustive object detection model investigation for detecting handwheels of gate-valves was presented. All models were trained and tested using a novel handwheels dataset containing 2447 images, nine handwheel types, and 6696 annotations. Results showed that YOLOv5s is the superior model for $mAP^{0.5}$, $mAP^{0.5:0.95}$, and inference time compared to R-CNN, Faster R-CNN, RetinaNet, and SSD. SpRAy analysis of R-CNN was inconclusive, while SpRAy analysis of YOLOv5s showed that meaningful features and classification strategies for certain handwheel types had been learned. Further research includes developing and improving XAI methods for CNNs, aiming for explainable decisions made by object detection and pose estimation algorithms when deployed in high-risk scenarios.

**Acknowledgments.** We gratefully acknowledge the support of the Austrian Research Promotion Agency (FFG), project SmartDis, and the Austrian Science Fund (FWF), under project No. I 6114, project iChores.

## References


1. Adadi, A., Berrada, M.: Peeking inside the black-box: a survey on explainable artificial intelligence (XAI). IEEE Access **6**, 52138–52160 (2018). https://doi.org/10.1109/ACCESS.2018.2870052
2. Anders, C.J., Neumann, D., Marin, T., Samek, W., Müller, K.R., Lapuschkin, S.: XAI for analyzing and unlearning spurious correlations in imagenet. In: ICML'20 Workshop on Extending Explainable AI Beyond Deep Models and Classifiers (XXAI) (2020)
3. Anders, C.J., Neumann, D., Samek, W., Müller, K., Lapuschkin, S.: Software for dataset-wide XAI: from local explanations to global insights with zennit, corelay, and virelay. CoRR abs/2106.13200 (2021). https://arxiv.org/abs/2106.13200
4. Anders, C.J., Weber, L., Neumann, D., Samek, W., Müller, K.R., Lapuschkin, S.: Finding and removing clever hans: Using explanation methods to debug and improve deep models. Inf. Fusion **77**, 261–295 (2022)
5. Arrieta, A.B., et al.: Explainable artificial intelligence (XAI): concepts, taxonomies, opportunities and challenges toward responsible AI. Inf. Fusion **58**, 82–115 (2020)
6. Bach, S., Binder, A., Montavon, G., Klauschen, F., Müller, K.R., Samek, W.: On pixel-wise explanations for non-linear classifier decisions by layer-wise relevance propagation. PLoS ONE **10**(7), e0130140 (2015)
7. Bengio, Y., Courville, A., Vincent, P.: Representation learning: a review and new perspectives. IEEE Trans. Pattern Anal. Mach. Intell. **35**(8), 1798–1828 (2013)
8. Doshi-Velez, F., Kim, B.: Considerations for evaluation and generalization in interpretable machine learning. In: Escalante, H.J., Escalera, S., Guyon, I., Baró, X., Güçlütürk, Y., Güçlü, U., van Gerven, M. (eds.) Explainable and Interpretable Models in Computer Vision and Machine Learning. TSSCML, pp. 3–17. Springer, Cham (2018). https://doi.org/10.1007/978-3-319-98131-4_1
9. Fung, A., Wang, L.Y., Zhang, K., Nejat, G., Benhabib, B.: Using deep learning to find victims in unknown cluttered urban search and rescue environments. Curr. Robot. Rep. **1**(3), 105–115 (2020)
10. Girshick, R., Donahue, J., Darrell, T., Malik, J.: Rich feature hierarchies for accurate object detection and semantic segmentation. In: 2014 IEEE Conference on Computer Vision and Pattern Recognition, pp. 580–587. IEEE (2013)
11. Goodfellow, I.J., Bengio, Y., Courville, A.: Deep Learning. MIT Press, Cambridge (2016)
12. Jocher, G., et al.: ultralytics/yolov5: v6.1 - tensorrt, tensorflow edge TPU and openvino export and inference (2022). https://doi.org/10.5281/zenodo.6222936
13. Karasmanoglou, A., Antonakakis, M., Zervakis, M.: Heatmap-based explanation of YOLOv5 object detection with layer-wise relevance propagation. In: 2022 IEEE International Conference on Imaging Systems and Techniques (IST), pp. 1–6. IEEE (2022)
14. Lapuschkin, S., Wäldchen, S., Binder, A., Montavon, G., Samek, W., Müller, K.R.: Unmasking clever hans predictors and assessing what machines really learn. Nat. Commun. **10**(1), 1–8 (2019)
15. Lin, T.Y., Goyal, P., Girshick, R., He, K., Dollar, P.: Focal loss for dense object detection. IEEE Trans. Pattern Anal. Mach. Intell. **42**(2), 318–327 (2020)
16. Lin, T.Y., et al.: Microsoft COCO: common objects in context. In: Fleet, D., Pajdla, T., Schiele, B., Tuytelaars, T. (eds.) ECCV 2014. LNCS, vol. 8693, pp. 740–755. Springer, Cham (2014). https://doi.org/10.1007/978-3-319-10602-1_48

17. Leibe, B., Matas, J., Sebe, N., Welling, M. (eds.): ECCV 2016. LNCS, vol. 9906. Springer, Cham (2016). https://doi.org/10.1007/978-3-319-46475-6
18. Miyakawa, K., Kanda, T., Ohya, J., Ogata, H., Hashimoto, K., Takanishi, A.: Automatic estimation of the position and orientation of stairs to be reached and climbed by a disaster response robot by analyzing 2D image and 3D point cloud. Int. J. Mech. Eng. Rob. Res. **9**(9), 1312–1321 (2020)
19. Nishikawa, K., Ohya, J., Matsuzawa, T., Takanishi, A., Ogata, H., Hashimoto, K.: Automatic detection of valves with disaster response robot on basis of depth camera information. In: 2018 Digital Image Computing: Techniques and Applications (DICTA). IEEE (2019)
20. Pearl, J., Mackenzie, D.: The Book of Why. Basic Books, New York (2018)
21. Redmon, J., Farhadi, A.: YOLOv3: An incremental improvement. arXiv preprint arXiv:1804.02767 (2018). https://arxiv.org/abs/1804.02767v1
22. Refaeilzadeh, P., Tang, L., Liu, H.: Cross-validation. Encycl. Database Syst. 1–7 (2016)
23. Ren, S., He, K., Girshick, R., Sun, J.: Faster R-CNN: towards real-time object detection with region proposal networks. IEEE Trans. Pattern Anal. Mach. Intell. **39**(6), 1137–1149 (2017)
24. Schneider, F.E.: ELROB - the European land robot trial (2022). https://www.elrob.org/
25. Schneider, F.E., Wildermuth, D.: Assessing the search and rescue domain as an applied and realistic benchmark for robotic systems. In: 2016 17th International Carpathian Control Conference (ICCC), pp. 657–662. IEEE (2016)
26. Selvaraju, R.R., Cogswell, M., Das, A., Vedantam, R., Parikh, D., Batra, D.: Grad-CAM: visual explanations from deep networks via gradient-based localization. In: Proceedings of the IEEE International Conference on Computer Vision, pp. 618–626 (2017)
27. Shim, V.A., Yuan, M., Tan, B.H.: Automatic object searching by a mobile robot with single RGB-D camera. In: 2017 Asia-Pacific Signal and Information Processing Association Annual Summit and Conference (APSIPA ASC), pp. 56–62. IEEE (2018)
28. Tjoa, E., Guan, C.: A survey on explainable artificial intelligence (XAI): towards medical XAI. IEEE Trans. Neural Netw. Learn. Syst. **32**(11), 4793–4813 (2019)
29. Turk, M.A., Pentland, A.P.: Face recognition using eigenfaces. In: Proceedings of the 1991 IEEE Computer Society Conference on Computer Vision and Pattern Recognition, pp. 586–587. IEEE Computer Society (1991)
30. Wang, J., Jing, X., Yan, Z., Fu, Y., Pedrycz, W., Yang, L.T.: A survey on trust evaluation based on machine learning. ACM Comput. Surv. **53**(5), 1–36 (2020)
31. Wöber, W.: Identifying geographically di erentiated features of ethopian nile tilapia (oreochromis niloticus) morphology with machine learning. PLoS ONE **16**, 1–30 (2021)
32. Wöber, W., Mehnen, L., Curto, M., Tibihika, P.D., Tesfaye, G., Meimberg, H.: Investigating shape variation using generalized procrustes analysis and machine learning. Appl. Sci. **12**(6) (2022)
33. Xiang, Y., Schmidt, T., Narayanan, V., Fox, D.: PoseCNN: a convolutional neural network for 6D object pose estimation in cluttered scenes. arXiv preprint arXiv:1711.00199 (2017). https://arxiv.org/abs/1711.00199v3
34. Zhang, J., Yin, B., Xiao, X., Yang, H.: 3D detection and 6D pose estimation of texture-less objects for robot grasping. In: 2021 6th International Conference on Control and Robotics Engineering, ICCRE 2021, pp. 33–38. IEEE (2021)

# Optimal Camera Placement for Maximized Motion Capture Volume and Marker Visibility

Jakob Ziegler(✉), Hubert Gattringer, and Andreas Müller

Institute of Robotics, Johannes Kepler University Linz,
Altenberger Str. 69, 4040 Linz, Austria
{jakob.ziegler,hubert.gattringer,a.mueller}@jku.at
https://www.jku.at/en/institute-of-robotics

**Abstract.** In optical motion capture, camera placement strongly effects the dimensions of the measurable area and the accuracy of measured 3D marker positions calculated via triangulation. Aligning a given set of cameras by hand according to spatial conditions and measurement requirements relies on expertise and experience. This paper introduces an approach to find optimal camera poses for maximizing the measurement volume. It is based on a formulation for the visibility of individual points and considers the convergence angles between the cameras. In addition, a straightforward methodology to adjust the real cameras according to the optimization result is presented.

**Keywords:** motion capture · optimal camera placement · visibility

## 1 Introduction

Most optical motion capture systems rely on marker-based approaches, where the position of active (light-emitting) or passive (purely reflective) markers is measured [5]. In general, markers must be visible by at least two cameras in order to calculate their respective 3D positions via triagulation. Robustness, as well as accuracy, of this calculation thereby improves with increasing number of cameras measuring a marker [3, 6, 8]. The overall performance of a motion capture system thus strongly depends on camera placement, especially as markers might be occluded by static or moving objects. Of course an excess of installed cameras all around the measured volume could cope with a multitude of related issues, but that is not reasonable in an economical sense. A limited amount of cameras of a motion capture setup is usually aligned by hand according to given spatial conditions and/or measurement scenarios. This requires expertise and experience to get the optimum in terms of accuracy and robustness. Even lacking detailed information about forthcoming measurements it still might be beneficial to have a motion capture volume as large as possible and to increase the accuracy in


T. Petrič et al. (Eds.): RAAD 2023, MMS 135, pp. 45–52, 2023.
https://doi.org/10.1007/978-3-031-32606-6_6

certain sectors of this area. In any case, the dimensions of and the accuracy within a measurement volume depend on the camera poses, which makes the optimization of a motion capture setup desirable.

Optimal camera placement has been investigated in the past, where propagation of triangulation errors, resolution degradation due to the distance between marker and camera, as well as static and dynamic occlusions were analyzed [1,4,7,8] and considered with optimization methods like simulated annealing or genetic algorithms. However, little attention seems to have been paid so far to the realization in terms of adjusting the real cameras according to the optimization result. Without that process being to a certain degree sufficiently simple and fast, it is at least questionable and needs to be properly considered if tailoring motion capture setups to detailed scenarios justifies the effort of possibly frequent camera realignment.

Adopting this thought, an approach for optimizing a motion capture setup considering its measurement volume dimensions together with the visibility of markers therein, as well as a method to implement the optimization result with the real system, is presented in this paper.

## 2 Optimization of Camera Poses

Goal of the optimization is to find camera poses so that the overall 'visibility' of the measurement volume is maximized. Relations between image resolution, marker size and dimensions of the measurement volume are assumed to allow for neglecting errors due to occurring distances from markers to cameras.

At first, a desired motion capture volume is defined and approximated by a 3D grid with a specified spatial resolution, resulting in a set of 3D points. With given poses and fields of view of the cameras it can then easily be calculated if a 3D point is visible to a certain camera of the setup or not. In the present case the camera field of view is given by a horizontal and a vertical angle. This results in a pyramid of vision which can be described by the camera center and four corner points at an arbitrary distance to the camera center. The vector $\mathbf{r}_{\mathrm{CP}}$ from the camera center to a point $P$ can now be decomposed into a vector $\mathbf{r}_{pl,i}$ parallel to any of the $i = 1 \ldots 4$ pyramid planes and a vector $\mathbf{r}_{pp,i} = (\mathbf{r}_{\mathrm{CP}}^{\mathsf{T}}\mathbf{n}_i)\mathbf{n}_i$ perpendicular to it, with $\mathbf{n}_i$ being the plane normal of the $i^{th}$ pyramid plane and defined to point towards the optical axis (the inside of the pyramid), as shown in Fig. 1. This means iff $\mathbf{r}_{\mathrm{CP}}^{\mathsf{T}}\mathbf{n}_i$ is positive for all four pyramid planes, the point lies inside the pyramid of vision of the camera and is thus visible.

An obvious approach to optimize the setup would now be to maximize the number of cameras to which a point is visible, for all points describing the desired measurement volume. However, many solvers require a continuous function describing the optimization objective where above binary decision per camera, i.e. a point is visible or invisible, is not appropriate. We therefore introduce a visibility index $\nu$ defined as

$$\nu = \sum_{i=1}^{4} \frac{1}{2}\left(\tanh \frac{\mathbf{r}_{\mathrm{CP}}^{\mathsf{T}}\mathbf{n}_i}{\epsilon\|\mathbf{r}_{\mathrm{CP}}\|} - 1\right), \tag{1}$$

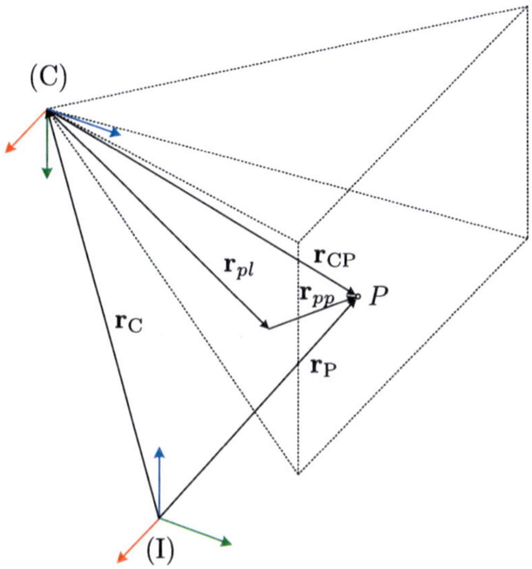


**Fig. 1.** $\mathbf{r}_{\mathrm{CP}}$ can be decomposed into $\mathbf{r}_{pl}$ (parallel to pyramid plane) and $\mathbf{r}_{pp}$ (perpendicular to it). Point $P$ is visible if $\mathbf{r}_{pp}$ is positive (pointing to the inside of the pyramid) for all four pyramid planes.

where $\mathbf{r}_{\mathrm{CP}}^{\mathsf{T}}\mathbf{n}_i$ describes how far point $P$ lies inside or outside the pyramid of view with respect to the direction of the plane normal $\mathbf{n}_i$, as described above.

This distance to the pyramid plane is normalized by $\|\mathbf{r}_{\mathrm{CP}}\|$, the distance from the point to the camera. $\epsilon$ influences the steepness of the tanh-function and should be set to a small number (e.g. 0.01). Eq. (1) results in a visibility index $\nu = 0$ for visible points and negative values for invisible points. For a camera $c$ the visibility indices of $M$ points are gathered in a vector $\boldsymbol{\nu}_c = [\nu_{c,1}, \dots \nu_{c,M}]^{\mathsf{T}}$. Considering a motion capture setup consisting of $N$ cameras a vector $\mathbf{v} = [\boldsymbol{\nu}_1^{\mathsf{T}}, \dots \boldsymbol{\nu}_N^{\mathsf{T}}]^{\mathsf{T}}$ can be generated.

To calculate the position of a 3D point via triangulation, at least two different camera views with non parallel optical axes are necessary. The convergence angle, i.e. the angle between the optical axes, plays an important role in terms of error propagation. It was shown that in an optimal configuration the optical axes of two cameras are perpendicular to each other, while the error of a triangulated position increases exponentially as the convergence angle approaches 0° or 180° [7,8]. Usually the body-fixed coordinate system of a camera is defined so that the $z$-axis matches the optical axis. Given two cameras $j$ and $k$ with known orientation, the convergence angle can be calculated directly from the according rotation matrices $\mathbf{R}_j$ and $\mathbf{R}_k$, or more specifically from their third columns, as $\Theta_{j,k} = \arccos \mathrm{col}_3(\mathbf{R}_j)^{\mathsf{T}} \mathrm{col}_3(\mathbf{R}_k)$. To find the optimal camera poses the nonlinear optimization problem can then be formulated as

$$\begin{aligned} \boldsymbol{\pi}^* = & \quad \arg\min_{\boldsymbol{\pi}} \frac{1}{2}\mathbf{v}(\boldsymbol{\pi})^\mathsf{T}\mathbf{v}(\boldsymbol{\pi}), \\ \text{s.t.} & \quad \Theta_{\min} \leq \Theta_{j,k} \leq \Theta_{\max} \end{aligned} \tag{2}$$

where $\boldsymbol{\pi}$ contains the positions and orientations (e.g. parameterized as Cardan angles or quaternions) of the cameras and $\Theta_{\min}$ and $\Theta_{\max}$ are the minimum and maximum convergence angles between the cameras.

## 3 Camera Alignment

Naturally, the real cameras have to be aligned with their calculated optimal poses. We introduce a straightforward approach in this regard, which is based on known 3D positions that are projected onto the 2D camera images.

It is assumed that the current setup is calibrated according to the advised procedure. Usually this involves a static reference object equipped with motion capture markers to define the origin and the axes of the inertial coordinate system (Fig. 2). As a result of the calibration the extrinsic parameters of the cameras, i.e. position and orientation with respect to this inertial coordinate system, as well as the intrinsic parameters are known. With this information it is simple to transform a vector represented in the inertial coordinate system (I) to the camera coordinate system (C). Let the position of a point $P$ represented in the inertial coordinate system be denoted as ${}_\mathrm{I}\mathbf{r}_\mathrm{P}$ and the position of a camera as ${}_\mathrm{I}\mathbf{r}_\mathrm{C}$. The vector from the camera center to point $P$ is then ${}_\mathrm{I}\mathbf{r}_\mathrm{CP} = {}_\mathrm{I}\mathbf{r}_\mathrm{P} - {}_\mathrm{I}\mathbf{r}_\mathrm{C}$. The transformation to the body-fixed camera coordinate system is performed by the rotation matrix $\mathbf{R}_\mathrm{CI}$ as ${}_\mathrm{C}\mathbf{r}_\mathrm{CP} = \mathbf{R}_\mathrm{CI}\,{}_\mathrm{I}\mathbf{r}_\mathrm{CP} = \mathbf{R}_\mathrm{CI}({}_\mathrm{I}\mathbf{r}_\mathrm{P} - {}_\mathrm{I}\mathbf{r}_\mathrm{C})$. This can be written as homogeneous transformation

$$\begin{bmatrix} {}_\mathrm{C}\mathbf{r}_\mathrm{CP} \\ 1 \end{bmatrix} = \begin{bmatrix} \mathbf{R} & \mathbf{t} \\ \mathbf{0}_{1\times3} & 1 \end{bmatrix} \begin{bmatrix} {}_\mathrm{I}\mathbf{r}_\mathrm{P} \\ 1 \end{bmatrix}, \tag{3}$$

with the rotation matrix $\mathbf{R} = \mathbf{R}_\mathrm{CI}$ and the translation vector $\mathbf{t} = -\mathbf{R}_\mathrm{CI}\,{}_\mathrm{I}\mathbf{r}_\mathrm{C}$.

With the known intrinsic camera matrix $\mathbf{K}$ a 3D point represented in the camera coordinate system can be mapped to pixel coordinates on the image plane by calculating the homogeneous vector

$$\begin{bmatrix} \hat{u} \\ \hat{v} \\ \hat{w} \end{bmatrix} = \mathbf{K}\,{}_\mathrm{C}\mathbf{r}_\mathrm{CP} = \begin{bmatrix} fk_x & -fk_x\cot\vartheta & u_0 \\ 0 & \frac{fk_y}{\sin\vartheta} & v_0 \\ 0 & 0 & 1 \end{bmatrix} {}_\mathrm{C}\mathbf{r}_\mathrm{CP}. \tag{4}$$

Neglecting lense distortion effects the required intrinsic parameters are thus the focal length $f$ in m, the pixel densities of the image sensor $k_x$ and $k_y$ in pixel/m, the skew angle $\vartheta$ in rad and the principle point (image center) in pixel coordinates $u_0$ and $v_0$. Note that Eq. (4) is ambiguous up to scale, which means all points on the ray defined by ${}_\mathrm{C}\mathbf{r}_\mathrm{CP}$ are projected to the same 2D point on the image plane. The final pixel coordinates are calculated by the homogenization $u = \hat{u}/\hat{w}$ and $v = \hat{v}/\hat{w}$.

Combining intrinsics and extrinsincs of the camera model in the projection matrix $\mathbf{P}$, a 3D point with known position ${}_{\mathrm{I}}\mathbf{r}_{\mathrm{P}}$ described in the inertial coordinate system can be projected to 2D pixel coordinates $\begin{bmatrix}\mathrm{u}\ \mathrm{v}\end{bmatrix}^{\mathsf{T}}$ of a camera image with

$$\begin{bmatrix}\hat{w}\mathrm{u}\\ \hat{w}\mathrm{v}\\ \hat{w}\end{bmatrix} = \mathbf{P}\ {}_{\mathrm{I}}\mathbf{r}_{\mathrm{P}}, \tag{5}$$

where $\mathbf{P} = \mathbf{K}\begin{bmatrix}\mathbf{R}\ \mathbf{t}\end{bmatrix}$.

Projecting known 3D positions, for instance the marker positions of the calibration tool, to the 2D camera images of the optimized setup gives the according "optimized" 2D pixel positions. By overlaying this desired 2D image and the actual camera image, and by adjusting the camera pose until the pixel positions match, it is now relatively easy to align the cameras according to the optimization result.

## 4 Results

The introduced methods were tested with a motion capture system, where the current setup consists of 10 infrared cameras (Qualisys, Sweden) measuring the 3D positions of passive (reflective) markers within a captured volume of approximately 4m width, 8m length and 3m height (Fig. 2). During the initial launch of the system the cameras were aligned by hand according to the recommendations. The CasADi framework [2] was used to implement the nonlinear optimization problem (2) which was solved with IPOPT [9]. All algorithms and tests were executed on a computer equipped with an i7 CPU (3.60GHz) and 8GB RAM.

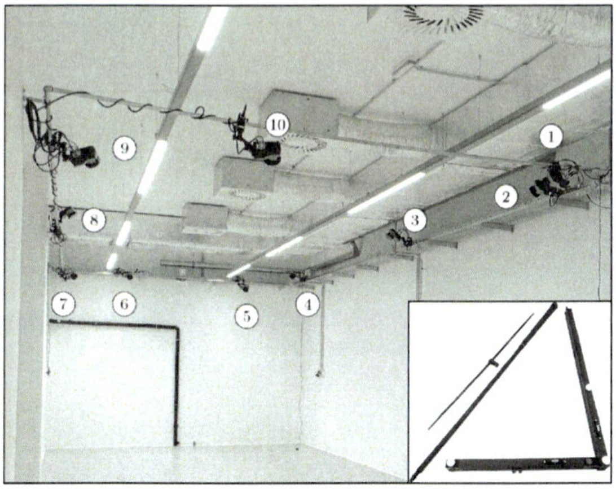


**Fig. 2.** Current motion capture setup with 10 cameras and the calibration kit (lower right corner) consisting of a frame with 4 markers and a wand

A calibration procedure, which is recommended before each measurement session, typically resulted in a standard deviation of around 0.6mm to 0.8mm regarding measured marker positions. This value is provided by the software of the motion capture system and is based on the reference distance of the markers on the calibration wand. An accompanying calibration file contains the identified extrinsic and intrinsic parameters of all cameras. The camera poses of this setup were used as initial guess for the optimization, while the intrinsics are assumed to be invariant. An optimization according to (2) with a minimum and maximum convergence angle of 20° and 160°, respectively, was carried out. In addition, restrictions regarding the possible camera positions according to the spatial conditions and mounting options in the lab, as well as a minimum distance of 50cm between the cameras, were defined.

After the optimization the known 3D positions of the calibration tool were then projected into the 2D images of the cameras with optimized poses, as described earlier. Figure 3 shows the initial and optimized pose of a camera, the measured 2D image with the identified markers (crosses), as well as the calculated projection of the 3D marker positions into the images of the initial (circles) and the optimized (diamonds) camera pose. Note that radial and/or tangential distortions are not considered here, as the software of the motion capture system already provides 2D images where these effects are compensated. The final step is now to adjust each camera by hand until the 2D images of the current and the optimized pose overlap, followed by a re-calibration of the system.

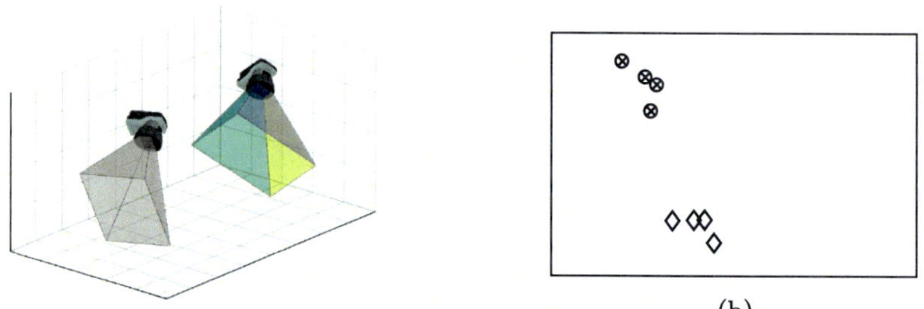

(a) (b)

**Fig. 3.** (a) initial (left, gray) and optimized (right, colored) pose of a camera and (b) measured 2D image with identified markers (crosses), 3D positions of the calibration frame projected according to the current pose (circles), projection according to the optimized pose (diamonds) (Color figure online)

Figure 4 visualizes the optimization result by exemplarily showing the volume captured by four or more cameras for the initial and the optimized setup. As can be seen, this volume increases drastically from approximately (25cm grid) 66.5% to 98.58% of the desired volume. Although not visualized, the volume measured by any other minimum number of cameras ($\geq 2$) increased in a similar manner. While this already indicates the functionality of the proposed methodology, as

(a) (b)

**Fig. 4.** Measurement volume visible by at least 4 cameras, (a) inital setup and (b) optimized setup (color indicates height)

the volume usable for motion capture measurements is enlarged, the typical accuracy for calibrations with the optimized setup also improved to a standard deviation of about 0.4mm to 0.6mm for measured marker positions.

The two main reasons for the increased measurement accuracy are assumed to be the following. On the one hand, measurement noise seems to be canceled out in a more effective way as the individual areas inside the measurement volume are viewed by a greater number of cameras. On the other hand, the convergence angles between the cameras are considered in an advantageous way. The convergence angles after the optimization were between 20.0° and 131.4° (76.19° on average) compared to the initial setup with values between 7.1° and 109.4° (62.5° on average). The computation time of the optimization increased with a rate slightly greater than linear with respect to the number of points describing the measurement volume, with about 50ms per point.

## 5 Conclusion

Intuitively aligning a motion capture system relies on expertise and experience. The dimensions of the recorded volume and accuracy of measured marker positions are strongly related to position and orientation of the individual cameras. This paper presents an approach for finding optimal poses of a limited amount of cameras maximizing the measurement volume and the visibility of markers.

To this end, a desired measurement volume is approximated by a 3D grid with defined spatial resolution. The amount of points visible by the cameras, depending on whether being inside the pyramids of view or not, is then maximized by optimizing the camera placement. Moreover, the convergence angles between the individual cameras are considered for increased accuracy. In addition, a straightforward methodology to adjust the real cameras according to the optimization result is introduced. Thereby, known 3D positions of motion capture markers are projected into the 2D images of the cameras with optimized poses and known intrinsic parameters. The individual cameras are aligned by adjusting position and orientation until the measured images and the projections match. An analysis of the optimization result showed a significant enlargement of the measurement volume measured by a defined minimum number of cameras. The functionality of the approach was further verified empirically. Calibration procedures with the optimized motion capture setup consistently showed improved measurement accuracy compared to the initial camera placement. However, it is not mandatory to maximize the overall measurement volume. Naturally, the presented approach also allows for a camera pose optimization according to specified areas inside the motion capture environment. This might be interesting when the motion sought to be captured takes place in a specific location or along a well known path, for instance gait measurements on a treadmill or mobile robots following a certain path. At this point it has to be mentioned that the occlusion of markers by static or moving objects, like the human body for instance, should probably be taken into account, which will be in the focus of future research.

**Acknowledgement.** This work has been supported by the "LCM - K2 Center for Symbiotic Mechatronics" within the framework of the Austrian COMET-K2 program.

## References

1. Aissaoui, A., Ouafi, A., Pudlo, P., Gillet, C., Baarir, Z.E., Taleb-Ahmed, A.: Designing a camera placement assistance system for human motion capture based on a guided genetic algorithm. Virtual Reality **22**, 13–23 (2018). https://doi.org/10.1007/s10055-017-0310-7
2. Andersson, J.A., Gillis, J., Horn, G., Rawlings, J.B., Diehl, M.: CasADi: a software framework for nonlinear optimization and optimal control. Math. Program. Comput. **11**(1), 1–36 (2019). https://doi.org/10.1007/s12532-018-0139-4
3. Aurand, A.M., Dufour, J.S., Marras, W.S.: Accuracy map of an optical motion capture system with 42 or 21 cameras in a large measurement volume. J. Biomech. **58**, 237–240 (2017). https://doi.org/10.1016/j.jbiomech.2017.05.006
4. Chen, X., Davis, J.: Camera placement considering occlusion for robust motion capture. Comput. Graph. Lab. Stanford Univ. Tech. Rep. **2**(2.2), 2 (2000)
5. Colyer, S.L., Evans, M., Cosker, D.P., Salo, A.I.: A review of the evolution of vision-based motion analysis and the integration of advanced computer vision methods towards developing a markerless system. Sport. Med. - Open **4**(1), 24 (2018). https://doi.org/10.1186/s40798-018-0139-y
6. Nagymáté, G., Kiss, R.M.: Motion capture system validation with surveying techniques. Materials Today: Proceedings **5**(13, Part 2), 26501–26506 (2018). https://doi.org/10.1016/j.matpr.2018.08.107, 34th Danubia Adria Symposium on Advances in Experimental Mechanics, 19 - 22 September 2017, Trieste, Italy
7. Olague, G., Mohr, R.: Optimal camera placement for accurate reconstruction. Pattern Recogn. **35**(4), 927–944 (2002). https://doi.org/10.1016/S0031-3203(01)00076-0
8. Rahimian, P., Kearney, J.K.: Optimal camera placement for motion capture systems. IEEE Trans. Visual Comput. Graph. **23**(3), 1209–1221 (2017). https://doi.org/10.1109/TVCG.2016.2637334
9. Wächter, A., Biegler, L.T.: On the implementation of an interior-point filter line-search algorithm for large-scale nonlinear programming. Math. Program. **106**(1), 25–57 (2006). https://doi.org/10.1007/s10107-004-0559-y

# Teaching Humanoid Robot Reaching Motion by Imitation and Reinforcement Learning

Kristina Savevska[1,2(✉)] and Aleš Ude[1]

[1] Humanoid and Cognitive Robotics Lab, Department of Automatics, Biocybernetics and Robotics, Jožef Stefan Institute, Jamova cesta 39, 1000 Ljubljana, Slovenia
kristina.savevska@ijs.si

[2] International Postgraduate School Jožef Stefan, Jamova cesta 39, 1000 Ljubljana, Slovenia

**Abstract.** This paper presents a user-friendly method for programming humanoid robots without the need for expert knowledge. We propose a combination of imitation learning and reinforcement learning to teach and optimize demonstrated trajectories. An initial trajectory for reinforcement learning is generated using a stable whole-body motion imitation system. The acquired motion is then refined using a stochastic optimal control-based reinforcement learning algorithm called Policy Improvement with Path Integrals with Covariance Matrix Adaptation ($PI^2$-CMA). We tested the approach for programming humanoid robot reaching motion. Our experimental results show that the proposed approach is successful at learning reaching motions while preserving the postural balance of the robot. We also show how a stable humanoid robot trajectory learned in simulation can be effectively adapted to different dynamic environments, e.g. a different simulator or a real robot. The resulting learning methodology allows for quick and efficient optimization of the demonstrated trajectories while also taking into account the constraints of the desired task. The learning methodology was tested in a simulated environment and on the real humanoid robot TALOS.

**Keywords:** Humanoids · Imitation Learning · Reinforcement learning

## 1 Introduction

Programming by Demonstration (PbD), also called Imitation Learning (IL) or Learning from Demonstration (LfD), is an effective way to program a robot without relying on a computer language, a complex graphical user interfaces, or expert knowledge. The idea is to provide a user-friendly approach to the use of robotic systems and to impart knowledge using a demonstration method inspired by the imitation capabilities of humans to acquire new skills [1,2].

IL is attracting a lot of attention in the field of humanoid robotics since it allows easy manipulation of the many degrees of freedom of the complex robotic


T. Petrič et al. (Eds.): RAAD 2023, MMS 135, pp. 53–61, 2023.
https://doi.org/10.1007/978-3-031-32606-6_7

system [8,14]. To that end, vision-based systems are primarily used for tracking the entire body of the demonstrator and translating it into a robot's whole-body motion [16]. However, these systems introduce the so-called correspondence problem where the differences in the kinematics and dynamics need to be addressed when transferring motion [6]. Taking the correspondence problem into account can lead to unsuccessful task execution, especially for tasks that require high precision as when manipulating objects.

Many recent approaches integrate imitation learning with other learning approaches [18]. In particular, human demonstrations have been exploited to initialize reinforcement learning (RL), which relies on multiple repetitions of the task performance to improve the learned behavior. Having a good initial demonstration impacts both the accuracy of the learned task and the learning time.

However, applying RL on high-dimensional systems such as humanoid robots can be a cumbersome and computationally expensive. The applied learning methods must be able to scale to high dimensional states and action spaces while also being computationally acceptable to work online. In most cases, the learning methods have to consider continuous spaces, which shows the necessity of parameterization through function approximation techniques [3,15].

While there are various applications of RL to humanoid robots, only a few deal with the high complexity of whole-body humanoid robot control [5,9,10,13, 18]. Stulp et al. [9] successfully applied an RL approach to a 34-DOF humanoid robot to learn full-body motor skills using a policy search algorithm called Policy Improvement with Path Integrals ($PI^2$), which is based on stochastic optimal control [12]. Theodorou et al. [13] discussed the $PI^2$ efficiency and robustness in highly dimensional spaces. The PoWER algorithm based on stochastic optimal control has also been shown to be applicable to high dimensions [5]. Another approach of RL in humanoids is presented in [18] who implemented a deep reinforcement learning framework along with imitation learning for teaching a human-style locomotion.

In this work we show how to improve the performance of a reaching task acquired by imitation learning and modified to account for the stability of a humanoid robot. To address the robot's stability within the imitation learning system, we employ a controller based on the linear inverted pendulum model approximation. An initially stable reproduction of a reaching task is then optimized using the $PI^2$-CMA algorithm while preserving the stability of a humanoid robot.

## 2 Methodology

### 2.1 Transfer of Human to Humanoid Robot Motion

We use a visual sensor to capture human motion. The acquired images are processed by Nuitrack Body Tracking SDK[1]. The estimated positions of human body segments and the orientations of coordinate frames attached to the segments are expressed in the global coordinate system attached to the sensor. For

[1] https://github.com/3DiVi/nuitrack-sdk.

each joint degree of freedom that is being imitated, we first compute the relative orientation between the two segments that are connected by the joint. For joints with three degrees of freedom (neck, waist, shoulders, hips, ankles), we convert the computed relative orientations into Euler angles around $z$, $y$, and $x$ axis, also known as yaw, pitch, and roll. The resulting Euler angles correspond to the humanoid robot joint angles around the 3 orthogonal joint axes. For joints with one degree of freedom (elbows and knees), we use the conversion into axis-angle representation [7]. The details regarding the specific joint angles of the robot are left out of this work as they are specific to each robotic platform.

As with every sensor, we must deal with noisy and difficult to process sensory data. The noisy and sometimes inaccurate data may violate the physical properties of the robot, such as exceeding joint or velocity limits, which may harm the robot if transferred without modification. Therefore, we apply additional joint angle modifications using a linear Kalman filter for each degree of freedom as well as safety check functions that prevent the execution of inaccurate or dangerous motions.

The above-described process addresses only the kinematic transformations between the human and the robot. To ensure the stability of the robot, the dynamics of the two subjects in the imitation process must also be considered.

### 2.2 Maintaining Stability with LIPM Based Control

The problem of keeping the stability of a humanoid robot is one of the fundamental challenges in humanoid robotics. The most popular stability concept, the Zero Moment Point (ZMP), was introduced by Vukobratović [17] and has been the basis for many approaches to stability control ever since.

Due to the complexity and high dimensionality of the humanoid robot's motion space, its highly nonlinear dynamics is often approximated using point mass models such as the linear inverted pendulum model (LIPM) [4]. This approach is based on the assumption that the whole robot mass is located at one point, i.e. at the robot's center of mass (CoM). In order to preserve the robot's stability, the dynamical relationship between the CoM and ZMP is employed to keep the ZMP within the support polygon. According to Sugihara et al., [11], we can indirectly manipulate the ZMP by controlling the CoM dynamics, i.e. $\dot{\boldsymbol{x}}_{CoM}$.

Sugihara et al. [11] also discuss how to apply constraints both in joint and in Cartesian space in addition to the CoM control. In this work, we are interested in generating whole-body motion for the robot that keeps both feet on the ground. Thus it makes sense to constrain the poses of both feet to be in contact with the ground and not moving. These constraints are expressed as follows

$$0 = \dot{\boldsymbol{x}}_{LF} = \boldsymbol{J}_{LF}\dot{\boldsymbol{\theta}}, \tag{1}$$

$$0 = \dot{\boldsymbol{x}}_{RF} = \boldsymbol{J}_{RF}\dot{\boldsymbol{\theta}}, \tag{2}$$

where $\dot{\boldsymbol{x}}_{LF}$ and $\dot{\boldsymbol{x}}_{RF}$ are the combined linear and angular velocity of the left and right foot, respectively. $\boldsymbol{J}_{LF}$ and $\boldsymbol{J}_{RF}$, expressed in the fixed world coordinate system, are the left and right foot Jacobians, respectively.

Finally, by combining the stable desired dynamics of the robot's CoM and the constant poses of the robot's feet, we can generate joint velocities of stable motion using the following relationship

$$\dot{\boldsymbol{x}} = \tilde{\boldsymbol{J}}\dot{\boldsymbol{\theta}}, \tag{3}$$

where

$$\dot{\boldsymbol{x}} = \left[\dot{\boldsymbol{x}}_{CoM},\ \boldsymbol{0},\ \boldsymbol{0}\right]^{\top},\ \tilde{\boldsymbol{J}} = \left[\boldsymbol{J}_{CoM},\ \boldsymbol{J}_{LF},\ \boldsymbol{J}_{RF}\right]^{\top}, \tag{4}$$

where $\boldsymbol{J}_{CoM}$ is the Jacobian of the CoM expressed in fixed world coordinate system. Furthermore, since humanoid robots are highly redundant robotic systems, we can consider solving multi-task inverse kinematics alongside the stability control. Thus, we can implement stable motion imitation by solving two tasks, of which the primary task is stability control and the secondary task is motion imitation, as follows

$$\dot{\boldsymbol{\theta}} = \tilde{\boldsymbol{J}}^{+}\dot{\boldsymbol{x}} + \boldsymbol{N}\dot{\boldsymbol{\theta}}_{\text{demonstrated}}, \tag{5}$$

where $\tilde{\boldsymbol{J}}^{+}$ is the Moore-Penrose pseudoinverse of $\tilde{\boldsymbol{J}}$, $\boldsymbol{N} = (\boldsymbol{I} - \tilde{\boldsymbol{J}}^{+}\tilde{\boldsymbol{J}})$ is the null space projection matrix of $\tilde{\boldsymbol{J}}$ and $\dot{\boldsymbol{\theta}}_{\text{demonstrated}}$ are the joint velocities acquired from the demonstrated motion and processed with the safety modules.

### 2.3 Motion Improvement by Reinforcement Learning

The process of stable motion imitation addresses successfully both the kinematics and dynamics of the robot. However, this results in a modified reproduction of the initially demonstrated trajectory. The extracted joint angle motion thus results in a suboptimal control policy for the reaching task. However, this suboptimal control policy can be used as an initial policy for reinforcement learning (RL). The aim of RL is to improve this initial policy with respect to specific task requirements and physical constraints of the robot.

We use Dynamic Movement Primitives (DMPs) [3,15] to represent the control policies for reaching movements. Thus encoded trajectories are then refined by RL using continuous trajectory rollouts (motion executions).

A DMP encodes the underlying motion as a dynamic system that contains a linear combination of $N$ nonlinear radial basis functions (RBFs), weighted by a finite vector $\boldsymbol{w}$. The weight vector $\boldsymbol{w}$ is then optimized using $\text{PI}^2$-CMA algorithm [10]. $\text{PI}^2$-CMA is based on $\text{PI}^2$ algorithm, which is a probabilistic RL algorithm derived from the principles of stochastic optimal control. The algorithm uses probability-weighted averaging to perform parameter updates instead of using an estimate of the gradient as is the case with many continuous RL algorithms. $\text{PI}^2$ has no open tuning parameters except for the exploration noise. $\text{PI}^2$-CMA [10] provides a mechanism to automatically adapt the exploration magnitude, which allows the algorithm to find a good exploration/exploitation trade-off regardless of the initial choice of the exploration noise.

We are interested in improving the imitation of a reaching task with respect to the stability of the robot, the ability to reach the final position of the reaching motion, and similarity to the initially demonstrated motion. Therefore, we define the cost function with the following terms

$$J(\boldsymbol{\tau}_k) = J(\boldsymbol{\tau}_k)_{stab} + w_{goal}J(\boldsymbol{\tau}_k)_{goal} + w_{traj}J(\boldsymbol{\tau}_k)_{traj}, \tag{6}$$

i.e. the overall cost function is a weighted sum of the cost functions of the individual cost functions addressing the desired constraints. The separate cost functions are defined as follows

$$J(\boldsymbol{\tau}_k)_{stab} = \frac{1}{N+1}\sum_{i=0}^{N} f_{stab}(\boldsymbol{x}_{ZMP}(t_i)), \tag{7}$$

$$J(\boldsymbol{\tau}_k)_{goal} = ||\boldsymbol{x}_{ee}(T) - \boldsymbol{g})||^2, \tag{8}$$

$$J(\boldsymbol{\tau}_k)_{traj} = \frac{1}{N+1}\sum_{i=0}^{N} \|\boldsymbol{\tau}(t_i) - \boldsymbol{\tau}_{demonstrated}(t_i)\|^2, \tag{9}$$

where

$$f_{stab}(\boldsymbol{x}_{ZMP}) = \gamma^2 \begin{cases} \left(\dfrac{\text{dist}_{SP}(\boldsymbol{x}_{SPC})}{\text{dist}_{SP}(\boldsymbol{x}_{ZMP})} - 1\right)^2, & \text{for } \boldsymbol{x}_{ZMP} \in \text{Support Polygon} \\ & \text{and } \text{dist}_{SP}(\boldsymbol{x}_{ZMP}) \geq \varepsilon > 0, \\ \left(\dfrac{\text{dist}_{SP}(\boldsymbol{x}_{SPC})}{\epsilon} - 1\right)^2, & \text{otherwise.} \end{cases} \tag{10}$$

Here $\text{dist}_{SP}(\boldsymbol{x})$ is a function that computes the Euclidean distance of the point $\boldsymbol{x}$ from the border of the support polygon. $\boldsymbol{x}_{SPC}$ is the constant center of the support polygon, i.e. the point within the support polygon that is the furthest away from the boundary. We defined $f_{stab}(\boldsymbol{x}_{ZMP})$ so that the cost function is bounded when $\boldsymbol{x}_{ZMP}$ comes very close to the border of the support polygon (within the $\epsilon$ neighborhood) or when the measured $\boldsymbol{x}_{ZMP}$ is outside of the support polygon. We assume that the feet are stationary and fixed to the ground.

The goal cost function given by Eq. (8) calculates the Euclidean distance between the robot's end-effector position and the desired goal position at the end of the trajectory, where $T$ is the duration of motion. The trajectory similarity cost (9) is calculated as the deviation of the performed rollout trajectory from the initially demonstrated trajectory.

# 3 Results

## 3.1 Learning in a Simulated Environment

As explained in Sect. 2.2, we start by capturing a reaching motion demonstrated by a human. The measured motion is converted into humanoid robot joint motion and processed to ensure the stability of a humanoid robot. The resulting motion is encoded as a DMP and fed to the reinforcement learning algorithm described in Sect. 2.3. The RL algorithm optimizes the obtained motion with respect to the stability, accuracy of reaching, and similarity to the initial trajectory.

Using the proposed framework, we computed the trajectories of 30 out of 32 degrees of freedom of a humanoid robot TALOS, which was used in our experiments. The learned DMP for each joint contained 20 RBFs. The learning with $PI^2$-CMA algorithm was performed in Gazebo simulated environment in 150 updates with 30 rollouts per update.

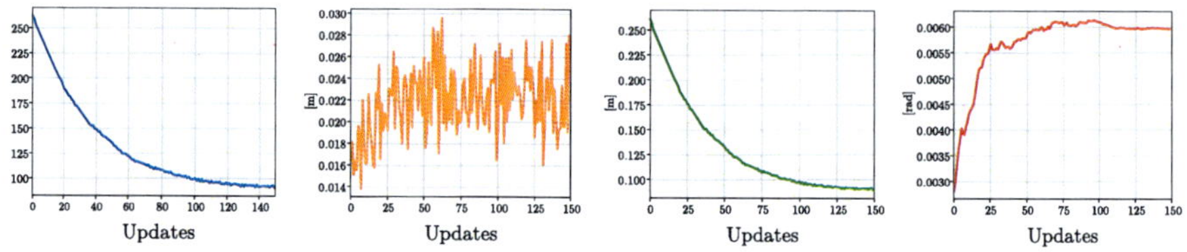


(a) Cost

(b) Distance from the center of the support polygon.

(c) Distance from the desired goal position.

(d) Deviation of the initial and the reproduced trajectory.

**Fig. 1.** Evaluation of the learning process conducted in Gazebo simulation.

The algorithm managed to improve the reaching motion without violating the stability of the robot. The learning curve is shown in Fig. 1a. It represents the weighted sum of all the cost terms relevant to the task. The performance of the learning process based on the separate cost terms is illustrated in Figs. 1b–1d.

### 3.2 Refinement of the Trajectory Learnt in Simulation

Because of the differences in the dynamics of the simulated and the real robot, the trajectories learnt in simulation often do not result in a successful task execution in the real world. To address this issue we propose to apply another round of learning to adapt the trajectory learnt in simulation to the dynamics of the real robot. Our goal is to adapt the motion with significantly fewer updates than when the learning is initialized with stabilized human demonstrated trajectories.

Due to the potential danger posed by the explorative nature of the reinforcement learning trials, the refinement of the learnt trajectory could not be implemented directly on the real robot. Instead, we used another simulator, recently developed by PAL Robotics, which closely corresponds to the robot's real dynamics, to further process the learned motion. This approach allowed us to verify the transferability and fast convergence of the learnt motion in a different dynamic environment, giving us confidence that the real robot will exhibit similar behavior during learning. The learning was performed in 50 updates with 30 rollout executions per update. The algorithm succeeded to improve the reaching accuracy and maintained the stability of the robot. The optimal solution was obtained after less than 20 updates. The learning curve of the task is shown in Fig. 2 along with the evaluation of each cost term.

In summary, we succeeded to adapt the learnt trajectory from one simulated dynamic environment to another in less than 20 updates. This is significantly less than what was needed when learning from scratch and proves that learning in one dynamic environment can benefit from the initialization with results obtained in another simulator. Thus the real robot should also be able to learn in significantly fewer rollouts by utilizing the optimal solution from the simulation environment. We are currently working towards implementing this learning procedure on a real robot.

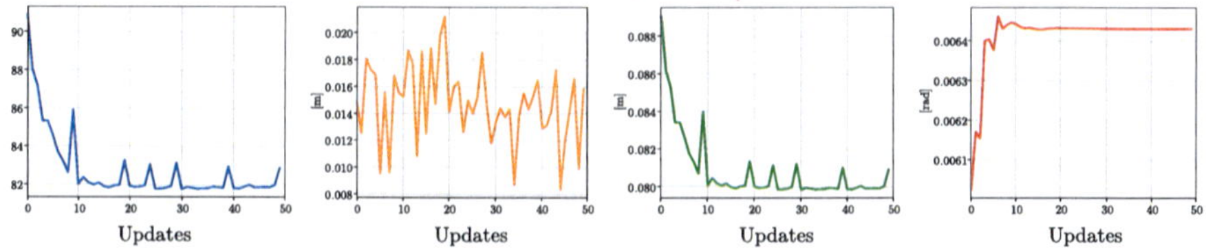


(a) Cost (b) Distance from the center of the support polygon. (c) Distance from desired goal position. (d) Deviation of the initial and the reproduced trajectory.

**Fig. 2.** Evaluation of the learning process conducted in the PAL Physics simulator

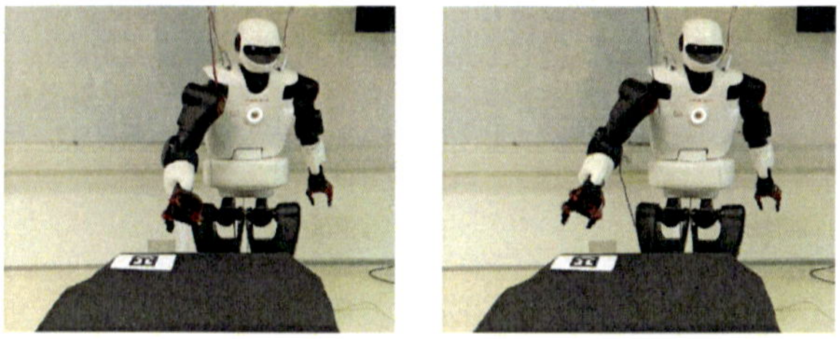

(a) The final robot posture on the initial, stable trajectory obtained from a human demonstration. (b) The final robot posture on the stable trajectory obtained by reinforcement learning.

**Fig. 3.** The outcome of learning on the real humanoid robot Talos.

The results of learning obtained in the PAL physics simulator were used to generate reaching motions on the real robot. An example of successful execution of the reaching task while maintaining the robot's stability is shown in Fig. 3.

## 4 Conclusions

In this work, we employed a combination of imitation learning and reinforcement learning to successfully program a humanoid robot to execute complex tasks while preserving stability. We established a framework to transfer human movements to the robot, taking into account the essential constraints of humanoid robots. The acquired motions were then improved using RL with a suitable cost function that accounts for the task-specific requirements. With this methodology, the robot was able to learn a reaching motion to a specific 3-D point using a single noisy visual observation. Our proposed algorithm was applied to a high degree of freedom humanoid robot and demonstrated efficient convergence within less than 100 updates.

As a next step, we aim to implement the refinement of the learnt trajectory on the real system. Additionally, by defining suitable cost functions, the proposed framework can be expanded to more complex tasks, including additional constraints and incorporating human-robot interaction.

## References


1. Argall, B.D., Chernova, S., Veloso, M., Browning, B.: A survey of robot learning from demonstration. Robot. Auton. Syst. **57**(5), 469–483 (2009)
2. Dillmann, R.: Teaching and learning of robot tasks via observation of human performance. Robot. Auton. Syst. **47**(2–3), 109–116 (2004)
3. Ijspeert, A.J., Nakanishi, J., Hoffmann, H., Pastor, P., Schaal, S.: Dynamical movement primitives: learning attractor models for motor behaviors. Neural Comput. **25**(2), 328–373 (2013)
4. Kajita, S., Kanehiro, F., Kaneko, K., Yokoi, K., Hirukawa, H.: The 3D linear inverted pendulum mode: a simple modeling for a biped walking pattern generation. In: IEEE/RSJ International Conference on Intelligent Robots and Systems (IROS), Takamatsu, Japan, pp. 239–246 (2001)
5. Kober, J., Peters, J.: Policy search for motor primitives in robotics. Mach. Learn. **84**, 171–203 (2010)
6. Koenemann, J., Burget, F., Bennewitz, M.: Real-time imitation of human whole-body motions by humanoids. In: IEEE International Conference on Robotics and Automation (ICRA), pp. 2806–2812 (2014)
7. Savevska, K., Simonič, M., Ude, A.: Modular real-time system for upper-body motion imitation on humanoid robot talos. In: Zeghloul, S., Laribi, M.A., Sandoval, J. (eds.) RAAD 2021. MMS, vol. 102, pp. 229–239. Springer, Cham (2021). https://doi.org/10.1007/978-3-030-75259-0_25
8. Schaal, S.: Learning from demonstration. In: 9th International Conference on Neural Information Processing Systems, pp. 1040–1046 (1996)
9. Stulp, F., Buchli, J., Theodorou, E., Schaal, S.: Reinforcement learning of full-body humanoid motor skills. In: 2010 10th IEEE-RAS International Conference on Humanoid Robots, pp. 405–410 (2010)
10. Stulp, F., Sigaud, O.: Path integral policy improvement with covariance matrix adaptation. In: Proceedings of the 10th European Workshop on Reinforcement Learning (EWRL 2012), vol. 1 (2012)
11. Sugihara, T., Nakamura, Y., Inoue, H.: Real-time humanoid motion generation through ZMP manipulation based on inverted pendulum control. In: IEEE International Conference on Robotics and Automation (ICRA), pp. 1404–1409 (2002)
12. Theodorou, E., Buchli, J., Schaal, S.: Learning policy improvements with path integrals. J. Mach. Learn. Res. **9**, 828–835 (2010)

13. Theodorou, E., Buchli, J., Schaal, S.: Reinforcement learning of motor skills in high dimensions: a path integral approach. In: IEEE International Conference on Robotics and Automation (ICRA), pp. 2397–2403 (2010)
14. Ude, A., Atkeson, C., Riley, M.: Programming full-body movements for humanoid robots by observation. Robot. Auton. Syst. **47**(2–3), 93–108 (2004)
15. Ude, A., Gams, A., Asfour, T., Morimoto, J.: Task-specific generalization of discrete and periodic dynamic movement primitives. IEEE Trans. Rob. **26**(5), 800–815 (2010)
16. Vakanski, A., Janabi-Sharifi, F.: Robot Learning by Visual Observation. Wiley, Hoboken (2017)
17. Vukobratović, M., Stepanenko, J.: On the stability of anthropomorphic systems. Math. Biosci. **15**(1), 1–37 (1972)
18. Yang, C., Yuan, K., Heng, S., Komura, T., Li, Z.: Learning natural locomotion behaviors for humanoid robots using human bias. IEEE Robot. Autom. Lett. **5**(2), 2610–2617 (2020)

# Deep Learning Prediction Models for the Detection of Cyber-Attacks on Image Sequences

Dusan Nedeljkovic(✉) and Zivana Jakovljevic

Faculty of Mechanical Engineering, University of Belgrade, Beograd, Serbia
{dnedeljkovic,zjakovljevic}@mas.bg.ac.rs

**Abstract.** With the introduction of Cyber Physical Systems and Industrial Internet of Things within Industry 4.0, vision systems, as indispensable element for robot cognition, become smart devices integrated into the control system using different communication links. In this control framework image streams are transferred between elements of distributed control system opening the possibility for various cyber-attacks that can cause changes in certain parts of images eventually triggering wrong decisions and negative consequences to the system performance. Timely detection of the attacks on communicated image streams is necessary to mitigate or completely avoid their negative effects. In this paper we propose a method for the prediction of the next image in the sequence which can be utilized for the development of anomaly-based cyber-attack detection mechanisms. For the model generation, we have explored the application of several deep learning architectures based on two-dimensional Convolutional Neural Networks and Convolutional Long Short-Term Memory Recurrent Neural Networks. Images obtained from the real-world experimental installation were utilized for model design. Our deep learning models proved to be effective in predicting the next frames according to the criteria of a discrepancy between pixels of the real and estimated images.

**Keywords:** Prediction models · Deep Learning · Convolutional Neural Networks · Long Short-Term Memory Recurrent Neural Networks

## 1 Introduction

Vision systems and image processing represent a rapidly growing field with numerous and diverse applications in robotics. The ability to accurately predict the contents of the next image in a sequence of images can have a profound impact on different spheres, such as manufacturing, healthcare, transportation, etc. In the field of autonomous robotic systems, image prediction can be used e.g., for anticipating the movements of other objects in the environment, enabling safer and more efficient navigation [1] and prevention of potential irregularities [2].

With the introduction of Industrial Internet of Things (IIoT) within Industry 4.0 framework internal and external vision systems are introduced into robots' control in


T. Petrič et al. (Eds.): RAAD 2023, MMS 135, pp. 62–70, 2023.
https://doi.org/10.1007/978-3-031-32606-6_8

the form of Cyber Physical Systems (CPS) that communicate image streams to other elements of robots' control system using different wired and/or wireless protocols. Widespread communication between CPS creates numerous opportunities for cyber-attacks with potentially catastrophic consequences. To mitigate or prevent negative effects, it is necessary to develop security mechanisms for timely detection of attacks.

Since adversaries can be very perfidious in their intention to harm the controlled process and launch different kinds of stealthy deception attacks, anomaly-based cyber-attacks detection methods represent a technique of choice for Industrial Control Systems (ICS) [3] including robots' control. In these detection methods the next value of the signal (image in the vision system) is predicted using analytical, regression or some other models, and the attack is detected as the discrepancy between predicted and value received through communication link. Considering the complexity of images obtained using vision systems, for these systems autoregression models are employed as a rule – the next image is predicted using a sequence of previous images.

Traditional methods for image prediction rely on hand-crafted features and a set of predefined rules. These methods can be effective, but they are limited in their ability to handle complex and varied data. Additionally, these techniques are usually less flexible and require more manual effort to design and tune the predictive models [4].

With the advent of Deep Learning (DL) techniques, significant progress has been made in this area. In contrast to the traditional techniques, DL methods can automatically learn hierarchical representations of visual world from data, allowing them to handle a wider range of data and produce more accurate predictions [5].

Convolutional Neural Networks (CNN) have found wide application in various forms, such as autoencoders [6] or deep CNN [7]. Long short-term memory networks (LSTM) based autoencoders were also employed [8] to predict the next frames. Since in some cases the LSTM are not able to consider well the spatial structure of data, in [9, 10] Convolutional LSTM (ConvLSTM) were introduced, where the internal transformations of an LSTM cell were replaced with convolutions. In this way, the model is capable to capture spatial dependencies through utilization of convolution operation, while still being able to capture temporal dependencies with the LSTM [11].

In this paper, we present a DL approach for predicting the next image in stream based on the sequence of previous images. Our method generates a model of system behavior utilizing a 2D-CNN, ConvLSTM, as well as their combination. We evaluate the performance of our method on images obtained from a real-world system. We also compare the results obtained using different Deep Neural Networks (DNN) architectures and choose the best model based on predefined criteria.

The remainder of the paper is structured as follows. Section 2 briefly outlines utilized DNN techniques and describes the image prediction process. Section 3 shows the experimental installation used for data acquisition and presents experimental results. Conclusions and future work guidelines are provided in Sect. 4.

## 2 Explored DNN Based Models

DNN based models that are generated in this paper estimate the next image – $\hat{I}_k$ from the sequence of previous $z$ images $(I_{k-z-1}, I_{k-z}, \ldots I_{k-1})$ in the stream, i.e. they perform spatiotemporal forecasting of the image through autoregression of two dimensional (2D) signal. We opted to explore the networks based on the following layers:

1. Two-dimensional CNN (2D-CNN) [5, 12],
2. Two-dimensional ConvLSTM (2D-ConvLSTM) [9, 10],
3. Combination of 2D-CNN with 2D-ConvLSTM layers in the network.

Namely, CNN bring into deep learning convolution – an operation that is frequently used in classical machine learning (ML) to extract the spectral features from the signals using different filters. In CNN at least one layer employs convolution of its input with a filter instead of matrix multiplication. Opposite to classical ML where the values of filter coefficients are predefined, in CNN they are obtained during network training aiming to extract the most prominent signal features for the application at hand. CNN have shown excellent results in 2D signals modelling [12].

LSTM, on the other hand, represent a class of Recurrent Neural Networks that generate current output based on the value of current and sequence of previous inputs and as such are suitable for modeling temporal characteristics of data. They consist of recurrently connected cells called memory blocks that contain internal cell state vector $\mathbf{c}(t)$ introduced to memorize important data during network functioning using several gates ($\mathbf{i}(t)$, $\mathbf{f}(t)$) that fine tune the current cell state and its output ($\mathbf{y}(t)$) [13]. Opposite to the conventional LSTM where matrix multiplication is used for transformations of input and recurrent transformations of output [5], in the ConvLSTM input and recurrent transformations are carried out using convolution [9]. In particular, the cell state $\mathbf{c}(t)$, hidden state $\mathbf{h}(t)$ and ConvLSTM output $\mathbf{y}(t)$ vectors are defined as follows (the input vector $\mathbf{x}(t)$ is given) [9]:

$$
\begin{aligned}
\mathbf{i}(t) &= \sigma(\mathbf{W_{xi}} * \mathbf{x}(t) + \mathbf{W_{hi}} * \mathbf{h}(t-1) + \mathbf{W_{ci}} \odot \mathbf{c}(t-1) + \mathbf{b_i}),\\
\mathbf{f}(t) &= \sigma(\mathbf{W_{xf}} * \mathbf{x}(t) + \mathbf{W_{hf}} * \mathbf{h}(t-1) + \mathbf{W_{cf}} \odot \mathbf{c}(t-1) + \mathbf{b_f}),\\
\mathbf{c}(t) &= \mathbf{f}(t) \odot \mathbf{c}(t-1) + \mathbf{i}(t) \odot tanh(\mathbf{W_{xc}} * \mathbf{x}(t) + \mathbf{W_{hc}} * \mathbf{h}(t-1) + \mathbf{b_c}),\\
\mathbf{o}(t) &= \sigma(\mathbf{W_{xo}} * \mathbf{x}(t) + \mathbf{W_{ho}} * \mathbf{h}(t-1) + \mathbf{W_{co}} \odot \mathbf{c}(t) + \mathbf{b_o}),\\
\mathbf{h}(t) &= \mathbf{o}(t) \odot tanh(\mathbf{c}(t))
\end{aligned}
\tag{1}
$$

where $\mathbf{W}_{\llcorner}$ and $\mathbf{b}_{\llcorner}$ represent weight matrices and bias vectors, $\odot$ denotes element-wise multiplication of the vectors and $\ast$ convolution operation. In this way ConvLSTM combine benefits of CNN in modeling spatial features and benefits of LSTM in modelling temporal characteristic of data and are a very useful tool for spatiotemporal modelling of 2D signals streams such as prediction of the next image in sequence.

The results of the predictions obtained using different architectures explored in this paper will be numerically compared using the mean absolute value of the summed discrepancy ($D$) between pixels of the ground truth ($I_k$) and image estimated using the developed model ($\hat{I}_k$) over $p$-$z$ images ($p$ is the number of images in the testing set):

$$
D = \frac{\sum_{k=1}^{p-z} \sum_{i=1}^{x} \sum_{j=1}^{y} I_k(i,j) - \hat{I}_k(i,j)}{p-z} \tag{2}
$$

In Eq. (2), $x$ and $y$ represent the number of image pixels along the horizontal and vertical axes, respectively.

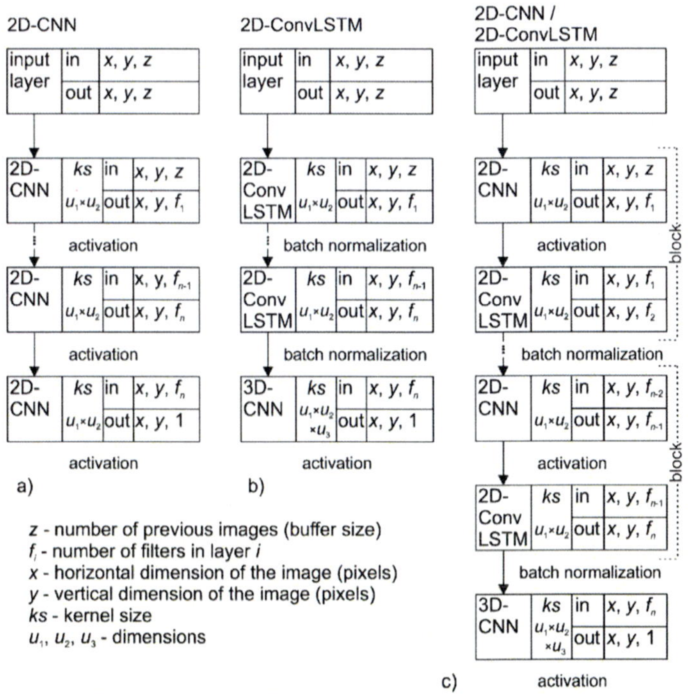


**Fig. 1.** Used DNN architectures: a) 2D-CNN, b) 2D-ConvLSTM, c) 2D-CNN/2D-ConvLSTM

The computational performances of the devices employed in ICS are as a rule relatively low and in these systems real-time applicability is essential. On the other hand, the computational complexity of the next image prediction algorithm is directly related to the number of parameters in DL model as can be observed from Eq. (1). In general, the real-time computational complexity of a model increases roughly linearly with the number of trainable parameters [14]. Thus, DL models should have as small number of parameters as possible, and this can be achieved through generation of models that are custom made for the images present in the application at hand without aiming general applicability that significantly influences the computational complexity of algorithm and its real-time applicability at computationally constrained devices. The focus should be on concrete industrial applications that as a rule contain simple scenes rather than on complex images such as in [15].

Figure 1 shows the architectures in general form for all types of DNNs considered for generation of custom-made models in this paper. Architectures are composed of an input layer, an output layer, and 2D-CNN and/or 2D-ConvLSTM layers in between, where $f_i$ denotes the number of filters in layer *i*, and *ks* denotes kernel size. The combined architecture (Fig. 1c) consists of sequentially placed blocks with 2D-CNN and 2D-ConvLSTM layers. CNN was utilized for the output layers in all three types of DNN architectures, 2D for the first type and 3D for the other architectures.

# 3 Experimental Results

## 3.1 Experimental Setup and Data Acquisition

The DNN based prediction models are generated for the sequences of images acquired using experimental setup presented in Fig. 2a. It consists of Cognex In-Sight 2000-120 camera, rotary table that carries the part whose images are obtained and PC for images acquisition. Cognex In-Sight 2000-120 is a monochrome camera with 640 × 480 px resolution [16]. It communicates with PC via Ethernet and images are acquired using In Sight Explorer software [17]. The drawing of the part that is photographed in different poses obtained using rotary table is presented in Fig. 2b.

During acquisition of streams of images different distances between camera and rotary table and offsets (translations) of part from table center were utilized. A total of 19,992 images structured in 18 sequences were acquired, where the smallest and largest sequences contain 293 and 3,000 images, respectively.

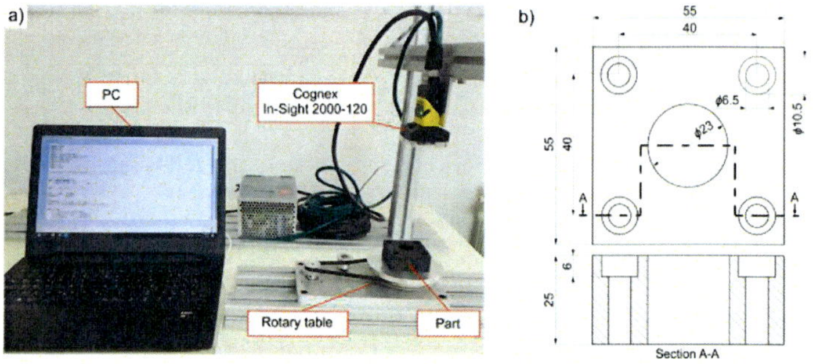


**Fig. 2.** A) Experimental installation; b) Part drawing (the dimensions are in mm)

## 3.2 Experimental Results

The results of the research will be presented for the sequence that contains 3,000 images. Before DNN training, the images were normalized to the range [0, 1]. To be able to carry out the training of the networks on the available computing platform, the images are downsampled from 640 × 480 to 160 × 120 px. 90% of data were used for training, and 10% for testing purposes. The buffer of previously streamed $z = 5$ images was used for estimation of the following image. The training is performed on workstation with double Intel Xeon Silver 4208 CPU, triple Nvidia Quadro RTX6000 GPU and 192GB DDR4 RAM for 100 epochs with 30% of training data utilized for validation. The models were trained in Python with Keras and TensorFlow at the background. Binary cross entropy loss function and Adam optimizer were employed.

All 2D-CNN and 2D-ConvLSTM layers in each DNN architecture used kernel size *ks* of 3 × 3 kernel, whereas kernel 3 × 3 × 3 was utilized in 3D-CNN layers. Rectified Linear Unit (ReLU) activation function was applied on each 2D-CNN layer, whereas sigmoid activation function was employed in 3D-CNN layers. Additionally, each 2D-ConvLSTM layer was followed by batch normalization.

Through the variation of the number of layers *n*, number of filters within each layer $f_i$, $i \in 1,\ldots n$, activation functions and other parameters of the network, a number of unique DNN models were generated. Table 1 shows a few best architectures by the criterion of minimum *D* for each type of DNN along with the number of trainable parameters. Discrepancy *D* is calculated over 295 images (300-*z*) from testing set. From this table, it can be observed that the networks 2D-CNN #1–3 were not able to adequately estimate the next image in the sequence. For these architectures, it is also interesting that the networks with a higher number of layers gave worse results. Furthermore, 2D-ConvLSTM has shown significantly better results than 2D-CNN. As expected, better estimation is obtained with the increase in architecture size and number of parameters. Nevertheless, the number of parameters for the best-performing 2D-ConvLSTM #3 is very high (161,185), which is not suitable for real-time work.

**Table 1.** An excerpt of trained architectures.

| DNN type | $n, f_{1\text{-}2\text{-}\ldots\text{-}n}$ | No. of param. | discrepancy *D* |
|---|---|---|---|
| 2D-CNN #1 | $n = 4, f_i = 8\text{-}16\text{-}16\text{-}32$ | 8,785 | 2,455.69 |
| 2D-CNN #2 | $n = 4, f_i = 64\text{-}64\text{-}128\text{-}128$ | 262,465 | 3,517.30 |
| 2D-CNN #3 | $n = 8, f_i = 8\text{-}8\text{-}16\text{-}16\text{-}32\text{-}32\text{-}64\text{-}64$ | 74,329 | 10,575.80 |
| 2D-ConvLSTM #1 | $n = 1, f_i = 16$ | 12,657 | 670.41 |
| 2D-ConvLSTM #2 | $n = 2, f_i = 16\text{-}16$ | 31,217 | 378.92 |
| 2D-ConvLSTM #3 | $n = 4, f_i = 16\text{-}16\text{-}32\text{-}32$ | 161,185 | 202.14 |
| Combined #1 | $n = 1, f_i = 16$ | 19,729 | 492.52 |
| Combined #2 | $n = 2, f_i = 16\text{-}16$ | 40,545 | 197.44 |

The model is selected based on the minimum discrepancy value *D*, considering the number of model parameters. According to the criterion of minimum value of discrepancy *D*, Combined #1-2 networks gave comparable results to 2D-ConvLSTM #1-3 networks with a significantly lower number of trainable parameters. By comparing the considered models in terms of the number of parameters and *D*, the conclusion is reached that the best architecture is Combined #2. In addition to the small value of *D*, this architecture has acceptable number of parameters for real-time implementation.

For graphical illustration of the performances of different DNN architectures that were tested, a buffer of $z = 5$ images that belong to testing set will be utilized (Fig. 3). Using this example of sequence, the results of the estimation for the considered architectures are graphically presented in Fig. 4. It can be noticed from this figure that the best prediction was given by the 2D-ConvLSTM #3 and Combined #2 architectures.

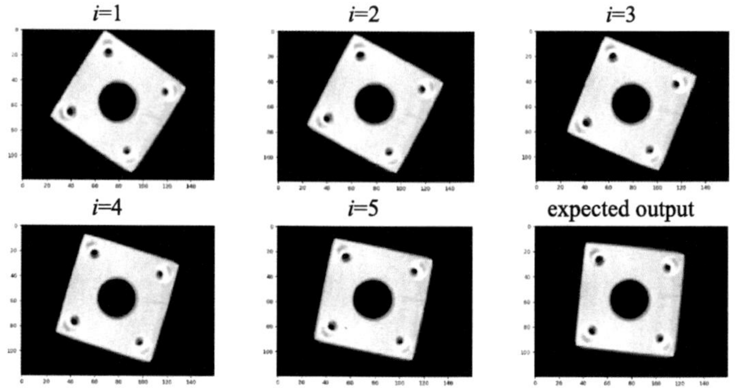


**Fig. 3.** A sequence of 5 images used for illustration of different architectures' performances

The developed models with the best results can be used successfully as a basis for the design of the cyber-attack detection method. In this case, the models will describe system behavior without attack, whereas potential attacks can be detected by introducing criteria based on the discrepancy between the real and estimated image.

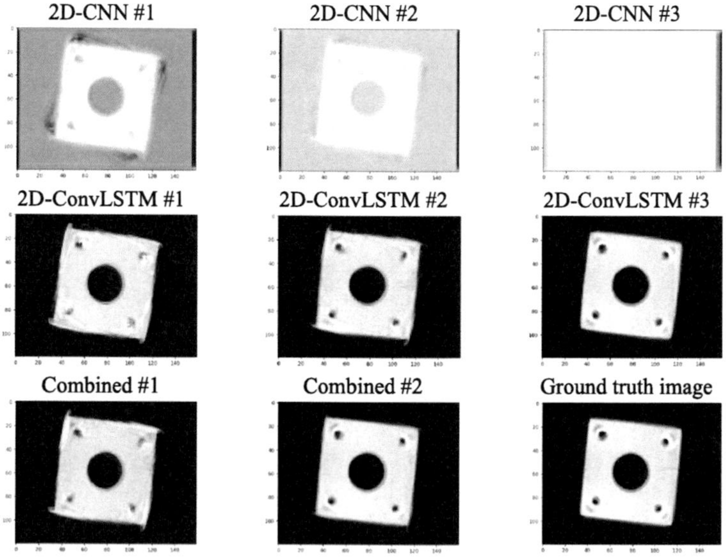


**Fig. 4.** Graphical representation of 2D-CNN models' image estimation results

## 4 Conclusion

In this paper we have explored different DL architectures based on 2D-CNN, 2D-ConvLSTM, and their combination for the generation of models for the prediction of the next image in the sequence based on a buffer of previous images. The best prediction model was chosen based on the minimum value of discrepancy between ground truth and predicted image, considering the model complexity and real-time applicability, and it is based on combination of 2D-CNN and 2D-ConvLSTM layers. The algorithm was evaluated using the images from the real-world experimental installation.

Our future work will focus on improving the algorithm generalization properties by introducing criteria to determine the maximum acceptable discrepancy. In addition, we will investigate the application of image pre-processing techniques to enhance the current performance. Finally, we will direct our attention to the implementation of the developed DL models on low-level controllers of CPS and their integration into the mechanisms for the detection of cyber-attacks on communicated image streams.

**Acknowledgment.** This research was supported by the Ministry of Education, Science and Technological Development of the Serbian Government under the contract No. 451-03-47/2023-01/200105.

## References

1. Hu, A., Cotter, F., Mohan, N., Gurau, C., Kendall, A.: Probabilistic future prediction for video scene understanding. In: European Conference on Computer Vision, pp. 767–785. Springer, Cham (2020)
2. Sultani, W., Chen, C., Shah, M.: Real-world anomaly detection in surveillance videos. In: Proceedings of the IEEE Conference on Computer Vision and Pattern Recognition (CVPR), pp. 6479–6488. IEEE, USA (2018)
3. Asghar, M.R., Hu, Q., Zeadally, S.: Cybersecurity in industrial control systems: Issues, technologies, and challenges. Computer Networks, 165, art. 106946 (2019)
4. Oprea, S., et al.: A review on deep learning techniques for video prediction. IEEE Trans. Pattern Anal. Mach. Intell. **44**(6), 2806–2826 (2020)
5. Goodfellow, I., Bengio, Y., Courville, A.: Deep Learning. MIT press (2016)
6. Hasan, M., Choi, J., Neumann, J., Roy-Chowdhury, A.K., Davis, L.S.: Learning temporal regularity in video sequences. In: Proceedings of the IEEE Conference on Computer Vision and Pattern Recognition, pp. 733–742. IEEE, USA (2016)
7. Paraskevoudis, K., Karayannis, P., Koumoulos, E.P.: Real-time 3D printing remote defect detection (stringing) with computer vision and artificial intelligence. Processes **8**(11), art. 1464 (2020)
8. Srivastava, N., Mansimov, E., Salakhudinov, R.: Unsupervised learning of video representations using lstms. In: Int. Conf. on Machine Learning, pp. 843–852. JMLR, France (2015)
9. Shi, X., Chen, Z., Wang, H., Yeung, D.Y., Wong, W.K., Woo, W.C.: Convolutional LSTM network: a machine learning approach for precipitation nowcasting. Adv. Neural. Inf. Process. Syst. **28**, 802–810 (2015)
10. Tsironi, E., Barros, P., Weber, C., Wermter, S.: An analysis of convolutional long short-term memory recurrent neural networks for gesture recognition. Neurocomputing **268**, 76–86 (2017)
11. Finn, C., Goodfellow, I., Levine, S.: Unsupervised learning for physical interaction through video prediction. Adv. Neural Inf. Process. Syst. **29** (2016)
12. Krizhevsky, A., Sutskever, I., Hinton, G.E.: Imagenet classification with deep convolutional neural networks. Commun. ACM **60**(6), 84–90 (2017)

13. Hochreiter, S., Schmidhuber, J.: Long short-term memory. Neural Comput. **9**(8), 1735–1780 (1997)
14. Thompson, N.C., Greenewald, K., Lee, K., Manso, G.F.: The computational limits of deep learning. arXiv preprint arXiv:2007.05558 (2020)
15. Lapuschkin, S., Wäldchen, S., Binder, A., Montavon, G., Samek, W., Müller, K.R.: Unmasking Clever Hans predictors and assessing what machines really learn. Nat. Commun. **10**(1), art. 1096 (2019)
16. In-Sight 2000 Vision Sensors: https://www.cognex.com/products/machine-vision/vision-sensors/in-sight-2000-vision-sensors. Last accessed 11 Jan 2023
17. In-Sight Explorer Software: https://www.cognex.com/products/machine-vision/2d-machine-vision-systems/in-sight-vision-software. Last accessed 1 Jan 2023

# Collaborative Robotics

# Variable Stiffness Joint Based on Pneumatic Relocation Mechanism

Luka Mišković[1,2(✉)] and Tadej Petrič[2]

[1] Department of Automation, Biocybernetics and Robotics, Jožef Stefan Institute, Jamova cesta 39, 1000 Ljubljana, Slovenia
luka.miskovic@ijs.si

[2] Jožef Stefan International Postgraduate School, Jamova 39, 1000 Ljubljana, Slovenia
https://abr.ijs.si/, https://www.mps.si/en/

**Abstract.** The concept of storing and reusing energy has emerged in quasi-passive exoskeletons as a solution to weight reduction and energy savings. These types of exoskeletons only store and reuse energy, but do not generate net positive mechanical work. In order to store different amounts of energy, the quasi-passive mechanism needs to be extended with the variable stiffness. In this paper, we present a newly designed variable stiffness joint that can change the geometry by exploiting the pneumatic artificial muscle (PAM). In addition to the PAM, the joint consists of a pneumatic cylinder and three fast-switching air valves. The PAM is the component that contracts in length when filled with compressed air and is inherently compliant. By filling the PAM with air from the cordless portable air pump, the pneumatic cylinder's effective length changes, resulting in a variation in torque, stiffness, and energy capacity. The joint is quasi-passive, meaning it only uses electricity to power the pump and the fast-switching valves. This implies that the joint can only store and reuse the energy, but not generate the torque. The article gives a system overview with the mathematical model, after which the experiment was performed to validate the hypothesis, and the results are discussed.

**Keywords:** Variable Stiffness Joint · Compliant Joint · Pneumatic · Exoskeletons

## 1 Introduction

A very frequently discussed topic today is intelligent energy distribution as it enables technological advances. The efficiency of the system can be greatly enhanced by having the ability to reuse some of the energy that would otherwise be lost. For instance, this can be found in prostheses that store and reuse energy in passive components or in industrial robotic and mechatronic systems

Supported by grants from the Slovenian Research Agency; PR-10489, and N2-0153.


T. Petrič et al. (Eds.): RAAD 2023, MMS 135, pp. 73–80, 2023.
https://doi.org/10.1007/978-3-031-32606-6_9

[1]. Similarly to this, higher energy efficiency enables advancements in robotic actuators. For instance, the series elastic actuator (SEA) can store energy in an elastic element placed between the gear and the actuator output and reuse it. However, the amount of energy that can be stored is limited by the constant stiffness and deflection of the spring. This led to an enhancement of the SEA with the possibility to vary the spring stiffness in variable stiffness actuators (VSA) to store various amounts of energy. Variable stiffness actuators [2] possess the ability to change both the actuator stiffness and the energy storage capacity. However, there is a trade-off between weight, size, and cost. Particularly when integrated into a moving system like an exoskeleton or humanoid robot, the assemblies must be very small. Each actuator is required to include a minimum of two motor units, frequently with two associated gearboxes, as well as a spring system for stiffness adjustment. There are usually three approaches to changing stiffness in electromechanical VSAs. These include 1) Changing the spring preload [3,4], 2) Changing the transmission ratio between the compliant element and the load [5], and 3) Changing the physical properties of the elastic element [6–8]. One solution to reduce the weight and size of actuators in exoskeletons and prostheses is to use quasi-passive mechanisms, as it has been shown that negative limb work can be accumulated and reused in the elastic element to reduce the metabolic cost of walking [9], running [10], or jumping. To achieve this, a variable stiffness spring is required where the energy storage capacity and stiffness can be adapted to the movement at the time, e.g., switching from walking to running. In addition to the design contradictions mentioned above, the aforementioned VSAs have an inverse relationship between stiffness and energy storage capacity, meaning that increasing stiffness decreases energy storage capacity. A notable exception was achieved by using air springs [7], where the authors demonstrated a way to achieve both the stiffness and energy storage capacity of a variable stiffness spring with a minimal amount of mechanical work. However, this strategy only worked in an experimental setup where the length of the pneumatic cylinder was changed using a press machine or a DC brush motor. If the length is to be changed during use on the exoskeleton, this must be done with additional components that are linearly actuated, such as actuated pinion and rack mechanism or an actuated ball screw. Since these components are usually heavy and do not fit the exoskeleton, we decided to use a pneumatic artificial muscle (PAM) whose length contraction is proportional to air pressure. The joint mechanism is built in a way that PAM's length change causes a shift in the length of the pneumatic cylinder and thus modifies the joint torque, stiffness, and energy storage capacity. The PAM is supplied with compressed air via a remote-controlled, chordless, rechargeable air pump, and once stiffness is set, the pump is no longer needed. Furthermore, no energy is required to maintain a stiffness setup. The air can be safely released by opening the valve, which means that the stiffness returns to its original setting. In addition, the PAM provides more safety because it is more compliant compared to other linear electromechanical actuators. The following chapter will provide a more thorough explanation of the joint mechanism.

# 2 System Description

The joint mechanism is described in detail in this section, as well as the method of stiffness modulation.

## 2.1 Overview

The joint mechanism is depicted in Fig. 1. When the rotating link is deflected from its starting angle ($\theta = 0°$), the mechanism resists rotation instead of actively producing torque in the joint. The pneumatic cylinder produces the force as a result of a change in pressure if all three air solenoid valves are closed. As explained previously [7], valve 1 has the function of a clutch, i.e., air can flow freely between two chambers without generating the force in the pneumatic cylinder. Thus, by opening valve 1, free movement is enabled. Conversely, closing valve 1 generates a force once the joint is deflected. The force sensor measures the force in the pneumatic cylinder. Valves 2 and 3 used to have the purpose of harvesting the energy, but in this work, we do not focus on energy harvesting, so valves 2 and 3 are considered passive components whose only task is to reset stiffness by releasing the air. By filling the PAM with compressed air, its length contracts, and the length of the cylinder is changed.

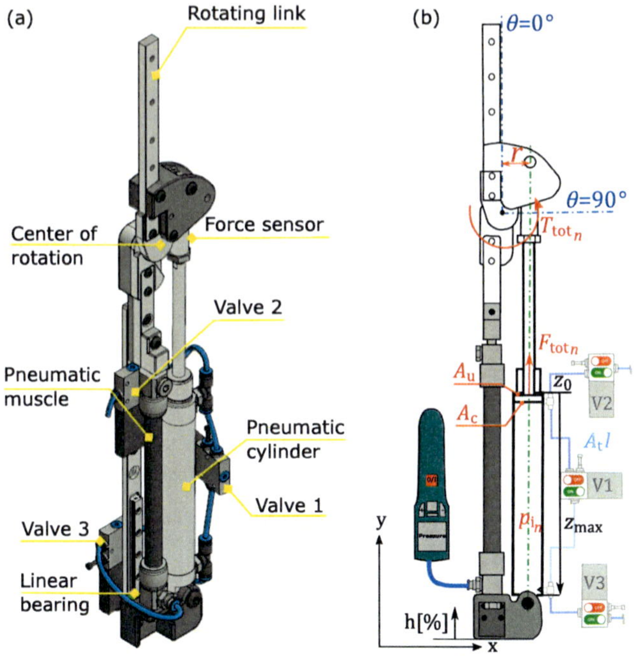


**Fig. 1.** (a) Variable Stiffness Joint Mechanism with specified parts, (b) Scheme of the mechanism

## 2.2 Mathematical Model

In the previous article [7] we developed the mathematical model and analyzed the pressure change in a pneumatic cylinder when operating with three air solenoid valves and using the passive dynamics of the body. For clarification, we briefly recapitulate the mathematical model here. The mechanism is shown in Fig. 1. The force in the pneumatic cylinder is modeled according to the pressure change in both cylinder chambers (see [7]), and is the function of the initial pressure in both cylinder chambers $p_{\mathrm{i}n}$, the initial piston position $z_0$ and the current piston position $z$. Thus, the equation can be expressed as follows:

$$F_{\mathrm{tot}n}(z, z_0, p_{\mathrm{i}n}) = p_{\mathrm{i}n}\left(\frac{A_{\mathrm{c}}(z_{\mathrm{max}} - z_0) + A_{\mathrm{t}}l}{A_{\mathrm{c}}(z_{\mathrm{max}} - z) + A_{\mathrm{t}}l}A_{\mathrm{c}} + \frac{A_{\mathrm{u}}z_0 + A_{\mathrm{t}}l}{A_{\mathrm{u}}z + A_{\mathrm{t}}l}A_{\mathrm{u}}\right). \tag{1}$$

The subscript $n$ refers to the number of steps in the previously described method of stiffness modulation. Although this method is not described here, the subscript is retained for consistency. The maximum piston position is denoted with $z_{\mathrm{max}}$, while $A_{\mathrm{t}}l$ is the volume inside the air hose and pneumatic fittings, where $A_{\mathrm{t}}$ is the hose cross-sectional area and $l$ is the length. Also, $A_{\mathrm{c}}$ is the larger cross-sectional area of the piston rod, and $A_{\mathrm{u}}$ is the smaller cross-sectional area as shown in Fig. 1. In order to calculate the theoretical torque of the joint, it is necessary to determine the lever arm length $\|r\|$, which is generated directly from the CAD model using Autodesk Inventor$^{\mathrm{TM}}$ for rotation angles from $\theta = 0°$ to $\theta = 90°$ (see Fig. 1 b). The lever arm is the perpendicular length connecting the joint's center of rotation with the cylinder's axis. Finally, with a known force and lever arm, the torque of the joint can be calculated as

$$T_{\mathrm{tot}n} = F_{\mathrm{tot}n}\|r\|. \tag{2}$$

When the PAM is filled with compressed air, the contraction $h[\%]$ increases proportionally to the air pressure, thereby adjusting the initial piston position $z_0$ as well as the maximum possible piston stroke $z_{\mathrm{max}}$, consequently affecting the piston cylinder force $F_{\mathrm{tot}n}$. The PAM contraction in relation to the pressure and force can be approximated by the FESTO data for the DMSP-10-140N pneumatic artificial muscle shown in Fig. 2, where the PAM force corresponds to the vertical component of the pneumatic cylinder force $F_{\mathrm{tot}n}$.

Moreover, changing the length of the PAM leads to a change in joint stiffness and energy storage capacity. Torsional stiffness can be expressed as

$$k_n = \frac{\partial T_{\mathrm{tot}n}}{\partial \theta}, \tag{3}$$

and the energy storage capacity is calculated by

$$H_n = \int_0^{\theta} T_{\mathrm{tot}n} d\theta. \tag{4}$$

Figure 3 depicts the force $F_{\mathrm{tot}n}$, torque $T_{\mathrm{tot}n}$, stiffness $k_n$, energy storage capacity $H_n$, and the lever arm $r$ versus angle deflection $\theta$ for different PAM

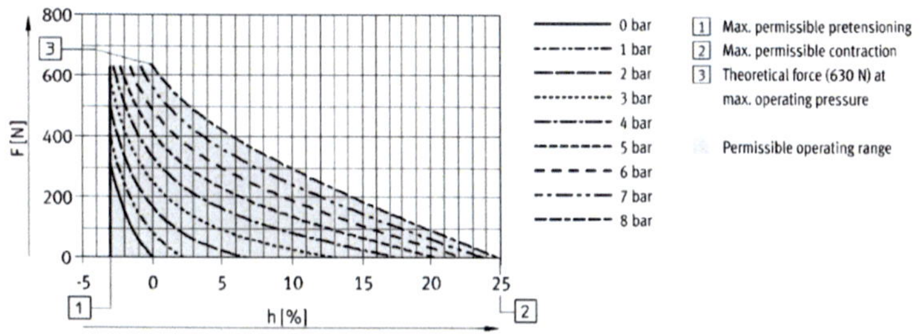


**Fig. 2.** Characteristic curves of the FESTO PAM DMSP-10-140N. Force as a function of contraction $h[\%]$ of the nominal length. The graphs are adopted from the datasheet of the PAM DMSP/MAS: www.festo.com

contraction lengths. The force, torque, stiffness, and energy storage capacity are affected by the PAM length contraction $h[\%]$, while the lever arm $r$ remains only a function of the angle $\theta$.

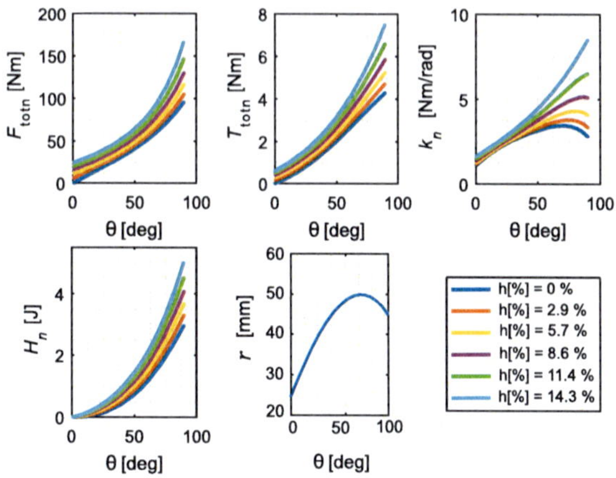


**Fig. 3.** Joint force, torque, stiffness, energy storage capacity and lever arm characteristics for different PAM contractions.

## 3 Experimental Evaluation

The experiment was carried out on the experimental platform shown in Fig. 4. The joint mechanism is coupled to and driven by a brushed DC motor to replace the human dynamics and verify the functionality of the mechanism. The motor was controlled in position mode, defining a sequence of seven motor rotations from 0° to 90° where the position control was done in Simulink$^{TM}$. After each joint rotation (except the first one), the PAM was filled with additional air by an external wireless pneumatic pump (Easy Pump Bosch). During the experiment, the pressure of both the cylinder and the PAM was measured. The rotation angle was measured by a rotary incremental encoder, while the linear contraction of the PAM was measured by a linear incremental encoder. The force in the cylinder was measured with a one-axis force transducer, while the torque of the joint was measured with a torque transducer placed between the DC motor and the joint. The measured torques and angles are used to calculate the torsional stiffness and energy in the post-analysis. All the components are detailed in Fig. 4.

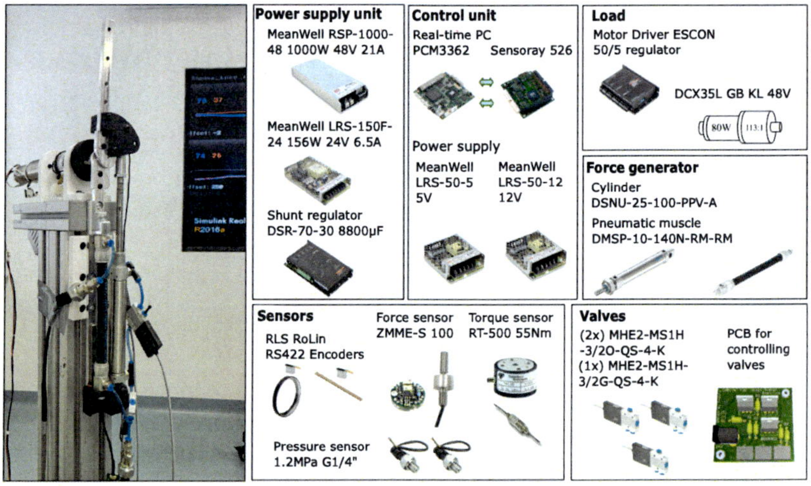


**Fig. 4.** Experimental platform with all components.

## 4 Results and Discussion

### 4.1 Results

The results of the experiment are shown in Fig. 5. The graphs show measured force, torque, calculated stiffness, and stored energy versus deflection, as well as the measured cylinder and PAM pressure along with the PAM length contraction percentage. The pressure in the cylinder increased from ~1.6 bar in the first rotation to ~2.3 bar in the last joint rotation due to the change in the effective length of the cylinder caused by the PAM shrinkage. The PAM was pumped to 5 bar, where the maximum contraction was ~10.5%. Between each rotation, the cylinder force and joint torque increased at the angle of 0°, as the PAM contraction compressed the air inside the cylinder without rotating the joint.

### 4.2 Discussion

The experimental results show a minor discrepancy with the mathematical model. This is due to the fact that the dynamic stretch of the PAM is not

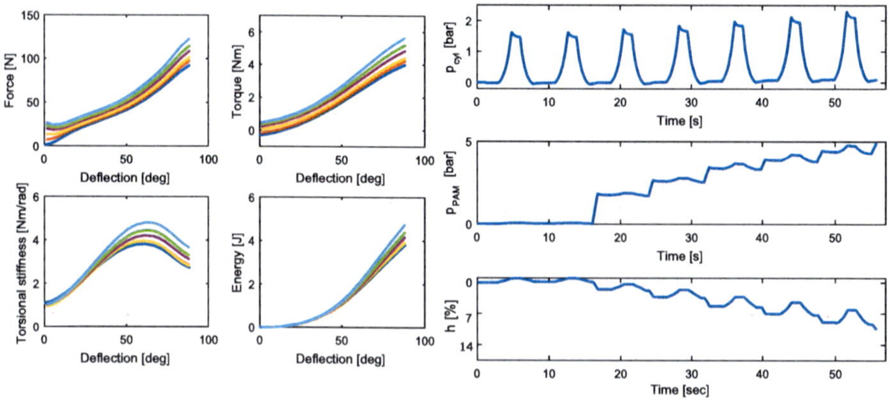


**Fig. 5.** Measured results of the experiment.

taken into account in the theoretical model, but is instead observed as if it were a rigid linear actuator with no compliance. This means that in the theoretical model, the pressure reaches higher values at the end of rotation. In reality, as the joint rotates, the PAM starts to stretch when the force exceeds the static value for the set pressure due to the inherent compliance of the PAM. Several dynamic PAM models exist, but they are still not accurate enough, and therefore future work will focus on a more thorough experimental dynamic characterization of the PAM and introduce additional terms in the model.

The great advantage of the joint mechanism is its low weight. The entire joint weighs only 0.76 kg. Secondly, the joint requires only a small amount of energy to maintain the set stiffness when the PAM is filled with air. To do this, the solenoid valve simply needs to be closed. Furthermore, few additional components are required to change and maintain the stiffness. The stiffness can also be easily reset by releasing the air. This can sometimes be problematic with electromechanical VSAs, if the fully tensioned mechanical spring is suddenly released, it will cause an energy burst. Finally, the main advantage is that it has a proportional relationship between the stiffness range and the energy storage capacity. It is possible to increase the stiffness range and the energy storage capacity simultaneously, which is not possible with electromechanical VSAs. Moreover, the joint can be fully portable and thus used in quasi-passive exoskeletons that store and reuse energy. However, electromechanical VSAs can produce much larger torques and achieve a wider range of possible stiffness configurations. Another drawback is that the presented joint mechanism does not generate the torque, but rather operates as a quasi-passive damper.

## 5 Conclusion

The paper presents a newly designed variable stiffness joint that varies stiffness based on pneumatic artificial muscle length contraction. The length contraction is achieved by filling the PAM with compressed air from a cordless portable pump. The length contraction of the PAM changes the effective length of the pneumatic cylinder and therefore changes the torque, stiffness, and energy storage capacity of the joint. The variable stiffness joint can be an alternative to electromechanical VSAs when lightweight applications are required (e.g. exoskeletons). However, the joint does not generate torque but acts as a quasi-passive damper that can store and reuse energy.

## References

1. Scalera, L., Palomba, I., Wehrle, E., Gasparetto, A., Vidoni, R.: Natural motion for energy saving in robotic and mechatronic systems. Appl. Sci. (Switz.) **9**(17) (2019)
2. Wolf, S., et al.: Variable stiffness actuators: review on design and components. IEEE/ASME Trans. Mechatron. **21**(5), 2418–2430 (2016)
3. Furnemont, R., Mathijssen, G., Van Der Hoeven, T., Brackx, B., Lefeber, D., Vanderborght, B.: Torsion MACCEPA: a novel compact compliant actuator designed around the drive axis. In: Proceedings - IEEE International Conference on Robotics and Automation, vol. 2015-June, no. June, pp. 232–237 (2015)
4. Catalano, M.G., et al.: VSA-CubeBot: a modular variable stiffness platform for multiple degrees of freedom robots. In: Proceedings - IEEE International Conference on Robotics and Automation, pp. 5090–5095 (2011)
5. Cestari, M., Sanz-Merodio, D., Arevalo, J.C., Garcia, E.: An adjustable compliant joint for lower-limb exoskeletons. IEEE/ASME Trans. Mechatron. **20**(2), 889–898 (2015)
6. Miakovic, L., Dezman, M., Petric, T.: Modular quasi-passive mechanism for energy storage applications: towards lightweight high-performance exoskeleton. In: 2021 20th International Conference on Advanced Robotics, ICAR 2021, pp. 588–593 (2021)
7. Miskovic, L., Dezman, M., Petric, T.: Pneumatic quasi-passive variable stiffness mechanism for energy storage applications. IEEE Robot. Autom. Lett. **7**(2), 1705–1712 (2022)
8. Miskovic, L., Dezman, M., Petric, T.: Pneumatic exoskeleton joint with a self-supporting air tank and stiffness modulation: design, modeling, and experimental evaluation. TechRxiv. Preprint. (2023)
9. Yandell, M.B., Tacca, J.R., Zelik, K.E.: Design of a low profile, unpowered ankle exoskeleton that fits under clothes: overcoming practical barriers to widespread societal adoption. IEEE Trans. Neural Syst. Rehabil. Eng. **27**(4), 712–723 (2019)
10. Simpson, C.S., et al.: Connecting the legs with a spring improves human running economy. J. Exp. Biol. **222**(17) (2019)

# Collision Avoidance in Collaborative Robotics Based on Real-Time Skeleton Tracking

Matteo Forlini(✉), Federico Neri, Cecilia Scoccia, Luca Carbonari, and Giacomo Palmieri

Department of Industrial Engineering and Mathematical Sciences, Polytechnic of Marche University, Ancona, Italy
m.forlini@pm.univpm.it

**Abstract.** Human-robot interaction in manufacturing is increasing more and more due to the spread of collaborative robots. However, it is important to ensure operator safety and improve the operator's physical and mental well-being. In this paper, an obstacle avoidance algorithm is implemented exploiting a vision system for human motion tracking. Once the positions of the human body joints relative to the robot's base are known in real time, the manipulator is able to avoid collisions by changing the trajectory and ensuring task execution at the same time. The motion tracking system is realized with three RGB-D cameras and customized software based on artificial intelligence and machine learning techniques. The whole system was tested in a real human-robot interaction scenario with the Universal Robot UR5e manipulator. The robot is able to avoid collisions over its entire kinematic chain. The distance between the human body and the manipulator always remains above a safety distance that is a settable parameter of the algorithm.

**Keywords:** Collaborative Robotics · Human Robot Interaction · Obstacle Avoidance · Machine Learning · Computer Vision · Deep Learning

## 1 Introduction

Today, collaborative robotics is becoming more and more common in manufacturing. Thanks to this new technology, humans and robots can work together, improving the physical and mental well-being of the operator [8]. The collaboration between humans and robots creates some important issues to pay attention to, such as operator ergonomics and his safety [6]. A new paradigm of production is rising: the human well-being is put at the center of attention. This new concept is bringing more productivity, more quality product, more flexibility in the task execution and less waste of materials [4,15]. Collaborative robots are able to stop in case of contact, however a better solution would be to detect any obstacle, like for example the human body that works next to it, and avoid


T. Petrič et al. (Eds.): RAAD 2023, MMS 135, pp. 81–88, 2023.
https://doi.org/10.1007/978-3-031-32606-6_10

it. In this way the robot is not forced to stop, whereas is able to complete its task by modifiyng its motion. This solution will increase the human's safety and physical and mental well-being, furthermore will make the interaction with the robot more profitable and usable. Lasota *et. al.* [9] argued that in order to have safe human-robot interaction there must be collision prevention methods such as safety zones, continuous monitoring of the distance between human and robot, and possible moving of the robot away from the operator whenever this distance becomes too small. Exteroceptive sensors can be used to know the position of the human respect to the robot base in real time [2]. For unwearable sensors, the most diffused for this kind of application are 3D camera, in particular depth camera sensors (RGB-D) like Microsot Kinect, Intel RealSense, Asus Xtion, etc. [7]. Thanks to commercial software or custom software these devices are able to identify and reconstruct the human skeleton and give the 3D position, respect to camera frame, of the human body joints. In a scene, where a single or more humans are present, it is important to use more than one camera to avoid occlusion problems [10,16]. Ferraguti *et al.* [5] implemented an algorithm based on controller barrier function around each body of the robot to ensure no collision with the human. Two RGB-D cameras were used to detect the position of the human; Kalman fliters were used to derive velocities and accelerations of the human, while data from the two cameras were merged using another bank of kalman filters. Morato *et al.* [11] presented a safety frame where humans and an industrial robot can work close together thanks to human tracking done by four Kinects sensors. The human's positions, velocities, and accelerations are known, and a roll-out strategy is used, in which the robot's trajectory in the near future is simulated. A precollision strategy is used that allows the human to operate in close proximity to the robot, pausing the robot's motion whenever an imminent collision is detected. Palmieri, Scoccia *et al.* [14,18] developed algorithms that for obstacle avoidance for different classes of collaborative robots. The algorithms are based on the potential fields method combined with Bézier curves fitting for off-line motion planning, while on-line motion control for obstacle avoidance is based on null space method (for redundant robots) and CLIK (Closed Loop Inverse Kinematics) control strategy. In [3], in particular, a specifc obstacle avoidance strategy is given for the Universal Robots class of manipulators. Safety regions around each link of the manipulator assume a cylindrical shape whose radius varies according to the speed of the colliding obstacle. The motion control law is based on two velocity tasks imposed to the manipulator: the first is imposed to the end-effector in order to follow the planned trajectory, the second is a repulsive velocity imposed at the point of the manipulator closest to the obstacle. Three different modalities for the collision avoidance control law are proposed, which differ in the type of motion admitted for the perturbation of the motion of the end-effector required to avoid the obstacle. In recent years Machine Learning (ML) techniques are becoming increasingly common in environments where human-robot interaction is expected [12,19]. ML is widely used for gesture recognition [17] and facial recognition and other applications that are able to facilitate the interaction between humans and robots.

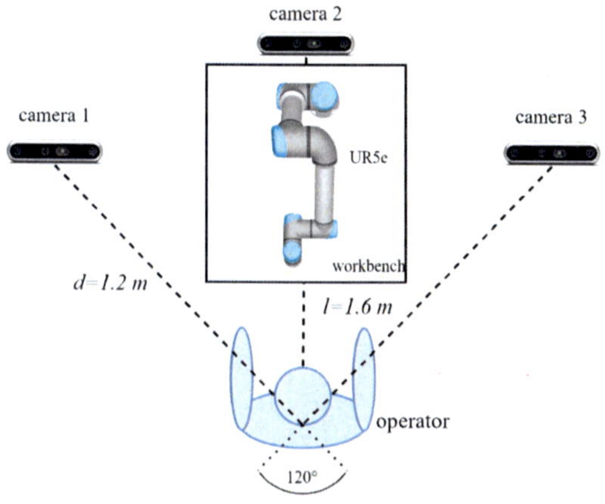


**Fig. 1.** System configuration

In this work an algorithm for obstacle avoidance developed in the past by the authors [3,13] is tested integrating a system for real-time obstacle detection. By the developed strategy the robot is able to avoid collisions with the human body, whenever it is getting close enough to the robotic arm. The detection of the human motion is done by three Intel Realsense D455 RGB-D cameras. A dedicated software framework in Python, which is able to recognize the human skeleton thanks to a machine learning technique, has been implemented. The skeleton detection provides the position of the human body joints to the control algorithm used to plan in real time the motion of the robot. An experimental case study of human robot interaction is performed to validate the algorithms and verify the accuracy of the system.

The manuscript is organized as follows: Sect. 2 describes the skeleton detection algorithm, Sect. 3 illustrates the experimental tests setup and Sect. 4 discusses the results obtained. Finally, concluding remarks and future works are summarised.

## 2 Skeleton Detection Algorithm

Human identification is achieved through three Intel Realsense D455 RGBD cameras. Their configuration is illustrated in Fig. 1. In this way it is possible to capture the entire scene from different points of view, mitigating the occlusion problem. The UR5e robot is positioned on top of the workbench at a height of 0.8 m above the floor, facing the operator with whom it shares the workspace.

Cameras are connected to a standard PC which executed the software for motion tracking. The software is developed in Python using the Mediapipe library from Google.

The stream of data from the depth sensor and from the RGB image sensor is set to a resolution of $424 \times 240$ pixels. An alignment of the two frames is performed to achieve a perfect match between the pixels of the depth image and those of the color image.

### 2.1 Artificial Intelligence for Skeleton Detection

MediaPipe Pose is a ML solution for high-fidelity body pose tracking, inferring 33 3D landmarks and background segmentation mask on the whole body from RGB video frames [1]. Mediapipe provides for each landmark $ID_i$ its $x, y$ pixels on the image. For the $z$ coordinate of $ID_i$ the value that comes from the depth frame for the correspondent pixel was used. An important output value is $ID_{visibility}$: Mediapipe is able to identify, for each landmark, its visibility in the scene. Visibility indicates the probability of a landmark not being occluded in the scene. Only landmarks with a $ID_{visibility}$ higher than 0.95 are taken into account, the other are discarded. $ID_{visibility}$ value was used to select among the three cameras (the maximum $ID_{visibility}$ value is chosen for each landmark), whereas the coordinates from the other two cameras. Furthermore, the Mediapipe parameters "min detection confidence" and "min tracking confidence" are both set to the high value of 0.90 in order to increase the robustness of the machine learning algorithm.

Once the intrinsic parameters of the cameras are determined by calibration, a deprojection that maps 2D pixel coordinates in 3D points in the camera frame space is performed exploiting the depth coordinate of the sensor. Camera calibration allows also to determine extrinsic parameters, i.e. the pose ${}_{cam}^{r}\mathbf{T}$ of the camera reference system relative to the robot base. Thus, human joint positions expressed relative to the camera reference system are transformed to the robot base reference system, as shown in Fig. 2:

$$ {}^{r}\mathbf{P} = {}_{cam}^{r}\mathbf{T}\ {}^{cam}\mathbf{P} \tag{1} $$

## 3 Experimental Tests

In this section experimental tests are presented. Skeleton detection is executed by a standard PC, human joints positions relative to the robot reference frame are sent via a TCP/IP 18 Hz to a second PC that executes the obstacle avoidance algorithm and sends velocity reference signals to the robot controller using UR RTDE interface [13]. For more information about the avoidance algorithm refer to [3]. Figure 3 shows an example with a graphical representation of the skeleton detection. It was possible to achieve an obstacle position communication speed 18 Hz, satisfactory for the purposes, by optimising the parameters of the human detection algorithm such as image resolution and so on, otherwise the communication speed would have been lower. Furthermore, by acting on intrinsic Mediapipe parameters, it was avoided that the human body was confused with the robot during detection and that the camera with occlusions was chosen. The results in terms of accuracy and speed in human detection obtained with the presented algorithm are comparable to the performance of other commercial skelton human detection software.

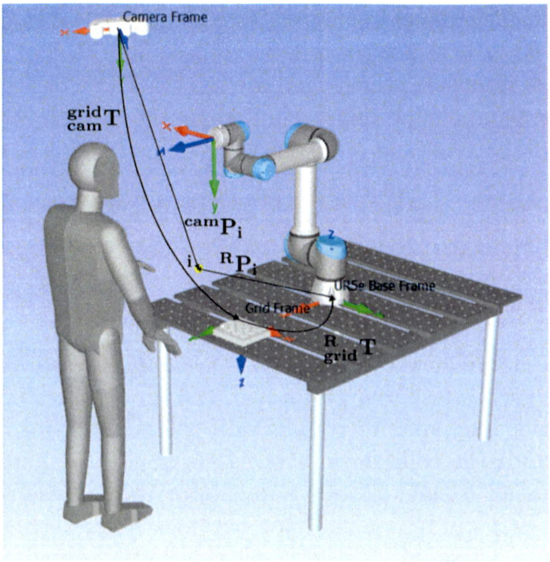


**Fig. 2.** Coordinates transformation from camera frame to robot base frame

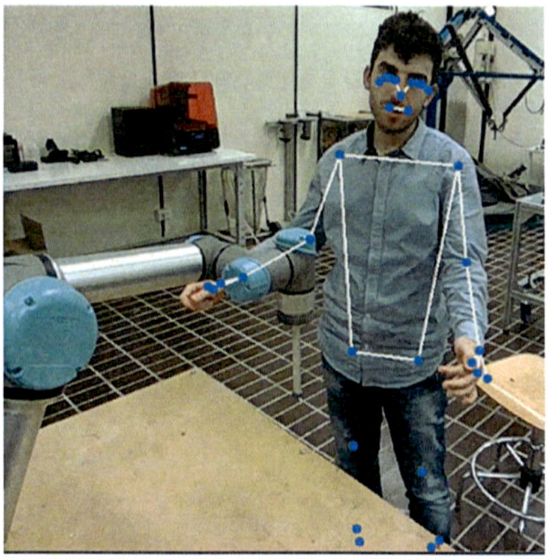

**Fig. 3.** Skeleton detection from camera 1

In the tests here presented the motion tracking system provides all visible joints coordinates, which are sent to the obstacle avoidance algorithm that considers for simplicity only the landmarks of the right and left wrists. The algorithm is capable of working with multiple obstacles, but in this case it was decided to use only the position of the two wrists, since they are the parts that can get closest to the robot given the test setup and since in a human-robot interaction scenario during an assembly task, they are the body parts that are most likely to come into contact with the robot. Some important parameters used in the obstacle avoidance algorithm were set with these values: $r = 0.20\,\text{m}$, $v_{rep} = 0.40\,\text{m/s}$, $\dot{\theta}_{max} = \pi/2\,\text{rad/s}$, where $r$ represents the safety distance from the robot, $v_{rep}$ the repulsive velocity imposed to the robot to avoid the obstacle and $\dot{\theta}_{max}$ the maximum joint speed value applicable to the robot. Parameter values were chosen by trial and error, ensuring that collision is avoided for the usual speeds of human movement, while at the same time preventing the robot from reacting too fastly, which can result in cognitive stress for the operator.

## 4 Results

To validate the functionality of the obstacle avoidance algorithm and skeleton detection, the following test was performed: the robot is stationary while the operator with his hands goes to collide with the tool and arm of the robot, which tries to avoids the collision in Mode I [3], i.e. with a perturbation of its pose that involves all 6 DoFs of the manipulator; when the distance between a hand and any part of the manipulator becomes lower than $r_{sup} = \frac{4}{3}r$, the avoidance control is activated and the repulsive velocity is gradually imposed to the robot in order to avoid the collision.

As it can be seen in Fig. 4, the operator tries to collide with his wrists against the robot three times. The distance falls below $r_{sup}$ value three times and the robot responds avoiding the obstacles as it can be seen in Fig. 5. For example the three peaks of $\dot{\theta}_1$ or $\dot{\theta}_2$ occurs at the same time that the wrists get closer to the robot. In alla cases the distance remains greater than the safety limit $r$.

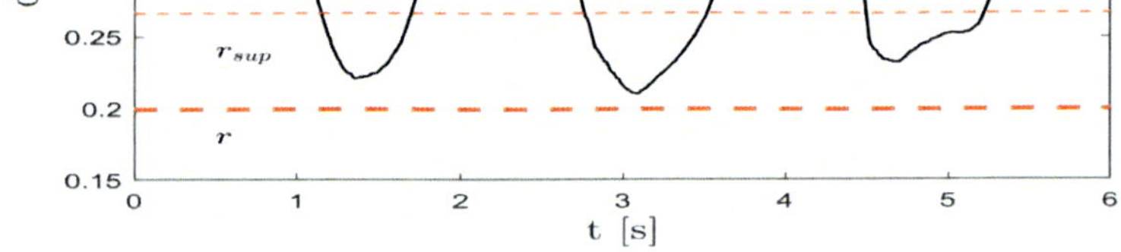


**Fig. 4.** Collision avoidance test; separation between human and robot.

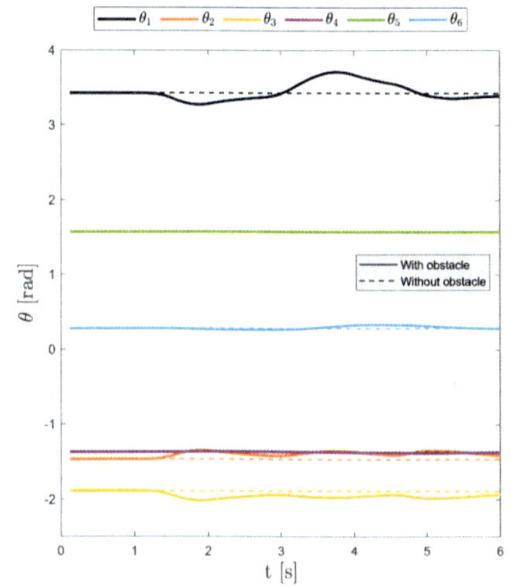


(a)

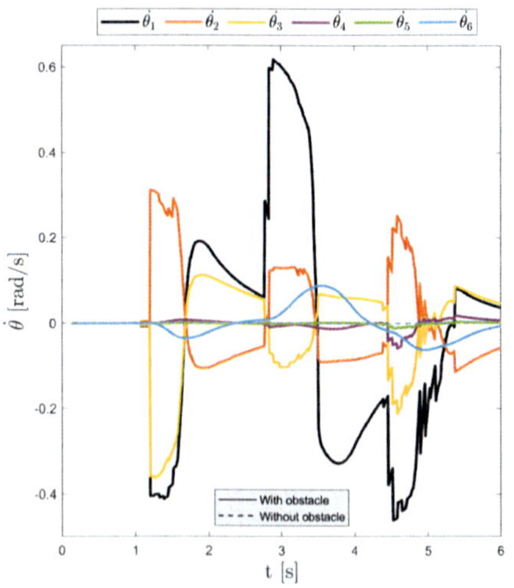


(b)

**Fig. 5.** Collision avoidance test; a) joint rotations, b) joint speeds.

## 5 Conclusions

In this work, a collaborative work cell was implemented in which human-robot interaction takes place. The robot is able to avoid contact with the human body, thus increasing the physical and cognitive well-being of the operator. A non-commercial skeleton detection system based on machine learning techniques was developed using three RGB-D cameras. Thanks to it, it was possible to test the avoidance algorithm with obstacles detected in real-time. The robot responds to any approach of the operator by changing its configuration in real-time in order to avoid the collision with the operator. Movements of the robot are smooth so as not to affect the cognitive stress of the operator. As further development it would be useful to take into account also the velocity of the human motions, as far as the relative velocity between robot and operator. Finally, a comparison between RGBD systems for obstacle detection and different systems, such as wearable IMU-based systems, will allow to evaluate the accuracy and robustness of the developed system.

## References


1. Mediapipe, pose detection (2022). https://google.github.io/mediapipe/solutions/pose.html
2. Cherubini, A., Navarro-Alarcon, D.: Sensor-based control for collaborative robots: fundamentals, challenges, and opportunities. Front. Neurorobot. **14**, 113 (2021)
3. Chiriatti, G., Palmieri, G., Scoccia, C., Palpacelli, M.C., Callegari, M.: Adaptive obstacle avoidance for a class of collaborative robots. Machines **9**(6), 113 (2021)

4. Ciccarelli, M., et al.: A system to improve the physical ergonomics in human-robot collaboration. Procedia Comput. Sci. **200**, 689–698 (2022)
5. Ferraguti, F., et al.: Safety barrier functions and multi-camera tracking for human-robot shared environment. Robot. Auton. Syst. **124**, 103388 (2020)
6. Gualtieri, L., Rauch, E., Vidoni, R.: Emerging research fields in safety and ergonomics in industrial collaborative robotics: a systematic literature review. Robot. Comput. Integr. Manuf. **67**, 101998 (2021)
7. Halme, R.J., Lanz, M., Kämäräinen, J., Pieters, R., Latokartano, J., Hietanen, A.: Review of vision-based safety systems for human-robot collaboration. Procedia CIRP **72**, 111–116 (2018)
8. Huang, J., et al.: An experimental human-robot collaborative disassembly cell. Comput. Ind. Eng. **155**, 107189 (2021)
9. Lasota, P.A., Fong, T., Shah, J.A., et al.: A survey of methods for safe human-robot interaction. Found. Trends Robot. **5**(4), 261–349 (2017)
10. Lenz, C., Grimm, M., Röder, T., Knoll, A.: Fusing multiple kinects to survey shared human-robot-workspaces (2012)
11. Morato, C., Kaipa, K.N., Zhao, B., Gupta, S.K.: Toward safe human robot collaboration by using multiple kinects based real-time human tracking. J. Comput. Inf. Sci. Eng. **14**(1) (2014)
12. Mukherjee, D., Gupta, K., Chang, L.H., Najjaran, H.: A survey of robot learning strategies for human-robot collaboration in industrial settings. Robot. Comput. Integr. Manuf. **73**, 102231 (2022)
13. Neri, F., Forlini, M., Scoccia, C., Palmieri, G., Callegari, M.: Experimental evaluation of collision avoidance techniques for collaborative robots. Appl. Sci. **13**(5), 2944 (2023)
14. Palmieri, G., Scoccia, C.: Motion planning and control of redundant manipulators for dynamical obstacle avoidance. Machines **9**(6), 121 (2021)
15. Probst, L., Frideres, L., Pedersen, B., Caputi, C.: Service Innovation for Smart Industry: Human-Robot Collaboration. European Commission, Luxembourg (2015)
16. Rybski, P., Anderson-Sprecher, P., Huber, D., Niessl, C., Simmons, R.: Sensor fusion for human safety in industrial workcells. In: 2012 IEEE/RSJ International Conference on Intelligent Robots and Systems, pp. 3612–3619. IEEE (2012)
17. Scoccia, C., Menchi, G., Ciccarelli, M., Forlini, M., Papetti, A.: Adaptive real-time gesture recognition in a dynamic scenario for human-robot collaborative applications. In: Niola, V., Gasparetto, A., Quaglia, G., Carbone, G. (eds.) Advances in Italian Mechanism Science (IFToMM Italy 2022). Mechanisms and Machine Science, vol. 122, pp. 637–644. Springer, Cham (2022). https://doi.org/10.1007/978-3-031-10776-4_73
18. Scoccia, C., Palmieri, G., Palpacelli, M.C., Callegari, M.: A collision avoidance strategy for redundant manipulators in dynamically variable environments: on-line perturbations of off-line generated trajectories. Machines **9**(2), 30 (2021)
19. Zamora, M., Caldwell, E., Garcia-Rodriguez, J., Azorin-Lopez, J., Cazorla, M.: Machine learning improves human-robot interaction in productive environments: a review. In: Rojas, I., Joya, G., Catala, A. (eds.) IWANN 2017. LNCS, vol. 10306, pp. 283–293. Springer, Cham (2017). https://doi.org/10.1007/978-3-319-59147-6_25

# Application of a Phase State System for Physical Human-Humanoid Robot Collaboration

Tilen Brecelj(✉) and Tadej Petrič

'Jožef Stefan' Institute, Jamova c. 39, 1000 Ljubljana, Slovenia
{tileb.brecelj,tadej.petric}@ijs.si
http://cobotat.ijs.si/

**Abstract.** Collaborative robotics is one of the fastest developing but at the same time one of the most challenging fields in robotics. The reason for the latter is that a good robot collaborator should precisely understand the intentions of the human coworker. In this paper, human-robot collaboration is addressed by the application of a dynamical phase state system guided by stable heteroclinic channel networks to the motion of a humanoid robot. With this approach, a person can control the underlying dynamical system by applying forces to the grippers of the humanoid robot and this way guide the robot. The robot motions presented in this paper are lifting up the forearms and squatting, but the dynamical model can be further extended to incorporate an arbitrary amount of motions. The presented approach, therefore, provides a suitable way for the realization of efficient human-robot collaboration and should be further explored and developed.

**Keywords:** Human–robot interaction · Collaborative robotics · Phase state system · Stable heteroclinic channel networks

## 1 Introduction

The presence of robots in our everyday lives is increasing on a daily basis as robots are being employed not only as home assistants but in many other fields such as science, industry, healthcare, medicine, tourism and hospitality, just to mention a phew. The goal of robotization is not only to substitute the human workforce but also to assist humans in executing different tasks. But robots perform their work mostly on their own following predefined trajectories or are teleoperated by humans. The reason lies in their incapability to fully understand human intentions to assist them in any situation in the desired way, which could lead to unwanted behavior of the robots and endanger the human

Supported by the Slovenian Research Agency grant N2-0153.


T. Petrič et al. (Eds.): RAAD 2023, MMS 135, pp. 89–96, 2023.
https://doi.org/10.1007/978-3-031-32606-6_11

coworker. Numerous strategies are therefore being developed for the interpretation of human behavior and the understanding of human intentions [11]. With the addition of grasping and object manipulation capabilities, robots could become suitable cooperators and assistants to human workers [6].

One way to make robots cooperate with humans is by mimicking human behavior. Humanoid robots are the most suitable robots for such a learning approach because of their human-like body shape which allows them to reproduce human motion. But humanoid robots have a relatively small support polygon in comparison to their body size, which makes them easily unstable. Performing fast movements and physically interacting with the environment, therefore represent a big challenge for them. For this reason, various stability parameters and balancing algorithms have been developed [3,4,12] which allow humanoid and other floating-based robots to walk [10], run and jump [9], ski [8], and complete many other tasks.

A suitable approach for the implementation of human-robot cooperation is based on the application of the stable heteroclinic channel (SHC) networks to the robot motion [5,7]. The SHC networks is a continuous dynamical phase state system (PSS), where the SHCs connect the saddle points, which can be interpreted as system states. Each saddle point lies on its coordinate axis, while the transitions between them are smooth. As each transition between two states has a phase, it can be coupled to the robot motion and make it transit from an initial to a final configuration.

In this paper, the motion of the humanoid robot Talos is coupled to a PSS consisting of three states, which makes Talos transit between three different configurations: standing up with its forearms oriented downwards, standing up with its forearms oriented upwards, and squatting. The behavior of the PSS is determined by the model parameters and by the forces applied to the grippers of Talos, which cause transitions between different states. Such an approach of coupling a dynamical PSS guided by the SHC networks is shown to be a suitable way of guiding a robot in the means of physical human-robot interaction in a safe and controlled way.

## 2 The SHC Networks

The robot motion presented in this paper is controlled by a continuous dynamical PSS, built on the SHC networks [5,7]. This dynamical system consists of saddle points that can be interpreted as system states and are connected with self-stabilizing limit cycles, each with its own activation time, velocity, and phase. Furthermore, each transition between two different states can be arbitrarily allowed or prohibited. When a transition is activated, it can be reversed and the system returns to the initial state. Furthermore, more transitions can be initialized simultaneously, as they are nonexclusive, but the PSS transits to only one final state, depending on the model parameters.

The PSS built on the SHC networks, applied to the robot motion and presented in this paper, is defined as

$$\dot{\boldsymbol{x}} = \boldsymbol{x} \circ \left(\boldsymbol{\alpha} + \left(\boldsymbol{\rho}_0 + \boldsymbol{\rho}_\Delta \circ (\boldsymbol{T} + \mathbf{G})\right)\boldsymbol{x}^\gamma\right)\eta(t) + \dot{\boldsymbol{\delta}}(t)\,, \tag{1}$$

where $\boldsymbol{x}$ is a $n$-dimensional state vector and $\circ$ denotes element-wise multiplication [5]. The state transition matrix $\boldsymbol{T}$ is defined as $T_{ji} = 1$ if the transition $i \to j$ exists and $T_{ji} = 0$ otherwise. The matrix

$$\boldsymbol{\rho}_0 = \left(\boldsymbol{\alpha}(\boldsymbol{\beta}^{-1})^T\right) \circ \left(\boldsymbol{I} - 1 - (\boldsymbol{\alpha}\boldsymbol{\alpha}^T)^{-1}\right), \tag{2}$$

determines the saddle points of the PSS, while the matrix $\boldsymbol{\rho}_\Delta$ is defined as

$$\boldsymbol{\rho}_\Delta = \left(\boldsymbol{\alpha} \circ (1 + \boldsymbol{\nu}^{-1})\right)\left(\boldsymbol{\beta}^{-1}\right)^T. \tag{3}$$

The $n$-dimensional vectors $\boldsymbol{\alpha}$, $\boldsymbol{\beta}$, and $\boldsymbol{\nu}$ determine the PSS transitions growth rates, the state's positions, and the shapes of the saddle points, respectively. The matrices $\boldsymbol{\rho}_0$ and $\boldsymbol{\rho}_\Delta$ are of dimensions $n \times n$ and maintain constant values. The greediness matrix $\boldsymbol{G}$ determines the probability of maintaining various transitions activated at the same time, but it can also reverse a transition. The locations of the PSS channels and the evolution rates are determined by the scalars $\gamma$ and $\eta(t)$, respectively. Finally, the parameter $\dot{\boldsymbol{\delta}}(t)$ activates the system transitions.

The system states lie each on its coordinate axis forming an orthonormal basis and therefore each transition between two states lies on the plane determined by the coordinate axis of the initial and the final state. The activation matrix $\boldsymbol{\Lambda}$ with the property $\sum \Lambda_{ji} = 1$, which contains all the state transitions and activations, can be defined as

$$\Lambda_{ji} = \begin{cases} \Lambda^{\mathrm{t}}_{ji}\,; & j \neq i \\ \lambda^{\mathrm{s}}_i\,; & j = i \end{cases}. \tag{4}$$

Here

$$\boldsymbol{\Lambda}^{\mathrm{t}} = \frac{16\boldsymbol{x}\boldsymbol{x}^T|\boldsymbol{x}^2|}{(\boldsymbol{x}\mathbf{1}^T + \mathbf{1}\boldsymbol{x}^T)^2 + |\boldsymbol{x}|^2} \circ \boldsymbol{T} \tag{5}$$

is the transition activation function and can have values from the interval $[0, 1]$,

$$\boldsymbol{\lambda}^{\mathrm{s}} = \frac{\boldsymbol{x}^2}{\sum \boldsymbol{x}^2}\left(\mathbf{1} - \sum \boldsymbol{\Lambda}^{\mathrm{t}}\right) \tag{6}$$

is the state activation matrix, while $\mathbf{1}$ is an $n-$ dimensional vector of ones.

Each transition from the state $i$ to the state $j$ has a phase determined by

$$\Phi_{ji} = \frac{|\boldsymbol{x}_j|}{|\boldsymbol{x}_j| + |\boldsymbol{x}_i|}\,. \tag{7}$$

with a value increasing from 0, at the beginning of the transition, to 1, when the system reaches the final state.

The PSS transition to the state $j$ can be triggered or blocked with the application of a small positive value or a small negative value to the parameter $\dot{\delta}_j(t)$, respectively. The vector $\dot{\boldsymbol{\delta}}(t)$ is defined as

$$\dot{\boldsymbol{\delta}}(t) = (\boldsymbol{\Lambda} \circ \boldsymbol{B})\boldsymbol{x}\,, \tag{8}$$

where the elements of the bias matrix $\boldsymbol{B}$ determine the velocity bias for the corresponding transition. In the case when multiple elements of the bias matrix $\boldsymbol{B}$ have a positive value, multiple transitions will be activated at the same time, but the system will converge to only one final state, according to the shape of the attractor.

The parameter $\eta(t)$ is defined as

$$\eta(t) = 2^{A \sum_{ji} \Lambda_{ji}} \tag{9}$$

and enables a continuous adjustment of the transition velocity at every iteration of the PSS by the modification of the parameter $A$. If $A$ is set to 0, the transition velocity is the default one, determined by the parameter $\alpha$.

# 3 The Application of the PSS Guided by the SHC Networks to a Humanoid Robot

## 3.1 Implementation

The maneuverability of a dynamical PSS guided by the SHC networks can be exploited to guide the motion of a robot. Each transition of the PSS has its own phase, which can be coupled to a certain predefined robot movement. At the beginning of a transition, its phase is 0 and the PSS is in its initial state, while the robot is in its initial position. But when the PSS transits towards its final state and the phase of the corresponding transition increases towards 1, the robot moves to reach its final configuration.

A dynamical PSS guided by the SHC networks, presented in Sect. 2, was applied to guide the motion of the humanoid robot Talos with 32 degrees of freedom [2]. The applied PSS consists of three states, each coupled to a different robot configuration. The possible PSS transitions are from state 1 to state 2 and vice versa, and from state 2 to state 3 and vice versa, as shown in Fig. 1. The transition matrix is therefore

$$\boldsymbol{T} = \begin{bmatrix} 0 & 1 & 0 \\ 1 & 0 & 1 \\ 0 & 1 & 0 \end{bmatrix}. \tag{10}$$

In state 1 the humanoid robot stands on its legs with its arms oriented downwards, in state 2 the humanoid robot stands on its legs with its forearms oriented

**Fig. 1.** A schematic representation of the PSS states with the corresponding configurations of the humanoid robot Talos and the possible transitions between them.

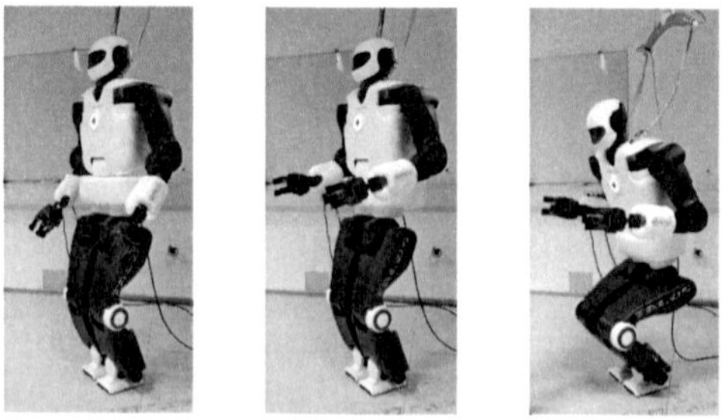

(a) (b) (c)

**Fig. 2.** Humanoid robot Talos in the standing-up position with its forearms oriented downwards (a) and horizontally (b), which correspond to states 1 and 2 of the PSS, respectively, and in the squatting position (c), which corresponds to state 3 of the PSS.

horizontally, while in state 3 the humanoid robot is in a squatting position, as shown in Fig. 2.

The transitions between different PSS states are triggered by the vertical pulling and pushing forces applied to the grippers of the humanoid robot, measured by the force-torque sensors mounted in its grippers. At every iteration of the PSS, the bias matrix $\boldsymbol{B}$ is evaluated as

$$\boldsymbol{B} = \boldsymbol{T} \circ \boldsymbol{F}_{\mathrm{d}} \circ \boldsymbol{s}, \tag{11}$$

where $\boldsymbol{F}_{\mathrm{d}}$ is the matrix corresponding to a certain force direction, $\boldsymbol{s} = (s_1, s_2, s_3)$ is the state vector determining the amount the PSS is in a certain state, and $s_{1,2,3} \in [0, 1]$. If the force is pulling up, pushing down, or if there is no force, d = up, d = down, or d = nf, respectively, while the corresponding force matrices are

$$\boldsymbol{F}_{\mathrm{up}} = 10^{-1} \begin{bmatrix} 0\,0\,0 \\ 1\,0\,1 \\ 0\,0\,0 \end{bmatrix}, \boldsymbol{F}_{\mathrm{down}} = 10^{-1} \begin{bmatrix} 0\,0\,0 \\ 0\,0\,0 \\ 0\,1\,0 \end{bmatrix}, \text{ and } \boldsymbol{F}_{\mathrm{nf}} = 10^{-3} \begin{bmatrix} 0\,1\,0 \\ 0\,0\,0 \\ 0\,0\,0 \end{bmatrix}. \tag{12}$$

If the PSS is in state 1, $\boldsymbol{s} = (1, 0, 0)$, Talos stands with its arms oriented downwards. If a pulling-up force is applied to its grippers, $F_{\mathrm{d}} = F_{\mathrm{up}}$, the only non-zero element of the bias matrix is $B_{21}$ and therefore the only allowed transition is from state 1 to state 2, when Talos lifts its forearms in a horizontal position.

If the PSS is in state 2, $\boldsymbol{s} = (0, 1, 0)$, Talos stands with its forearms oriented horizontally. If a pushing-down force is applied to its grippers, $F_{\mathrm{d}} = F_{\mathrm{down}}$, the only non-zero element of the bias matrix is $B_{32}$, which allows only the transition from state 2 to state 3, when Talos squats. But if on the other hand, there is no force applied to Talos while standing with its forearms oriented horizontally, $F_{\mathrm{d}} = F_{\mathrm{nf}}$, the only non-zero element of the bias matrix, $B_{12}$, allows the transition from state 2 to state 1 and Talos puts its forearms down.

Finally, if the PSS is in state 3, $\boldsymbol{s} = (0, 0, 1)$, Talos is in a squatting position. If a pulling-up force is applied to its grippers, $F_{\mathrm{d}} = F_{\mathrm{up}}$, the only non-zero element of the bias matrix, $B_{23}$, triggers the transition from state 3 to state 2, which makes Talos stand up.

All the other combinations of forces and PSS states produce a null matrix (matrix with only zeros) and no PSS transitions and therefore Talos' movements occur. In every step of the iteration, all the zero values of the matrix $B$ are set to $-10^{-3}$ in order to block all the PSS transitions that should not occur.

The motion of the forearms of Talos from the vertical to the horizontal position was obtained from a simulation in the Gazebo simulator [1]. On the other hand, the squatting motion was obtained by applying the squatting motion of a person, recorded with a motion capture system, to the robot Talos.

### 3.2 Results

The results presented in this paper were obtained by setting all the elements of the greediness matrix $\boldsymbol{G}$ to zero and this way blocking all the simultaneous multiple transitions. Furthermore, all the values of the vector $\boldsymbol{\alpha}$ were set to 20, and therefore the growth rates of all the channels were the same, while all the values of the vectors $\boldsymbol{\beta}$ and $\boldsymbol{\nu}$ were set to 1, which makes all the saddle points of the same shape and positioned on their own coordinate axis at the value of 1. To make the transitions directly leave the initial states and approach their final states, the parameter $\gamma$ was set to 1.

To fix the PSS in the initial state 1, in the beginning, all the values of $\boldsymbol{B}$ were set to $-10^{-3}$. But after a force was applied, the matrix $\boldsymbol{B}$ was modified according to Eq. (11). To make the PSS transitions and therefore Talos' motion force dependent, the transition velocity, determined by the Eq. (9), was modulated by the parameter $A = -1.5 + 10^{-2} {F_{\mathrm{grip}}}^{1.4}$ in each step of the iteration, where $F_{\mathrm{grip}} = 0.5(F_{\mathrm{L}} + F_{\mathrm{R}})$, and $F_{\mathrm{L}}$ and $F_{\mathrm{R}}$ are the vertical forces applied to the left and right gripper of Talos, respectively.

Figure 3 shows the PSS states and the transitions between them, along with the vertical force, applied to the grippers of Talos, as a function of time. The vertical force was obtained as the average vertical force applied to both grippers of Talos. As it can be seen, the PSS was initially in state 1 and Talos was standing on its feet with its arms oriented downwards. A pulling-up force, exerted on the grippers of Talos, made the PSS transit at time $t = 6.7$ s to state 2 and Talos lifted its forearms. Further on, a pushing-down force made the system transit at time $t = 17.8$ s to state 3 and Talos squatted. At time $t = 34.8$ s a pulling-up force made the PSS transit back to state 2 and Talos stood up. Further on,

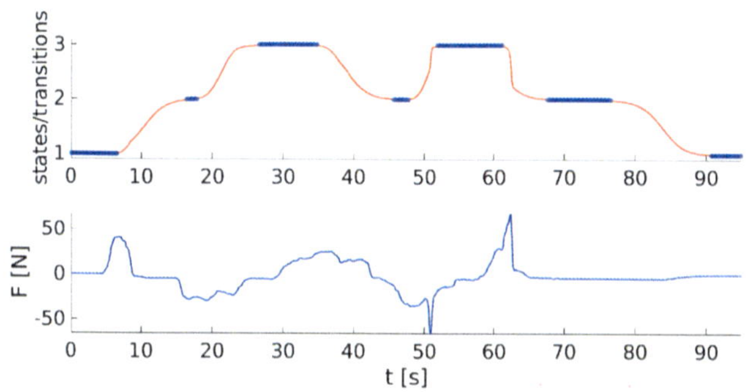


**Fig. 3.** PSS states and the transitions between them, along with the average vertical force, applied to the grippers of Talos, as a function of time.

larger vertical forces than previously were applied to the grippers of Talos. The forces were first oriented downwards and later oriented upwards, which made the PSS transit at time $t = 47.7$ s from state 2 to state 3, and at time $t = 61.2$ s from state 3 back to state 2. This made Talos squat and stand back up. Larger forces caused the PSS transitions and therefore also the movements of Talos to be faster, which can be clearly seen in Fig. 3. In the end, when the PSS was in state 2 and Talos was standing with its forearms oriented horizontally, no more force was applied to its grippers and at time $t = 76.5$ s the PSS slowly transited from state 2 to state 1 and Talos put its forearms down.

## 4 Discussion and Conclusions

Robots are being employed in many different areas to physically assist humans at their work. Due to their human-like appearance, humanoid robots are of special interest, as they can perform human-like motion and mimic human behavior. But efficient cooperation requires a thorough understanding of human intentions, which makes human-robot collaboration very challenging. Nevertheless, human-robot collaboration has lately experienced a tremendous advance combining knowledge from many different fields such as psychology, computer science, and engineering.

In this paper, human-robot collaboration is addressed by the application of a dynamical PSS guided by the SHC networks to the motion of the humanoid robot Talos. The application of vertical forces to the grippers of Talos triggers the transitions of the PSS and causes the motion of the robot. This is possible as each PSS transition has a phase, that is coupled to a different predefined movement of Talos. With this approach, a person is able to dynamically control Talos and make it lift its forearms, squat, and stand back up. Moreover, by applying forces with different amplitudes, the speed of the PSS transitions is altered, and consequently, the velocity of the motion of Talos is controlled.

The implementation of this algorithm is simple, as in every iteration of the loop, only Eqs. (9) and (11) need to be evaluated to compute the evolution of the

PSS. Moreover, such an approach includes the understanding of human intentions from the directions and amplitudes of the forces applied on the grippers of Talos. But despite the fact that all the robot movements were predefined, the discussed method represents a step towards efficient human-robot collaboration. As the presented dynamical PSS can be further extended, many more tasks could be added to the discussed method.

The next step in the development of the proposed approach for human-robot collaboration could be the inclusion of different learning algorithms to make the robot's movements smoother and adapt them to specific situations. This could further contribute to the evolution of techniques for efficient autonomous human-robot collaboration.

## References


1. Gazebo. https://gazebosim.org/home
2. Talos. https://pal-robotics.com/robots/talos/
3. Brecelj, T., Petrič, T.: Zero moment line — universal stability parameter for multi-contact systems in three dimensions. Sensors **22**(15), 5656 (2022). https://doi.org/10.3390/s22155656, https://www.mdpi.com/1424-8220/22/15/5656
4. Caron, S., Pham, Q.C., Nakamura, Y.: Leveraging cone double description for multi-contact stability of humanoids with applications to statics and dynamics (2015). https://doi.org/10.15607/RSS.2015.XI.028
5. Deimel, R.: Reactive interaction through body motion and the phase-state-machine. In: 2019 IEEE/RSJ International Conference on Intelligent Robots and Systems (IROS), pp. 6383–6390 (2019)
6. Harada, K., Kajita, S., Kaneko, K., Hirukawa, H.: Dynamics and balance of a humanoid robot during manipulation tasks. IEEE Trans. Rob. **22**(3), 568–575 (2006). https://doi.org/10.1109/TRO.2006.870649
7. Horchler, A.D., Daltorio, K.A., Chiel, H.J., Quinn, R.D.: Designing responsive pattern generators: stable heteroclinic channel cycles for modeling and control. Bioinspiration Biomimetics **10**(2), 026001 (2015). https://doi.org/10.1088/1748-3190/10/2/026001
8. Lahajnar, L., Kos, A., Nemec, B.: Skiing robot - design, control, and navigation in unstructured environment. Robotica **27**, 567–577 (2009). https://doi.org/10.1017/S0263574708004955
9. Nagasaka, K., Kuroki, Y., Suzuki, S., Itoh, Y., Yamaguchi, J.: Integrated motion control for walking, jumping and running on a small bipedal entertainment robot. In: IEEE International Conference on Robotics and Automation, 2004. Proceedings. ICRA 2004, vol. 4, pp. 3189–3194 (2004). https://doi.org/10.1109/ROBOT.2004.1308745
10. Petrič, T., Žlajpah, L., Garofalo, G., Ott, C.: Walking control using adaptive oscillators combined with dynamic movement primitives (2013)
11. Strabala, K., .: Toward seamless human-robot handovers. J. Hum.-Robot Interact. **2**(1), 112–132 (2013). https://doi.org/10.5898/JHRI.2.1.Strabala
12. Vukobratovič, M., Borovac, B.: Zero-moment point - thirty five years of its life. I. J. Humanoid Robot. **1**, 157–173 (2004). https://doi.org/10.1142/S0219843604000083

# Input-Observer-Based Estimation of the External Torque for Single-Link Flexible-Joint Robots

Nikola Knežević[1], Maja Trumić[1,2](✉), Kosta Jovanović[1], and Adriano Fagiolini[2]

[1] School of Electrical Engineering, University of Belgrade, Belgrade, Serbia
[2] MIRPALab, Department of Engineering, University of Palermo, Palermo, Italy
maja.trumic@etf.rs

**Abstract.** To enable safe interaction between a robot and its environment, it is necessary to monitor in real-time the external torque that acts on the robot. This paper proposes a strategy that exploits solely encoder and current data to estimate the external torque acting on the single-link flexible-joint robot driven by a variable stiffness actuator in the antagonistic setup. The strategy comprises two steps: first, the elastic torques generated by motors within a variable stiffness actuator are reconstructed, then the external torque is estimated leveraging the data on the previously reconstructed elastic torques and robot dynamic model. The effectiveness of the estimator is tested in the simulation environment with respect to the time-varying external torque.

**Keywords:** Estimation · Input observer · Variable Stiffness Actuators · External torque

## 1 Introduction

Ever since a global shift to the paradigm of collaborative robotics, which empowered the research on the physical human-robot interaction, it has become necessary to ensure that robots are aware of their unstructured and time-varying environment. In such close-collaborative circumstances, there is a higher probability of an unintentional collision between robots and the environment, or even an intentional collision initiated by humans [3]. Naturally, when a collision occurs, it becomes of the utmost importance to choose the appropriate robot's behavior such that a safe environment is ensured [6], and then accordingly control its position and impedance [11,13].

This work has been partially supported by the Science Fund of the Republic of Serbia, PROMIS, #6062528, ForNextCobot and by the Ministry of Science, Technological Development and Innovation of the Republic of Serbia under contract number: 451-03-47/2023-01/200103.


T. Petrič et al. (Eds.): RAAD 2023, MMS 135, pp. 97–105, 2023.
https://doi.org/10.1007/978-3-031-32606-6_12

The initial steps in determining a robot's response to a collision include detecting whether a collision has occurred, isolating the link affected by the collision, and estimating the amplitude of the external torque, as presented in a collision pipeline in [4]. In this paper, our focus is on detecting the external torque and estimating its amplitude.

There are several solutions for detecting and estimating external force/torque, which include electronic skin [7] and solutions based on collecting information from force/torque sensors mounted in robot joints and/or robot's end-effector [1]. Furthermore, there are solutions such as energy, velocity, and momentum observers, which reconstruct the information about the external torque indirectly, by leaning on the information from the position and current sensors [4]. Recently, learning-based methods have been introduced with the aim to achieve generalization of results [5] and to avoid friction modeling and parameter tuning [8].

Compared to the state-of-the-art literature, this paper focuses on external torque estimation for flexible-joint robots driven by variable stiffness actuators (VSAs) in an antagonistic setup. We build upon the theory developed for stiffness estimation [2,12], but this time generalize the previous work by assuming the existence of the external torque.

First, the linear and delayed Unknown Input Observer (UIO) is used to reconstruct the local elastic torques, which are by nature highly nonlinear and uncertain due to the existence of elastic elements inside the VSA. This step allows achieving a higher accuracy of external torque estimation, since there is no need to use a generic and less precise model of the elastic torque. Then, an additional delayed Unknown Input Observer leverages the knowledge about robot link dynamics and inner-UIO information about the local elastic torques to estimate the external torque acting on the robot. The verification of this method has been performed in the simulation environment on a single-link robot and the results show that the external torque can be accurately estimated.

The use of a delayed UIO is suitable due to its several advantages. This type of observer is capable of simultaneously estimating the system states and reconstructing the unknown inputs, which allows the end-user to avoid mounting additional sensors, such as velocity and force/torque sensors. Moreover, the UIO is composed of pre-calculated matrices whose values do not depend on the nature of the unknown signal, thus no manual tuning of the observer is needed. Finally, it has been shown that the use of UIO leads to better estimation results compared to similar types of observers [9].

## 2 System Model and Problem Statement

Consider a 1-DoF articulated soft robot consisting of a rigid link and a flexible revolute joint, actuated by a VSA device including a pair of motors that operate in decoupled antagonistic setup, as shown in Fig. 1.

Having denoted with $q$ the robot's link angle, $\tau_{\text{ext}}$ an external torque, $q_a$ and $q_b$ the angles of the agonistic and antagonistic motors, respectively, $\tau_a$ and $\tau_b$

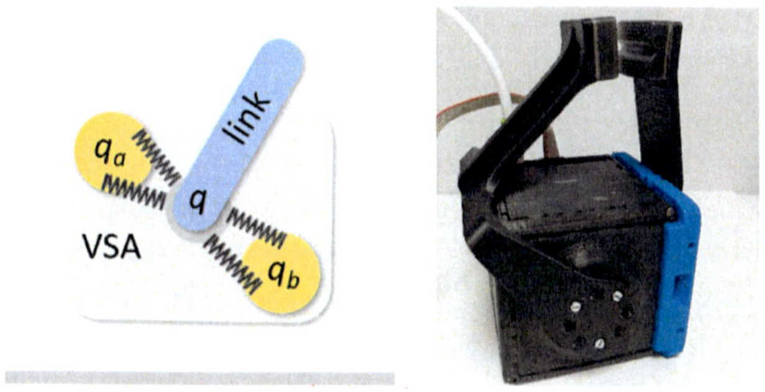


**Fig. 1.** Depiction of the single-degree-of-freedom robot driven by electromechanical antagonistic variable stiffness actuator (left) and picture of the laboratory setup consisting of *qbmove* actuator

their commanded torques, and $\phi_a = q - q_a$ and $\phi_b = q - q_b$ the VSA internal transmission deflections, the robot's dynamics is

$$I\,\ddot{q} + \beta\,\dot{q} + mgl\sin q = \tau_e(\phi_a, \phi_b) + \tau_{\text{ext}}\,, \tag{1a}$$

$$I_a\,\ddot{q}_a + \beta_a\,\dot{q}_a - \tau_{e,a}(\phi_a) = \tau_a\,, \tag{1b}$$

$$I_b\,\ddot{q}_b + \beta_b\,\dot{q}_b - \tau_{e,b}(\phi_b) = \tau_b\,, \tag{1c}$$

where $m$ and $I$ are the link's mass and inertia, respectively, $\beta$ is a viscous damping coefficient, $g$ is the gravity acceleration, $l$ is the distance of the link's center of mass from the joint, $I_a$, $I_b$ and $\beta_a$, $\beta_b$ are the motors' inertias and damping coefficients, $\tau_{e,a}$ and $\tau_{e,b}$ are their flexibility torques, and $\tau_e$ is the total flexibility torque at the link, being given by

$$\tau_e(\phi_a, \phi_b) = \tau_{e,a}(\phi_a) + \tau_{e,b}(\phi_b)\,. \tag{2}$$

Within this framework, we aim at finding a solution to estimate the external torque $\tau_{ext}$ by using information about the link's position $q$, its first time-derivative $\dot{q}$, the two motor's angles $q_a$ and $q_b$, their dynamic models and motor currents. The sought solution is *sensorless* in the sense that it is required to exploit no information from force/torque sensors mounted on the joint.

## 3 The Proposed External Torque Estimator for Variable Stiffness Actuator

The estimator of the external torque is graphically presented in Fig. 2. Essentially, it consists of two-layered UIOs, of which the first one addresses the reconstruction of the local elastic torques generated by the DC motors and the second one targets the estimation of the external torque, leaning on the knowledge of link dynamics.

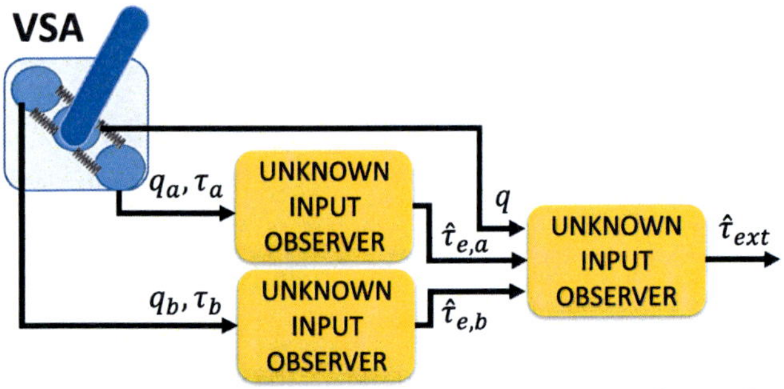


**Fig. 2.** Block diagram of the proposed external torque estimator. The diagram shows the solution for a single link robot actuated by a variable stiffness actuator with the highlighted inputs and outputs of unknown input observers.

### 3.1 The Delayed Unknown Input Observer

The principal framework of Unknown Input Observers is based on the equivalence between a linear system's invertibility and unknown input observability. When it comes to the *delayed* UIOs, the introduction of delayed output samples endows this observer with the capability to estimate both states and unknown inputs with a constant and a-priori determined delay, under looser existence conditions compared to zero-delay observers.

For a linear, time-invariant, discrete-time system in the state form

$$\dot{x}_{k+1} = Ax_k + Bu_k\,, \quad y_k = Cx_k + Du_k\,, \tag{3}$$

where $k$ is a discrete-time step, $x_k \in \mathbb{R}^n$ is a state vector, $u_K \in \mathbb{R}^m$ is a vector of unknown inputs, $y_k \in \mathbb{R}^p$ is an output vector, $A$, $B$, $C$, and $D$ are system matrices, the linear discrete-time delayed Unknown Input Observer has a form

$$\begin{aligned} \hat{x}_{k+1} &= \quad E\,\hat{x}_k + F\,\mathbb{Y}_k\,, \\ \hat{u}_k \;&= G\begin{pmatrix} \hat{x}_{k+1} - Ax_k \\ y_k - Cx_k \end{pmatrix}, \end{aligned} \tag{4}$$

where $\hat{x}_k$ and $\hat{u}_k$ are the state and input estimates, respectively, $\mathbb{Y}_k = (y_k, \cdots, y_{k+L})^T$ is the history of the latest $L$ output samples, $E$ and $F$ are matrices of suitable dimension that ensure $\hat{x}_K \to x_k$ and $\hat{u}_k \to u_k$, and $G = (B^T, D^T)^{T\dagger}$, where $M^\dagger$ is the pseudoinverse of matrix $M$. More details on the UIO's existence conditions, design, and various applications are covered in [2,10,12].

### 3.2 Reconstruction of Elastic Torques

Observing the whole system dynamics in Eq. 1, it is apparent that there are two perspectives towards estimating the local elastic torques $\tau_{e,a}$ and $\tau_{e,b}$. One approach is to target the motor dynamics (Eq. (1b) and Eq. (1c)) and consider the local elastic torques as the unknown inputs to the motor system (similar to [2]). The other approach would be to assume that the total elastic torque $\tau_e = \tau_{e,a} + \tau_{e,b}$ is the unknown input to the link dynamics (similarly to [12]). It is evident, however, that in the latter case, elastic torque estimation cannot be decoupled from the external torque estimation, therefore, we proceed with the former approach and focus on the agonistic part (Eq. (1b)) since the same procedure can be repeated for the antagonistic part as well.

By choosing the state vector $x = (q_a, \dot{q}_a)^T$ and suitably defining an input variable $u = (\tau_{e,a}, \tau_a)$, the relation in Eq. 1b can be written as a linear state form:

$$\dot{x} = A_c\, x + B_c\, u\,, \quad y = C\, x + D\, u\,, \tag{5}$$

where $A_c = \begin{pmatrix} 0 & 1 \\ 0 & -\frac{\beta_a}{I_a} \end{pmatrix}$, $B_c = \begin{pmatrix} 0 & 0 \\ \frac{1}{I_a} & \frac{1}{I_a} \end{pmatrix}$, $C = \begin{pmatrix} 1 & 0 \\ 0 & 0 \end{pmatrix}$, and $D = \begin{pmatrix} 0 & 0 \\ 0 & 1 \end{pmatrix}$.

This allows us to easily design an input-state observer of the dynamic system in Eq. 5, described by the state form in Eq. (4), where $A = I_n + T_s A_c$ is a discrete-time form of a matrix $A_c$ obtained by Euler's method, $T_s$ is a sampling period, and matrices $E$ and $F$ are given by

$$E = \begin{pmatrix} 1 & T_s \\ -\frac{1}{T_s} & -1 \end{pmatrix}, \quad F = \begin{pmatrix} 0 & 0 & 0 & 0 & 0 & 0 \\ 0 & 0 & 0 & 0 & \frac{1}{T_s} & 0 \end{pmatrix}.$$

Once an estimated $\hat{u}_k$ of the input vector is obtained, the current value of the local elastic torque, $\tau_{e,a}$ is its first element $\hat{u}_k(1)$ at the instant $t = k\,T_s$.

*Remark 1.* It is worth noticing that the proposed observer requires the use of the motor's link position $q_a$, its inertia $I_a$ and the damping coefficient $\beta_a$, and knowledge about motor torque $\tau_a = k_m i_a$ that can be obtained through the datasheet and an embedded current sensor, while no force/torque sensor is needed.

*Remark 2.* To make a UIO feasible, the necessary delay is equal to $L = 2$ samples. This value satisfies the statement of propositions 1 and 2 in [2].

### 3.3 Estimation of the External Torque

Once the local elastic torques are estimated, the total elastic torque can be calculated as in Eq. (2). Therefore, considering Eq. (1a), the only unknown term that remains is the external torque $\tau_{ext}$. To achieve the final goal and estimate its value, another linear and delayed Unknown Input Observer can be constructed, this time adopting an unknown input that collects the nonlinear term $mgl \sin q$, total elastic torque and the external torque. Hence, the Eq. (1a) gains the linear form

$$I\ddot{q} + \beta\dot{q} = u_{ext}\,, \tag{6}$$

where $u_{ext} = \tau_{ext} - \tau_e - mgl \sin q$ or equivalently, in the state-space form

$$\dot{x} = A_c\, x + B_c\, u\,, \quad y = C\, x + D\, u\,,$$

where $A_c = \begin{pmatrix} 0 & 1 \\ 0 & -\frac{\beta}{I} \end{pmatrix}$, $B_c = \begin{pmatrix} 0 \\ \frac{1}{I} \end{pmatrix}$, $C = \begin{pmatrix} 1 & 0 \end{pmatrix}$, and $D = 0$. The observer matrices are given by

$$E = \begin{pmatrix} 1 & T_s \\ -\frac{1}{T_s} & -1 \end{pmatrix}, \quad F = \begin{pmatrix} 0 & 0 & 0 \\ 0 & 0 & \frac{1}{T_s} \end{pmatrix}.$$

Such a linear system can be straightforwardly used to design an Unknown Input Observer, described by the state form already presented in Eq. (4). Once an estimated value of the unknown input $u_{k,ext}$ is obtained, the external torque is reconstructed as

$$\hat{\tau}_{ext} = \hat{u}_{ext} + \hat{\tau}_e + mgl \sin q .$$

*Remark 3.* It is worth noticing that the proposed observer uses the information about the link position $q$ and link dynamics.

*Remark 4.* To estimate the unknown input, i.e. the external torque, the designed unknown input observer needs to have a two-sample delay. Hence, taking into account that a two-sample delay is needed to reconstruct the total elastic torque (Remark 2), and now additional two samples to find the external torque, the total delay of estimation is equal to four samples.

## 4 Simulation Results

The previously described method is verified in the simulation of a single degree-of-freedom robot. For this purpose, we considered the dynamical model of a 1-DoF robot with a flexible joint actuated by a *qbmove MakerPro* variable stiffness actuator in the antagonistic setup.

The parameters of the system's model have the following values. The mass of the link is $m = 0.26\,\mathrm{kg}$, the link's length is $l = 0.09\,\mathrm{m}$, the link's inertia is $I = 0.15\,\mathrm{kgm}^2$, and the damping coefficient is $\beta = 10^{-6}\,\mathrm{Nms/rad}$. Local elastic torques are described by functions provided in the datasheet of the actuators: $\tau_{e,a} = k_1 \sinh\left(a_1(q - q_a)\right)$, $\tau_{e,b} = k_2 \sinh\left(a_2(q - q_b)\right)$, where $a_1 = 6.7328\,\mathrm{rad}^{-1}, a_2 = 6.9602\,\mathrm{rad}^{-1}, k_1 = 0.0227\,\mathrm{Nm}, k_2 = 0.0216\,\mathrm{Nm}$. We considered an external torque of amplitude $0.003\,\mathrm{Nm}$ that acts on the link from the 8-th till the 12-th s. The link position and joint stiffness are commanded to follow a sinusoidal trajectory, that is for the position $q_{des} = 0.25 \sin \pi t$ and for the stiffness $\sigma_{des} = 6 + 2 \sin \pi t$.

Results of the simulation run are reported in Figs. 3, 4, and 5. Figure 3 shows the evolution of the position and stiffness with respect to their desired values. As expected, it can be observed that the position changes during the period when the external torque acts on the link. The estimation of the total elastic torque is depicted in Fig. 4, which clearly shows a successful reconstruction of its value. Finally, the estimated external torque and its real value are shown in Fig. 5. After the initial transient, the results show an accurate estimation of the

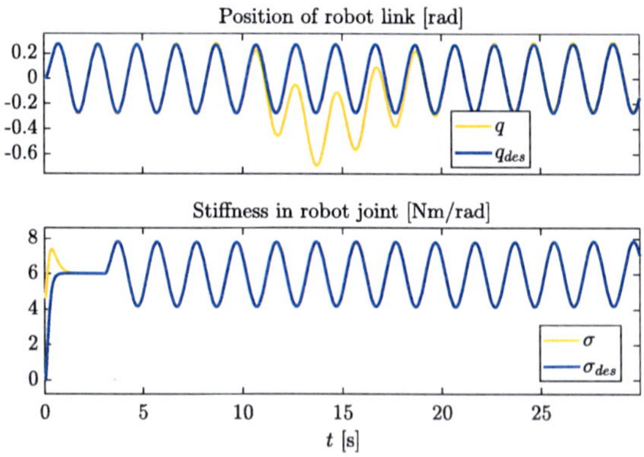


**Fig. 3.** Simulation - Position (top) and stiffness (bottom) desired and achieved trajectories for a single flexible-joint robot.

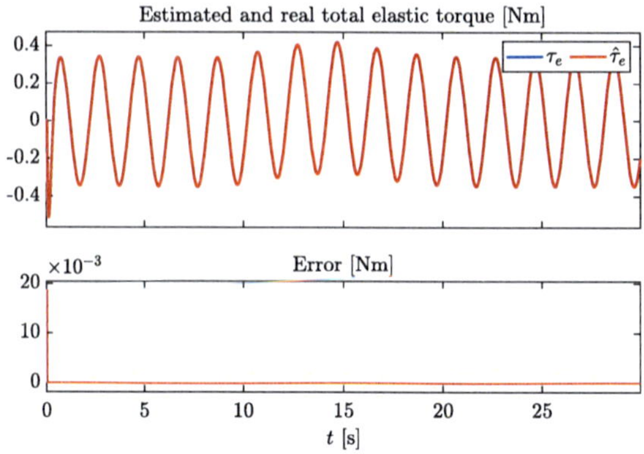


**Fig. 4.** Simulation - Reconstruction of the total elastic torque $\hat{\tau}_e$ by the UIO for a single flexible-joint robot.

unknown external torque value, until the 12th s when an error becomes non-zero for a short period. The reason for such an error is due to the abrupt change of the external torque. Since the estimator lags four samples (Remark 4) and the abrupt change occurred with higher frequency, the estimator performed worse during the short time span.

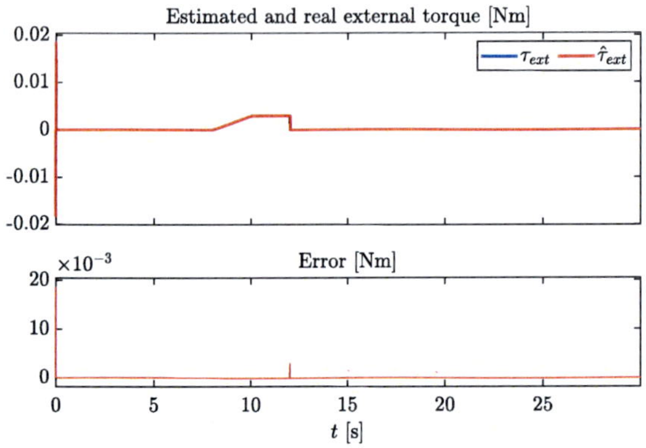


**Fig. 5.** Simulation - Reconstruction of the external torque $\hat{\tau}_{ext}$ by the external UIO for a single flexible-joint robot.

## 5 Conclusion

The presented work has shown a new approach to estimating the external torque that acts on a single-link robot driven by a variable stiffness actuator. The proposed solution included the two-level delayed unknown input observers, which were first used to find the total elastic torque and then to reconstruct the external torque. The simulation results have shown good performance, making a bounded error only when the external torque is changing faster than the estimator's delay. Future work will extend the proposed method to a robot with multiple degrees of freedom, perform the experimental validation and achieve robustness with respect to model uncertainties and measurement noise.

## References

1. De Luca, A., Albu-Schaffer, A., Haddadin, S., Hirzinger, G.: Collision detection and safe reaction with the DLR-III lightweight manipulator arm. In: 2006 IEEE/RSJ International Conference on Intelligent Robots and Systems, pp. 1623–1630. IEEE (2006)
2. Fagiolini, A., Trumić, M., Jovanović, K.: An input observer-based stiffness estimation approach for flexible robot joints. IEEE Robot. Autom. Lett. **5**(2), 1843–1850 (2020)
3. Galin, R., Meshcheryakov, R.: Review on human–robot interaction during collaboration in a shared workspace. In: Ronzhin, A., Rigoll, G., Meshcheryakov, R. (eds.) ICR 2019. LNCS (LNAI), vol. 11659, pp. 63–74. Springer, Cham (2019). https://doi.org/10.1007/978-3-030-26118-4_7
4. Haddadin, S., De Luca, A., Albu-Schäffer, A.: Robot collisions: a survey on detection, isolation, and identification. IEEE Trans. on Robot. **33**(6), 1292–1312 (2017)
5. Kim, D., Lim, D., Park, J.: Transferable collision detection learning for collaborative manipulator using versatile modularized neural network. IEEE Trans. Robot. **38**, 2426–2445 (2021)

6. Lippi, M., Gillini, G., Marino, A., Arrichiello, F.: A data-driven approach for contact detection, classification and reaction in physical human-robot collaboration. In: 2021 IEEE International Conference on Robotics and Automation (ICRA), pp. 3597–3603. IEEE (2021)
7. Pang, G., et al.: CoboSkin: soft robot skin with variable stiffness for safer human-robot collaboration. IEEE Trans. on Ind. Electron. **68**(4), 3303–3314 (2020)
8. Park, K.M., Kim, J., Park, J., Park, F.C.: Learning-based real-time detection of robot collisions without joint torque sensors. IEEE Robot. Autom. Lett. **6**(1), 103–110 (2020)
9. Pedone, S., Trumić, M., Jovanović, K., Fagiolini, A.: Robust and decoupled position and stiffness control for electrically-driven articulated soft robots. IEEE Robot. Autom. Lett. **7**(4), 9059–9066 (2022)
10. Sundaram, S., Hadjicostis, C.N.: Delayed observers for linear systems with unknown inputs. IEEE Trans. on Autom. Control **52**(2), 334–339 (2007)
11. Trumić, M., Della Santina, C., Jovanović, K., Fagiolini, A.: Adaptive control of soft robots based on an enhanced 3D augmented rigid robot matching. In: 2021 American Control Conference (ACC), pp. 4991–4996. IEEE (2021)
12. Trumić, M., Grioli, G., Jovanović, K., Fagiolini, A.: Force/torque-sensorless joint stiffness estimation in articulated soft robots. IEEE Robot. Autom, Lett. **7**, 7036–7043 (2022)
13. Trumić, M., Jovanović, K., Fagiolini, A.: Decoupled nonlinear adaptive control of position and stiffness for pneumatic soft robots. Intl. J. Robot. Res. **40**(1), 277–295 (2021)

# A Multi-feature Augmented Reality-Based Robot Control

Mohammad-Ehsan Matour(✉) and Alexander Winkler

Mittweida University of Applied Sciences, Mittweida, Germany
{matour,alexander.winkler}@hs-mittweida.de

**Abstract.** This paper presents an augmented reality-based interface that provides a set of tools needed for full control of a collaborative robot. A mixed reality head-mounted display (HMD) serves as an interface for the operator by superimposing virtual content, allowing interaction with the robotic system. In addition to providing robot state information, such as the joint position, velocity and the force applied to the robot flange, motion commands can also be sent to the robot, which significantly facilitates robot programming. In order to move the robot, the operator can either adjust the joint angles with the visualized sliders or simply drag a cursor that represents the position and orientation of the robot tool center point (TCP). Moreover, the proposed user interface (UI) allows the operator to preset the gripping force while the robot carries out hybrid force position control. Considering that the safety aspect should always be taken into account when moving the robot, the virtual model of the robot is also superimposed on the working environment so that the operator can perform a preview maneuver indicating the current and final position of each joint, before the motion is executed on the real system. The human-robot interface and virtual content are created on top of the Unity3D game engine, while the Robot Operating System (ROS) provides reliable data transmission and processing between the HMD and the robot controller.

**Keywords:** Augmented reality · Robot control · Intuitive robot programming · Human robot interface · Microsoft HoloLens

## 1 Introduction

As the complexity of systems in industry and workplaces increases, so does the need for user-friendly interfaces and intuitive robot programming. To handle evolving manufacturing, a new, simplified workflow is needed that allows users to focus on the task at hand rather than the complexity of the system. Augmented reality (AR) is a rapidly growing field that provides real-time interactive experiences by combining the real world with computer-generated visual elements. The content is blended seamlessly with the physical world, being perceived as a


T. Petrič et al. (Eds.): RAAD 2023, MMS 135, pp. 106–114, 2023.
https://doi.org/10.1007/978-3-031-32606-6_13

feature of the real environment. AR has been explored in many fields over the past few decades, from design and modeling to military and industry. In the industrial sector, AR capabilities can be leveraged in assembly, training, maintenance and repair, and robotics. It can also significantly increase productivity and worker safety in factories by improving digital simulation and enhancing the efficiency of human-robot interaction.

Taking advantage of AR in robotics opens up new opportunities for applications such as robot control. Since the operator is able to retrieve the robot's state information through the HMD without being distracted by the physical environment, the collaborative tasks can be performed more efficiently. One approach to enhance the level of automation in assembly tasks is to employ collaborative robots. Given that such robots play an important role in industry, this paper proposes a set of tools for controlling a cobot through HMD.

## 2 Augmented Reality in Robotics

In recent years, the application of AR in robotics has increased enormously due to its great potential. Scientists have already developed many applications that make the operation of robots much more convenient.

Liu [1] proposes an AR-based approach to abstract the functionality, decision-making process, and diagnosis of a robotic system while carrying out a task. By means of the temporal graph network, the user can intuitively comprehend and interpret the inner workings of the robot and easily diagnose any errors in performance. Moreover, by parsing the graph, the user is able to oversee the robot's action planner and visually track the robot's latent states. While this method provides usable information about the robot's state and intentions, it does not go beyond a visualization framework and does not program the robot directly through the HMD, rather through human demonstration. Similar to this work, [2] developed an adaptive AR system to display faults in industrial robots using a HMD. [3] introduces an intuitive programming method through marker-based tracking AR to speed up the execution of assembly tasks, where the operator first assembles components virtually via an HMD using the CAD models. Once the process is acquired, the robot is able to reproduce the task in reality. This method has been successfully validated and tested and has an accuracy of about 1 mm to 2 mm. In the automotive and manual assembly context, [4] outlined a new approach towards providing instruction information using simulation data, reducing the cognitive load of manual assembly tasks. In [5], a multi-feature system is presented that enables the user to program a 7-DoF robot through an augmented trajectory. In addition to visualizing robot parameters, the robot can follow a user predefined path both in free space and in contact with the surface. The first mode can be applied to a pick-and-place task, while the latter involves planning the trajectory using a MYO armband.

An HMD is not always required to realize an augmented reality experience. [6] and [7] proposed and demonstrated a smartphone-based AR application with a 4-DoF manipulator that can effortlessly perform a pick-and-place task. [8]

illustrates an interactive fabrication process based on ARHMD, where the user plans a robot path based on CAD information using a simple user interface. [9] features two control modes for effective and intuitive control of a robotic manipulator via Microsoft HoloLens. The interesting feature of this work is the operator's ability to drag and move the virtual end effector. After successful path planning, the robot moves to the desired position. In addition, the torques experienced by the manipulator's joints are visualized via color media. A similar work to ours is [10], however, the authors have developed a handheld device with a wireless mouse and infrared markers that allows the operator to move around the work cell to define waypoints for the real robot to follow. The path and waypoints drawn by the user are displayed via the HMD. However, in our work, we use a virtual pointer integrated with the virtual robot model. [11] suggests an AR-based interface with the ability to set waypoints that the robot needs to follow. The discussed system has collision detection that makes path planning safe. However, the orientation of the pointer and thus the orientation of the end effector is not considered because the pointer has a circular shape. Since we used an arrow as a pointer, the user can determine not only the position, but also the orientation of the end effector. In a recent work, [12] presented a novel method on visualizing the robot's internal intentions for human-to-robot handover through HMD. In this method, the user is aware of the pose at which the robot intends to grasp the object. The user study conducted on this method reveals significant improvements in the participants' subjective experience of the fluency of the interaction, trust towards the robot, perceived safety, mental load, and predictability.

## 3 Structure of the System

In this project we utilize a collaborative robot alongside a Microsoft HoloLens 2 as augmented reality HMD. Since the robot is intended to be in direct contact with humans, safe interaction is of great concern. The robot manipulator used in this work is an economical, lightweight robot from Universal Robots of type UR3e. The deployment of such a robot enables the realization of the concept of human-robot collaboration as well, in which the robots are not completely separated from humans anymore. The robot of type URe are equipped with an internal six-axis force/torque sensor (FTS). This feature makes the sensor guarded motion possible and setting up the threshold for acceptable collision forces or maximal velocities allows safe human robot collaboration. This type of robot fits perfectly to hybrid force-position control, in which the contact forces can be adjusted along a certain axis. On the other hand, world-anchoring, hand tracking, large field-of-view, built-in simultaneous localization and mapping and precision are powerful feature of the untethered HoloLens 2 that make it a proper AR device for robot applications. The graphical user interface is created using Unity3D game engine. It has the infrastructure to render engineering concepts in a virtual world and supports Mixed Reality Toolkit (MRTK), a powerful platform with a set of building blocks and features for spatial interactions and UI.

With the help of this toolkit the user is able to interact with the virtual objects created in Unity through hand gesture, speech and gaze. To enable the communication between HMD and robot controller, robot operating system (ROS) is employed. As a middleware it offers the opportunity to combine different hardware and software and makes communication between system units and an external computer possible. ROS-middleware is built on top of Ubuntu Linux and is responsible for handling the communication and message-passing between processes and devices in a distributed system and has evolved to a mature framework in recent years providing powerful tools for robot engineers and researchers. For the connection and control of the robot via ROS, the *Universal Robots Ros Driver*, a stable and efficient ROS interface, is deployed. This driver provides topics and action interfaces like the *FollowJointTrajectory* action, which asynchronously invokes the functionality of another node to communicate with the robot. For establishing a connection between Unity and ROS, *ROS TCP Connector* package should be included in the Unity platform. On the ROS side *ROS TCP Endpoint* package running as a ROS node is necessary. One of the useful tools of this package is the URDF importer, which can be integrated into Unity environment and allows the user to import robot model defined in URDF format into Unity scene. This format defines the geometry, kinematics, and dynamics attributes of a robot. The imported robot may serve as a simulated robot, before sending the motion commands on the real one, to make sure whether a given path is safe or not. The connection between ROS and Unity takes place through RosBridge, which provides a JSON interface to ROS and allows any client to send JSON, in order to communicate to ROS. It uses standard TCP/IP sockets for transporting message data. Figure 1 shows the overall system structure. After creating the AR application in Unity and making it compatible with ROS by

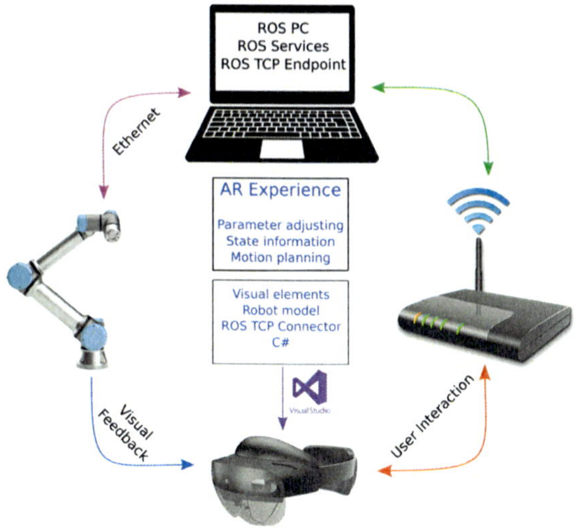


**Fig. 1.** Components and the structure of the system

attaching *C#* functionalities to each feature, it should be compiled and built as Universal Windows Platform (UWP). The obtained file is going to be deployed to the HoloLens. Alternative, Holographic remoting feature of MRTK can be used, to stream holographic content from an external PC to Microsoft HoloLens in real-time. Due to the reduced workload on the HoloLens, this feature may considerably enhance the developer's productivity.

## 4 Human-Robot-Interface Layout

Figure 2 illustrates the proposed UI created in Unity. It consists of texts, buttons, and sliders, which all give the possibility to control the manipulator. The robot model is also imported and placed in the scene. The functionality of each feature is described below:
By clicking on buttons 1 and 2, Unity subscribes to ROS and displays the current joint angles and velocities of the real robot. The operator can hide this information at any time by clicking again on the buttons. Clicking on button 3 opens

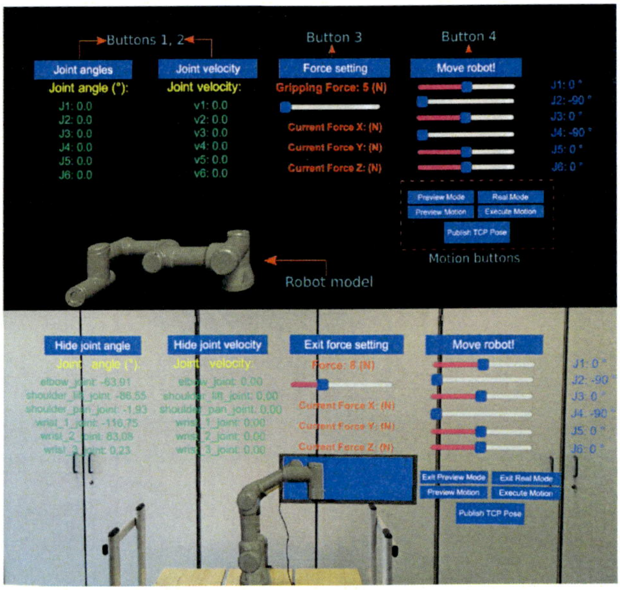


**Fig. 2.** The proposed user interface

the force setting and indicators. Here the information of the internal FTS of the manipulator is called and the force in free space is displayed in three forms. First, the current force as a vector in free space; second, as a graph consisting of the user-configured reference value of the gripping force and the current force curve, which can be useful to derive the force progression, especially when the manipulator hits an object during the force-controlled movement; and third, as a color media. The sixth joint of the robot uses a color spectrum and represents the force applied to the robot as a color. This gives quick visual feedback about

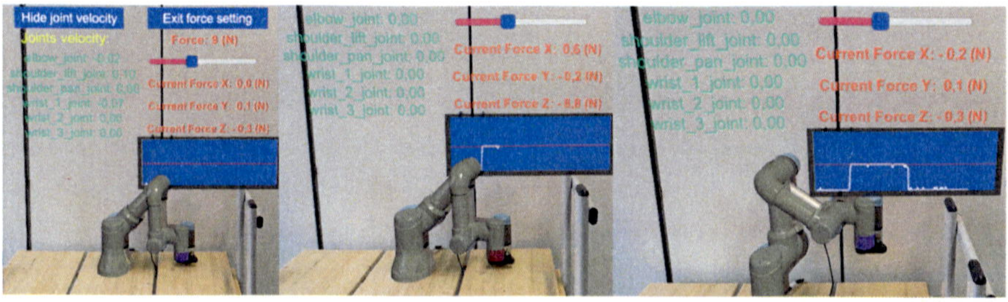


**Fig. 3.** Force indicators during force-controlled motion

current force. The operator is also able to set the gripping force through the slider. In our case the robot is intended to grasp an object using an adhesive gripper. To this end, the force applied to the object is a major factor to make the object stick to the gripper. Hence, the gripping process takes place in hybrid force-position control and the force exerted on the object can be dynamically adjusted and easily configured by the operator. The force value is displayed at the top of the slider and published to the server automatically. Figure 3 shows the robot during the gripping process and the indicators which are showing the current force. Button 4 is related to the motion of the manipulator. It allows the user to set the joint angles manually, give a TCP pose and to preview the robot motion before executing. The user can switch between two modes, preview and real mode. The former serves as a pure simulation to monitor the trajectory and ensure that the movement of the robot does not cause any danger. In this mode the robot model does not receive any data from the real one. The latter mode must be active, to enable the communication between ROS and Microsoft HoloLens and to send motion commands. The six sliders on the right side of the UI are supposed to adjust the joint angles manually. The current angle of each joint is displayed under button 1. After adjusting the angles, the user may verify the given position. By pressing the button *Preview Motion* the specified values are assigned to the simulated robot and the user can get an overview of the final position of the robot (see Fig. 4). After ensuring a collision-free position the angles can be sent to robot controller through ROS by pressing *execute motion*. The other feature of the proposed UI to move the robot, is the TCP pose adjusting. To this end, an arrow-like pointer is realized and added to the end effector of simulated robot whose position and orientation corresponds to the TCP-pose of the real one. After repositioning the arrow through HMD and pressing *Publish TCP Pose* button, its position in Unity reference frame is published to ROS. The obtained position in robot base frame is then sent to inverse

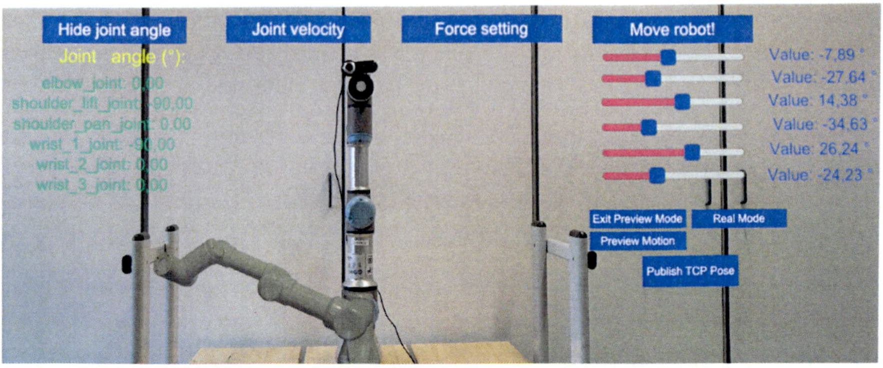


**Fig. 4.** Preview mode: the simulated model shows robot's final position given through sliders by the user

kinematic solver. Some joint constraint is also defined to calculate the inverse kinematic under joint constraints. This may be crucial to avoid unexpected joint positions. Figure 5 shows the robot positioning via the arrow.

**Fig. 5.** TCP repositioning via the virtual pointer

## 5 Conclusion

The demand for fast, effective, and reliable robot programming has led researchers to explore the possibilities of intuitive approaches such as AR-based programming. In this context, we have provided a set of tools that facilitate the operator to get visual feedback on the internal parameters of the robot and to dynamically adjust the control parameters. To illustrate this concept, we have attempted to adjust the gripping force using a HMD while the robot performs a force-controlled gripping operation by means of an adhesive gripper. In this instance, the gripping force should be adjustable as a function of the weight of the object. In the presented user interface, the operator can effortlessly configure this parameter and receives visual feedback about the current force being applied to the object. In addition, the UI provides the ability to adjust the TCP position using a simulated pointer, configure the robot's joint angle, and preview the robot's motion before it is executed on the real part of the system. Going forward, the implemented pointer can be leveraged to define some waypoints and create a path by interpolating through the points, along with the configurable force, leading in a fully force-controlled motion driven by HoloLens.

## References


1. Liu, H., Zhang, Y., Si, W., Xie, X., Zhu, Y., Zhu, S.: Interactive robot knowledge patching using augmented reality. In: 2018 IEEE International Conference on Robotics and Automation (ICRA) (2018). https://doi.org/10.1109/ICRA.2018.8462837
2. Avalle, G., De Pace, F., Fornaro, C., Manuri, F., Sanna, A.: An augmented reality system to support fault visualization in industrial robotic tasks. IEEE Access **7**, 132343–132359 (2019). https://doi.org/10.1109/ACCESS.2019.2940887
3. Blankemeyer, S., Wiemann, R., Posniak, L., Pregizer, C., Raatz, A.: Intuitive robot programming using augmented reality. Procedia CIRP **76**, 155–160 (2018). https://doi.org/10.1016/j.procir.2018.02.028
4. Lampen, E., Teuber, J., Gaisbauer, F., Bär, T., Pfeiffer, T., Wachsmuth, S.: Combining simulation and augmented reality methods for enhanced worker assistance in manual assembly. Procedia CIRP **81**, 588–593 (2019). https://doi.org/10.1016/j.procir.2019.03.160
5. Perez Quintero, C., Li, S., Pan, M., Chan, W., Van der Loos, H.: Robot programming through augmented trajectories in augmented reality. In: 2018 IEEE/RSJ International Conference on Intelligent Robots and Systems (IROS) (2018). https://doi.org/10.1109/IROS.2018.8593700
6. Frank, J., Moorhead, M., Kapila, V.: Mobile mixed-reality interfaces that enhance human-robot interaction in shared spaces. Front. Robot. AI (2017). https://doi.org/10.3389/frobt.2017.00020
7. Chacko, S., Kapila, V.: An augmented reality interface for human-robot interaction in unconstrained environments. In: 2019 IEEE/RSJ International Conference on Intelligent Robots and Systems (IROS) (2018). https://doi.org/10.1109/IROS40897.2019.8967973
8. Kyjaneka, O., Al Bahar, B., Vasey, L., Wannemacher, B., Menges, A.: Implementation of an augmented reality AR workflow for human robot collaboration in timber prefabrication. In: 36th International Symposium on Automation and Robotics in Construction (ISARC 2019) (2019). https://doi.org/10.22260/ISARC2019/0164
9. Manring, L., et al.: Augmented reality for interactive robot control. In: Dervilis, N. (ed.) Special Topics in Structural Dynamics & Experimental Techniques, Volume 5. CPSEMS, pp. 11–18. Springer, Cham (2020). https://doi.org/10.1007/978-3-030-12243-0_2
10. Ong, S., Yew, A., Thanigaivel, N., Nee, A.: Augmented reality-assisted robot programming system for industrial applications. Robot. Comput. Integr. Manuf. **61**, 101820 (2020). https://doi.org/10.1016/j.rcim.2019.101820

11. Shamaine, C., Qiao, Y., Henry, J., McNevin, K., Murray, N.: RoSTAR: ROS-based telerobotic control via augmented reality. In: 2020 IEEE 22nd International Workshop on Multimedia Signal Processing (MMSP) (2020). https://doi.org/10.1109/MMSP48831.2020.9287100
12. Newbury, R., Cosgun, A., Crowley-Davis, T., Chan, W., Drummond, T., Croft, E.: Visualizing Robot Intent for Object Handovers with Augmented Reality (2021). https://doi.org/10.48550/arXiv.2103.04055

# Use Case Driven Active Teaching in Robotics for Education and Continuous Professional Development Through a 6-DOF Open-Source Robotics Platform

Lukáš Mácha(✉), Roman Šmíd, Filip Šáfr, and František Petrů

Inomech s.r.o, Tábor 390 02, Czech Republic
{macha,smid,safr,petru}@inomech.com

**Abstract.** Robotics is an ever-evolving field that is predicted to generate demand for high-quality, well-paid roles and generate billions in revenue in the coming years. The impact of introducing robotics into education to fight the scarcity of required talent has been widely studied and surveyed with different aims ranging from improving the wellbeing of the population to continuous professional development. Experience however shows that learning retention rate following lecturing, or demonstration, falls far behind that of active teaching methods. To enable active teaching of Robotics, in education as well as continuous professional development, we introduce INNOBOT, an open-source, 3D-printable, 6 degrees-of-freedom (DOF) robotics platform built using off-the-shelf components, to support the development of talent that might otherwise be scarce. Topics covered by INNOBOT platform exceed the traditional areas such as systems thinking, and problem solving and cover wide range of practically oriented topics required for "full stack" robotics development.

**Keywords:** robotics education · active teaching · project-based learning · science and technology · open-source · ROS 2 · MoveIt 2

## 1 Introduction

Robotics, [1], is considered a multidisciplinary field that brings together scientists and engineers from various backgrounds. Originating from the Czech word "robota", meaning difficult and forced labor, the role of robots has since evolved into more than just fulfilling the forced labor that is can be dangerous or unsuitable for humans. Robots have, in one form or another, been introduced into our everyday lives [4].

Report [2] estimates that the current total market for robot systems accounts for enormous USD 48 billion with about 30% revenue created by the robot itself, 25% revenue created by accessories and about 45% of the total revenue created by services such as auxiliary hardware, installation, programming, and software. In [3], prediction is made that the global robotics market value will reach between USD 160–260 billion. In [2] it is further predicted that by 2030 the number of new roles that will need to


T. Petrič et al. (Eds.): RAAD 2023, MMS 135, pp. 115–122, 2023.
https://doi.org/10.1007/978-3-031-32606-6_14

be filled will increase to 50 million globally and about 375 million workers globally will be required to master new skills due to their current jobs developing along with the developments in automation, robotics, and artificial intelligence. With [2] and [3] highlighting that talented workforce with robotics-related capabilities is already scarce, this scarcity is predicted to severely increase and potentially even slow the predicted growth of robotics.

To manage the rate at which science and innovation is changing the field, introduction of robotics [10] into curriculum is needed to develop future talent. Traditionally, the use of robotics in education has been primarily focused on improving students Science, Technology, Engineering and Mathematics skills [9]. Effective methods are through Project-Based Learning (PBL), where groups of students are assigned a specific problem that needs to be further investigated and refined. Skills taught through these PBL assignments often involved critical thinking, thinking outside of the box, problem solving and working in a team, or communicating their findings to others. These traditional PBLs are considered to be a powerful teaching strategy that can promote motivation and self-directed learning.

The technical skills traditionally taught at schools and universities however begin to fall behind of the technical skills required by companies from fresh students and graduates [11]. To tackle the growing variety of robotics applications one must possess the knowledge and understating beyond critical thinking, system integration, kinematics, motion planning and robot control, common topics covered by traditional texts on robotics. Choice of skills that are regularly sought after by companies however are not included in a traditional academic curriculum. These skills include programming languages, continuous improvement/ continuous delivery, agile methodologies, version control systems and project management tools.

This contribution aims to fill this gap through providing a robotics development platform for use case driven active teaching in an accessible form factor. This approach enables learners to gain holistic overview of the requirements for developing a robotics application as well as lower-level implementation detail for those wanting to explore further. The rest of this paper is split as follows: part two introduces projects in similar form factor available in the public domain, part three introduces the developed and now online available[1] INNOBOT platform and 3+1 layers of skills and knowledge requirements, part four describes a Pick and Place application and how it contributes to active teaching, discussion and conclusion is provided in section five.

## 2 Related Work

With the goal of bringing digital manufacturing technology to everyone and lowering the costs of the materials used by undergraduate students for their internships, [5] introduced fully open-source, 3D printed, 5-dof robotic arm, the Moveo. Providing robotic arm, that can be modified and reproduced by students at a lower cost than its industrial counterpart allows students to benefit from all the activities they undertake when building the arm. This covers several topics such as mechanical design, electrical, electronics, automation,

[1] https://github.com/inomech/innobot.

and programming. Whilst the electronics includes available solutions such as Arduino and RAMPS, the control is built on Marlin, firmware primarily intended for 3D printers that might pose challenges for expansion of the arm's capabilities and integration of new robotics use cases.

Another 6DOF, 3D printed manipulator, RBX1, [6], builds on top of the existing Moveo robot and adds an extra DOF to rotate the end effector. This contributes spherical wrist and allows for 6DOF manipulation, bringing the platform closer to industrial robotics. Whilst majority of the kit is open sourced, RBX1 is powered using Raspberry Pi and a custom stepper motor driver which however are no longer available, and one thus cannot use the RBX1 platform without putting significant effort into custom electronics and control software.

In [7], open-source, 6DOF robotic arm, the MOTUS, was introduced winning the entry for the 2022 Digital Fabrication Student Design Challenge. MOTUS contributes to the design of educational robotic arms using a 3D printed planetary gearbox to increase the torque of the stepper motors. This eliminates the need for belts making its form factor yet closer to industrial robot. Another advancement is introduced in the form of a Quick Release system developed for rapid change of the end effector tool. No electronics and control software were however developed for the MOTUS arm.

It is shown that the field of educational robotics has wide range of contributors from manufacturers to students themselves. Whilst building standalone robotic arm is a great experience to have, we identified several shortfalls of the existing solutions that would provide full experience of developing a robotics application. These include 1) clear use case definition, 2) reliance on custom-build electronics, 3) use of production-grade, future-ready open-source software frameworks.

## 3 INNOBOT Platform

INNOBOT, Fig. 1, is a robotics platform developed for upskilling talent in robotics education in schools as well as research and continuous professional development that enables low-cost, rapid prototyping. The entire platform is built using the Robot Operating System 2, framework designed to meet the challenges of modern robotics [12], building on production grade software that can be easily adapted and extended to suite different applications. Estimated cost of the platform is shown in Table 1.

**Table. 1.** Cost Estimates

| Category | Description | Cost (EUR) |
|---|---|---|
| Mechanical | Actuators, bearings, belts, screws, heat inserts | 650 |
| 3D printing | Robot links | 350 |
| Electronics | Arduino, Raspberry Pi, power source, motor drivers, wiring | 300 |
| Total | | 1300 |

It is a 6DOF, 3D printed robotic arm with improved mechanical design through removed complex geometries, increased wall thickness, and removed overhangs that may cause difficulties using lower end 3D printers. The arm consists of 6 rigid bodies (links) connected using joints that enable the relative rotational motion between the adjacent links. Belts and pulleys are used to scale the torque available at the joint actuators to reduce their power requirements, keep the platform operating under safe voltage and current ratings and to keep the cost to minimum. All components used in the platform are available off-the-shelf eliminating the need for custom build electronics. Range of topics is introduced through a 3+1-layer approach – Hardware Layer, Firmware Layer, Robot Software Layer and Day-to-Day Layer. This allows the separation of concern between hardware components (the robot and gripper assembly), electrical and electronics driving the robot movement at the low-level, high-level logic implemented using ROS 2 and open-source robotics manipulation platform MoveIt 2. Finally, the so-called "day-to-day" layer, introduces the ecosystem of tools required on daily basis to be able to plan, code, build, test, release, deploy, operate, and monitor the application. This layer might at first seem insignificant in contributing to the robotics application itself but is equally as important as the rest and requires significant amount of effort to grasp and understand to effectively start building robotics applications.

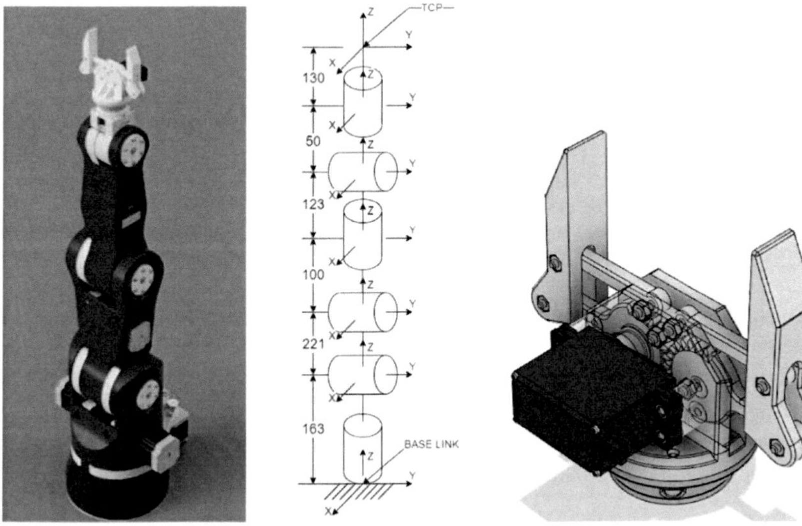


**Fig. 1.** INNOBOT (from left: rendered view, kinematic structure, gripper mechanism)

Individual layers are now introduced in more detail:

**The Hardware Layer** includes the physical components of the robotic arm and gripper that interact with the environments during manipulation tasks. The kinematic structure of the arm is shown in Fig. 1. (Dimensions in mm). The description follows right-hand notation for coordinate systems and positive rotations about joint axes. The gripper mechanism is opposed-finger, actuated through single servo motor which moves both fingers through set of levers and gears.

**The Firmware Layer** is the bridge between the Hardware and Robot Software Layers that provides services such as communication, feedback control, and data processing.

It receives setpoints from the Robot Software Layer at specified interval. PID controller is written for each joint that controls the position of each stepper which are actuated using the AcelStepper[2] library. The state of individual joints is sent back to the Software Layer via serial to update joint_state_broadcaster[3] controller feeding the information back to MoveitCpp for motion planning.

**The Robot Software Layer** is implemented using ROS 2 and MoveIt 2. This provides software application that runs on top of Ubuntu 20.04 and provides functionalities such as decision making and manipulation. The manipulation is implemented through high level MoveitCpp C++ interface that plans the arm's motion during PickNPlace application as shown in Fig. 3. Execution of the planned trajectory is implemented through joint_trajectory_controller[4] of ros2_control[5] package. Serial communication is used between the Software and Firmware Layers where setpoints from the joint_trajectory_controller are sent to the PIDs at frequency specified in the configuration file (e.g. 100 Hz). Inverse kinematics for the gripper is computed at grasp and release poses based on the width of the manipulated object and executed using the forward_command_controller[6].

**The "Day-to-day"** layer is added to the platform to cover the extras that are not discussed elsewhere. This layer covers topics that help in the day-to-day job to write the robotics software itself in an adequate text editor (e.g., VSCode), enable setting up the Linux development environment and finally software tools such as PlatformIO to be able to flash the robot firmware onto the microcontroller board.

This development environment with the tools and processes listed above was selected and catalogued[7] based on our own needs for setting up an efficient and where possible zero-cost working environment that allows robotics application development between remote teams.

## 4 Active Teaching with INNOBOT

Canonical example in robotics is picking an object up from a specific location A and placing it in another specific location B. This use case is known as Pick and Place (PnP). Whilst picking and placing an object might be an easy task for a human being, tackling the PnP application for a manipulator presents a series of challenges that must be addressed and serve as a clearly defined use case for a wide range or learners. The rest of this section introduces the PnP use case, and how INNOBOT can contribute to the delivery of active teaching methods.

### 4.1 Pick and Place Use Case

Pick and Place is often the first application where the focus of begins to move away from controlling the robot's motion (e.g., robot kinematics or dynamics) to where the robot

[2] https://github.com/waspinator/AccelStepper.

[3] https://github.com/ros-controls/ros2_controllers/tree/foxy/joint_state_broadcaster.

[4] https://github.com/ros-controls/ros2_controllers/tree/foxy/joint_trajectory_controller

[5] https://github.com/ros-controls/ros2_control/tree/foxy.

[6] https://github.com/ros-controls/ros2_controllers/tree/foxy/forward_command_controller.

[7] https://www.innobot.eu/.

interacts with its environment to move an object [8]. The focus is therefore shifted from the manipulator itself (the robotic arm), to the manipulation task (manipulating target object).

To support PnP use case in the INNOBOT platform, manipulation board, shown in Fig. 2, was developed which consists of 2 × 8 matrix of possible pick and place locations, each defined by a 50 × 50 mm square. The object being manipulated is a 3D printed 40 × 40 × 40 mm cube with X, Y and Z planes marked for simple orientation validation. The position and orientation of the centre of each cell of the 2 × 8 matrix relative to the robot is known and its pose can therefore, after its initial placement, be computed in the robot's base coordinate system. The PnP application consist of the following steps:

1) launching the INNOBOT applications
2) placing the cube on a cell on the PnP board,
3) entering the Pick Cell ID and Place Cell ID
4) calling the manipulation client to execute PnP.

**Fig. 2.** Early prototype of the INNOBOT Manipulation Board

The control flow of the PnP implementation is modelled as a state machine in Fig. 3. Using a state machine. This PnP use case provides a great starting point for learning robotics with a tangible objective for reasons such as:

1) **Provides clear task definition.** Pick and Place is a simple application that involved picking an object up from specific "Pick" location and placing it in another specific "Place" location. As such, it is easy to define and understand task making it easier for learner to conceptually develop solution and implement it.
2) **Exercises basic robotics skills.** The task involves computation of intermediate approach, grasp and retreat transformation poses in the Robot Software Layer using tf2. This requires understanding of the robot's configuration space, kinematics, and rigid body motions such as translation and rotation. It further requires controlling flow of the application to ensure robot does not first attempt to place the object before it is picked demanding the use of means to control the flow of the execution.

3) **Provides a hands-on experience.** Despite its simplicity, this PnP Use Case provides a hands-on experience in building and programming robot that must meet a set of requirements to fulfil its use case. When executed on the physical robot using the set of tools described in Sect. 3 one still must account for things like latency in the communication and inaccuracies in the original robot pose. This allows the learner to develop and gain confidence in their practical skills applicable to a variety of robot platforms from educational platform such as INNOBOT to industrial robots.
4) **Creates opportunity for adding complexity.** Whether it is through increasing the complexity of objects (e.g., different shapes and sizes) being manipulated or through adding more functionality to the application. This simple PnP use case can be easily adapted to meet increasing needs as the learner progresses with their robotics development and is therefore considered a great starting point for learning robotics.

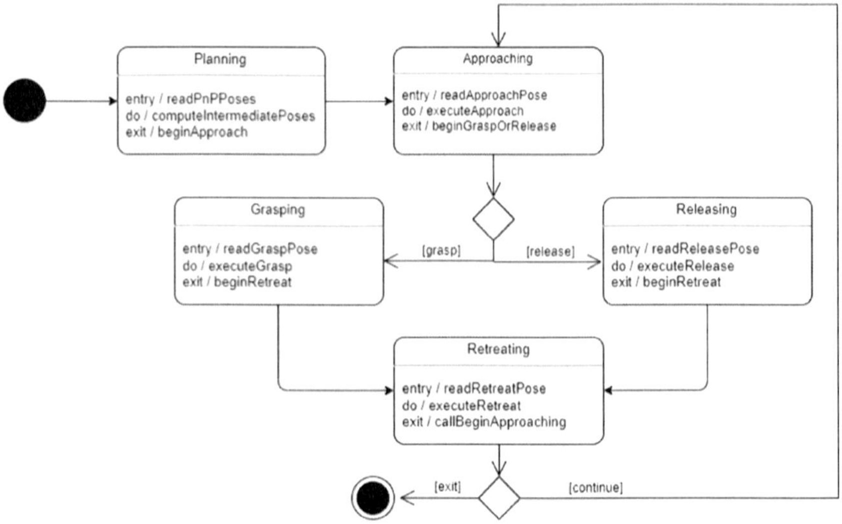


**Fig. 3.** Pick and Place Application State Machine

## 5 Conclusion

Robotics, as it was identified, is an ever-evolving field predicted to create billions in revenue with growing demand for skilled talent. The lack of an active teaching platform with a clear definition of task that needs to be achieved, was identified during our own upskilling efforts, as the main cause of low learning retention rate and poor self-motivation. We thus developed an open-source, 3D printed, robotics platform with an accessible form factor which on top of providing hands-on experience through clearly defined initial use case, allows learners, with online guidance, to grasp concepts beyond what is traditionally taught in school curriculum despite it being sought after by current employers. Whilst INNOBOT and its software is implemented with low-level entry prerequisites to be accessible to wider range of learners it remains an active and live platform with new topics being introduced and explained on top of the base as we continue efforts in our own talent development and share the practices with the community.

## References

1. Future Learn: The future of robotics: How will robots change the world?, https://www.futurelearn.com/info/blog/general/introduction-robotics-future-robots. Last accessed 29 Jan 23
2. McKinsey & Company, Industrial robotics: Insights into the sector's future growth dynamics. https://www.mckinsey.com/. Last accessed 2 Feb 23
3. BCG How intelligent and mobility will shape the future of the robotics industry. https://www.bcg.com/publications/2021/how-intelligence-and-mobility-will-shape-the-future-of-the-robotics-industry. Last accessed 15 Jan 23
4. Dimitrova, M., et al.: Identifying needs of robotic and technological solutions in the classroom. In: 26$^{th}$ International Conference on Software, Telecommunications and Computer Networks, pp. 444–450, IEEE, Split (2018)
5. BCN3D Moveo: https://www.bcn3d.com/bcn3d-moveo-the-future-of-learning-robotic-arm/. Last accessed 18 Jan 23
6. Roboteurs homepage: https://roboteurs.com/products/rbx1-remix-3d-printed-6-axis-robot-arm-kit?variant=40314908751. Last accessed 12 Dec 22
7. Instructables MOTUS: https://www.instructables.com/MOTUS-Open-Source-3D-Printed-Robotic-Arm/. Last accessed 15 Jan 23
8. Lynch, K.M., Park, C.F.: Modern Robotics: Mechanics Planning and Control. Cambridge University Press, Cambridge (2017)
9. Karahoca, D., et al.: Robotics teaching in primary school education by project-based learning for supporting science and technology courses. Proc. Comp. Sci. **3**, 1425–1431 (2011)
10. Music, J., et. al.: Robotics and information technologies in education: four countries from Alpe-Adria-Denube Region Survey. Int. J. Technol. Design Educ. 749–771 (2022)
11. Entrepreneur What School Doesn't Teach Us about the Workforce. https://www.entrepreneur.com/leadership/what-school-doesnt-teach-us-about-the-workforce/356318. Last accessed 15 Jan 23
12. Macenski, S., et. al.: Robot Operating System 2: Design, architecture, and uses in the wild. Sci. Robot. **7**(66) (2022)

# Impact of Robot Related User Pre-experience on Cognitive Load, Trust, Trustworthiness and Satisfaction with VR Interfaces

Laurent Frering[1(✉)], Clemens Koenczoel[2], Jochen A. Mosbacher[2], Martina Kames[2], Matthias Eder[1], Peter Mohr-Ziak[1], Sebastian Gabl[1], Denis Kalkofen[1], Dietrich Albert[2], Bettina Kubicek[2], and Gerald Steinbauer-Wagner[1]

[1] Graz University of Technology, 8010 Graz, Austria
{laurent.frering,matthias.eder,steinbauer}@ist.tugraz.at,
{mohr,gabl,kalkofen}@icg.tugraz.at

[2] University of Graz, 8010 Graz, Austria
{clemens.koenczoel,jochen.mosbacher,martina.kames,dietrich.albert,
bettina.kubicek}@uni-graz.at

**Abstract.** This paper investigates how user pre-experience impacts the quality of Human-Robot Interaction (HRI) when using a virtual reality (VR) interface to control a robot. The benefit lies in the fact that creating interfaces for these complex systems is a challenging process, and our aim is to shed light on one of the contributing factors. We aim to complement existing works by conducting a user study exploring the link between multiple psychological factors and users' pre-experience with robot systems, video games and gamepads while interacting with a real robot through a VR interface. Results showed that experienced users reported less cognitive load, but also less trustworthiness of and satisfaction with the system and that these factors are correlated with system usability and satisfaction. We assume that this is based in pre-experienced users, having higher expectations and a more critical perception of new systems. These findings could guide the development of future interfaces tailored to specific user groups.

**Keywords:** Human-Robot Interaction · Virtual Reality · Trust · Trustworthiness · Cognitive Load · Satisfaction

## 1 Introduction

When developing an interactive robotics system, it is crucial to not only optimize the performance and reliability on a technical level but also to consider HRI

Supported by the Austrian Research Promotion Agency (FFG) - KIRAS Program - Project EASIER (FO999886357).


T. Petrič et al. (Eds.): RAAD 2023, MMS 135, pp. 123–131, 2023.
https://doi.org/10.1007/978-3-031-32606-6_15

aspects to ensure efficient human-robot teams. An often-considered option to improve the interaction between users and robotics systems are Virtual Reality (VR) interfaces [5].

Literature suggests multiple psychological factors that are important to investigate in order to assess the efficiency of a VR interface for robots: Trust is a major factor in HRI and is closely related to the trustworthiness of the robot, which is determined by robot attributes (ability, benevolence and integrity) [13]. It is well recognized that trust can affect people's willingness to accept [12] and utilize robots [10]. Additionally, trust is a prerequisite for successful and efficient human-robot cooperation [10], which also affects users satisfaction with the system. Cognitive load is a predictor of performance in Human-Computer Interaction (HCI) [15] and has a direct relationship to trust in robots, with lower load being linked to higher trust and vice versa [1]. The perceived cognitive load is influenced by system and interface design aspects (e.g. type of steering) [19]. Another interesting aspect is satisfaction, which refers to the discrepancy between individual expectations and the actual perceived quality of a system [16] and plays a critical role in the acceptance. In this paper, we are interested in the role of prior experience for these factors and their associations with other important aspects like system usability. System usability has been found to relate to many relevant factors in HCI [2], but in the VR context, it also relates to the users' preferences and possible VR sickness [11]. The use of VR headsets can lead to unusual perceptions among users and different effects such as signs of eye fatigue, disorientation, and nausea can occur [4]. According to the review in [4], causes of VR sickness can lie in the hardware (e.g. display type), content (e.g. optical flow), or person (e.g. previous experience). As a result of current research, it is evident that the above variables often influence each other, implying that all variables should be considered when examining the impact of a tool or approach for HRI. Furthermore, it is shown that these psychological variables depend on the personal factors of the users interacting with a UI [8]. As robotic systems are often specified to be used by users with a given pre-experience with such systems, be it novice or trained users, we hypothesized that user pre-experience with the different technologies involved may impact the quality of user-robot interaction in the context of VR for robotics. Based on the research mentioned, we therefore examined the following research question:

*RQ1. How does an operator's pre-experience with robots, technical input devices or video games impact trust, trustworthiness, cognitive load, usability, and virtual sickness when operating a robot in VR?*

To examine this RQ, we conducted a user study as part of the larger EASIER project [7], where users were asked to control a real mobile robot using a VR interface, analyzing the influence of users pre-experience on these variables. The methods use for the experimental procedure, technical setup and measurements are described below.

## 2 Methods

### 2.1 Experimental Procedure

The study started with a set of questionnaires assessing demographic data, personality, and others. Afterward, participants were introduced to the robot and were made familiar with the displays and controls. The next part consisted of different exercises relating to HRI for disaster response scenarios. In this paper, we focus on the VR part, where participants were asked to explore the room for a set duration of 10 min (or until they wished to stop) through the use of a VR interface allowing teleoperation and waypoint navigation control of a mobile robot. Finally, this was followed by trust and cognitive load evaluation.

### 2.2 Robot Presentation

A Clearpath Husky robot was used during this study. It was equipped with two 2D Lidars (Hokuyo UTM-30LX) for navigation and a wide-angle camera (RICOH THETA S) for user teleoperation. The robot navigated in a pre-recorded incomplete map of the environment using a combination of Dijkstra's algorithm and DWA planner [6] implemented using a Behaviour Trees architecture powered by the Move Base Flex ROS package [18]. A picture of the robot is shown in Fig. 1(a).

### 2.3 VR Interface

The VR interface, shown in Fig. 1(b), was implemented in Unity3D and viewed on an Oculus Quest 2 headset using Quest Link. The user can navigate within the extruded map of the area while observing a digital twin of the robot. In addition, we show the confidence of the robot to maneuver as color-coded overlay on the floor of the 3D environment (see the orange area in Fig. 1(b)).

We show a real-world camera view and a control panel attached to the left controller. The user can place navigation waypoints, confirm or abort the planned movement, and teleport to the location of the robot. The right controller is used to operate the control panel buttons via raycasting, while its joystick allows the user to teleoperate the robot (y-axis for forward/backward motion and x-axis for left/right rotation). The left controller is used for user movement and teleportation.

### 2.4 Participants Demographics and Ethical Statement

For the study, subjects were approached and invited by the researchers. A total of 39 participants (21 female, 18 male) between the ages of 19 and 33 ($M = 22.54$, $SD = 3.04$) took part in the study. The study was approved by the ethics committees of the authors' universities and all participants gave written informed consent. Psychology students got a compensation in form of a "participation certificate", a kind of course credit.

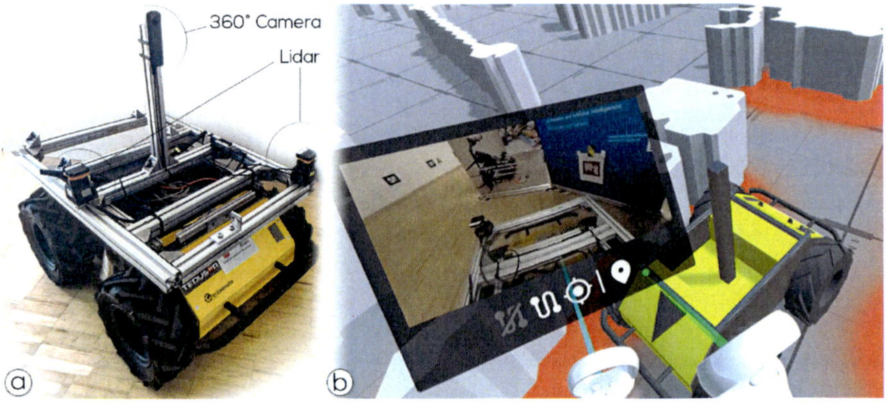


**Fig. 1.** Setup: (a) The Husky robot used during the experiment. (b) The VR interface showing the user's view.

### 2.5 Measures

German versions of questionnaires were used, as the study was conducted with German native speakers.

**Participants Pre-experience** with robots was assessed by a single item with the options "yes" or "no". Of the 39 participants, eight reported previous experience with robots prior to the study and two are currently dealing with it in their work or studies. Experience with video games and gamepads was assessed with single-item questions each on a five-point Likert scale from 1 (never) to 5 (very often). For analysis as independent variables, data on pre-experience with video games and gamepads was dichotomized, yielding a variable separating the participants into a group with and one without experience with video games or gamepads. Of the 39 participants, 25 reported video game experience and 19 reported gamepad experience.

**Trust, Trustworthiness, and Satisfaction** were assessed with one item each. The trust question was adapted from Körber [13]. To evaluate trustworthiness, participants answered the self-developed item: "The robot seemed trustworthy in the last situation". Satisfaction was measured based on the Post Study System Usability Questionnaire (PSSUQ) from Lewis [14]. Answers to all three items were given on a five-point Likert scale ranging from 1 (Strongly disagree) to 5 (Strongly agree).

**Cognitive load** was assessed using the mental demand dimension of the NASA-Task Load Index [9]. Here, a slider scale function with extreme endpoints from "very low" on the left, to "very high" on the right was used.

System Usability was measured using the 10-item System Usability Scale (SUS) by Brooke [3]. Participants rated their subjective assessments of usability on a five-point rating scale, ranging from 1 (Strongly disagree) to 5 (Strongly agree). Items were summed to obtain a standardized SUS-score.

**VR Sickness** was evaluated by three single items addressing the symptoms eye fatigue, disorientation, and nausea mentioned in [4]. Participants rated their symptoms right after taking off the VR headset on a 10-point rating scale from 0 (no symptoms) to 10 (strong symptoms).

## 3 Results

### 3.1 Effects of Pre-Experience

Analysis of the influence of video games, gamepads, and robot experience on trust, trustworthiness, and satisfaction was conducted using Mann-Whitney-U-Tests. Results showed that persons with pre-experience with video games reported the robot as less trustworthy ($p = .032$). Similarly, persons with pre-experience with gamepads reported the robot as less trustworthy ($p = .011$) and were less satisfied with the robot ($p = .004$). Additionally, a trend towards a significant difference can be seen, as users with pre-experience with video games as well as pre-experience with robots show tendentially less satisfaction with the system. Moreover, a trend towards a significant difference can be seen for users with pre-experience with robots, that show tendentially less trustworthiness. The medians, mean-ranks, and detailed results are given in Table 1.

The influence of video games, gamepads, and robot experience on cognitive load, system usability, and VR sickness was examined using analysis of variance. Results showed that persons with experience with video games ($p = .046$), gamepads ($p = .044$), and robots ($p = .016$) reported less cognitive load. The means, standard deviation, and detailed results are given in Table 2.

### 3.2 Correlations

Correlations were calculated using Pearson Correlation Analysis and Kendall's rank correlation analysis where appropriate. Results showed significant relations between cognitive load and system usability ($r = -.429$; $p = .006$) and VR sickness ($r = .470$; $p = .003$), as well as between system usability and VR sickness ($r = -.450$; $p = .004$). Furthermore, trust, trustworthiness, and satisfaction are all positively correlated (all $p < .001$), and so are satisfaction and system usability ($p = .013$). The correlation between system usability and trust only showed a trend to a positive relation ($r = .248$; $p = .063$). The correlation results are summarized in Table 3.

## 4 Discussion

From the perspective of *RQ1*, the main results demonstrate that experience, or more precisely familiarity, with robots, input devices, and the "game-like" setting of the interface setup decreased cognitive load, but also the reported trustworthiness of and satisfaction with the VR system.

**Table 1.** Pre-Experience Analysis on Trust, Trustworthiness and Satisfaction.

| *Independent Variables* | *Dependent Variables* | *Experience* | *n* | *Mean Ranks* | *U* | *p* |
|---|---|---|---|---|---|---|
| *Experience with...* | | | | | | |
| Video Games | Trust | no | 14 | 21.43 | 155.0 | 0.501 |
| | | yes | 25 | 19.20 | | |
| | Trustworthiness | no | 14 | 24.75 | 108.5 | 0.032 |
| | | yes | 25 | 17.34 | | |
| | Satisfaction | no | 14 | 24.07 | 118.0 | 0.063 |
| | | yes | 25 | 17.72 | | |
| Gamepads | Trust | no | 20 | 22.58 | 141.0 | 0.114 |
| | | yes | 19 | 17.55 | | |
| | Trustworthiness | no | 20 | 24.32 | 108.0 | 0.011 |
| | | yes | 19 | 15.90 | | |
| | Satisfaction | no | 20 | 24.82 | 98.5 | 0.004 |
| | | yes | 19 | 15.43 | | |
| Robots | Trust | no | 31 | 20.52 | 108.0 | 0.523 |
| | | yes | 8 | 18.00 | | |
| | Trustworthiness | no | 31 | 21.61 | 74.0 | 0.056 |
| | | yes | 8 | 13.75 | | |
| | Satisfaction | no | 31 | 21.42 | 80.0 | 0.088 |
| | | yes | 8 | 14.50 | | |

Results on satisfaction indicate that operators with pre-experience with gamepads might have different expectations in the interface than novices. Tendentially significant results on possible effects of pre-experiences with robots and video games point in the same direction and give a hint for further research focusing on operators' expectations. Further insights might be beneficial in terms of e.g. interface design, optimizing information flow, or further development of automated functions, especially as satisfaction also correlates positively with system usability. Operators with pre-experience rate trustworthiness lower than those without pre-experience. This indicates, that either novices overestimate the trustworthiness and probably the capabilities of the robot or that operators with experience have a more critical point of view on the robot. To get a more valid evaluation of the actual trustworthiness of the system, an expert rating on the robot's trustworthiness could serve as a baseline, to further develop the robot and relating aspects such as the explanation of the robot's capabilities and limits to the user. This might also be relevant for the development of specific training to enhance operators' expertise on the system.

The impact of pre-experience on cognitive load when using a new system might seem obvious. But, especially regarding the gaming experience results

**Table 2.** Pre-Experience Analysis on Cognitive Load, System Usability and VR sickness.

| *Independent Variables* | *Dependent Variables* | *Experience* | *n* | *M* | SD | *F* | df | *p* | $\eta p2$ |
|---|---|---|---|---|---|---|---|---|---|
| *Experience with...* | | | | | | | | | |
| Video Games | Cognitive Load | no | 14 | 13.36 | 2.47 | 4.26 | 1.37 | 0.045 | 0.104 |
| | | yes | 25 | 10.48 | 4.83 | | | | |
| | System Usability | no | 14 | 68.93 | 15.28 | 0.08 | | 0.775 | 0.002 |
| | | yes | 25 | 70.60 | 18.43 | | | | |
| | VR Sickness | no | 14 | 3.50 | 4.05 | 0.11 | | 0.745 | 0.003 |
| | | yes | 25 | 3.08 | 3.71 | | | | |
| Gamepads | Cognitive Load | no | 20 | 12.95 | 3.55 | 4.34 | 1.37 | 0.042 | 0.107 |
| | | yes | 19 | 10.15 | 4.65 | | | | |
| | System Usability | no | 20 | 70.26 | 14.62 | 0.01 | | 0.927 | 0.000 |
| | | yes | 19 | 69.75 | 19.68 | | | | |
| | VR Sickness | no | 20 | 3.63 | 3.56 | 0.41 | | 0.527 | 0.011 |
| | | yes | 19 | 2.85 | 4.04 | | | | |
| Robots | Cognitive Load | no | 31 | 12.36 | 3.63 | 6.40 | 1.37 | 0.015 | 0.150 |
| | | yes | 8 | 8.25 | 5.50 | | | | |
| | System Usability | no | 31 | 68.55 | 16.85 | 1.08 | | 0.305 | 0.028 |
| | | yes | 8 | 75.63 | 18.41 | | | | |
| | VR Sickness | no | 31 | 3.48 | 4.07 | 0.67 | | 0.419 | 0.018 |
| | | yes | 8 | 2.25 | 2.31 | | | | |

have not been unambiguous [17]. Here, however, we found clear evidence for the role of not only pre-experience with robots but also with video games and gamepads. These results indicate that using common tools for teleoperation, like standard gamepads, can be advantageous for the implementation of new systems. Especially as the perceived cognitive load correlated negatively with the perceived system usability and positively with the amount of experienced VR sickness. Both aspects are critical for a successful design and implementation of an interface.

**Table 3.** Pearson's and Kendall's Rank Correlation Analysis. Asterisks highlight significant relations.

| Pearson | Cognitive Load | System Usability | VR Sickness |
|---|---|---|---|
| System Usability | −0.429** | – | / |
| VR Sickness | 0.470** | −0.450** | – |
| Kendall's Tau | Trust | Trustworthiness | Satisfaction |
| Trustworthiness | 0.613** | – | / |
| Satisfaction | 0.598** | 0.644** | – |
| Cognitive Load | −0.082 | −0.079 | −0.135 |
| System Usability | 0.248 | 0.185 | 0.329* |
| VR Sickness | −0.027 | −0.022 | −0.172 |

# 5 Conclusion

Results indicate that pre-experience with robots and similar but unrelated devices affects users' cognitive load when using a new system and their appraisal thereof. We assume that pre-experienced users have higher expectations and are therefore more critical of new systems, also in terms of systems trustworthiness. On the other hand, even when working with a new system they show less cognitive load, which is of special interest for use in safety-critical applications where trained users and high cognitive load are expected. Taking this into account, one might think of several implications to optimize e.g. the UI design, input devices, special information elements such as transparency modules, or the choice of a system's shared autonomy level with respect to the expected user base.

# References


1. Ahmad, M.I., Bernotat, J., Lohan, K., Eyssel, F.: Trust and cognitive load during human-robot interaction (2019). arXiv:1909.05160 [cs]
2. Aubert, B.A.: The impact of interface quality on trust in web retailers. Cahier du GReSI no **1**(05) (2001)
3. Brooke, J., et al.: SUS-A quick and dirty usability scale. Usabil. Eval. Ind. **189**(194), 4–7 (1996)
4. Chang, E., Kim, H.T., Yoo, B.: Virtual reality sickness: a review of causes and measurements. Int. J. Hum.-Comput. Interact. **36**(17), 1658–1682 (2020)
5. Dianatfar, M., Latokartano, J., Lanz, M.: Review on existing VR/AR solutions in human-robot collaboration. Proc. CIRP **97**, 407–411 (2021)
6. Fox, D., Burgard, W., Thrun, S.: The dynamic window approach to collision avoidance. IEEE Robot. Autom. Mag. **4**(1), 23–33 (1997). https://doi.org/10.1109/100.580977
7. Frering, L., et al.: Enabling and assessing trust when cooperating with robots in disaster response (EASIER) (2022). http://arxiv.org/abs/2207.03763
8. Gasteiger, N., Hellou, M., Ahn, H.S.: Factors for personalization and localization to optimize human–robot interaction: a literature review. Int. J. Soc. Robot. 1–13 (2021)

9. Hart, S.G., Staveland, L.E.: Development of NASA-TLX (task load index): results of empirical and theoretical research. In: Advances in Psychology, vol. 52, pp. 139–183. Elsevier (1988). https://doi.org/10.1016/S0166-4115(08)62386-9
10. Khavas, Z.R.: A review on trust in human-robot interaction. arXiv preprint arXiv:2105.10045 (2021)
11. Khundam, C.: A study on usability and motion sickness of locomotion techniques for virtual reality. ECTI Trans. Comput. Inf. Technol. (ECTI-CIT) **15**(3), 347–361 (2021)
12. Kim, Y.D., Kang, J.H., Sun, D.H., Moon, J.I., Ryuh, Y.S., An, J.: Design and implementation of user friendly remote controllers for rescue robots in fire sites. In: Proceedings of SICE Annual Conference 2010, pp. 875–880. IEEE (2010)
13. Körber, M.: Theoretical considerations and development of a questionnaire to measure trust in automation. In: Bagnara, S., Tartaglia, R., Albolino, S., Alexander, T., Fujita, Y. (eds.) IEA 2018. AISC, vol. 823, pp. 13–30. Springer, Cham (2019). https://doi.org/10.1007/978-3-319-96074-6_2
14. Lewis, J.R.: Psychometric evaluation of the PSSUQ using data from five years of usability studies. Int. J. Hum.-Comput. Interact. **14**(3–4), 463–488 (2002)
15. Longo, L.: Experienced mental workload, perception of usability, their interaction and impact on task performance. PLoS One **13**(8), e0199661 (2018). https://doi.org/10.1371/journal.pone.0199661
16. Michalos, A.C.: Satisfaction and happiness. Soc. Indic. Res. **8**, 385–422 (1980)
17. Nenna, F., Gamberini, L.: The influence of gaming experience, gender and other individual factors on robot teleoperations in VR. In: 2022 17th ACM/IEEE International Conference on Human-Robot Interaction (HRI), pp. 945–949. IEEE, Sapporo (2022). https://doi.org/10.1109/HRI53351.2022.9889669
18. Pütz, S., Simón, J.S., Hertzberg, J.: Move base flex: a highly flexible navigation framework for mobile robots. In: 2018 IEEE/RSJ International Conference on Intelligent Robots and Systems (IROS) (2018)
19. Wang, Z., Reed, I., Fey, A.M.: Toward intuitive teleoperation in surgery: human-centric evaluation of teleoperation algorithms for robotic needle steering. In: 2018 IEEE International Conference on Robotics and Automation (ICRA), pp. 5799–5806. IEEE, Brisbane (2018). https://doi.org/10.1109/ICRA.2018.8460729

# Physical Education Exercises Validation Through Child-Humanoid Robot Interaction

Tara Knežević[1](✉), Marija Radmilović[2], Jefimija Borojević[3], Jovan Šumarac[2], Marko Švaco[1], and Mirko Raković[3]

[1] Faculty of Mechanical Engineering and Naval Architecture, University of Zagreb, Ul. Ivana Lučića 5, 10000 Zagreb, Croatia
{tk223365,marko.svaco}@fsb.hr

[2] Institute Mihailo Pupin, University of Belgrade, Volgina 15, 11060 Belgrade, Serbia
{marija.radmilovic,jovan.sumarac}@pupin.rs

[3] Faculty of Technical Sciences, University of Novi Sad, Dositej Obradovic Square 6, 21102 Novi Sad, Serbia
{borojevic.mh9.2019,rakovicm}@uns.ac.rs

**Abstract.** This paper presents the concept of integrating a robot into the children's environment as an instructor for physical education exercises. The research includes four phases: recording human physical education movements, mapping the recorded movements to the movements of the Pepper robot representing the instructor, analyzing the children's movements while it repeats the movements of the robot, and making a decision about the quality of the children's movements. The goal of this research is to motivate children to perform physical education exercises as well as to make it easier for them to accept the robot in their environment through play. The concept was tested on 10 adults and 2 children. The results show that the children were very stimulated to do the movements with the robot. They performed the exercises well, with a small lack of concentration that can be improved through interaction with the robot.

**Keywords:** Pepper robot · Physical education exercise · Child-robot interaction · Human motion estimation

## 1 Introduction

It is evident that the decreased physical activity and sedentary lifestyle of children have led to the disruption of the development of basic movement skills and the emergence of problems related to proper body posture and physical development. Research about the basic developmental skills of movement in preschool and primary school-age children has shown that the motor skills of children decrease year by year. The results have shown that children in the first school year had the highest level of knowledge of gross motor skills compared to all their peers from other age groups [1]. In recent decades, an unsatisfactory level of coordination development among school-aged children has been observed, which is manifested in a significant delay in the development of fine and gross


T. Petrič et al. (Eds.): RAAD 2023, MMS 135, pp. 132–140, 2023.
https://doi.org/10.1007/978-3-031-32606-6_16

motor skills, as well as a serious and widespread problem of obesity [2]. The robot-human interaction has been studied in many researches whose concepts are based on physical exercises presented by robots in order to increase children's motivation in performing exercises. They differ in who the movements are intended for, as well as in the methods used to establish interaction with children. The proof of numerous researchers which have implemented their robot assisting interaction precisely to autistic children is given in [3]. The implementation of our study has a general application on humans, regardless of their age or physical state and in this way its usage could be more represented. Furthermore, researches such as [4, 5] have achieved the robot's role not only to show, but also to track a child's motions and to correct them if they are wrong by using Inertial Measurement Units for capturing children's movements. However, IMUs can influence children to feel unnatural and forced to do exercises rather than fluently enjoy them.

It is common for physical education exercises to require a certain number of repetitions of the same movements demonstrated by the instructor. For children with reduced motor skills, process of imitation can be useful for the improvement of basic gross motor skills [6]. However, the motivation to repeat the same movements in children decreases over time. Therefore, it is necessary to additionally motivate children to properly do exercises to strengthen and develop the body.

The goal of this research is to encourage children's interaction with robots through creative methods that also motivate them to acquire new and/or improve existing gross motor skills. The inclusion of robots in the environment of children and their interaction (mostly verbal) has been described in many papers [7–9]. The method of including robots for therapeutic purposes in children with developmental disorders is presented in the paper [10, 11]. Unlike industrial robots, humanoid robots are somewhat aware of the social obstacles and can build a meaningful interaction. In this research, the robot is presented as a physical education instructor for healthy children who demonstrates arm and back exercises and gives an assessment of the quality of the performed exercise. The advantages of this method are multiple:

1. The inclusion of robots in children's environments, taking into account that robots are increasingly present in everyday life.
2. The robot as an instructor of physical education exercises is one of the creative and game-based ways of interaction between children and robots with the aim of improving gross motor skills.

# 2 Materials and Methods

## 2.1 Experimental Design

One healthy volunteer without sensorimotor disorder participated in the experiment for the purposes of this study. A set of five basic physical education exercises was defined and they are presented in the Fig. 1. At the beginning of each exercise, the participant was in the initial position, with arms horizontally extended in front of him and palms facing the floor. The equipment used during the experiments is the Vicon Motion Capture System, which includes eight infrared Vicon cameras from the T-series, the Vicon MoCap

a) b) c) d) e)

**Fig. 1.** The set of five basic physical education exercises for the arms and back.

Suit, and 30 fluorescent MoCap markers [12]. During the experiments, the X, Y, and Z coordinates of the fluorescent markers affixed to the MoCap suit worn.

by the participant were recorded. The fluorescent markers are placed one on each hand, elbow, shoulder, neck, and torso joint; six markers on the torso, one marker on each upper arm, forearm, shoulder, hand, and head segment; and one marker on each index finger, thumb, and middle finger. The global coordinate system based on which the X, Y, and Z positions of the markers are estimated was placed on the ground level, mid-ways in the between of the participant's feet.

### 2.2 Robot Pepper

The humanoid robot Pepper has been developed by Aldebaran Robotics [13] (see Fig. 2a). Pepper is 1.2 m tall and has 15 degrees of freedom (DoFs) in the body, including 5 DoFs per arm, 3 DoFs for the leg, and 2 DoFs for the head, as well as an omnidirectional mobile base (3DoFs) on three wheels and open/close hands. The kinematic model of the Pepper robot with mDH parameters was calculated. For the purposes of research in this paper arms (10 DoFs), LCD display and loudspeakers were used.

### 2.3 The Imitation Process

The information provided by the motion capture system was used to define the characteristics of the human body and its motion. Since the information obtained from the Vicon motion capture system is defined for the model of the human, the kinematic model of the Pepper robot was scaled to the size of the human and extended to the standard kinematic model of the human's arm with 2 additional joints in the wrist (yaw and pitch). The size of the robot's segments was scaled to the size of the actor's limbs. In that manner, it was possible to calculate human motion in joint space using the imitation process.

An analytical imitation algorithm based on the Jacobian matrix capable of real-time extraction of Cartesian motions and joint motions is used [14]. The imitation process is formulated as an optimization algorithm. The imitation algorithm is based on minimization function given with Eq. 1:

$$q_{imitation}(t_i) = \min_q(\zeta) \tag{1}$$

where $t_i$ is each time sample with $i = [1 \ldots N]$. The criterion function was comprised of the error between the positions of the frames attached to the actor's joint $P_{aj}$ and those of the scaled and extended model of the Pepper robot $P_{rj}$, as well as the error between the positions of the segment's markers attached to the actor $P_{am}$ and those of the scaled and extended model of the Pepper robot $P_{rm}$, as given by Eq. 2:

$$\zeta = \left\| \begin{bmatrix} \alpha\left(\vec{P}_{am}(t_i) - \vec{P}_{rm}(q)\right) \\ \beta\left(\vec{P}_{aj}(t_i) - \vec{P}_{rj}(q)\right) \end{bmatrix}^2 \right\| \tag{2}$$

where $q$ is the current joint configuration of the scaled and extended model of the Pepper robot, $\alpha = 1$ and $\beta = 1.3$ are the weight factors for the error in markers and joint positions, respectively; $\vec{P}_{aj}(t_i)$ is the vector of the 3D position of the shoulder, elbow and wrist joints; $\vec{P}_{am}(t_i)$ are vectors of the 3D position of the recorded upper arm, forearm, hand, index finger, thumb and middle finger markers; $\vec{P}_{rj}(q)$ is the vector of the 3D position of the shoulder, elbow and wrist joints, and $\vec{P}_{rm}(q)$ are the vectors of the 3D position of the upper arm, forearm, hand, index finger, thumb and middle finger markers of scaled and extended model of the Pepper robot in the configuration $q$.

In the initial position, the 3D positions of each marker of the scaled and extended model of the Pepper robot $P_{rm}$ are defined for the initial configuration and are the same as the position of the same actor's marker $P_{am}$. The kinematics of the scaled and extended model of the Pepper robot is calculated analytically with SYMORO + software [15].

The paper [14] in detail presents the calculation of robot marker frames and mapping the initial configuration of the scaled and extended robot model and the actor's initial configuration. The motion obtained with imitation algorithm $q_{imitation}$ could not be used directly by the Pepper robot. We used the algorithm presented in the paper [16] to calculate Pepper robot joint coordinates $q_{robot}$ in order to adjust the estimated human movement $q_{imitation}$ with Pepper robot joint limits and the robot's size.

### 2.4 Human Pose Estimation and Evaluation of the Quality of the Realized Movement

Human body pose estimation is a crucial step in any task of human-robot interaction. Besides that, it also plays an important role in other applications such as quantifying physical exercises, sign language recognition, and gesture control.

In order to estimate human motion in real time, we decided to use the MediaPipe library which utilize the BlazePose model [17]. An Intel D435i camera is placed in front of the actor (see Fig. 2a). In order to adequately validate the concept, ten adults (18–22 years) and two children (4 and 8 years) participated in the experiment. Adults were asked to strictly follow the robot's movements, while children spontaneously interacted with the robot. Adults gave verbal consent to participate in the experiment, while parental approval was obtained for children's participation. The robot performed the recorded physical education exercise according to the joint motion estimation $q_{robot}$ obtained by the imitation algorithm. The child/adult repeats the robot motion while the camera records it.

The complete landmark map is shown in Fig. 2b) with key points annotated. Each key point is represented with position and orientation in the body coordinate system also shown in Fig. 2b).

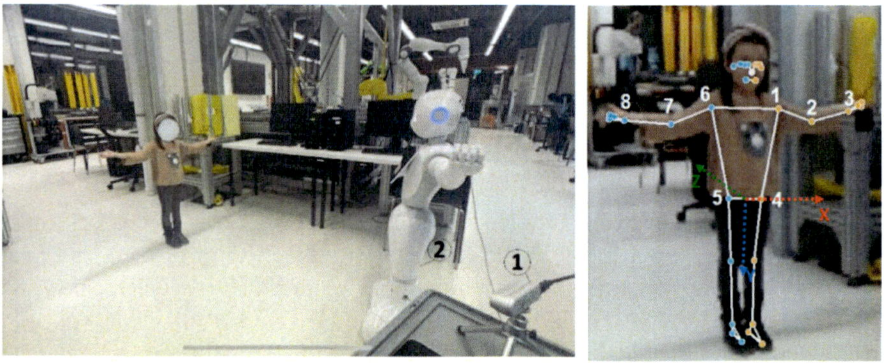


a) b)

**Fig. 2.** Pepper robot and child during the experiment: **a)** 1 - Intel D435i camera, 2 – Pepper robot; **b)** Pose landmark model: 1 – Left shoulder, 2 – Left elbow, 3 – Left wrist, 4 – Left hip, 5 – Right hip, 6 – Right shoulder, 7 – Right elbow, 8 – Right wrist.

Based on the 3D positions of the hip, shoulder, elbow and wrist landmarks, the shoulder pitch, shoulder yaw, and elbow yaw joints are calculated using the Eq. 4:

$$q_{human} = \begin{bmatrix} q_{LShPitch} \\ q_{LShYaw} \\ q_{LElbowYaw} \\ q_{RShPitch} \\ q_{RShYaw} \\ q_{RElbowYaw} \end{bmatrix} = \begin{bmatrix} \arctan(\left|\vec{v}_{4\_1} x \vec{v}_{1\_2}\right| / \vec{v}_{4\_1} \cdot \vec{v}_{1\_2}) \\ \arctan(\left|\vec{v}_{6\_1} x \vec{v}_{1\_2}\right| / \vec{v}_{6\_1} \cdot \vec{v}_{1\_2}) \\ \arctan(\left|\vec{v}_{2\_1} x \vec{v}_{2\_3}\right| / \vec{v}_{2\_1} \cdot \vec{v}_{2\_3}) \\ \arctan(\left|\vec{v}_{5\_6} x \vec{v}_{6\_7}\right| / \vec{v}_{5\_6} \cdot \vec{v}_{6\_7}) \\ \arctan(\left|\vec{v}_{1\_6} x \vec{v}_{6\_7}\right| / \vec{v}_{1\_6} \cdot \vec{v}_{6\_7}) \\ \arctan(\left|\vec{v}_{6\_7} x \vec{v}_{7\_8}\right| / \vec{v}_{6\_7} \cdot \vec{v}_{7\_8}) \end{bmatrix} \quad (4)$$

where $q_{human}$ is the vector of the estimated human (child's/adult's) joint angles $q_{LShPitch}$,$q_{LShYaw}$, $q_{LElbowYaw}$, $q_{RShPitch}$, $q_{RShYaw}$, and $q_{RElbowYaw}$. The operation $x$ represents cross product of the vectors, while operation $(\cdot)$ represents dot product of the vector.

The shoulder roll and wrist roll joints cannot be estimated using the estimated joint positions using MediaPipe, as this software does not detect movement in landmarks for rotations in these joints. The estimated overall trajectory of the child's shoulder and elbow joints should be segmented according to the duration of each exercise. The intersection points in movement segmentation were the times of the occurrence of pauses between movements. Dynamic Time Wrapping [18] algorithm was utilized on $q_{robot}$ and the estimated human joint angle $q_{human}$ in order to quantify the estimation deviation between child's/adult's and robot's joint angle.

# 3 Results

A comparative analysis of the angles in all joints of the robot, adult, and child arms for all experiments is presented in Fig. 3. The execution time of each exercise varies and depends on the participant. Therefore, robot, adult and child joint trajectories for each experiment were time normalized to be mutually comparable.

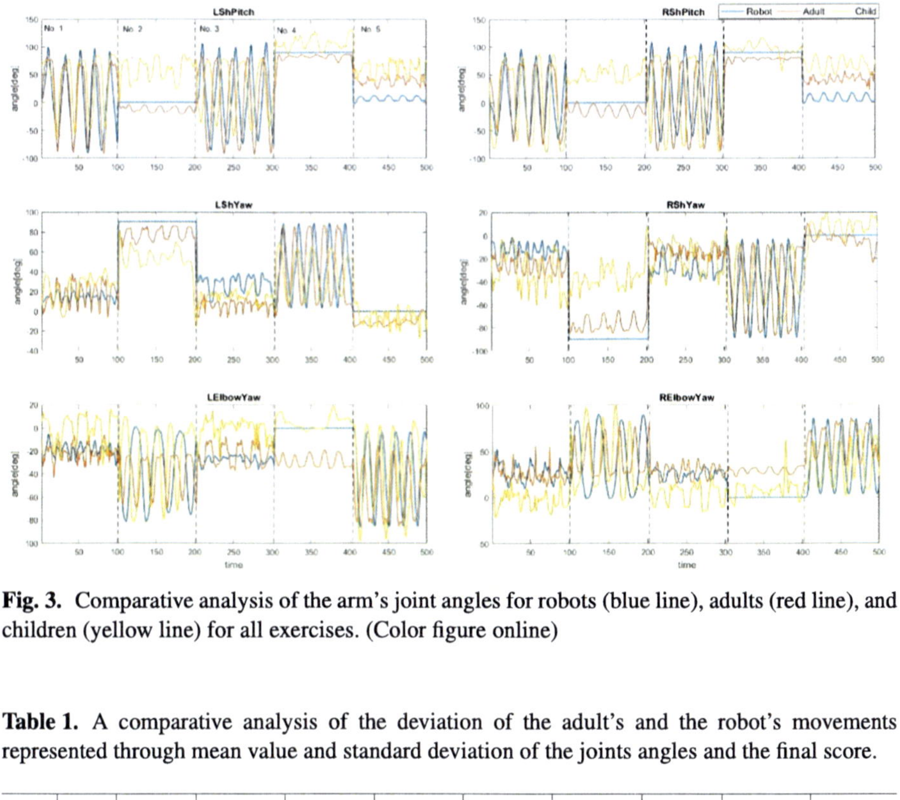


**Fig. 3.** Comparative analysis of the arm's joint angles for robots (blue line), adults (red line), and children (yellow line) for all exercises. (Color figure online)

**Table 1.** A comparative analysis of the deviation of the adult's and the robot's movements represented through mean value and standard deviation of the joints angles and the final score.

| | Adult 1 | Adult 2 | Adult 3 | Adult 4 | Adult 5 | Adult 6 | Adult 7 | Adult 8 | Adult 9 | Adult 10 |
|---|---|---|---|---|---|---|---|---|---|---|
| *Ex.*1 | 6.5 | 10.5 ± 5 | 17.1 ± 3 | 9.7 ± 4 | 13.7 ± 4 | 6.83 ± 1 | 7.5 ± 2 | 9.2 ± 4 | 7 ± 3 | 11.8 ± 6 |
| *Ex.*2 | 19.4 ±12 | 16.1 ± 7 | 24 ± 11 | 16.7 ± 9 | 13.9 ± 6 | 14.1 ± 7 | 16.3 ± 8 | 14.3 ± 9 | 12.8 ± 1 | 24.3 ± 10 |
| *Ex.*3 | 9.9 ±4 | 9.1 ± 4 | 14.3 ± 8 | 9.7 ± 6 | 11.1 ± 4 | 8.9 ± 4 | 9.4 ± 6 | 9.5 ± 6 | 10.6 ± 5 | 9.7 ± 2 |
| *Ex.*4 | 12.7 ± 10 | 14.5 ± 12 | 16.8 ± 13 | 14.2 ± 12 | 13.4 ± 12 | 13.2 ± 12 | 13.5 ± 13 | 13.3 ± 9 | 12.9 ± 10 | 11.6 ± 9 |
| *Ex.*5 | 11.9 ±6 | 6.2 ± 2 | 12.3 ± 5 | 10.4 ± 6 | 19.9 ± 5 | 7.1 ± 4 | 9.3 ± 4 | 15.6 ± 8 | 14.4 ± 10 | 9.8 ±6 |
| *score* | **5** | **5** | **3** | **3** | **4** | **4** | **4** | **5** | **5** | **4** |

**Table 2.** A comparative analysis of the deviation of the child's and the robot's movements, the values of the criterion function for each exercise, and the final score.

| | Child 1 | | | | | Child 2 | | | | |
|---|---|---|---|---|---|---|---|---|---|---|
| Exercise/joint | No.1 | No.2 | No.3 | No.4 | No.5 | No.1 | No.2 | No.3 | No.4 | No.5 |
| *LShPitch* | 14.25 | 34.41 | 14.24 | 16.18 | 34.66 | 21.29 | 49.32 | 17.35 | 11.4 | 43.88 |
| *LShYaw* | 6.908 | 51.12 | 8.12 | 6.08 | 3.87 | 8.16 | 38.02 | 9.25 | 9.73 | 5.19 |
| *LElbowYaw* | 9.90 | 15.39 | 18.57 | 8.88 | 15.56 | 13.42 | 15.81 | 23.58 | 3.64 | 13.29 |
| *RShPitch* | 23.88 | 53.83 | 22.75 | 14.77 | 44.53 | 24.87 | 48.22 | 22.74 | 6.61 | 51.71 |
| *RShYaw* | 17.57 | 57.99 | 8.84 | 6.84 | 7.47 | 10.57 | 49.35 | 9.42 | 10.09 | 5.43 |
| *RElbowYaw* | 13.68 | 22.39 | 13.89 | 8.99 | 12.78 | 16.25 | 15.14 | 16.77 | 7.77 | 8.84 |
| *criteria* | **1** | **0** | **1** | **1** | **0** | **1** | **0** | **1** | **1** | **0** |
| *score* | **3 (good)** | | | | | **3 (good)** | | | | |

The quality of the match between the robot's movements and the realization of the exercise by ten adults and two children is presented in Tables 1 and 2. The angles of the robot joints $q_{robot}$ and the human (adult or child) joint angles $q_{human}$ were compared using the DTW value, divided by the number of samples, and represented in degrees. The *criteria* value is equal to 1 if the quality of matching for each joint per exercise is less than 25 degrees, otherwise, the value of the criteria is equal to 0. The final *score* is the sum of the *criteria* values for all exercises.

## 4 Discussion and Further Works

The joint trajectories presented in Fig. 3 show a satisfactory matching of the adult's and robot's movements. The biggest deviation in these trajectories is in the examples of exercises . 2 and No. 5. It is evident that the arms of the adult 1 had a flexion in the *ShPitch* joint of ~ 23 deg more than the robot's arms. It is also observed that the elbow trajectories *ElbowYaw* match with the error ~ 21deg, which is a consequence of imprecise estimation of the elbows landmark while adults in which they performed the movement precisely. Comparative analysis of matching of the adult's and robot's movements for 10 adults is presented in Table 1 and shows that small deviations in tracking the robot's movements caused the score to be lower for some adults. The children motion deviates more from the robot motion for tasks No. 2 and No. 5, especially in the shoulder joints. The deviation in the *ShPitch* and *ShYaw* is ~ 50deg and evident in the results presented in Table 2. The reason for this is a lack of concentration and holding the elbow at shoulder height, which is one of the more difficult poses.

The results presented in this paper are part of our proof-of-concept research. Our research has multiple implications and can be applied to the analysis of the movements of adults as well as healthy and sick children. In addition to the basic goal of motivating children to do physical education exercises, this research can be very applicable for rehabilitation purposes. While our results provide valuable information, they are limited by the fact that they do not reflect a general conclusion about using robots to motivate children to do physical exercises because of the low number of participants. We plan on expanding our research in the future to include a larger participant sample (>30) and a comparison of the motivation gained from actual human instructors compared to the Pepper robot. The results of the presented research can be improved in several ways. The estimation of children's skeletons and the calculation of joint angles can be improved by using a more sophisticated system for recording and analyzing human movements, such as an inverse kinematics algorithm in real time. In this way, the estimation of the shoulder yaw joint and wrist joints, which were not taken into consideration in this research due to the characteristics of the equipment, would be included in the analysis.

**Acknowledgement.** The authors would like to acknowledge the support of the project CRTA - Regional Centre of Excellence for Robotic Technologies, funded by the ERDF fund. Also, the authors would like to thank all participants in the experiments.

## References

1. Sgrò, F., Quinto, A., Messana, L., Pignato, S., Lipoma, M.: Assessment of gross motor developmental level in Italian primary school children. J. Phys. Edu. Sport **17**(3), 1954–1959 (2017). https://doi.org/10.7752/jpes.2017.03192
2. D'Elia, F., Tortella, P., Sannicandro, I., D'Isanto, T.: Design and teaching of physical education for children and youth. Journal of Human Sport and Exercise **15**(4proc), S1527-S1533 (2020)
3. Ismail, L.I., Verhoeven, T., Dambre, J., Wyffels, F.: Leveraging robotics research for children with autism: a review. Int. J. Soc. Robot. **11**, 389–410 (2019)
4. Guneysu, A., Arnrich, B.: Socially assistive child-robot interaction in physical exercise coaching. In: 2017 26th IEEE international symposium on robot and human interactive communication (RO-MAN), pp. 670–675. IEEE (2017)
5. Guneysu, A., Arnrich, B., Ersoy, C.: Children's rehabilitation with humanoid robots and wearable inertial measurement units. In: 2015 9th International Conference on Pervasive Computing Technologies for Healthcare (PervasiveHealth), pp. 249–252. IEEE (2015)
6. Sutapa, P., Suharjana, S.: Improving gross motor skills by gross kinesthetic-and contemporary-based physical activity in early childhood. Jurnal Cakrawala Pendidikan **38**(3), 540–551 (2019)
7. Ioannou, A., Andreou, E., Christofi, M.: Pre-schoolers' interest and caring behaviour around a humanoid robot. TechTrends **59**(2), 23–26 (2015)
8. Meghdari, A., Alemi, M., Zakipour, M., Kashanian, S.A.: Design and realization of a sign language educational humanoid robot. J. Intell. Rob. Syst. **95**(1), 3–17 (2019)
9. Kennedy, J., et al.: Child speech recognition in human-robot interaction: evaluations and recommendations. In: Proceedings of the 2017 ACM/IEEE International Conference on Human-Robot Interaction, pp. 82–90 (2017)
10. Borovac, B., et al.: Robotics as Assistive Technology for Treatment of Children with Developmental Disorders—Example of Robot MARKO. In: International Conference on Machine and Industrial Design in Mechanical Engineering, pp. 69–76. Springer, Cham (2021)
11. Barakova, E.I., Bajracharya, P., Willemsen, M., Lourens, T., Huskens, B.: Long-term LEGO therapy with humanoid robot for children with ASD. Expert. Syst. **32**(6), 698–709 (2015)
12. Vicon Motion Systems, Inc., Oxford, UK. Nexus 2.11 Documentation: Kinetic Modeling, https://docs.vicon.com/display/Nexus211/Kinetic+modeling (2021)

13. Soft Bank Robotics: Pepper. Technical Datasheet 1.8 Version 2 (2019)
14. Tomić, M., Chevallereau, Ch., Jovanović, K., Potkonjak, V., Rodić, A.: Human to humanoid motion conversion for dual arm manipulation tasks. Robotica, 1–21 (2018). https://doi.org/10.1017/S0263574718000309, ISSN: 0263-5747
15. Khalil, W., Creusot, D.: SYMORO+: A system for the symbolic modelling of robots. Robotica **15**, 153–161 (1997)
16. Safonova, A., Pollard, N., Hodgins, J.K.: Optimizing human motion for the control of a humanoid robot. In: Proceedings of International Conference on Robotics and Automation, pp. 1390–1397 (2003)
17. Bazarevsky, V., Grishchenko, I., Raveendran, K., Zhu, T., Zhang, F., Grundmann, M.: Blazepose: On-device real-time body pose tracking. arXiv preprint arXiv:2006.10204 (2020)
18. Sakoe, H., Seibi, C.: Dynamic programming algorithm optimization for spoken word recognition. IEEE Transactions on Acoustics, Speech, and Signal Processing 1978. Vol. ASSP-26, No. 1, pp. 43–49 (1987)

# A Singularity-Free Approach for Safe Operation of a Parallel Robot for Lower Limb Rehabilitation

Paul Tucan(✉), Bogdan Gherman, Adrian Pisla, Alin Horsia, Calin Vaida, and Doina Pisla

CESTER, Technical University of Cluj-Napoca, 400114 Cluj-Napoca, Romania
{paul.tucan,doina.pisla}@mep.utcluj.ro

**Abstract.** This paper presents a singularity analysis of a parallel robot for lower limb rehabilitation with the scope of creating a safe environment for the patient and physical therapist during the rehabilitation procedure. The singularities are identified by analyzing the determinants of Jacobi matrices and avoidance solutions using mechanical constraints and control algorithm are proposed. The solutions are validated by performing two sets of experimental laboratory tests, one test was performed without any human subjects to check the functionality of safe operation mode and the second experimental test was performed using 5 healthy subjects to validate the functionality of the entire robotic system.

**Keywords:** Singularity analysis · parallel robot · kinematics · lower limb rehabilitation · safe operation

## 1 Introduction

Worldwide, stroke is one of the main reasons for limb impairment [1]. Any limb disability affects the activities of daily living (ADLs), but lower limb impairment may seriously affect the quality of life of the patient [2]. The rehabilitation requires a physical therapist to manage the repetitive motion and to record the evolution of the patient between sessions. The increased life expectancy is leading to a continuously increasing number of patients requiring physical rehabilitation, and even today the medical system is unable to provide efficient treatment for everyone; a predicted deadline for an international crisis in healthcare is 2030 [3]. Thus, the European Union [4, 5] is encouraging the setup of strategic partnerships between different research domains leading to new paradigms that should create an optimized and sustainable healthcare system and improved health outcomes for citizens and patients.

As a possible solution to this crisis some robotic devices for lower limb rehabilitation have been developed in the last years [6–8]. In most cases, patients need rehabilitation in the acute phase of the disease as the effectiveness of the treatment is highest. When dealing with patients that are unable to move their lower body or in some cases actively communicate with the medical personal, robotic rehabilitation must take place in strict


T. Petrič et al. (Eds.): RAAD 2023, MMS 135, pp. 141–149, 2023.
https://doi.org/10.1007/978-3-031-32606-6_17

safety conditions. Safety in robotic assisted rehabilitation may be achieved by analyzing a series of factors regarding the control system of the robot and the mechanical design. From the control system point of view, a robotic system for rehabilitation should contain external and internal sensor systems. External sensor system is responsible of monitoring the patient bio-signals, especially when the patient has little to none feeling and a surpass of the motion capability may create trauma for the patient. The internal sensor system is responsible for monitoring the robot position during the entire procedure. Any singularity from the robot workspace must be avoided in the control algorithm. From mechanical design point of view, a singularity free operation workspace may be achieved through efficient design [9].

There are different approaches to assess the safety of robotic assisted rehabilitation procedure. He et al. [10, 11] proposes a hazard analysis combined with FMEA (Failure Modes & Effects Analysis) to identify any aspect that may influence the efficiency of the rehabilitation. Pan et al. [12] propose a safety supervisory fuzzy controller to monitor the forces within the system and using a decision-making section to adapt the force or halt the robot if necessary. Wu et al. [13] propose achievement of safer environment through singularity avoidance using different methods (gradient projection and Jacobian optimization) and use of vision systems to externally check the proximity of singular areas. Gherman et al. [14] uses singularity analysis for analytical workspace generation to create a safer, singularity-free workspace of a parallel robot.

This paper presents a singularity-free approach for developing a parallel robotic system for lower limb rehabilitation able to operate in safe conditions. This approach uses a detailed singularity analysis carried on in the designing stage of the robotic system and implemented during the prototyping of the robot. The singularities areas are avoided by mechanical constrains and by using the mathematical model implemented within the control system and programed to avoid the singularities.

After the Introduction, the Second Section contains data regarding the medical task of lower limb rehabilitation, Third Section illustrates data regarding the robotic system, Fourth Section presents singularity analysis, singularity avoidance methods and experimental validation of the robotic system followed by Conclusions.

## 2 Lower Limb Rehabilitation Task

Lower limb rehabilitation implies repetitive mobilization of hip joint, knee joint or ankle joint (Fig. 1). For the hip joint the rehabilitation motions are flexion ($H_F$)-extension ($H_E$), adduction ($H_{Ad}$)-abduction ($H_{Ab}$) and internal rotation ($H_{IR}$)-external rotation ($H_{ER}$). For the knee joint the rehabilitation motions are flexion ($K_F$) and extension ($K_E$). The rehabilitation motions of the ankle joint are flexion ($A_F$)-dorsiflexion ($A_D$) and eversion ($A_E$)-inversion ($A_I$). Range of motion (ROM) for each rehabilitation motion is presented in Fig. 2, relative to a healthy human [15].

The rehabilitation procedure may present different challenges depending on the health state of the patient. If the patient is able to maintain the vertical position, some gait training devices may be used and some harnesses to avoid any harm of the patient, but if the patient is bedridden then the ergonomics of the rehabilitation procedure change entirely, and a robotic rehabilitation solution must contain bed attachment device or be designed with a special ergonomic bed.

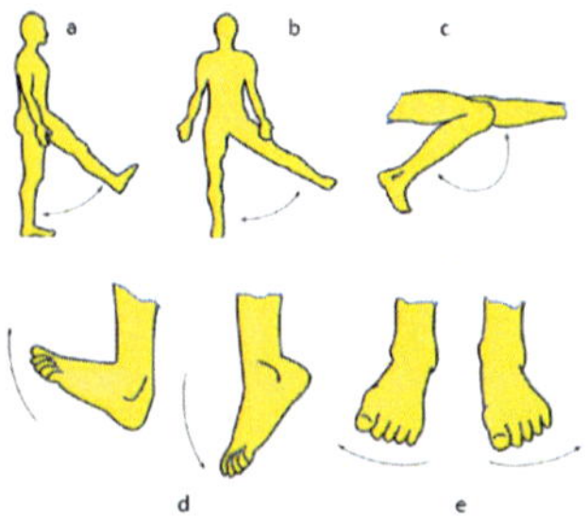


**Fig. 1.** Lower limb rehabilitation motion a) $H_F$-$H_E$; b) $H_{Ad}$-$H_{Ab}$; c) $K_F$-$K_E$; d) $A_F$-$A_D$; e) $A_I$-$A_E$

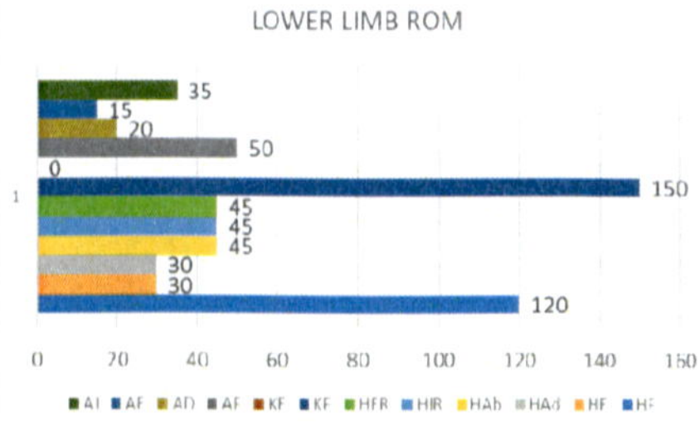


**Fig. 2.** Lower limb ROM

## 3 RAISE-Parallel Robot for Lower Limb Rehabilitation

Starting from the required characteristics of a lower limb rehabilitation robot, RAISE parallel robot was developed and patented [16]. The robotic structure is composed of two mechanisms, one for hip-knee rehabilitation which performs the hip motion in two perpendicular planes (vertical and lateral) and the knee motion in the vertical plane and the other one for ankle rehabilitation which performs the angular motion in two perpendicular planes using the ankle joint as center of rotation. The targeted lower limb rehabilitation motions are: $H_E$-$H_F$ (0°–80°), $H_{Ad}$-$H_{Ab}$ (0°–30°), $K_E$-$K_F$ (0°–80°), $A_D$–$A_F$ (20°–35°), $A_E$-$A_I$ (−20°–20°). The values have been chosen in such a way for these to be lower than the motion amplitudes of a healthy subject.

Figure 3a presents the kinematic scheme of hip-knee module which uses three active translational joints and several passive joints (seven rotational and two prismatic). As the targeted joints (hip and knee) perform rotational motions, the passive joints are used to connect the robot to the patient lower limb, using two rotational joints for the hip (to allow the motion in two planes) and a rotational joint for the knee.

The mobile elements of the hip-knee module are illustrated along with the seven (7) rotational joints ($R_i$, $i = 1{:}7$) and the five (5) translational joints ($T_i$, $i = 1{:}5$). The active joints, $q_1, q_2$ and $q_3$ and the corresponding angular values with respect to the lower limb $\alpha_{hip\_l}, \alpha_{hip\_v}$ and $\alpha_{knee}$ are also illustrated to point out the mechanism behavior. The kinematic scheme of the second module is presented in Fig. 3b. Using the same conventions for notations the mobile elements and the rotational joints of the ankle mechanism are defined. The mechanism has two active joints $q_4$ and $q_5$ and

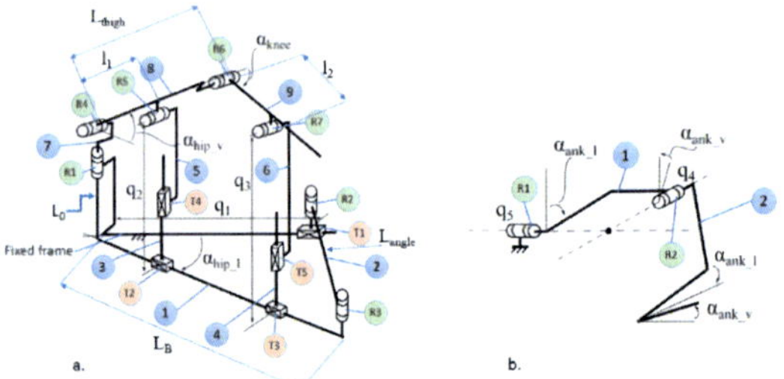

**Fig. 3.** The kinematic model for: a) hip-knee mechanism; b)ankle mechanism [17]

modify the angular position of the ankle with respect to the lateral and vertical plane ($\alpha_{ank\_l}$ and $\alpha_{ank\_v}$). The structural analysis of the mechanism yields 3 degrees of mobility for the hip-knee module and 2 degrees of mobility for the ankle module.

## 4 Singularity Analysis

Singularities, or singular configurations are poses where a mechanism may lose or gain a degree of freedom (DOF). These singular configurations should be avoided during the medical procedure to assure the safety of the patient, operator and of the robotic system. One method to determine the singularities of a mechanism is to analyze the determinants of the Jacobi matrices, computed from the implicit functions of the geometric model (Eq. 1) [18]. The Jacobi matrix for the inverse model (A) is presented in Eq. (2) and the expressions of the terms $A_{11}$ and $A_{32}$ are given in the Eq. (3).

$$\begin{cases} f_1 \Rightarrow q_1 - L_B \cdot \cos\left(\alpha_{hip\_l}\right) + \sqrt{L_B^2 \cdot \cos(\alpha_{hip\_l})^2 - L_B^2 + L_{angle}^2} = 0 \\ f_2 \Rightarrow q_2 - L_O - l_1 \cdot \sin\left(\alpha_{hip\_v}\right) = 0 \\ f_3 \Rightarrow q_3 - L_O - L_{thigh} \cdot \sin\left(\alpha_{hip\_v}\right) - l_2 \cdot \sin(\alpha_{hip\_v} + \alpha_{knee}) = 0 \\ f_4 \Rightarrow q_4 - \alpha_{ank\_v} = 0 \\ f_5 \Rightarrow q_5 - \alpha_{ank\_l} = 0 \end{cases} \tag{1}$$

The Jacobi matrix B, for the forward model, is the identity matrix because when computing the matrix of the joint space the implicit functions are differentiated by values of $q_i$, where $i = 1{:}6$ yielding 1 for the main diagonal, rest of the terms being zero and its determinant always yields 1. The determinant of the Jacobi matrix A is given in Eq. (4). Considering Eq. (4), three cases are distinguished: the first case when $\sin\left(\alpha_{hip\_l}\right) = 0 \rightarrow \alpha_{hip\_l} = 0$, the second case: $\cos\left(\alpha_{hip\_v}\right) = 0 \rightarrow \alpha_{hip\_v} = \pm\frac{\pi}{2}$ and the third case: $\cos\left(\alpha_{hip\_v} + \alpha_{knee}\right) = 0 \rightarrow \alpha_{hip\_v} + \alpha_{knee} = \pm\frac{\pi}{2}$.

$$A = \begin{bmatrix} A_{11} & 0 & 0 & 0 & 0 \\ 0 & -l_1 \cdot \cos(\alpha_{hip\_v}) & 0 & 0 & 0 \\ 0 & A_{32} & -l_2 \cdot \cos(\alpha_{hip\_v} + \alpha_{knee}) & 0 & 0 \\ 0 & 0 & 0 & -1 & 0 \\ 0 & 0 & 0 & 0 & -1 \end{bmatrix} \tag{2}$$

$$A_{11} = L_B \cdot \sin\left(\alpha_{hip\_l}\right) - \frac{L_B^2 \cdot \cos(\alpha_{hip\_l}) \cdot \sin(\alpha_{hip\_l})}{\sqrt{L_B^2 \cdot \cos(\alpha_{hip\_l})^2 - L_B^2 + L_{angle}^2}}$$
$$A_{32} = -l_2 \cdot \cos\left(\alpha_{hip\_v} + \alpha_{knee}\right) - L_{thigh} \cdot \cos\left(\alpha_{hip\_v}\right) \tag{3}$$

$$\det(\mathrm{A}) = L_B \cdot \sin\left(\alpha_{hip\_l}\right) \cdot \left[1 - \frac{\cos(\alpha_{hip\_l}) \cdot L_B}{\sqrt{\cos(\alpha_{hip\_l})^2 \cdot L_B^2 + L_{angle}^2}}\right] \cdot \cos\left(\alpha_{hip\_v}\right) \cdot l_1 \cdot l_2 \cdot \cos\left(\alpha_{hip\_v} + \alpha_{knee}\right) \tag{4}$$

First case of singularity (Fig. 4a) is given by the condition where the patient's leg is straight which means that the robot will have a singular starting position. The second singularity (Fig. 4b) appears when the patient's thigh is maintained at 90° in any direction, position that is improper for the rehabilitation exercises according to the medical specialists' recommendations. The third singularity (Fig. 4c) represents the situation in which the lower part of the limb is vertical, parallel with the OX axis of the fixed coordinate frame.

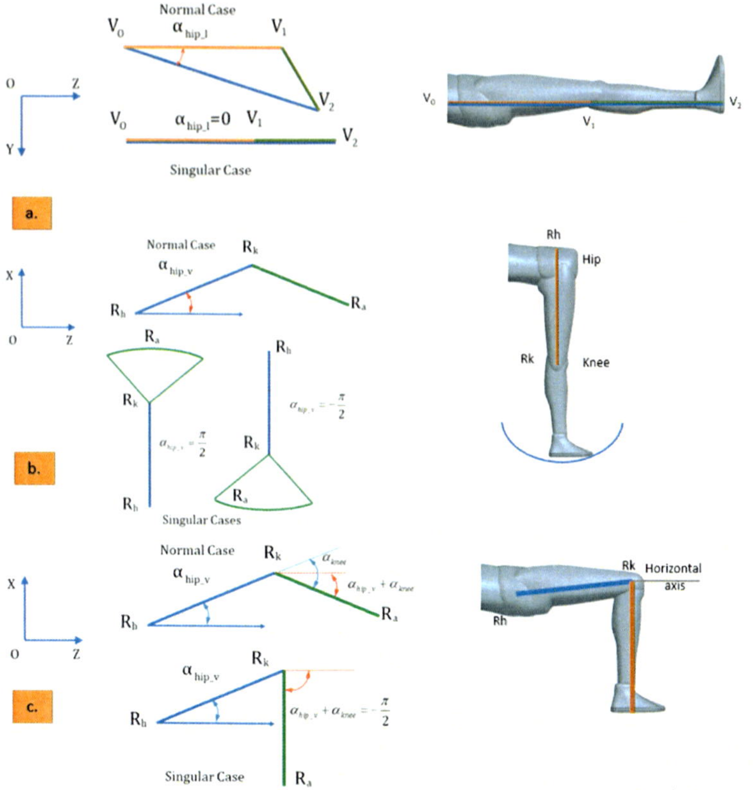


**Fig. 4.** Singularity cases: a)singular pose one; b) singular pose two; c) singular pose three

The singularity analysis points out that there are singular configurations only in the case of the first mechanism (hip-knee module). To avoid the first singularity case, a mechanical distancing element has been designed and implemented on the robot prototype. The CAD model and the mechanical part integrated within the prototype can be seen in Fig. 5. The mechanical part contains a buffer plastic component that constraints the amplitude of the motion and does not allow $\alpha_{hip\_v}$ angle to achieve zero value. The mechanical component is only a final safety assurance method, the amplitude of motion being always verified by the control system (see Fig. 6). The second singularity presents a position that should not be achieved as is it exciding the hip ROM of a bedridden patient, which means that this condition has no negative influence on the robot performance with respect to the medical task and will be avoided by limiting the actuator strokes in the robot control algorithm.

Due to mechanical configuration, the hip rehabilitation mechanism is unable to reach the amplitude of 90 degrees (max $H_F$ROM = 80°) and at the same time due to the fact that the rehabilitation motion is performed with a bedridden patient, there is no extension motion for the hip.

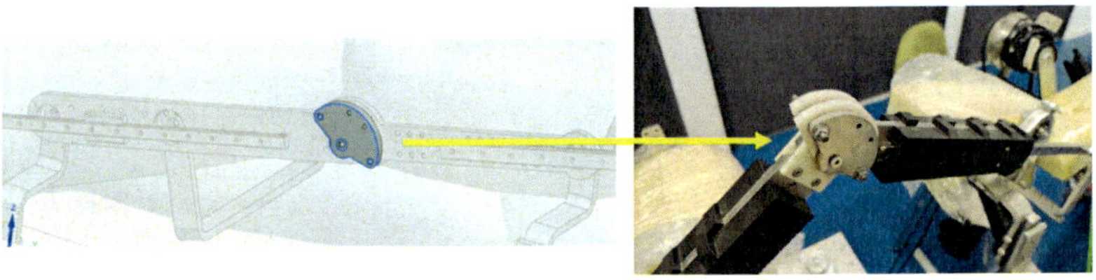

**Fig. 5.** Mechanical avoidance of the first case of singularity (using of plastic damper to eliminate the mechanicl elements to reach the singularity pose)

The third singularity implies vertical position of the $l_2$ link. To ensure the safety of the procedure, this particular condition is avoided within the robot control system. In the Fig. 4c only the $-\pi/2$ case is represented as the other one leads to an anatomically impossible position of the leg.

The control algorithm to avoid the three singular poses is presented in Fig. 6. The algorithm continuously monitors the values of three angles ($\alpha_{hip\_v}$, $\alpha_{hip\_l}$ *and* $\alpha_{knee}$) and check them against some predefined values (values defined to avoid using the mechanical components for stopping the motion), if any of the monitored values is not in the defined interval then the variable "SAFE MODE" changes its value to "FALSE" and the "HALT" command is transmitted to the drivers controlling the motors. Decision to stop the motion is given by the fact that the singularity is placed at the end of the reach for the joint. If the values of the angles are within the specified intervals, then the rehabilitation motion is not interrupted.

The experimental model of the robotic system RAISE is presented in Fig. 7. The robotic system has been subject of two experimental tests. First test was performed without human subjects (dry run) to test the functionality of the safe operation module, where the robot was "pushed" to the limits and the control algorithm monitoring the singularities was validated. The second experimental test was performed using healthy human subjects to check the functionality of the entire robotic system. The test was carried on with the written consent of the participants and involved 5 healthy subjects (1 female, 4 males, mean age 28 years ± 3.098 years, mean weight 73.7 kg ± 14.49 kg, mean height 176.8 cm ± 10.35 cm).

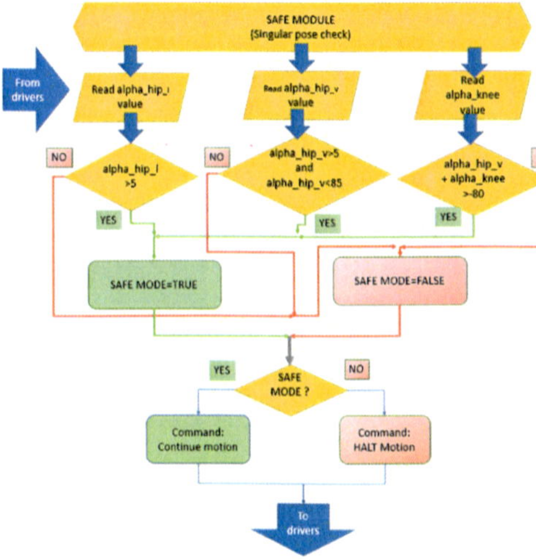


**Fig. 6.** Singularity avoidance control algorithm

The robotic system was able to reach the maximum imposed amplitude for each subject validating the functionality of the system, also the control algorithm for safe operation was validated using the same subjects.

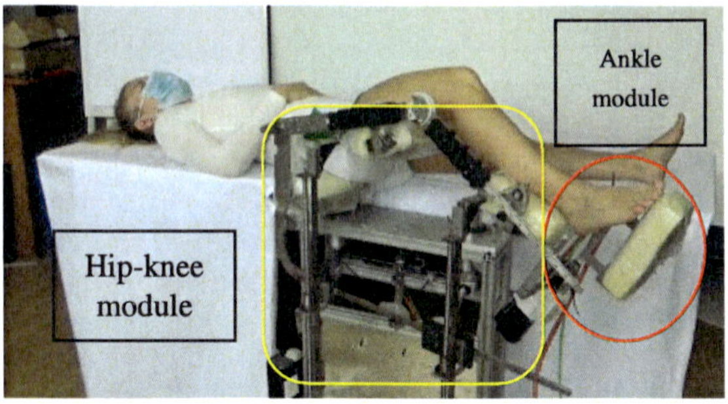


**Fig. 7.** The experimental model of RAISE

## 5 Conclusions

This paper presents a singularity analysis of the parallel robot RAISE for lower limb rehabilitation. The analysis is used to determine the situations where the robot may become unstable for performing the rehabilitation procedure, thus threating the safety of the patient during the procedure. Three cases of singular poses were determined and further avoided using mechanical design and control algorithms. The control algorithm and the functionality of the robotic system were validated using laboratory experimental tests. Future work regarding the RAISE robotic system implies mechanical and control preparation for functional validation using patients.

**Acknowledgement.** This work was supported by a grant from the Romanian Ministry of Research and Innovation, CCCDI—UEFISCDI, project number PN-III-P2-2.1-PED-2021-3430/608PED/2022 (Hope2Walk) within PNCDI III.

## References

1. Kim, Y.W.: Update on stroke rehabilitation in motor impairment. Brain Neurorehabilitation **15**(2), e12 (2022)
2. Jiang, D.W.et al.: Design and research of a lower limb cycling rehabilitation robot. J. Mech. Med. Biol. **22**(6) (2022)
3. Robotics 2020 multi-annual roadmap for robotics in Europe, SPARC Robotics, Eu-Robotics AISBL, The Hauge, The Netherlands (2018)
4. Strategic Research, Innovation and Development Agenda. AI, Data and Robotics Partnership, Third release (2020)
5. European Partnership on Personalised Medicine, Draft proposal (2022)
6. Li, W., Liu, K., Li, C., Sun, Z., Liu, S., Gu, J.: Development and evaluation of a wearable lower limb rehabilitation robot. J. Bionic Eng. **19**(3), 688–699 (2022). https://doi.org/10.1007/s42235-022-00172-6
7. Dong, M., et al.: A lower limb rehabilitation robot with rigid-flexible characteristics and multi-mode exercises. Machines **10**, 918 (2022)
8. Eimmanussakul, T., et al.: A lower limb rehabilitation robot in sitting position with a review of training activities. J. Healthc. Eng. **2018**, 1927807 (2018)
9. Birlescu, I.: Studies regarding the safe operation of innovative medical parallel robots, Doctoral Thesis, Technical University of Cluj-Napoca (2019)
10. He, Y., et al.: Risk management and regulations for lower limb medical exoskeletons: a review. Med Devices (Auckl) **10**, 89–107 (2017)
11. Tucan, P., et al.: risk-based assessment engineering of a parallel robot used in post-stroke upper limb rehabilitation. Sustainability **11**(10), 2893 (2019)
12. Pan, L., et al.: Safety supervisory strategy for an upper-limb rehabilitation robot based on impedance control. Int. J. Adv. Robotic Syst. **10** (2013)
13. Wu, K., et al.: Safety -enhanced model-free visual servoing for continuum tubular robot through singularity avoidance in confined environments. IEEE Access **7**, 21539–21558 (2019)
14. Gherman, B., et al.: Singularities and workspace analysis for a parallel robot for minimally invasive surgery. In: International Conerence. on Automation, Quality and Testing, Robotics (2010)
15. O'Rahilly, R., Muller, F.: Basic Human Anatomy: A Regional Study of Human Structure. Dartmouth Medical School, National Library of Medicine (2004)
16. Pisla, D., et al.: Development of a control system and functional validation of a parallel robot for lower limb rehabilitation. Actuators **10**(10), 277 (2021). https://doi.org/10.3390/act10100277

17. Vaida, C., et al.: RAISE - an innovative parallel robotic system for lower limb rehabilitation. In: Carbone, G., Ceccarelli, M., Pisla, D. (eds.) New Trends in Medical and Service Robotics. MMS, vol. 65, pp. 293–302. Springer, Cham (2019). https://doi.org/10.1007/978-3-030-00329-6_33
18. Vaida, C., et al.: Systematic design of a parallel robotic system for lower limb rehabilitation. IEEE Access **8**, 34522–34537 (2020)

# Blood Serum Recognition Method for Robotic Aliquoting Using Different Versions of the YOLO Neural Network

L. A. Rybak[1(✉)], V. V. Cherkasov[1], D. I. Malyshev[1], and G. Carbone[1,2]

[1] Belgorod State Technological University N.a. V.G. Shukhov, Belgorod, Russia
rlbgtu@gmail.com

[2] Department of Mechanical, Energy and Management Engineering, University of Calabria, 87036 Rende, Italy

**Abstract.** This article discusses vision systems based on neural networks to optimize various processes in robotics. To perform aliquotation tasks, it is necessary to determine the volume of liquid available for sampling from a test tube. This problem can be solved by using vision systems. To determine the coordinates along which the delta manipulator should move, it was proposed to use a vision system based on the YOLO neural network. The algorithm, as well as the structure of the YOLO neural network for the operation of the control system of a robotic complex for aliquoting is considered. Conducting this study allowed us to choose the most appropriate version of the neural network for subsequent work.

**Keywords:** automated aliquoting · blood serum · computer vision · YOLO · machine learning

## 1 Introduction

Aliquoting is the process of manipulating a precisely measured portion of the sample (volume of solution) taken for analysis, which preserves the properties of the main sample. The process of aliquoting of blood of various patients involves the excavation of biological material from a test tube with a divided fraction into small-volume test tubes. To do this, move the test tube from the tripod to the dirty area (working area) where aliquots are dispensed. The quality of sample preparation is ensured at all stages, including aliquoting [1].

Currently, technical vision systems based on neural networks are widely used. The article [2] trains five deep neural networks CNN, VGG19, VGG16, ResNet50, InceptionNet v3 and YOLO v5, for processing chest X-ray data and making a diagnosis of pneumonia, using the RSNA Pneumonia Detection Challenge dataset. The study consists of comparing and evaluating deep learning methods based on convolutional neural networks for identifying pneumonia. According to the tests, VGG16 had the highest recognition accuracy (88%) and AUC-ROC (91.8%), while YOLO v5 was used for localization of inflammation with a confidence level of 99%. The articles [3–5] show


T. Petrič et al. (Eds.): RAAD 2023, MMS 135, pp. 150–157, 2023.
https://doi.org/10.1007/978-3-031-32606-6_18

examples of training convolutional neural networks for recognizing and detecting diseases such as bone and brain cancers and leukemias in images. The results of speech recognition accuracy were also presented. The article [6] describes a method for optimizing the X-ray inspection process for tracking dangerous goods. Using the dataset, several versions of the YOLO deep learning network model were evaluated YOLO and the results were compared with those of the YOLO-T model. The proposed YOLO-T method demonstrated a higher accuracy in detecting prohibited goods. In [7, 8], YOLOv5s-KE and Mask R-CNN neural networks were trained to recognize surface defects in aircraft engine components and blades. The efficiency of speech recognition was compared with other neural networks. In the article [9], it is proposed to use a cascading convolutional neural network for segmentation of organs affected by radiation therapy. The performance of the cascading neural network was compared with the U-Net and FSU neural networks in medical segmentation. The result showed excellent segmentation accuracy, and the two-step neural network reduced the average segmentation time per patient by 48% compared to FSU and by 5% compared to U-Net. In [10], it is proposed to use нейронную сеть the YOLO neural network, which allows solving problems of determining the fraction of blood serum and calculating the volume available for performing aliquoting operations. And an open source network such as the COCO data set is used, as it has been particularly successful in performing similar tasks: object segmentation, object detection, and classification. The use of the YOLO neural network is appropriate due to the higher recognition quality compared to other networks. The advantage of YOLO is its high speed compared to R-CNN and its variations, which work at a speed of 155 frames per second. This makes it effective in solving real-time object detection problems.

## 2 Structure of the Aliquoting Process Management System

A pipette is used to collect blood serum from the test tube. The need to make full use of the available volume of serum, as far as possible, forces the laboratory assistant to carry out manipulations in the area of its lower border. As a result, red blood cells can be drawn into the pipette from the surface of the clot, which will lead to contamination of the sample. In automated aliquoting systems, the determination of the height of the upper and lower boundaries of serum within the test tube, as well as the volume of liquid for excavation using neural networks, is an urgent research task. A robotic system is proposed, which includes a delta robot, with a dosing head installed on it for collecting and pipetting liquid (Fig. 1a) The algorithm of operation of the robotic system together with a neural network is shown in Fig. 1b.

To implement the presented algorithm, we will consider the use of a neural network for image recognition.

## 3 Structure and Method of Training the YOLO Neural Network

For automated aliquoting using a robot, it is necessary to accurately determine the fraction of blood serum to determine the volume available for pipette, so it is important to develop an effective recognition model. Thanks to the latest developments in deep

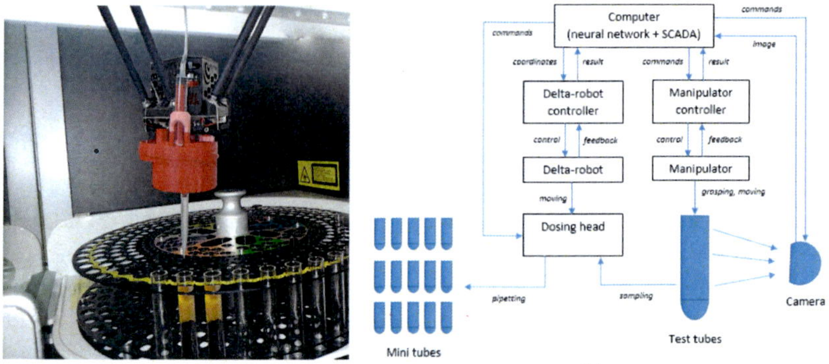


a) b)

**Fig. 1.** a) robotic aliquoting system b) functional diagram of the aliquoting control system.

learning, the YOLO family of models provides scalable tools for classifying objects in images. Taking into account the accuracy and efficiency of object detection, версии YOLOv5 and YOLOv 7 versions were chosen, which consist of three main architectural blocks: Backbone, Neck, and Head, as shown in Fig. 2. YOLO Backbone uses CSPDarknet as the basis for extracting features from images consisting of partial networks between stages. YOLO Neck uses PANet to create a feature pyramid network to aggregate features and pass them to Head for prediction.

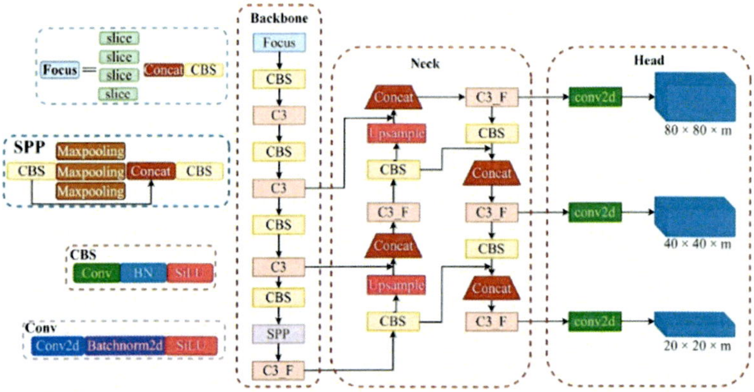


**Fig. 2.** Architecture of the YOLOv5 model.

The combination of FPN and PAN can aggregate parameters of different detection levels from different backbone layers, which significantly enhances he network's ability to combine functions [11]. YOLO Head consists of layers that generate forecasts from anchor fields for object detection. Head includes two parts: the loss function and non-maximal suppression (NMS). In YOLO, the binary cross-entropy loss function is used to calculate classification loss and confidence loss, while the full *IoU* loss function [12]. The neural network is trained on a prepared dataset. The amount of data for training was

257 images, but with the use of augmentation, the number of images for training was 1020. During the processing of the data set, each image was marked up: areas containing the serum fraction in each test tube were highlighted (Fig. 3).

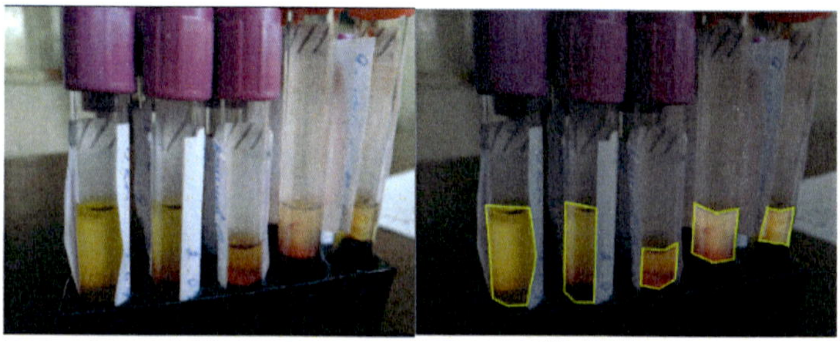

a) b)

**Fig. 3.** Formation of the training sample. a) the original image. b) selection of the serum contour during data set processing.

Two versions of neural networks were trained: YOLOv5 and YOLOv7. Training was conducted on the same data set and over the same number of epochs, which allowed us to objectively compare the results obtained. The training sample size was divided into a training sample and test images in the ratio of 70/30. Training was performed over 100 epochs for both versions of the neural network. To filter multiple bounding boxes that define a single object, the NMS algorithm is used-not maximum suppression. The selection criteria are $IoU$ - intersection over the union $IoU = S/A$, where S is the total area of overlap, and A is the area of the union. $IoU$ is used to determine whether the bounding box was correctly predicted. If there is an exact match, the ratio of bounding rectangles becomes exactly 1, and if there is no overlap, the ratio is 0. After training, the YOLO neural network versions V5 and V7 successfully cope with the task of recognizing blood serum in the image.

## 4 Speech Recognition Results and Comparative Analysis

The Google Colab platform is used as the development environment Google Colab because of its accessibility and ease of writing Python code. To select the most effective version, let's compare the recognition quality indicators.

**Accuracy Level:** Accuracy, also called a positive predictive value, is the proportion of matching instancesв among the extracted instances $precision = \frac{Tp+Tn}{TS}$, where $Tp$ is truly positive, $Tn$ is truly negative, and $Ts$ is the total sample. The calculation results are shown in Fig. 4.

After training, the accuracy of YOLOv5 was 0.998, and the accuracy of YOLOv7 was 0.999. The following indicators were used to analyze speech recognition quality:

Recall effect: a recall, also called sensitivity, which is the proportion of recognized instances of an object in the total number of instances of an object in the image $Recall =$

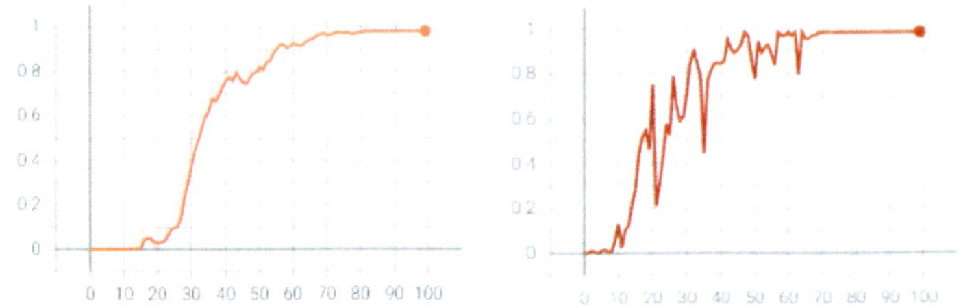


a) b)

**Fig. 4.** Accuracy achieved after training: a) YOLOv5, b) YOLOv7.

$\frac{\sum^{T} p}{\sum^{(} Tp+Fn)}$, where Tp is true positive and Fn is false negative. The calculation results are shown in Fig. 5.

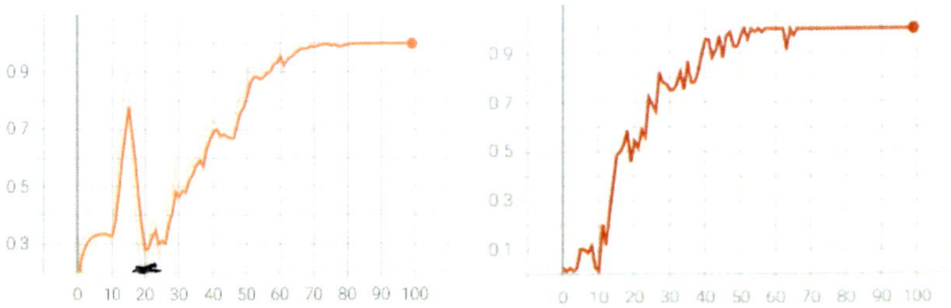


a) b)

**Fig. 5.** Recall rate: a) YOLOv5, b) YOLOv7.

As can be seen from the obtained graphs, therecall efficiency achieved by YOLOv5 and YOLOv7 is 0.978 and 0.989, respectively.

**Average Accuracy mAP_0. 5:** mAP (Average Accuracy) is a metric that is used to evaluate object detection models such as YOLO. Calculating mAP requires the intersection by union, precision, recall, exact recall curve, and mAP. Average accuracy $mAP = \frac{\sum_{q-1}^{Q} AP(q)}{Q.}$. The calculation results are shown in Fig. 6.

The mAP_0.5 accuracy index for YOLOv5 and YOLOv7 is 0.99.

**Average Accuracy mAP_0.5: 0.95.95:** mAP_0.5: 0.95 means the average accuracy map for various *IoU* thresholds ranging from 0.5 to 0.95 in increments of 0.05. Thus, the average accuracy of blood serum recognition for YOLOv5 was 63%, and for YOLOv7, this indicator is 75%. The results are shown in Fig. 7.

If the predicted block during recognition has a significant overlap with the main block of the true location of the object, then box_lossthe box_loss regression loss in the bounding box (root-mean-square error) is calculated to see how efficiently the model is able to identify objects. Figure 8 shows the box_loss graphs box_loss obtained during training 100 iterations for версий the YOLOv5 and YOLOv7 versions.

The mean square error based on the training results was 0.03 for YOLOv5, and 0.018 for YOLOv7.

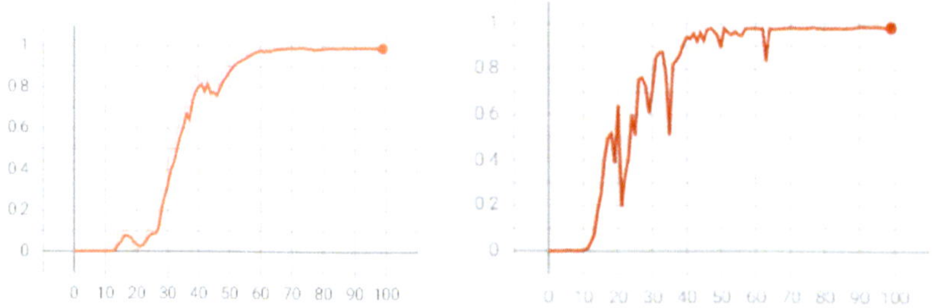


a) b)

**Fig. 6.** mAP_0. 5 accuracy indicator for versions: a) YOLOv5, b) YOLOv7.

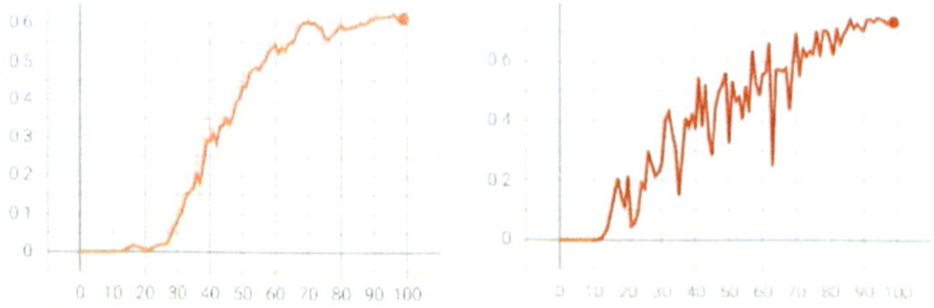


a) b)

**Fig. 7.** Map of average accuracy mAP_0. 5:0.95 after 100 epochs: a) YOLOv5, b) YOLOv7.

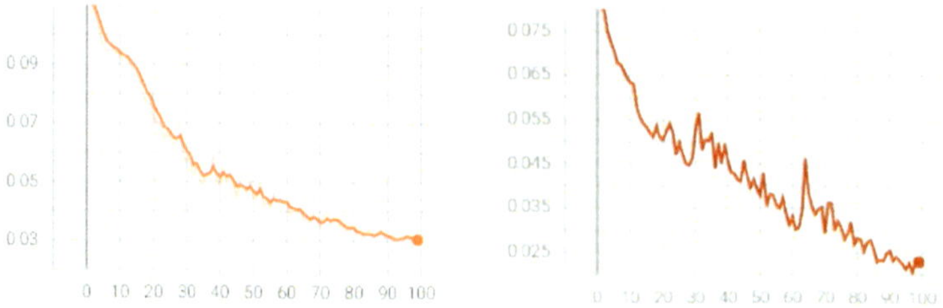


a) b)

**Fig. 8.** Training mean square error graph: a) YOLOv5, b) YOLOv7.

## 5 Algorithm for Determining the Levels of Media Boundaries

The purpose of the conducted determination of the serum fraction in the image is to determine the volume available for aliquoting.

The calculation of this volume and the corresponding depth of immersion of the pipette is performed in accordance with the following algorithm:

1. Calculates the height $h$ (in pixels).
2. If we assume 1 pixel to be 0.26 mm, then the calculated height will be $m = 0.26 \cdot h$.
3. The volume is calculated by the formula $V = m \cdot \pi \cdot r^2$, where r is the known test tube radius of 13 mm.

4. The number of aliquots $n = \lfloor V/V_0 \rfloor$ is determined (Fig. 9).
   where $V_0$ is the aliquot volume, and square brackets indicate taking the integer part of the number (rounding down to the nearest integer).
5. If $n < 1$, automated aliquoting is not allowed, otherwise, aliquoting is performed taking into account the number $n$ of aliquots.

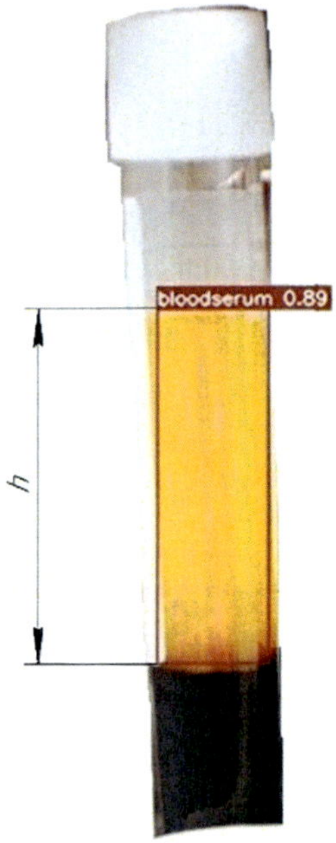


**Fig. 9.** Determination of serum volume and pipette immersion depth.

## 6 Conclusion

In conclusion, we can note that the considered algorithms YOLOv5 and YOLOv7 were trained on the same data set, which allowed us to compare the effectiveness of their work. The parameters of average accuracy, root-mean-square error, and loss of objectivity were also determined, which allowed us to conclude that the YOLOv7 algorithm is the most suitable for the task of determining the blood serum fraction. This study allowed us to select the most appropriate version of the neural network for subsequent work. To perform aliquoting tasks, it is necessary to determine the volume of liquid available for sampling from the test tube. This problem can be solved by image segmentation. The result of image segmentation is the contour of the liquid available for pipetting, which can be used to determine the actual volume.

**Acknowledgements.** This work was supported by the state assignment of Ministry of Science and Higher Education of the Russian Federation under Grant FZWN-2020-0017.

## References

1. Malm, J., et al.: Developments in biobanking workflow standardization providing sample integrity and stability. J. Proteomics **95**, 38–45 (2013)
2. Chiwariro, R., Julius, B.: Comparative analysis of deep learning convolutional neural networks based on transfer learning for pneumonia detection. Int. J. Res. Appl. Sci. Eng. Technol. **11**(1), 1161–1170 (2023)
3. Zimbalist, T., Rosen, R., Peri-Hanania, K., Caspi, Y., Rinott, B., Zeltser-Dekel, C., Bercovich, E., Eldar, Y., Bagon, S.: Detecting Bone Lesions in X-Ray Under Diverse Acquisition Conditions (2022)
4. Abass, S.: Diagnosing the leukemia using faster region based convolutional neural network. J. Appl. Sci. Technol. Trends **3**, 35–38 (2022)
5. Zine-Dine, I., Riffi, J., Fazazy, K., Mahraz, M., Tairi, H.: Brain Tumor Classification using Machine and Transfer Learning (2022)
6. Wang, M., et al.: YOLO-T: multitarget intelligent recognition method for x-ray images based on the YOLO and transformer models. Appl. Sci. **12**, 11848 (2022)
7. Li, X., Wang, C., Ju, H., Li, Z.: Surface defect detection model for aero-engine components based on improved YOLOv5. Appl. Sci. **12**, 7235 (2022)
8. Shang, H., Sun, C., Liu, J., Chen, X., Yan, R.: Deep learning-based borescope image processing for aero-engine blade in-situ damage detection. Aerosp. Sci. Technol. **123**, 107473 (2022)
9. Men, K., et al.: More accurate and efficient segmentation of organs-at-risk in radiotherapy with Convolutional Neural Networks Cascades. Medical Physics **46**, 286–292 (2018)
10. Woo, S., Park, J., Lee, J.-Y., Kweon, I.: CBAM: Convolutional block attention module. In: European Conference on Computer Vision, pp. 3–19. Munich, Germany (2018)
11. Bochkovskiy, A., Wang, C.-Y., Liao, M.: YOLOv4: Optimal speed and accuracy of object detection (2020)
12. Jiang, P., Ergu, D., Liu, F., Ying, C., Ma, B.: A review of YOLO algorithm developments. Procedia Comput. Sci. **199**, 1066–1073 (2022)

# Mobile Robotics and Localisation

# Scenario Concatenation for Vehicle Testing by Solving a Vehicle Routing Problem with Skill and Temporal Synchronization Constraints

Robert Fina[1(✉)], Johannes Wenninger[2], Daniel Reischl[2], Hubert Gattringer[1], Andreas Müller[1], Thomas Schlömicher[3], and Heiko Scharke[3]

[1] Institute of Robotics, Johannes Kepler University Linz, Linz, Austria
{robert.fina,hubert.gattringer,a.mueller}@jku.at
[2] Linz Center of Mechatronics, Linz, Austria
{johannes.wenninger,daniel.reischl}@lcm.at
[3] AVL List GmbH, Graz, Austria
{thomas.schloemicher,heiko.scharke}@avl.com
http://robotik.jku.at , http://www.lcm.at , http://www.avl.com

**Abstract.** In the automotive industry new technology has to be tested thoroughly before a market launch. In addition, many regulations hinder a short time-to-market, as well as the manually operated proving grounds (PG). Therefore, a toolchain is introduced to be able to automate and to increase the efficiency of test scheduling. Three steps are proposed. In the first step, common standards, which represent the road network and the test workflow, are converted into a graph representation. The second step combines a customer request and the collected data to a suitable base for a Vehicle Routing Problem with temporal synchronisation and skill constraints (VRPSS) for the test participants. In the third step, the VRPSS is modelled and solved by Metaheuristic methods. It is shown, that the proposed toolchain can automate the scenario concatenation for PGs.

**Keywords:** Heterogeneous fleet scheduling · Vehicle Routing Problem · Scenario concatenation · Proving ground automation

## 1 Introduction

The market roll-out of a novel automotive technology, e.g. functions for autonomous driving (AD), requires thorough testing and validation. For such technology, where there is a high risk of injury, simulation results are not sufficient [18], which is why regulatory requirements call for thorough testing in a realistic environment – a proving ground (PG) (e.g. General Motors PG in Michigan, Applus IDIADAs' PG in China and Spain, ZalaZONE in Hungary).


T. Petrič et al. (Eds.): RAAD 2023, MMS 135, pp. 161–169, 2023.
https://doi.org/10.1007/978-3-031-32606-6_19

Planning and conducting all regulatory tests at a PG in the shortest possible time is a challenge, mainly because it is still done manually. In addition, the urge to minimise time-to-market and the need to increase the variety of tests due to the high level of technology, precludes the possibility of manual processing if efficient and time-saving use of the PG is to be ensured.

Most tests involve other vehicles such as bicycles and other road users in addition to the vehicle-under-test (VUT). These entities consist of a mobile platform called a carrier and a corresponding crash dummy. The heterogeneity of the vehicle fleet introduces a high complexity for planning the test series, as the carrier batteries need to be recharged, scenarios need different dummies, etc.

Therefore, a toolchain is introduced that determines a time-efficient scenario concatenation at a PG. For a proper description of the PG and tests, open and common standards are used. *OpenDrive* [1] and *OpenScenario* [2] describe the vehicle's environment, such as the geometry of roads, as well as the path of each vehicle, and the workflow of scenario can be described, respectively.

The problem of creating a time-efficient test plan is modelled as a Vehicle Routing Problem with skill and temporal synchronisation constraints (VRPSS). Due to the nature of the VRP, the problem is NP-hard and the time sensitivity of the scheduling task, (Meta-)heuristics are used for solving [13]. The toolchain is implemented using Python 3.10 and C++ 17.

## 2 State-of-the-Art

Beringhoff et al. conducted interviews with experts around the world and observed 31 challenges for testing automated/autonomous vehicles [3]. Two of those are addressed in this work. The lack of test automation and the concatenation of scenarios, which is only scarcely dealt with and not covered in any literature, respectively.

A graph representation of a route network, especially a PG, comes natural and simplifies the definition of the VRPSS [5]. Thus, the toolchain uses a complete graph, whose edges represent the shortest routes between test locations and the vertices represent test locations.

In the specific application of vehicle testing, not all locations at a PG need to be visited - unlike the classic VRP definition. This means that the necessary places for the required tests have to be found first. This can be done with the help of comparative algorithms.

One of the more unconventional trajectory/path comparison algorithm is *Partial Curve Mapping* (PCM). It allows the detection of a suitable test location as a partial element of a road - or the road itself [19].

The classic Vehicle Routing Problem (VRP) can be described as the problem to visit multiple locations once as quickly as possible, starting and ending on the same position, called depot [13]. A large variety of VRPs are investigated in the last decades in research, including VRP with time window constraints, heterogeneous fleet VRP, VRP with pickup-and-delivery, and others [7,14].

A comprehensive overview on methods to solve a VRP can be found in [13,15], ranging from (Mixed) Integer Linear Programming models, using Metaheuristic methods to AI based methods. Whereby the majority of researchers have chosen single-solution based Metaheuristic methods to solve VRPs [7], since it is NP-hard. These methods start with an initial solution which is modified step-by-step – traversing a trajectory in the solution space until (close to) optimality [9]. *Local Search* (LS), *Guided Local Search* (GLS), and *Large Neighbourhood Search* (LNS) are examples for single-solution based Metaheuristics [9]. The individual methods are not presented in more detail, since it would exceed the scope of this paper.

Scenarios and a heterogeneous fleet introduces two additional challenges to the classic VRP. Firstly, a synchronisation of many vehicles at a specific location at a specific time is necessary - called temporal synchronisation [6]. Secondly, vehicles with specific attributes, further called skill, are required for a test. Bredström et al. [4] present a branch-and-price algorithm for the routing and scheduling with synchronisation, but only focus on pairwise synchronisation. Further work is done by Parragh et al. [16] in the field of service technician routing using adaptive LNS to solve the problem. In the home-healthcare sector similar challenges arise for scheduling, which are addressed by Polnik et al. [17] with a multistage approach using GLS, LNS and LS. Both have a limitation in their model only 10% of all jobs are pairwise synchronised. But, the introduced test scheduling does need a 100% temporal synchronisation of all vehicles, which is investigated in the following sections.

## 3 Toolchain

A customer can select the necessary scenarios from a collection of predefined scenarios and road networks, available as *OpenScenario* and *-drive* files, '*.osc' and '*.odr', respectively. The customer request set $\mathcal{R} = \{\langle S, \hat{m}, \mathbf{w}\rangle_i \mid i \in \{1, \dots, L\}\}$ stores $L \in \mathbb{N}$ scenarios $S_i$, which has to be performed $\hat{m}_i \in \mathbb{N}$ times. The skills required for the participants of a scenario are stored in $\mathbf{w}_i = \left(w_1 \dots w_{\hat{n}}\right)^T$, with $w \in \mathcal{W}$. In this paper, the skill set $\mathcal{W} = \{C, P, B\}$ includes the skills: car $C$, pedestrian $P$ and bicycle $B$. All scenarios and PGs are known in advance and change rarely. Therefore, the data can be processed, stored and used later on for any VRP, see Fig. 1.

The procedure of the toolchain is divided into three steps.

- Step 1: All information of the PGs and scenarios is extracted and stored in a global database. This is done with the help of the simulator *esmini*[1]. In addition, all possible scenarios at each graph vertex are determined by the scenario location matching (SLM).
- Step 2: A search for a suitable area at the PG is performed (Scenario location optimisation - SLO) and the final form of a suitable graph is generated.
- Step 3: The VRPSS model is formulated and the VRPSS is solved.

A detailed description of step 1 and step 2 can be found in [8].

[1] https://github.com/esmini/esmini.

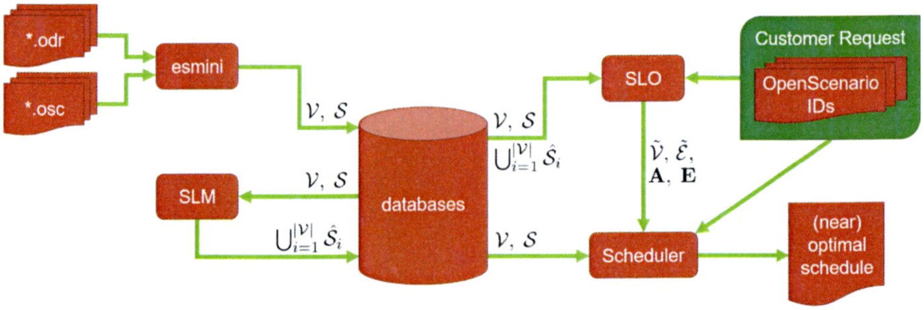


**Fig. 1.** Structure of the proposed toolchain and its environment.

### 3.1 Step 1: Data Representation

In this step, a PG description is transformed into a graph representation $P = \langle \mathcal{V}, \mathcal{E}, d \rangle$. All roads are part of the set $\mathcal{V} = \{v_i \mid v_i = \langle \sigma_i, p_i(\cdot) \rangle, \; i \in \{1, \ldots, N\}\}$, $N \in \mathbb{N}$, where $\sigma_i$ represents the arc length of the reference line of the road and the road geometry is described by $p_i(\cdot) : \mathbb{R} \rightarrow \mathbb{R}^2$.

The set of edges $\mathcal{E} = \{e_{ij} \mid e_{ij} = (v_i, v_j) = \langle \xi, p(\cdot) \rangle, v_i, v_j \in \mathcal{V}, i, j \in \{1, \ldots, N\}, i \neq j\}$, $N \in \mathbb{N}$ represents the direct connections between roads. The function $d : \mathcal{E} \rightarrow \mathbb{R}$, $d(e_{ij}) = t_{min}$ determines a cost for each edge $e \in \mathcal{E}$. $t_{min}$ is the minimal time in minutes which is needed to traverse $\sigma_{min}$ - the minimal distance between two vertices. The shortest path leading to $\sigma_{min}$ is determined with the *Dijkstra* algorithm by the framework *NetworkX*. In this step, an unweighted and undirected graph is generated and the python package *NetworkX* [11] is used for the graph representation.

In order to schedule a set of tests $\mathcal{S} = \{S_1 \cdots S_M\}$, $M \in \mathbb{N}$, which are rigorously defined by organisations such as EURO NCAP, a unique location at the PG must be found for each scenario. For example a Car-to-Car Rear braking test which requires straight roads. The locations are determined by the Scenario Location Matching (SLM) algorithm, which compares the road geometry with the geometric path of the VUT using PCM. The comparison of geometric paths is not constrained to individual vertices only, but can also include adjacent vertices.

PCM is provided by the Python package *similaritymeasures* [12]. The SLM determines a set of possible scenarios $\hat{\mathcal{S}}$ for each vertex, resulting in the full set of possible scenarios at the PG $\tilde{\mathcal{S}} = \bigcup_{i=1}^{|\mathcal{V}|} \hat{\mathcal{S}}_i$ at the end of this step.

### 3.2 Step 2: Optimal Allocation Graph

The next procedure is an optimisation, called Scenario Location Optimiser (SLO), which allocates each scenario to one location only. The objective is to minimise the sum of time needed to visit all locations, see Eq. (1), but each requested scenario has to be allocated to a location (2) and a scenario can be located on exactly one vertex, or on many vertices (3).

The cost matrix $\mathbf{C} \in \mathbb{R}^{n \times n}$, $C_{ij} = d(e_{ij})$, with $n = |\mathcal{V}|$, is determined beforehand. Thereby, direct neighbour costs are $C_{ij} = 0$, or else $C_{ij} = d(e_{ij})$.

The elements of the matrix $\mathbf{A} \in \mathbb{B}^{m \times n}$, with $n = |\mathcal{V}|$ and $m = |\mathcal{R}|$, are the decision variables to determine which scenarios are allocated to which vertices.

To be able to define a simple objective the matrix $\mathbf{E} \in \mathbb{B}^{n \times n}$ is introduced. Its form is an upper triangle matrix, with zeros in its diagonal and holds the information which edge is used by a vehicle. Therefore, two more constraints have to be introduced. A constraint that guarantees that an edge is used between two selected scenario-vertex-allocations (4). The second constraint (5) ensures that not more scenarios are visited than requested in $\mathcal{R}$.

$$\min_{\mathbf{A},\mathbf{E}} \sum \sum C_{ij} E_{ij} \quad (1)$$

$$\text{s.t.} \sum_{j=1}^{n} A_{ij} > 0 \quad (2)$$

$$\sum_{i=1}^{m} A_{ij} \leq 1 \quad (3)$$

$$E_{ij} \leq \frac{\sum_{k=1}^{m} A_{ki} + \sum_{k=1}^{m} A_{kj}}{2} \quad (4)$$

$$\sum_{\forall i,j} E_{ij} = \frac{m(m-1)}{2} \quad (5)$$

$$\text{with } m = \sum_{i=1}^{|\mathcal{R}|} \hat{m}_i,\ \hat{m}_i \in \mathcal{R} \quad (6)$$

The result is used to build a new graph $\tilde{P} = \langle \tilde{\mathcal{V}}, \tilde{\mathcal{E}}, \tilde{\mathcal{S}}, d \rangle$. The new set of vertices $\tilde{\mathcal{V}} = \{v_i \mid i \in \{1, \ldots, |\mathcal{V}| + \sum_{\forall S \in \mathcal{R}} \hat{n}_i + \hat{m}_i\}$ contains all original vertices $v_i \in \mathcal{V}$ and the vertices multiplied for each participant in all scenarios in $\tilde{\mathcal{S}}$ $(\hat{n}_i)$, as well as all repetitions of all scenarios $(\hat{m}_i)$. If a vertex is not a member of the new set $v_i \in \mathcal{V} \setminus \tilde{\mathcal{V}}$, it is degraded to an edge with a certain weight. However, no connection between vertices may be lost in the process.

### 3.3 Step 3 - Model of VRPSS and Strategy to Solve the VRPSS

In order to solve the test-scheduling problem, a Mixed Integer Linear Problem (MILP) is proposed. The routing is required to be time-optimal, meaning that the sum of the time to travel shall be minimal (7). Thereby, the decision variables $x_{ijk} \in \mathbb{B}$ define if an edge $e_{ij} \in \tilde{\mathcal{E}}$ is traversed (=1) or not (=0) by the vehicle $a_k$. The classic constraints of a VRP are covered by (12) and (13).

The defined participant requirements $w_i \in \mathcal{W}$ of a scenario $S_i \in \mathcal{R}$ and the actual fleet of vehicles $\mathcal{F} = \{a_i = \langle w_i \rangle \mid i \in \{0, \ldots, K\}\}$, with $w_i \in \mathcal{W}$, introduce the skill matching constraints (10) to the classic VRP. The elements of the matrix $\mathbf{N} \in \mathbb{B}^{M \times K}$ are decision variable and define which vehicle participates in which scenario. The elements are used to model three constraints. First, the skill matching constraint (10). Second, the constraint (8) which guaranties that the VUT participates in all scenarios and (9) to fulfil the carrier requirements of a scenario.

The temporal synchronisation (11) requires additional definitions. Firstly, the set $\mathcal{T}_k = \{(v_i, v_j) \mid v_i, v_j \in \tilde{\mathcal{V}}\} \subset \tilde{\mathcal{V}} \times \tilde{\mathcal{V}}, k \in \{1, \ldots, |\mathcal{R}|\}$, which contains all vertices allocated to the scenario $S_k$, is proposed. Therefore, all pairs in $\mathcal{T}$ have to be considered for a temporal synchronisation.

Secondly, a positive travel time $t_{ik}$ which is the sum of all preceding travel times and scenario execution times of the vehicle $a_k$ at the vertex $v_i$ and a positive waiting time $\alpha_i$ for each vertex $v_i$ is introduced to soften the hard synchronisation constraint. It allows that a vehicle can wait for others at a vertex. The MILP is defined as

$$\min_{x_{ijk}, \mathbf{N}, \alpha_j} \sum_{k=1}^{|\mathcal{F}|} \left( \sum_{i=1}^{|\tilde{\mathcal{V}}|} \sum_{j=1}^{|\tilde{\mathcal{V}}|} x_{ijk} d(e_{ij}) + \alpha_j \right) \tag{7}$$

$$\text{s.t.} \quad \sum_{i=1}^{|\mathcal{R}|} N_{i0} = |\mathcal{R}| \tag{8}$$

$$\sum_{j=0}^{|\mathcal{R}|} N_{ij} = \hat{m}_i \tag{9}$$

$$N_{ij} w_j = N_{ij} w_i \quad w_j \in a_j, \, w_i \in S_i \tag{10}$$

$$\sum_{k \in K} t_{ik} + \alpha_i = \sum_{k \in K} t_{jk} + \alpha_j, \quad \forall (v_i, v_j) \in \mathcal{T}_l, \, l \in \{1, \ldots, |\mathcal{R}|\} \tag{11}$$

$$\sum_{k=1}^{|\mathcal{F}|} \sum_{i=1}^{|\tilde{\mathcal{V}}|} x_{ijk} = 1 \tag{12}$$

$$\sum_{k=1}^{|\mathcal{F}|} \sum_{j=1}^{|\tilde{\mathcal{V}}|} x_{ijk} = 1 \tag{13}$$

The problem is formulated using *Google Operations Research Tools* (OR-Tools) [10] in C++ 17. It is an open source software suite for solving combinatorial optimisation problems.

## 4 Toolchain Example

In this section, the set-up of the graph is depicted based on a simple case study. In this example, six vertices and six edges with weight zero represent a PG, as can be seen in Fig. 2(A). Using the known specifics of the PG, all locations are identified where individual tests can be executed, and allocated to unique vertices, marked in green in Fig. 2(B).

The allocation indicates, that the vertices 1, 3, 5, and 6 are the optimal locations, whereas vertices 2 and 4 can be degraded to edges. The customer request includes the scenarios 1 and 4 once and the others $\hat{m}_2$ and $\hat{m}_3$ times. The final graph, with some edges with weights not equal to zero, is shown in Fig. 3.

## 5 Results

The test sets which are discussed in this section include a symmetric distance matrix, and five vehicles including the VUT, two cars, one pedestrian, and one

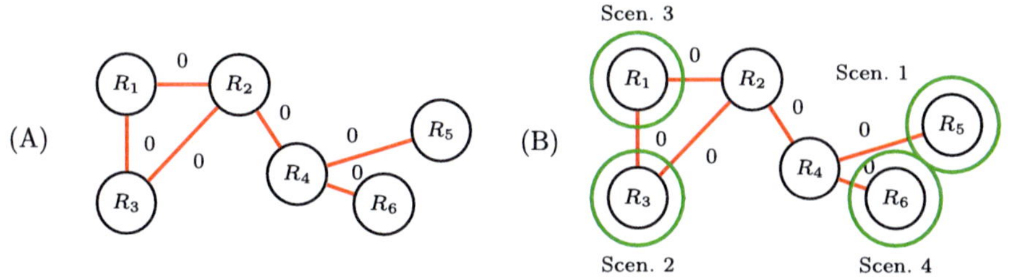


**Fig. 2.** (A) Undirected, weighted graph constructed with an *OpenDrive* description of a road network. (B) Unique assignment of scenarios to locations on the PG.

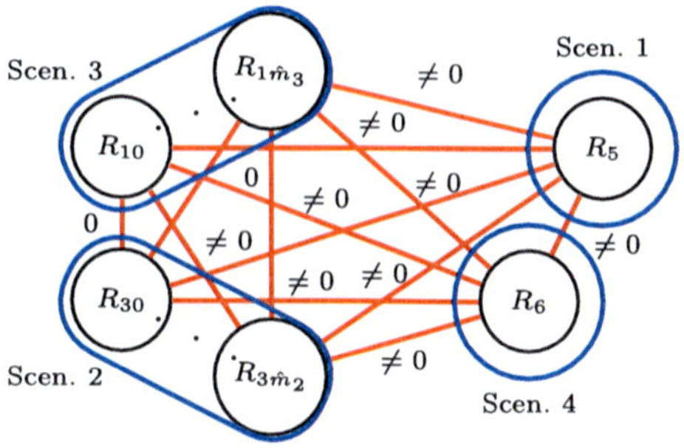


**Fig. 3.** Graph $\tilde{P}$ including the original and the duplicated vertices for the VRP.

bicycle. The first test set covers four scenarios which include the following skill requirements: VUT + car, VUT + pedestrian, VUT + bicycle, VUT + car + pedestrian + bicycle. These scenarios lead to a complete graph with 10 vertices. The second and third test sets contain the first test set multiplied two times and three times, respectively.

The Google OR-Tools settings are as follows. The first solution strategy is the *path cheapest arc search*, which generates an initial route by iteratively adding the cheapest arc. As for the Metaheuristics all methods as of 2022 are used and compared. That includes Greedy Descent (GD), Simulated Annealing (SA), Tabu Search (TS), Generic Tabu Search (GTB), and GLS. The individual methods are not presented in more detail, since it would exceed the scope of this work.

The maximum computational time is restricted for all methods to 300 s. The calculations are done on a PC with an Intel® Core$^{TM}$ i5-9500 (3 GHz) CPU and 16 GB RAM (DDR4). The results are listed in Table 1.

All Metaheuristics require the entire allowed computation time, except GD due to its greediness. Therefore, GD gives the worst results, and as the number of vertices increases - the complexity of the problem - the travel time gets worse. In addition, GD is one of three (SA, GTB) methods which do not include the fourth vehicle to determine a better solution of the more complex test set.

The results also show that the max. waiting time is always related to a carrier and not the VUT, in each solution. Moreover, a carrier is always the last one in

**Table 1.** List of all test sets and the results given by different Metaheuristic implementations of the Google OR-Tools.

| Scenario Count | Vertex Count | Method | Computation Time | No. of carriers | VUT time | Max. waiting time |
|---|---|---|---|---|---|---|
| 4 | 10 | GD | 5 ms | 3 | 25 min | 12 min |
| 4 | 10 | GLS | 300 s | 3 | 25 min | 12 min |
| 4 | 10 | SA | 300 s | 3 | 25 min | 12 min |
| 4 | 10 | TS | 300 s | 3 | 25 min | 12 min |
| 4 | 10 | GTB | 300 s | 3 | 25 min | 12 min |
| 8 | 20 | GD | 66 ms | 3 | 46 min | 12 min |
| 8 | 20 | GLS | 300 s | 3 | 44 min | 9 min |
| 8 | 20 | SA | 300 s | 3 | 46 min | 12 min |
| 8 | 20 | TS | 300 s | 3 | 44 min | 9 min |
| 8 | 20 | GTB | 300 s | 3 | 44 min | 9 min |
| 12 | 30 | GD | 5.779 s | 3 | 95 min | 49 min |
| 12 | 30 | GLS | 300 s | 4 | 85 min | 30 min |
| 12 | 30 | SA | 300 s | 3 | 93 min | 34 min |
| 12 | 30 | TS | 300 s | 4 | 85 min | 30 min |
| 12 | 30 | GTB | 300 s | 3 | 95 min | 49 min |

the depot at the end of the tests. This behaviour is not included in the model of the VRPSS, but should be emphasised and investigated in future work.

Overall, the GLS and TS give the best solutions for all test sets in terms of max. waiting time, VUT travel time, overall travel time and the amount of carriers used. Hereby it is shown that the implementations of these methods are sufficiently efficient for small test sets. However, a more intensive investigation needs to be carried out - with several VUTs and a higher number of vertices - to make a complete statement about the usability of the Google OR-Tools for VRPSS.

## 6 Conclusions and Outlook

A toolchain to determine a time-efficient scenario concatenation is introduced - in terms of planning and execution time. In three steps data is arranged according to a customer request and a VRPSS is solved. The data is extracted from descriptions of a PG and a variety of scenarios defined in open standards. A variety of Metaheuristics, accessible by Google OR-Tools, is investigated to solve the VRPSS; the most promising are GLS and TS. This result requires further investigation in more detail for larger numbers of scenarios and a larger fleet size.

However, the here applied approach still lacks some important model extensions, like carrier recharging, or road capacities. The problem can also be formulated as a job shop scheduling model – beneficial when including more carriers and VUTs. This needs to be investigated in further studies.

**Acknowledgement.** This work has been supported by the "LCM - K2 Center for Symbiotic Mechatronics" within the framework of the Austrian COMET-K2 program.

## References

1. ASAM OpenDrive (2022), v1.7.0, www.asam.net
2. ASAM OpenScenario (2022), v1.1.1, www.asam.net
3. Beringhoff, F., Greenyer, J., Roesener, C., Tichy, M.: Thirty-one challenges in testing automated vehicles: interviews with experts from industry and research. In: Proceedings of the IEEE Intelligent Vehicles Symposium (2022)
4. Bredström, D., Rönnqvist, M.: A branch and price algorithm for the combined vehicle routing and scheduling problem with synchronization constraints. NHH Department of Finance & Management Science Discussion Paper (2007)
5. Deo, N.: Graph Theory with Applications to Engineering and Computer Science. Courier Dover Publications, New York (2017)
6. Drexl, M.: Synchronization in vehicle routing-a survey of VRPs with multiple synchronization constraints. Transp. Sci. (2012)
7. Elshaer, R., Awad, H.: A taxonomic review of metaheuristic algorithms for solving the vehicle routing problem and its variants. Comput. Ind. Eng. (2020)
8. Fina, R., et al.: A toolchain definition to solve a vehicle routing problem for a time-efficient testing on a proving ground. In: Proceedings of the Austrian Robotics Workshop (2023)
9. Gendreau, M., Potvin, J.Y., et al.: Handbook of Metaheuristics. Springer, Heidelberg (2010)
10. Google OR-Tools (2022), https://developers.google.com/optimization, v9.5.2237
11. Hagberg, A.A., Schult, D.A., Swart, P.J.: Exploring network structure, dynamics, and function using NetworkX. In: Proceedings of the 7th Python in Science Conference (2008)
12. Jekel, C.F., Venter, G., Venter, M.P., Stander, N., Haftka, R.T.: Similarity measures for identifying material parameters from hysteresis loops using inverse analysis. Int. J. Mater. Forming (2019)
13. Koç, Ç., Bektaş, T., Jabali, O., Laporte, G.: Thirty years of heterogeneous vehicle routing. Eur. J. Oper. Res. (2016)
14. Lahyani, R., Khemakhem, M., Semet, F.: Rich vehicle routing problems: from a taxonomy to a definition. Eur. J. Oper. Res. (2015)
15. Nazari, M., Oroojlooy, A., Snyder, L., Takác, M.: Reinforcement learning for solving the vehicle routing problem. Adv. Neural Inf. Process. Syst. (2018)
16. Parragh, S.N., Doerner, K.F.: Solving routing problems with pairwise synchronization constraints. Cent. Eur. J. Oper. Res. (2018)
17. Polnik, M., Riccardi, A., Akartunalı, K.: A multistage optimisation algorithm for the large vehicle routing problem with time windows and synchronised visits. J. Oper. Res. Soc. (2021)
18. Wang, F.Y., Wang, X., Li, L., Mirchandani, P.: Creating a digital-vehicle proving ground. IEEE Intell. Syst. (2003)
19. Witowski, K., Stander, N.: Parameter identification of hysteretic models using partial curve mapping. In: 14th ISSMO Multidisciplinary Analysis and Optimization Conference (2012)

# Gaining Insights into a Robot Localization Monitor Using Explainable Artificial Intelligence

Matthias Eder(✉), Laurent Frering, and Gerald Steinbauer-Wagner

Graz University of Technology, 8010 Graz, Austria
{matthias.eder,laurent.frering,steinbauer}@ist.tugraz.at

**Abstract.** Monitoring the state of a localization component in robotic systems has received increasing attention in recent years, as navigation behaviors of robots rely on a reliable pose estimation to a large extend. Nowadays, research focuses on the development of new approaches to monitor the localization state of a robot. Many of those approaches use Machine Learning techniques which do not provide direct insight into the decision making process and are thus often handled as a black box. In this work, we aim to open this black box by making use of an Explainable Artificial Intelligence (XAI) framework that allows us to improve the understanding of a machine learning based localization monitor. To gain insights into the machine learning model, we make use of the open-source framework SHapley Additive exPlanations (SHAP). Results show that investigations in the model structure of a localization monitor using XAI helps to improve the model's transparency. Overall, XAI proves to be useful in understanding the decision-making process of a localization monitor and can even help to improve the model's design quality.

**Keywords:** Explainable Artificial Intelligence · XAI · Localization Monitoring · Machine Learning · SHAP

## 1 Introduction

Robot localization is a fundamental capability of all intelligent mobile robots, aiming to determine the pose of a robot in space. There is a huge corpus of research that has been conducted to allow reliable estimation of the current pose using uncertain sensor data and action execution [8]. Because of uncertainties in sensor data and environmental factors such as people flocking around robots there is always the risk that the error of the pose estimation increases significantly or that it diverges completely without being noticed by the robot. Usually, navigation and other behaviors of robots rely on a valid pose estimation. Thus,

This work was partially supported by the Austrian Research Promotion Agency (FFG) with the project Easier.


T. Petrič et al. (Eds.): RAAD 2023, MMS 135, pp. 170–177, 2023.
https://doi.org/10.1007/978-3-031-32606-6_20

for performance as well as dependability of robot systems it is of great interest for the system to know the state of its localization component [4]. Moreover, if a robot gets delocalized near an object this can result in undesirable behaviors, such as a collision. To avoid this, it is aimed to detect the delocalization before it occurs to be able to perform appropriate counter measures. Such an event can be detected by a localization monitor which observes the behavior of the used localization technique and assesses if the localization is still correct [5]. Localization monitors are typically based on machine learning (ML) methods and trained on localization data to predict a delocalization. However, such methods are typically hard to understand, lack interpretability and are therefore often seen as a black-box [1]. The field of Explainable Artificial Intelligence (XAI) aims to open the black box of Machine Learning and to understand its internal decisions better [1]. This work extends the work from [5] by investigating the decisions coming from a ML-based localization monitor using SHapley Additive exPlanations (SHAP) [7] with the aim to understand the decision making process. SHAP is an approach arising from game theory to explain the trained output of a ML model by allocating credit to the model input [7] using Shapley values. A Shapley value is defined to be the average expected marginal contribution of an input after all possible combinations have been considered [11]. The localization monitor is developed as proposed in [5] which observes the behavior of a particle filter used for localization [15]. In this method, the estimated pose and features arising from the output of the particle filter are extracted and recorded together with a ground truth in a simulation environment. Due to the divergence of the estimated and real robot pose, the localization state of a robot can be calculated and used as training target for a ML model. As input, the extracted features from the particle filter are used. While [5] uses AdaBoost to predict the robot's localization state, we use a different boosting technique, namely XGBoost [3] as it has been proven to a be powerful model for interpretation with SHAP [6]. By extracting Shapley values from the trained XGBoost, explanations on how a decision is made and how various input features influence the decision can be generated. These explanations are then used to identify the most influential features for a further refinement of the localization monitor.

The remainder of this paper is as follows. Section 2 briefly discusses related research in the field of localization monitoring and XAI. Section 3 presents the used localization monitor in more detail which is investigated using SHAP in Sect. 4. Section 5 concludes the paper.

## 2 Related Research

### 2.1 Robot Localization Monitoring

The authors of [5] present a localization monitor which uses deep learning to identify temporal patterns from the particle filter in the case of losing/lost localization. These patterns are then combined with extracted features from the particle set and sensor perception to boost a localization state estimator. While the results from this work are promising, the extracted features were not thoroughly analyzed and neither was the boosted localization estimator. In [17], the authors present an optimal two-stage volume Kalman filter to estimate the state of IMU measurements and detect faults. They show that they can obtain an optimal estimation of the system state and thus improve the accuracy of the IMU measurements, leading to an improved localization. Similar to that, the authors of [2] propose a method to monitor a robot's localization integrity risk using an Extended Kalman Filter. The authors of [4] presents an approach to assess the performance of a localizer by embedding a spatial model of the expected localization performance into a map. To predict the localization performance at points in the map without data, a combination of Gaussian Process model and physical sampling is applied.

### 2.2 Explainable Artificial Intelligence

There are two types of XAI: data-driven and goal-driven XAI [1]. While data-driven XAI focuses on the explainability of data-driven methods such as machine learning, goal-driven approaches aim to generate explanations for intelligent agents who's behavior arise from symbolic AI. In this work, we apply data-driven XAI, as we train a localization monitor using pre-recorded data. A common method to extract explanations from models trained on images is Class Activation Mapping (CAM) which not only gets the prediction and its certainty score of a neural network but also provides the information on what regions of the input the result comes from by overlaying the input with a heatmap indicating the regions of interest [18]. [13] presented an XAI approach called *Grad-CAM* which extends *CAM* by including gradient information. Another approach available under the open-source license is SHapley Additive exPlanations (SHAP) [7]. The framework provides additive feature attribution methods and visualization, by connecting optimal credit allocation with local explanations. Other currently used techniques are *LIME* [10], *DeepLIFT* [14] and *CXplain* [12].

## 3 Robot Localization Monitor

This section describes the setup and data generation of the localization monitor which is based on the concept presented in [5]. The localization monitor is designed to solve a binary classification problem: either the robot is well localized or delocalized. We declare the robot delocalized if the estimated pose diverged from the true pose more than a given threshold. In order to have access to a ground truth of the robot's pose and to be able to provoke delocalizations we simulate the robot's motion and sensing (i.e. laser scans) as well as the environment using a simulation environment based on Stage [16]. The data from the perception as well as the data from the particle filter-based pose estimation (i.e. particles) are recorded and used to extract statistical features that are used as training data for classification. The approach follows the assumption that statistical features derived from the perception and the particle set can be used to learn a classifier, since the particle set represents a belief distribution of the pose in the environment. Following the approach in [5], a total of 33 statistical features are extracted from the localization approach holding information about the robot's state. Example features which are extracted are the size and the center of gravity of the particle cloud, as well as the number of identified particle clusters. A detailed list and description of the extracted features can be found in [5]. To generate training data, the robot is randomly driven trough the simulated environment which allows to add random obstacles within the environment that disturb the scan data and thus provoke delocalizations. Using a delocalization threshold $d = 1.2$ m, the extracted data points are labeled. Samples with propoer localization are assigned to class 1 and delocalized samples to class 0. A total of 150.000 data points are extracted where 50% are assigned to each class. 25% (37.500) data points are excluded from training and used for validation.

For classification, XGBoost [3] is used on the generated training data as it has proven to be a powerful machine learning model when paired with interpretability tools such as SHAP [6]. Since a single feature by itself is a rather bad classifier for estimating the localization state, the idea is to combine multiple weak features which do not provide enough information about certain classes themselves and combine them into a strong classifier that can be used for identifying the localization state of a robot. With an accuracy of 92.05%, XGBoost shows good overall performance and outperforms AdaBoost (85.44%) used in [5] by more than 6%, demonstrating that XGBoost is also a good model for localization monitoring. Table 1 gives an overview on the resulting metrics of XGBoost and compares it to AdaBoost as proposed in [5].

## 4 Gaining Insights into the Localization Monitor

Since it is hard to understand how the trained localization monitor comes to its predictions, we investigated the trained classifier using SHAP. SHAP helps to identify informative relationships between the given input features and the prediction of the classifier which can be used to explain how the model's decision process is trained [6]. This is done by assigning each input feature a shapley value which can be used to make correlations from the trained model transparent using different visualization techniques [7]. Figure 1 displays explanations for four individual predictions. $f(x)$ shows the aggregated SHAP value for a given prediction. A negative value represents a prediction of class 0 (delocalized) while a positive value indicates prediction for class 1 (localized). High SHAP values (positive or negative) indicate a high certainty in the prediction. The visualization also shows how strong and in which direction each feature impacts the prediction. Blue features have a negative impact on the SHAP value while red features a positive one. Figure 1a represents a correctly predicted delocalized state (true negative), Fig. 1b a falsely classified delocalized state (false negative), Fig. 1c shows the SHAP values for a false prediction for a localized state (false positive) and Fig. 1a represents a correctly classified prediction for the localized state (true positive). It can be seen that correct predictions have higher (absolute) SHAP values and fewer opposing feature values than incorrect predictions. Similar observations can be seen when generating explanations for other data samples. It is interesting to see that the features *ray_inlier_percent* (percentage of scan points that were found before an obstacle occured in the map) and *cog_mean_abs_dev* (mean over the absolute deviation of the particles from the center of gravity) seem to have the biggest impact on the prediction. However, since the plots only show a single data sample, we cannot see the impact of these features on all data points. This can be done by visualizing a scatter plot over all data samples as exemplary visualized in Fig. 2. The scatter plot shows how changing the value of a feature impacts the model's SHAP values and thus its predictions. The $x$-axis represents the feature value and the $y-axis$ the SHAP value as trained by the model. The light grey area the bottom depicts the distribution of the data values. While the vertical dispersion of the plot shows that similar feature values have a different impact on the model's prediction, having the dots follow and increasing pattern indicates that there is a direct correlation between feature value and prediction. This allows to estimate how a change in the feature value affects the feature's SHAP value and thus the model's prediction. Similar observations, but with less impact on the prediction, can be made when extracting explanations for other features. To retrieve a general explanation on the importance of features, SHAP provides a visualization technique where the overall importance of each feature is represented as the mean absolute SHAP value over all data samples. An example visualization is given in Fig. 3 for 16 out of the 33 features. From there, the overall importance of a feature on the outcome can be observed. From there it can be seen, that the feature *angle_inliers* and *point_dist* have the overall highest impact on a prediction, followed by the features presented above in Fig. 2. Additionally, partial or full redundant features can be identified using SHAP by building a hierarchical cluster of the features which were trained with XGBoost. A redundant feature for a trained model indicates that any feature could be used for prediction while still getting the same results. The results of the clustering is represented as a distance measure in $[0, 1]$, where a distance of 0 between two features indicates perfect redundancy and 1 indicates complete independence. The redundant features can be visualized as shown in Fig. 3 on the right. The visualization is done up to a defined distance, also called the clustering cutoff ($co \in [0, 1]$). This information can be used to simplify the localization monitor by removing redundant features without sacrificing performance significantly. Additionally, removing features from training can significantly improve the training time. Table 1 shows the performance of the trained model and its corresponding training time when removing redundant features with a clustering cuttof $co = 0.1$ and $co = 0.25$. For $co = 0.1$, a total of ten features are removed, improving the training time by roughly 20% without forfeiting the model's prediction performance. Similar results can be observed in the model with $co = 0.25$. It has to be noted that while removing redundant features can be done without sacrificing performance, manipulating a trained model based on the insights gained from explanations generated with SHAP should only be conducted with care. SHAP makes correlations trained within a model transparent, but making them transparent does not make them causal [6]. To understand causal relationship within the data, further investigation such as the creation of a causal graph [9] is required. However, such functionality is not offered by any known open-source XAI framework so far.

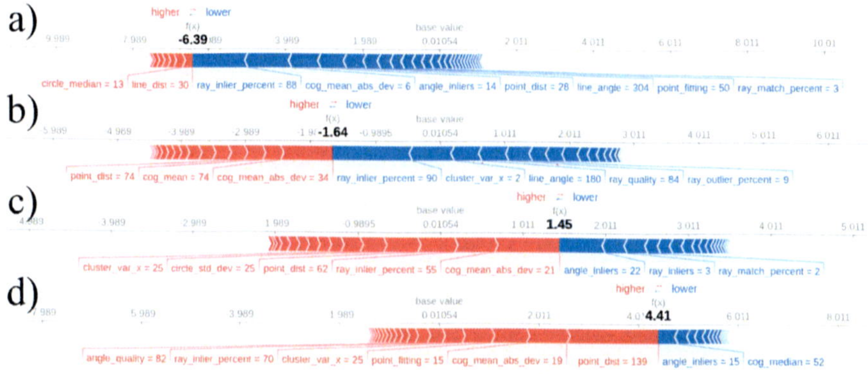


**Fig. 1.** Example visualization of the contribution of each feature to the single prediction. a) true negative b) false negative c) false positive d) true positive. While true positive and negative show high values, false predictions seem to have lower shap values.

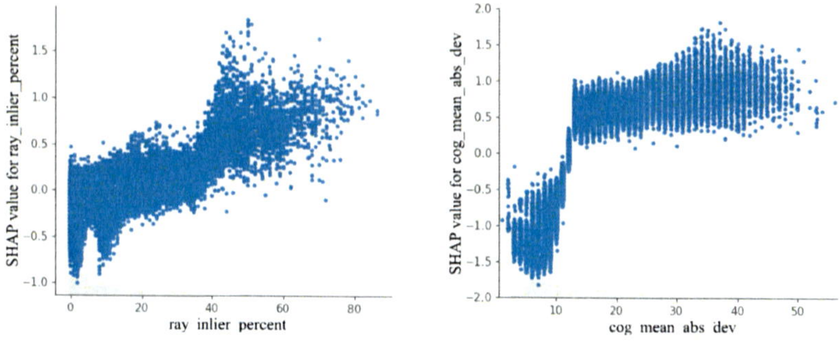


**Fig. 2.** Example visualization of the impact of SHAP values on the feature values for *ray_inlier_percent* (left) and *cog_mean_abs_dev* (right).

## 5 Conclusion

This paper uses SHAP, a data-driven XAI framework for generating explanations for a machine learning based localization monitor, which was trained on extracted localization features from a particle filter. The localization monitor is trained using XGBoost on the task to classify the robot's localization state in a

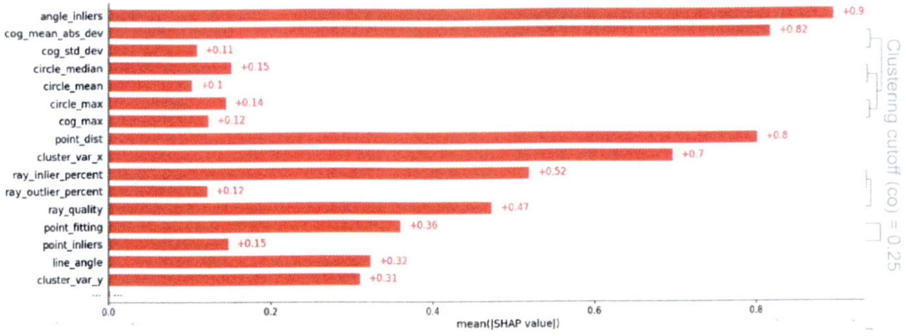


**Fig. 3.** Feature importance plot representing the (absolute) global importance of each feature over all data samples.

**Table 1.** Performance results of two different models used as localization monitor.

| Method | # Features | Precision | Recall | F1-Score | Accuracy | Training time |
|---|---|---|---|---|---|---|
| AdaBoost | 33 | 85.26% | 85.69% | 85.48% | 85.44% | 13.67 min |
| XGBoost | 33 | 91.21% | 93.01% | 92.11% | 92.05% | 3.28 min |
| XGBoost, $co = 0.1$ | 23 | 90.96% | 93.17% | 92.05% | 91.97% | 2.61 min |
| XGBoost, $co = 0.25$ | 19 | 90.65% | 92.95% | 91.78% | 91.70% | 2.32 min |

binary localized/delocalized value. A direct comparison of an existing localization monitor trained with AdaBoost shows that XGBoost provides acceptable results and even outperforms the existing method. The explanations generated from SHAP are used to gain insights into the predictions of the monitor and to improve the understanding of the trained model. The generated explanations provide insights to the developer about which features have the biggest impact on a prediction for a given data sample and how they contribute to the prediction. SHAP also provides explanations on how the prediction changes when changing the value of a single feature. By examining the most important features and identifying redundant features, the ML model can be simplified without sacrificing the prediction performance. However, SHAP only provides insights into correlations by making them transparent and is not to be used to draw casual conclusions. To do so, a method is required which is able to understand causal relationships within the features.

## References

1. Anjomshoae, S., Najjar, A., Calvaresi, D., Främling, K.: Explainable agents and robots: results from a systematic literature review. In: Proceedings of the 18th International Conference on Autonomous Agents and MultiAgent Systems, AAMAS 2019, Richland, SC, pp. 1078–1088 (2019)
2. Arana, G.D., Hafez, O.A., Joerger, M., Spenko, M.: Recursive integrity monitoring for mobile robot localization safety. In: 2019 International Conference on Robotics

and Automation (ICRA), pp. 305–311 (2019). https://doi.org/10.1109/ICRA.2019.8794115
3. Chen, T., Guestrin, C.: XGBoost: a scalable tree boosting system. In: Proceedings of the 22nd ACM SIGKDD International Conference on Knowledge Discovery and Data Mining, KDD 2016, pp. 785–794. Association for Computing Machinery, New York (2016). https://doi.org/10.1145/2939672.2939785
4. Churchill, W., Tong, C.H., Gurău, C., Posner, I., Newman, P.: Know your limits: embedding localiser performance models in teach and repeat maps. In: 2015 IEEE International Conference on Robotics and Automation (ICRA), pp. 4238–4244 (2015). https://doi.org/10.1109/ICRA.2015.7139783
5. Eder, M., Reip, M., Steinbauer, G.: Creating a robot localization monitor using particle filter and machine learning approaches. Appl. Intell. **52**(6), 6955–6969 (2022). https://doi.org/10.1007/s10489-020-02157-6
6. Lundberg, S., Dillon, E., LaRiviere, J., Roth, J., Syrgkanis, V.: Be careful when interpreting predictive models in search of causal insights. Towards Data Sci. 1–15 (2021)
7. Lundberg, S.M., Lee, S.I.: A unified approach to interpreting model predictions. In: Guyon, I., et al. (eds.) Advances in Neural Information Processing Systems, vol. 30, pp. 4765–4774. Curran Associates, Inc. (2017). https://doi.org/10.48550/arXiv.1705.07874
8. Panchpor, A.A., Shue, S., Conrad, J.M.: A survey of methods for mobile robot localization and mapping in dynamic indoor environments. In: 2018 Conference on Signal Processing and Communication Engineering Systems (SPACES), pp. 138–144 (2018). https://doi.org/10.1109/SPACES.2018.8316333
9. Pearl, J.: Graphs, causality, and structural equation models. Sociol. Methods Res. **27**(2), 226–284 (1998). https://doi.org/10.1177/0049124198027002004
10. Ribeiro, M.T., Singh, S., Guestrin, C.: "Why should i trust you?": explaining the predictions of any classifier (2016). https://doi.org/10.48550/arXiv.1602.04938
11. Roth, A. (ed.): The Shapley Value: Essays in Honor of Lloyd S. Shapley. Cambridge University Press, Cambridge (1988). https://doi.org/10.1017/CBO9780511528446
12. Schwab, P., Karlen, W.: CXPlain: causal explanations for model interpretation under uncertainty (2019). https://doi.org/10.48550/arXiv.1910.12336
13. Selvaraju, R.R., Cogswell, M., Das, A., Vedantam, R., Parikh, D., Batra, D.: Grad-CAM: visual explanations from deep networks via gradient-based localization. Int. J. Comput. Vision **128**(2), 336–359 (2020). https://doi.org/10.1007/s11263-019-01228-7
14. Shrikumar, A., Greenside, P., Kundaje, A.: Learning important features through propagating activation differences. CoRR **abs/1704.02685** (2017). https://doi.org/10.48550/arXiv.1704.02685. Accessed 12 Oct 2019
15. Thrun, S., Fox, D., Burgard, W., Dellaert, F.: Robust Monte Carlo localization for mobile robots. Artif. Intell. **128**(1), 99–141 (2001)
16. Vaughan, R.: Massively multi-robot simulation in stage. Swarm Intell. **2**(2–4), 189–208 (2008). https://doi.org/10.1007/s11721-008-0014-4
17. Wei, H., Qizhi, H., Weiguo, Z., Yu, L.: The optimal state estimation and IMU fault diagnosis of aircraft based on optimal two-stage CKF. In: 2021 China Automation Congress (CAC), pp. 447–452 (2021). https://doi.org/10.1109/CAC53003.2021.9728458
18. Zhou, B., Khosla, A., Lapedriza, A., Oliva, A., Torralba, A.: Learning deep features for discriminative localization (2015). https://doi.org/10.48550/arXiv.1512.04150

# LiDAR-Based Scene Understanding for Navigation in Unstructured Environments

Hamid Didari(✉) and Gerald Steinbauer-Wagner

Institute of Software Technology, Graz University of Technology, 8010 Graz, Austria
{hamid.didari,steinbauer}@ist.tugraz.at

**Abstract.** Reliable scene understanding is crucial for autonomous off-road navigation. This work proposes a perception framework based on multiple LiDARs and odometry that is able to analyze a robot's environment to generate an occupancy grid map for the navigation task. A gradient-based approach separates obstacles and ground points. The exact position of negative obstacles (cliffs and holes) is corrected using geometrical relation. Then, obstacle points are used to create an occupancy grid map for the robot. Observing obstacles are propagated to the next frame to cover blinds spot in the sensor setup, and temporary misclassification and dynamic obstacles are handled using ground points. The proposed framework is tested on a robot with two LiDARs to evaluate the performance. The results show successful navigation in the presence of positive and negative obstacles.

**Keywords:** Mobile Robots · Scene Understanding · Off-Road Navigation

## 1 Introduction

Autonomous off-road navigation has a wide range of applications, from space exploration [5,6] to terrestrial search and rescue [8]. One of the main challenges of these applications is reliable scene understanding, which aims to enable vehicles to perceive the surrounding environments and move safely. A reliable scene understanding should be able to detect positive obstacles (trees, cars, etc.), negative obstacles that include cliffs and holes, which can severely damage the vehicle if they are not avoided, and impassable slopes (steep uphill or downhill) where it could tip over. Negative obstacles make scene understanding harder since they can not be detected directly. They can be modeled as physical objects with specific sizes, shapes, positions, and motion parameters. Therefore, vehicle sensors can use those parameters information as a baseline to estimate the position and

This work was partially supported by the Austrian Research Promotion Agency (FFG) with the project RoboNav.


T. Petrič et al. (Eds.): RAAD 2023, MMS 135, pp. 178–185, 2023.
https://doi.org/10.1007/978-3-031-32606-6_21

orientation of the obstacle. For instance, a laser beam measures the distance between the laser and where the laser beam hits, but when we have a negative obstacle, It hits inside and not the beginning of it.

Papadakis [1] summarises the prior works on traversability analysis by classifying them into proprioceptive and exteroceptive methods. Diaz et al. [2] developed an algorithm to separate the obstacle and ground points in the point cloud. They use a voxel grid map to sort points in a point cloud. After that, the normal vectors for each point are extracted and labeled using Principal Component Analysis (PCA). Thenceforth, a neighborhood analysis of each point is utilized to correct the mislabeled points. Finally, a plane is fitted to the ground points. Although this method offers promising results, the assumption of the robot moving on a plane is unrealistic for an off-road environment. Additionally, their work did not address negative obstacles explicitly.

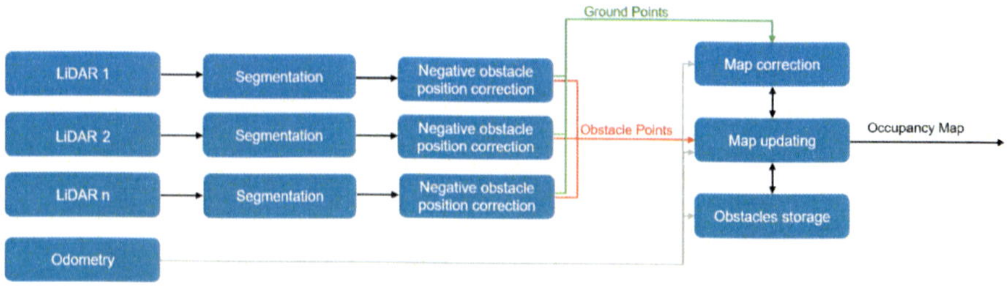


**Fig. 1.** A flowchart of the proposed method.

Huang et al. [4] assume the LiDAR is installed horizontally and calculate the distance between two sequences of LiDAR rings by using the horizontal angle between the lasers in LiDAR. If the distance between two points in the two sequence rings is less than a threshold, it is labeled as an obstacle. Then, a Markov random field (MRF) is used for fine-tuning the classification of points. The assumption of horizontally installed LiDAR limits the method. Additionally, their work does not consider slopes and negative obstacles.

Shang et al. [3] proposed using two tilted LiDARs to detect negative obstacles. Initially, a number of negative and positive obstacle samples are manually selected from LiDAR data, and a feature vector for each sample is computed. Then, a Support Vector Machine (SVM) is trained on these features. After that, the SVM can automatically label the points. This method relies heavily on the initial sensor setup due to the required data collection step.

The method proposed in this work relies on multiple LiDARs and odometry to estimate an occupancy grid map of negative and positive obstacles in generic environments. The main differences between the proposed method and previous works are:

- the proposed method is not limited to a specific number of LiDARs and can utilize multiple sources for better accuracy.
- the proposed method corrects the position of negative obstacles to enhance the quality of the final occupancy map.
- the proposed method uses the ground points to maintain a consistent final occupancy map.

The remainder of the paper is structured as follows. Section 2 presents the proposed method in detail. After that, Sect. 3 describes the experiments to show the proposed method's effectiveness and the analysis of the result. Finally, Sect. 4 concludes the work.

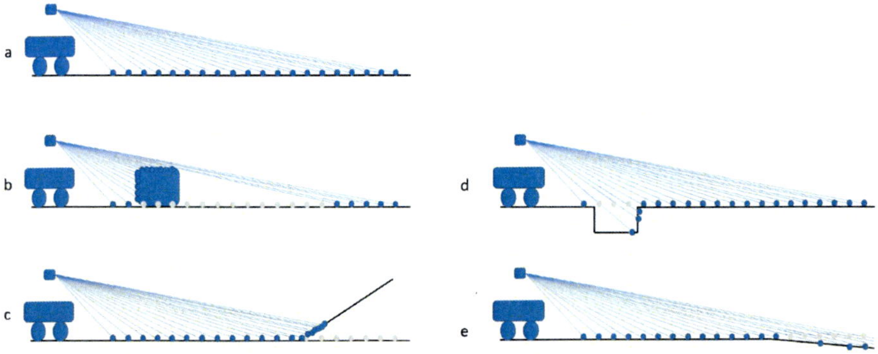


**Fig. 2.** LiDAR points arrangement in different situations. (a) no obstacle in the path, (b) a positive obstacle in the path, (c) a positive slope in the path, (d) a negative obstacle in the path, and (e) a negative slope in the path.

## 2 Proposed Method

The proposed method is executed in multi-stages. First, for each LiDAR, a gradient-based segmentation is used in parallel. In this step, the ground and obstacle points are separated, and the position of negative obstacles is calculated. In order to mitigate the effect of blind spots around the robot, a queue of observed obstacles is stored and propagated to the current map. Finally, the ground points are used to reduce the effect of mislabeling and corruption caused by noisy data and dynamic obstacles. Figure 1 shows a flow chart of the proposed method. Following subsections describe each stage of it.

### 2.1 Point Cloud Segmentation

There are a number of rotating lasers inside of a LiDAR that measures the distance between the laser and where the laser beam hit. LiDAR converts the measured distance to 3D points and sends them as a point cloud. These points can belong to a traversable surface(ground points) or an obstacle that needs to be avoided. To have safe navigation, the obstacle points need to be separated.

The lasers inside a LiDAR are arranged straight or staggered so that the relative horizontal angle between the lasers is fixed. When there are no obstacles in the scanning direction, the horizontal distance between the point and the origin will increase with the vertical rise of its corresponding laser, as shown in Fig. 2(a). A negative or positive obstacle will affect the arrangement of these points. Thus, the distance relationship will change accordingly, as shown in Fig. 2(b, d). Furthermore, a slope in the path will also affect the arrangement of points, as shown in Fig. 2(c, e). To separate obstacle points, first, we sort points inside point clouds into a 2D array. Each row belongs to one laser of LiDAR, and each cell contains the measured point in one scan.

Due to the behavior of laser points, the gradient over the z-axis can be used to separate obstacle points from the ground. This value can also indicate whether the robot can traverse a specific slope. The gradient can be calculated as follows:

$$\nabla_{i,j} = \max_{-1 \leq m,n \leq 1} \frac{p_z^{i,j} - p_z^{i-n,j-m}}{\sqrt{(p_x^{i,j} - p_x^{i-n,j-m})^2 + (p_y^{i,j} - p_y^{i-n,j-m})^2}} \tag{1}$$

where $\nabla_{i,j}$ is the gradient over the z-axis, $p^{i,j}$ is the coordinates of a point in the 2D array, $i$ indicates laser number, and $j$ is the place of measured point by LiDAR in one scan. Depending on the robot's ability, this gradient can be compared to a threshold to find the candidates for an obstacle. Since our robot can pass slopes that are less than 15°, We used $sin(15) = 0.25$ as a threshold. Smaller obstacles, like pebbles, may have a big gradient, but they can be traversed easily. Thus, the height of the obstacle is also examined to determine whether that robot can pass it or not. The robot that is used in this work can pass obstacles that have a height of less than 20 cm. Since a tilted installation of the LiDAR results in a better perspective in front of the robot, a rotation is required to bring the point clouds to the robot frame for the proposed method to function correctly. This rotation matrix can be calculated from the sensor setup.

## 2.2 Correction of Negative Obstacles

The detected laser point for a negative obstacle rests at the end of the obstacle. Thus, a correction method is required to find the starting point of the obstacle (the virtual grey point in Fig. 3) to create an occupancy map. The location of this point can be derived by using the basic geometrical properties of similar triangles:

$$\frac{\Delta h}{\Delta d} = \frac{\Delta h'}{\Delta d'} \tag{2}$$

where $\Delta h$ is the height of the sensor, $\Delta d$ is the distance between the negative obstacle and the robot, $\Delta d'$ is the width of the negative obstacle, $\Delta h'$ is the

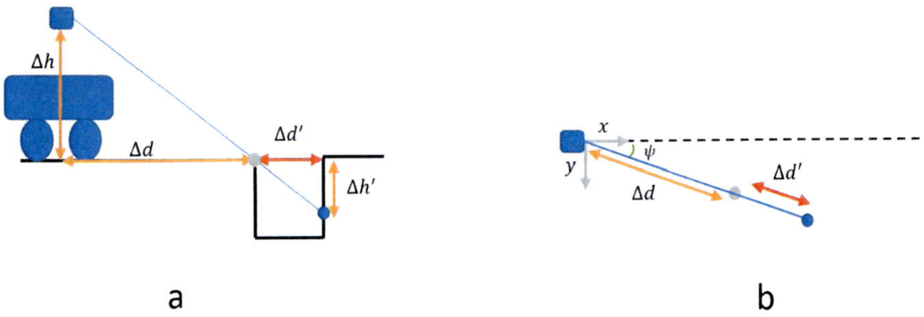


**Fig. 3.** The virtual grey point that indicates the position of the negative obstacle. (a) Is side view, and (b) is the top view of a negative obstacle.

depth of the negative point measured by $p$. The information from $p$ can be used to calculate the missing values in the Eq. 2:

$$\begin{aligned} d = \Delta d + \Delta d' &= \sqrt{p_x^2 + p_y^2} \\ \Delta h' &= -p_z \end{aligned} \tag{3}$$

By utilizing the values of Equation.3, the solution for $\Delta d$ is obtained as $\Delta d = \frac{\Delta h}{\Delta h' + \Delta h} d$ .The position of the virtual grey point can be calculated as follows:

$$\begin{aligned} \psi &= sin^{-1} \frac{p_y}{\sqrt{p_x^2 + p_y^2}} \\ p_x' &= \Delta d\ cos\psi = \frac{\Delta h}{\Delta h' + \Delta h} p_x \\ p_y' &= \Delta d\ sin\psi = \frac{\Delta h}{\Delta h' + \Delta h} p_y \end{aligned} \tag{4}$$

where $\psi$ also indicates the yaw angle of the laser beam in the laser scanner (Fig. 3(b)).

### 2.3 Creating the Local Occupancy Map

The occupancy map is represented as a grid map of a robot's environment, and the robot is located in the center of it. Each cell contains the probability of being occupied or unoccupied. A local planner can utilize this grid map to optimize the path for the robot. The map starts as an empty grid map where the probability ($P(C)$) of being occupied for each cell is 0. In each scan, the detected obstacle points are integrated into this map by updating the probability of each cell with the following equation:

$$P_t(C_{i,j}) = P_{t-1}(C_{i,j})e^{\frac{n}{w_b}} \tag{5}$$

where $n$ is the number of obstacle points integrated into $C_{i,j}$ in the current scan, $w_b$ is the weight of each obstacle point, and $t$ is the scanning step. If the $P_t(C_{i,j})$ is more than 1, it caps at 1.

The sensor setup usually has blind spots that make navigation unsafe. The current blind spots can be seen during navigation. For instance, if we install a LiDAR horizontally on a robot, depending on its height, it may not see around of robot in a short distance, but this spot can be seen when the robot is moving toward that. So, we can use previous information to cover blind spots. One way to do this is to transport the previous occupancy grid map to the current one. However, this solution results in a mismatch of cells when the robot moves around in the environment and it requires that each cell be integrated into four cells in each step. Another way is to store all obstacle points and import them to the current map, but this needs a huge amount of memory. To reduce memory usage, in each step, the projected obstacle points in each cell are considered to be part of the same obstacle. The position of this new obstacle is the average of all the points that contribute to it. Once the robot moves, this obstacle is transferred to the appropriate cells in the grip map, where the new probability of each cell is calculated as follows:

$$P(C_{i,j}) = \prod_{i=1}^{k} P(obs_i) \tag{6}$$

where $obs$ indicates the newly defined obstacle and $k$ is the total number of obstacles that are projected into the cell.

While projecting the obstacles into subsequent scans helps cover the blind spots, it may result in the corruption of the occupancy map when the obstacles are dynamic. Moreover, any mislabel will be projected into future scans and further corrupt the occupancy map without any mechanism to detect it. To counter these effects, the ground points are used to recalculate the probability for each cell as follows:

$$P_t(C_{i,j}) = P_t(C_{i,j})e^{-\frac{g}{w_g}} \tag{7}$$

where $g$ is the number of ground points in the current scan that is projected into $C_{i,j}$ and $w_g$ is the weight of each ground point.

## 3 Evaluation and Analysis

The proposed method was tested on a robot equipped with two 32-layer Hesai laser scanners on the university campus and in a highway tunnel at Zentrum am Berg in Eisenerz, Austria.. In these two scenarios, the robot must navigate safely.

In the highway tunnel test, the robot needed to go inside a tunnel simulated for a disaster, make a 3D map environment, and find victims. Figure 4(a) shows

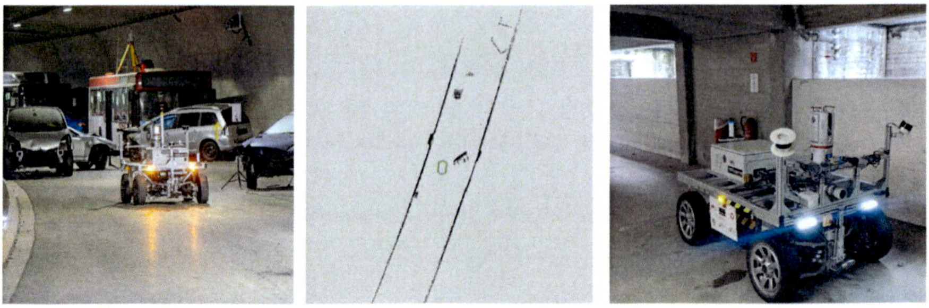

a b c

**Fig. 4.** (a) A photo of the robot in the highway tunnel. (b) The produced occupancy grid map. Black cells indicate obstacles, and the green hexagon shows the robot's footprint. (c) The sensor setup in the campus test.

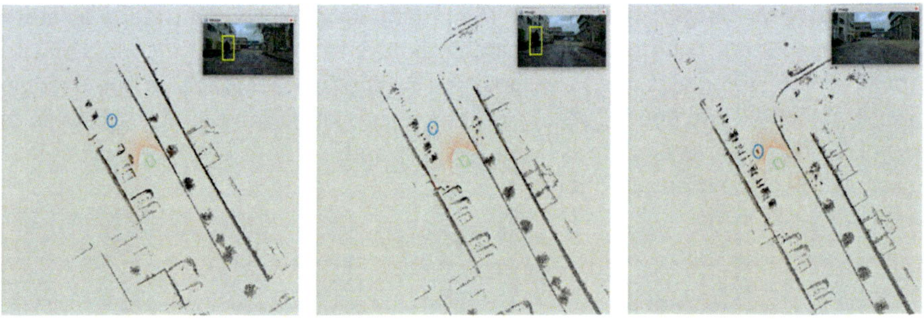

$t_1$ $t_2$ $t_3$

**Fig. 5.** The produced occupancy grid map for the campus test at different time steps. The observed obstacles are integrated into the current map, and the walking person in front of the robot didn't corrupt the map.

the robot setup in the highway tunnel. The previous work [7] was utilized to fuse LiDAR, IMU, and wheel odometry. The LiDARs are installed horizontally on the front-left and front-right of the robot to provide a better view of the environment. In this test, the robot went inside the tunnel for about 200 m to create a 3D map of the environment and returned safely. Figure 4(b) shows the occupancy grid map produced during this mission.

In the campus test, GPS was used as the primary odometry source. The data from GPS was fused with IMU and wheel odometry by Unscented Kalman Filter (UKF) to achieve better accuracy for navigation and scene understanding. To improve the detection of negative obstacles and slopes, each LiDAR is tilted in pitch and roll axis by 15°. While the setup provides better frontal information, it has blind spots on the sides and back of the robot. Figure 4(c) shows the robot and sensor setup in the campus test. In this test, the robot needed to navigate safely inside the campus at a speed of 2 m/s. Figure 5 shows the created occupancy map for the campus environment during the test. Red points on the map show the current LiDAR point cloud. As seen in the figure, the observed obstacles in the previous scan propagated successfully inside the current map. The yellow box shows a walking person in front of the robot, and the blue box shows its position inside the occupancy map. The person had a different position inside the occupancy map during the time, and it did not corrupt the map.

## 4 Conclusion

This paper proposes a perception framework based on multiple LiDARs and odometry to produce an occupancy grid map for the navigation task. The proposed method utilizes a gradient-based approach to separate obstacles and ground points. Geometrical relations are used to rectify the position of negative obstacles. To compensate for the potential problem of blind spots in the setup, a storing algorithm was proposed to cover the spots while moving in the environment. To evaluate the performance of the proposed method, two locations with different criteria were selected for testing: a university campus and a highway tunnel. The results indicate the effectiveness of the proposed method when dealing with natural or man-made environments.

In the future, a deep neural network could be utilized on the organized point cloud to segment obstacles and ground instead of the gradient method. Moreover, more information could be added to the cost map, such as terrain type, slope degree, and navigation risk.

## References

1. Papadakis, P.: Terrain traversability analysis methods for unmanned ground vehicles: a survey. Eng. Appl. Artif. Intell. **26**(4), 1373–1385 (2013)
2. Diaz, N., Gallo, O., Caceres, J., Porras, H.: Real-time ground filtering algorithm of cloud points acquired using terrestrial laser scanner (TLS). Int. J. Appl. Earth Obs. Geoinf. **105**, 102629 (2021)
3. Shang, E., An, X., Li, J., He, H.: A novel setup method of 3D LIDAR for negative obstacle detection in field environment. In: 17th International IEEE Conference on Intelligent Transportation Systems (ITSC) (2014)
4. Huang, W., et al.: A fast point cloud ground segmentation approach based on coarse-to-fine Markov random field. IEEE Trans. Intell. Transp. Syst. **23**(7), 7841–7854 (2022)
5. Didari, H., et al.: The AMADEE-20 robotic exploration cascade: an experience report. In: Müller, A., Brandstötter, M. (eds.) RAAD 2022. Mechanisms and Machine Science, vol. 120, pp. 477–484. Springer, Cham (2022). https://doi.org/10.1007/978-3-031-04870-8_56
6. Haruyama, J., et al.: Lunar holes and lava tubes as resources for lunar science and exploration. In: Badescu, V. (ed.) Moon, pp. 139–163. Springer, Heidelberg (2012). https://doi.org/10.1007/978-3-642-27969-0_6
7. Didari, H., Reitbauer, E., Schmied, C., Steinbauer-Wagner, G.: A low-drift LiDAR-based odometry for subterranean areas. In: Proceedings of the Austrian Robotics Workshop (ARW) (2022)
8. Jennings, J., Whelan, G., Evans, W.: Cooperative search and rescue with a team of mobile robots. In: 8th International Conference on Advanced Robotics (1997)

# Understanding Why SLAM Algorithms Fail in Modern Indoor Environments

Linus Nwankwo(✉) and Elmar Rueckert

Chair of Cyber-Physical Systems, Montanüniversität, Leoben, Austria
linus.nwankwo@unileoben.ac.at
https://cps.unileoben.ac.at/

**Abstract.** Simultaneous localization and mapping (SLAM) algorithms are essential for the autonomous navigation of mobile robots. With the increasing demand for autonomous systems, it is crucial to evaluate and compare the performance of these algorithms in real-world environments. In this paper, we provide an evaluation strategy and real-world datasets to test and evaluate SLAM algorithms in complex and challenging indoor environments. Further, we analysed state-of-the-art (SOTA) SLAM algorithms based on various metrics such as absolute trajectory error, scale drift, and map accuracy and consistency. Our results demonstrate that SOTA SLAM algorithms often fail in challenging environments, with dynamic objects, transparent and reflecting surfaces. We also found that successful loop closures had a significant impact on the algorithm's performance. These findings highlight the need for further research to improve the robustness of the algorithms in real-world scenarios.

**Keywords:** SLAM algorithms · Mapping · SLAM evaluation

## 1 Introduction

SLAM algorithms have been widely researched in the past decade and have shown significant progress in simulated and laboratory environments [1]. However, the real-world deployment of these algorithms is still a challenging task due to factors such as poor odometry, dynamic objects, transparent surfaces e.g., glass walls, etc. In real-world environments, SLAM algorithms face several challenges such as illumination variations, geometric-free regions, limited sensor range, etc. These challenges often cause the algorithm to fail in providing accurate and consistent estimates of the robot's position and the environment's structure. For example, in Fig. 1a–b, the algorithms failed to create a consistent metric map of the environments due to the glass walls reflectivity, different lightening conditions, unpredictable obstacles, no geometric features at the open entrance, and problems attributed to odometry error. The glass walls (Fig. 1b) reflect the laser beams which therefore return inaccurate measurements.


T. Petrič et al. (Eds.): RAAD 2023, MMS 135, pp. 186–194, 2023.
https://doi.org/10.1007/978-3-031-32606-6_22

These challenges limits the applicability of the SLAM algorithms in various fields such as autonomous navigation, robotic mapping, and augmented reality. Therefore, it is important to understand the sources of failure and to develop new methods to overcome them.

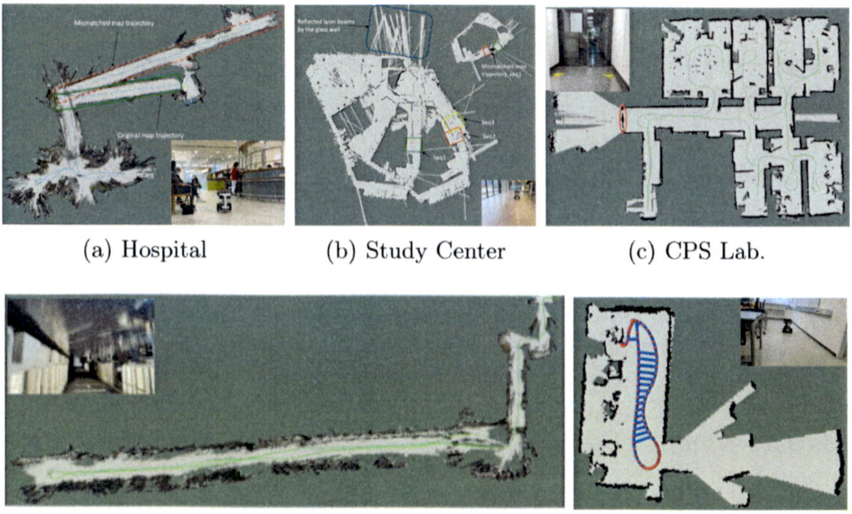

(a) Hospital (b) Study Center (c) CPS Lab.

(d) Long Passage (e) Controlled Env.

**Fig. 1.** Illustration of the environments. In the areas marked in red in (a) and (b), the algorithm could not track the robot's trajectory. The algorithm failed to create a consistent metric map as a result of the accumulation of odometry error and insufficient geometric features for the pose estimation due to the reflectivity of glass walls.

## 2 Related Work

Most SLAM algorithms are based on vision (with monocular, stereo, or RGBD cameras) or lidar (2D or 3D) [12]. Previous studies have shown that traditional SLAM algorithms, such as Kalman Filter-based [16], particle filter-based [15], and graph-based [3] approaches struggles in challenging environments due to their reliance on accurate sensor data and assumptions about the environment. Recent research has focused on developing more robust SLAM algorithms that can better handle such environments. For example, Xiangwei Dang et al. [2] proposed incorporating sensor fusion techniques, where multiple sensors are used to improve the robustness and accuracy of the algorithm.

Additionally, a study [8] have proposed using visual-based SLAM algorithms, which rely on visual information from cameras, and can be more robust in dynamic environments. However, these algorithms still have limitations in certain scenarios such as low lighting conditions or when the camera is occluded.

# 3 Method

To ascertain the root cause of failures of SLAM algorithms, we set up a series of tests in controlled and challenging real-world scenarios that included factors such as reflecting surfaces, different lightening conditions, high levels of noise, and dynamic or unpredictable obstacles. In this section, we describe the hardware platforms, the features of the SLAM algorithms used for this work. Thereafter, we provide details of the test environments, datasets, evaluation metrics, and our experimental setup.

## 3.1 Hardware Platforms

We used our developed low cost mobile robot [9] with two standard high-resolution RGB-D cameras, a 360° horizontal field of view (FOV) lidar sensor, and a 9-axis inertial measurement unit (IMU) to record novel challenging datasets used to evaluate the robustness and accuracy of the selected SLAM algorithms. Other hardware platforms used for this test include a Lenovo Think-Book Intel Core i7, Intel iRIS Graphics running on Ubuntu 20.04, ROS Noetic distribution with 8GB RAM. These were used for on-the-site (OTS) data-set acquisition. Additionally, since the recorded datasets are processed offline on each of the considered algorithms, we used a ground station PC with Nvidia Geforce RTX 3060 Ti GPU, 8GB GDDR6 memory running Ubuntu 20.04 and ROS Noetic distribution.

## 3.2 SLAM Algorithms

We evaluated SOTA open-source SLAM algorithms actively supported and maintained by the long-term end-of-life (EOL) distribution of the robot operating system (ROS) [13]. Due to the large availability of open-source SLAM algorithms, we limited our selection to the most popular and well-supported (a) three filter and graph-based SLAM approaches namely: Hector-SLAM [5], Gmapping [4], and Karto-SLAM [6] and (b) two visual-based techniques namely: real-time appearance-based mapping (RTAB-Map) [7] and ORB-SLAM2 [11].

These algorithms were run on the same datasets recorded in each of the test environments, and their performances are assessed based on the evaluation metrics in Subsect. 3.4.

## 3.3 Environment and Data Collection

Our datasets were recorded in five complex indoor environments of different geometric and feature characteristics depicting realistic operations scenarios for autonomous mobile robots. Figure 1 shows the exemplary layout of the environments. The CPS Lab (Fig. 1c) depicts a typical office layout with multiple occlusions, several pieces of furniture, and different lighting conditions (i.e., natural daylight and artificial lights). The Long Passage (Fig. 1d) is a typical passageway of approximately $6 \times 120$ m with several tables and natural day lighting conditions. Figure 1b is the student study centre at the University of Leoben with large reflecting surfaces and many moving objects (people) which depicts a typical static and dynamic environment. The hospital environment (Fig. 1a) includes both static, dynamic characteristics (i.e., people moving around, some sitting, and several pieces of furniture) as well as multiple reflecting glasses. The last datatset was collected in a room with limited obstruction and dynamics which we considered as controlled environment (Fig. 1e).

We captured the dataset by teleoperating the mobile robot around the environments leveraging the ROS-Mobile framework developed by [14]. We recorded specific topics of interest, e.g., for lidar-based SLAM algorithms, the robot's transformation, odometry, and scan data are required. For vision-based technique, the RGB-D dataset contains the colour and depth images of an Intel Realsense D435i camera along the ground-truth trajectory of the sensor. The data was recorded at a full frame rate of 30 frames per second (fps), RGB and depth resolution of $1920 \times 1080$ and $1280 \times 720$ respectively. The ground-truth trajectory was obtained from both Intel Realsense $T265$ tracking camera and a high-accuracy external 9 degree-of-freedom (DOF) motion tracking IMU device mounted on the robot base.

### 3.4 Metrics

We assessed the performance and the accuracy of the algorithms based on the metrics described in this subsection. To ensure a detailed performance analysis, we performed both quantitative and qualitative evaluations. For the quantitative assessment, we defined various metrics relating to the robot's trajectory and translation (Eqs. 1, 2, 3, and 4). Furthermore, for the qualitative evaluation, we performed a visual inspection of the generated map and compared it to the ground truth map.

**Absolute Trajectory Error (ATE):** We utilised this metric to measure the overall alignment between the estimated trajectory and the ground truth trajectory. The ATE error directly reflects the accuracy of the estimation algorithm and the consistency of the global trajectory. More formally, given a sequence of poses from the estimated trajectory $s_i \triangleq [s_i^{trans}, s_i^{rot}]^T \in SE(3)$ and the ground truth trajectory $g_i \triangleq [g_i^{trans}, g_i^{rot}]^T \in SE(3)$ at the $i-th$ timestamp, we computed the ATE as shown in Eq. 1. $SE(3)$ denotes the homogeneous transformation of the robot in 3D space [10]. The superscripts $trans$ and $rot$ denotes the translational and rotational components of the transformation matrix respectively.

$$ATE_{trans} \triangleq \sqrt{\frac{1}{n}\sum_{i=1}^{n} \|g_i^{trans} - s_i^{trans}\|^2} \tag{1}$$

where $||.||$ is the $L(2)$ norm of the position difference of the Euclidean distance between ground truth poses $g_i$ and the estimated poses $s_i$. The variable $n$ is the total number of poses in the trajectory sequence.

**Relative Pose Error (RPE):** We choose this error metric as complementary to ATE since it does not account for the accumulated error of the robot's trajectory. However, it enabled us to independently estimate the drift in each of the SLAM algorithms. We computed RPE as the average of the Euclidean distance between the $g_i$ and the $s_i$, normalized by the $g_i$ to ensure that the RPE is expressed as a ratio, rather than an absolute value i.e.,

$$RPE_{trans} \triangleq \sqrt{\frac{1}{m}\sum_{i=1}^{m}\left\| \frac{g_i^{trans} - s_i^{trans}}{g_i^{trans}} \right\|^2} \tag{2}$$

where $m$ is the number of relative poses in the trajectory. We used Eq. 2 and 3 to compute the difference in position and orientation between the consecutive estimated and ground truth poses respectively.

$$RPE_{rot} \triangleq \frac{1}{n}\sum_{i=1}^{n}(cos^{-1}(g_i^{rot} - s_i^{rot})) \tag{3}$$

**Scale Drift (SD):** We defined this metric to measure the deviation of the algorithm's estimated scale of the environment to the ground truth scale over time. It enabled us to measure the consistency of the estimated trajectory with respect to the true scale. We computed the scale drift by comparing the cumulative distance of the estimated pose with the cumulative distance of the ground truth pose i.e.,

$$SD = \frac{1}{n-1}\sum_{i=1}^{n-1}\frac{\|(s_{i+1}^{trans} - s_i^{trans})\|}{\|(g_{i+1}^{trans} - g_i^{trans})\|} \tag{4}$$

From 4, if the SLAM algorithm is performing optimally (i.e., with no deviation), then SD $= 1$, and when it deviates from the actual values, SD values less than or greater than 1 indicate underestimation or overestimation of the distances.

**Map Quality:** We used this metric to measure the quality of the map generated by each SLAM algorithm. The map quality was evaluated by comparing the generated map to a ground truth map, and by assessing the consistency, correctness and completeness of the map through visual inspection.

## 4 Results

In this section, we present the statistical results of the performance of the different algorithms on our datasets. Each algorithm was evaluated on the same recorded dataset with its default settings to ensure a fair comparison.

### 4.1 Lidar-Based and Visual-Based SLAM Techniques

Table 1 presents the evaluation results of each of the SLAM algorithms considered in this work. The performances are benchmarked on our recorded datasets from the five indoor environments. We grouped the results into lidar and visual based. For each group and environment, we represent the best results for the metric in **bold** font. From Table 1, all the algorithms had difficulty dealing with the complexity of the environments, resulting in large ATE, RPE, and RMSE (as evident in the hospital environment for visual-based SLAM technique, and CPS Lab for lidar-based approach).

**Table 1.** Statistical results of the algorithms performance with the different recorded datasets. $\mu \rightarrow$ mean, $\sigma \rightarrow$ standard deviation, $t/a \rightarrow$ translation/angular RMSE.

| Metrics | | Lidar-Based | | | Visual-Based | |
|---|---|---|---|---|---|---|
| | | Hector | Gmapping | Karto | ORBSLAM2 | RTAB |
| Hospital | ATE($\mu/\sigma$) | **0.12/0.09** | 0.28/0.26 | 0.29/0.26 | 8.51/8.76 | **8.49/8.76** |
| | RPE($\mu/\sigma$) | **0.10/0.35** | 0.12/0.78 | 0.13/0.79 | 4.80/29.36 | **0.53/0.45** |
| | SD($\mu/\sigma$) | 1.50/0.38 | **1.02/0.05** | 1.03/ 0.06 | **1.59/3.50** | **1.59/3.50** |
| | RMSE(t/a) | **0.16/0.38** | 0.39/0.13 | 0.39/0.14 | **12.22/1.48** | 12.23/1.49 |
| St. Center | ATE($\mu/\sigma$) | **0.15/0.01** | 0.19/0.01 | 0.17/0.03 | 3.59/2.37 | **0.25/0.20** |
| | RPE($\mu/\sigma$) | **0.06/0.64** | 0.08/0.82 | 0.07/0.73 | 0.32/0.91 | **0.08/0.84** |
| | SD($\mu/\sigma$) | 1.01/0.02 | 1.04/0.02 | **0.99/0.02** | 0.67/0.16 | **0.81/0.05** |
| | RMSE(t/a) | **0.15/0.13** | 0.19/0.12 | 0.17/0.12 | 4.30/0.51 | **0.32/0.08** |
| CPS Lab. | ATE($\mu/\sigma$) | 1.73/1.14 | 1.77/1.36 | **1.54/1.18** | **0.75/0.60** | 0.87/0.53 |
| | RPE($\mu/\sigma$) | 0.25/0.59 | **0.21/0.56** | 0.22/0.59 | **0.10/0.16** | 0.32/2.68 |
| | SD($\mu/\sigma$) | 1.05/0.16 | 1.03/0.07 | **0.97/0.13** | **1.03/0.03** | 1.13/0.99 |
| | RMSE(t/a) | 2.08/0.34 | 2.23/0.36 | **1.94/0.71** | **0.96/0.60** | 1.02/0.53 |
| Long Pass. | ATE($\mu/\sigma$) | **0.06/0.04** | 1.32/0.77 | 1.57/0.92 | 0.13/0.05 | **0.06/0.06** |
| | RPE($\mu/\sigma$) | **0.01/0.03** | 0.08/0.23 | 0.09/0.23 | 0.48/4.79 | **0.01/0.06** |
| | SD($\mu/\sigma$) | 0.96/0.01 | **0.98/0.02** | **0.98/0.02** | 1.18/2.14 | **1.17/2.14** |
| | RMSE(t/a) | **0.07/0.12** | 1.53/0.39 | 1.82/0.43 | 0.13/0.11 | **0.08/0.11** |
| Ctrld. Env. | ATE($\mu/\sigma$) | 0.29/0.24 | **0.01/0.00** | 0.02/0.00 | **0.31/0.25** | 0.32/0.30 |
| | RPE($\mu/\sigma$) | 0.46/0.62 | **0.03/0.08** | 0.10/0.25 | 0.43/0.67 | **0.35/0.62** |
| | SD($\mu/\sigma$) | 1.75/0.49 | **1.00/0.00** | **1.00/0.00** | 0.70/0.21 | **0.71/0.20** |
| | RMSE(t/a) | 0.38/0.47 | **0.01/0.00** | 0.02/0.00 | **0.40/0.31** | 0.44/0.22 |

Comparing the individual lidar-based algorithms, Hector-SLAM outperforms all other methods in the Long Passage environment with low mean and low standard deviation. This is attributed to its unreliance on the error-prone odometry data to perform accurate scan matching. Also, the cluttered static obstacles at environment enabled it to perform accurate scan matching and generate consistent occupancy grid map of the environment.

Gmapping performed sufficiently at the controlled environment. We attributed its poor performance in other environments to its reliance on perfect odometry to maintain a set of hypotheses about the robot's pose. Odometry is inherently uncertain and prone to error due to the complexity of the operating environments and factors such as wheel slippage, sensor noise, and drift over time. The performance in the controlled environment could also be attributed to the fact that the robot covered less distance and thus accrued fewer odometry errors.

Karto-SLAM had near optimal SD values in the majority of the environments (fourth column) due to its graph-based SLAM approach, which adapts well to changes in the environment. However, the high mean and standard deviation indicates that the algorithm still struggles to estimate the robot's pose accurately, leading to larger errors.

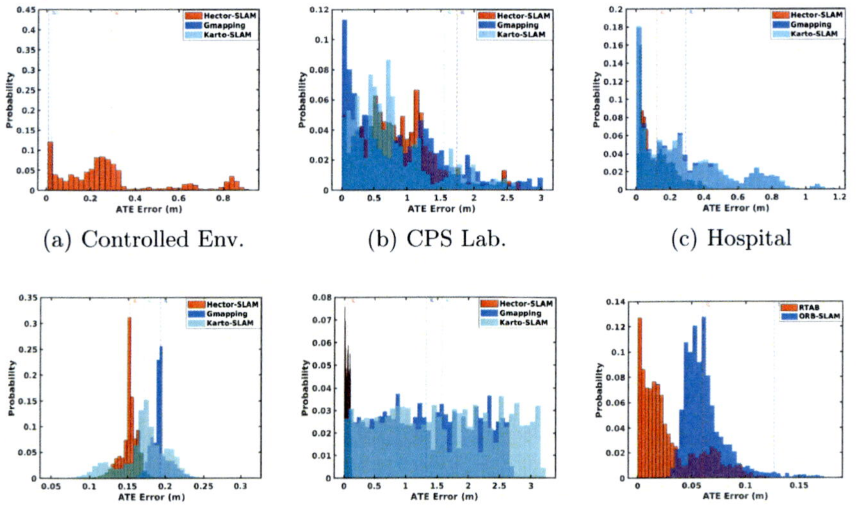


(a) Controlled Env. (b) CPS Lab. (c) Hospital

(d) Study Center (e) Long Passage (f) Visual Long Passage

**Fig. 2.** ATE result for all the environments. In each plot, the dashed vertical lines represent the mean ATE for each algorithm.

Figure 2 shows the distribution of the ATE for the algorithms. In the hospital environment (Fig. 2c), the ATE for Gmapping algorithm has a probability of 0.18 of being between 0.00 m and 1.18 m. This means that for 18% of the time, the error in the estimated trajectory is within that range. Furthermore, the ATE and the mean translational and angular root mean square error $RMSE(t/a)$ for the CPS Lab are high for all SLAM algorithms due to frequent turning, loop closures, unpredictable obstacles, and the limited geometric features in the environment. The mean ATE of 1.77 m e.g., indicates that if at any moment, the robot has to return to a specific location in the environment, there is at least 1.77 m of error to reach it, without considering the accumulation of such error.

## 5 Conclusion

In this work, we provided an evaluation strategy and real-world datasets to evaluate SOTA SLAM algorithms in controlled and in challenging indoor environments. In the controlled environment, the majority of the algorithms performed adequately, with relatively low ATE, RPE, and RMSE values e.g., ATE mean of the lidar based Gmapping algorithm was 0.01 m and of the visual SLAM approach ORBSLAM2 was 0.31 m. However, in challenging environments, the ATE, RPE, RMSE and SD values were higher e.g., the ATE mean of Gmapping was 1.77 m and of RTAB 0.87 m in the CPS lab environment. The complete evaluation results of five algorithms in five environments is shown in Table 1.

These results demonstrate the limitations of the algorithms in complex and dynamic environments, where the robot has to navigate through obstacles (see Fig. 1a), changing lighting conditions, and dealing with dynamic objects. The results were influenced by factors such as the reflecting surfaces (see Fig. 1b), sensor noise, odometry errors, and the large-scale of the environment. In summary, we have shown that the SLAM algorithms face significant challenges in challenging environments and that further research and development are needed to improve their robustness in real-world scenarios.

To improve the performance, we are exploring the incorporation of floor plans as prior to complement the SLAM map in current ongoing work.

## References

1. Cadena, C., et al.: Simultaneous localization and mapping: present, future, and the robust-perception age. IEEE Trans. Robot. **32** (2016)
2. Dang, X., Rong, Z., Liang, X.: Sensor fusion-based approach to eliminating moving objects for SLAM in dynamic environments. Sensors **21**, 230 (2021)
3. Grisetti, G., Kümmerle, R., Stachniss, C., Burgard, W.: A tutorial on graph-based SLAM. IEEE Trans. Intell. Transp. Syst. Mag. **2** (2010)
4. Grisetti, G., Stachniss, C., Burgard, W.: Improved techniques for grid mapping with rao-blackwellized particle filters. IEEE Trans. Robot. **23**(1) (2007)
5. Kohlbrecher, S., Meyer, J., Graber, T., Petersen, K., Klingauf, U., von Stryk, O.: Hector open source modules for autonomous mapping and navigation with rescue robots. In: Behnke, S., Veloso, M., Visser, A., Xiong, R. (eds.) RoboCup 2013. LNCS (LNAI), vol. 8371, pp. 624–631. Springer, Heidelberg (2014). https://doi.org/10.1007/978-3-662-44468-9_58
6. Konolige, K., Grisetti, G., Kümmerle, R., Burgard, W., Limketkai, B., Vincent, R.: Efficient sparse pose adjustment for 2D mapping. In: 2010 IEEE/RSJ International Conference on Intelligent Robots and Systems, pp. 22–29 (2010)
7. Labbé, M., Michaud, F.: RTAB-map as an open-source lidar and visual simultaneous localization and mapping library for large-scale and long-term online operation. J. Field Robot. **36** (2018)
8. Li, G., Liao, X., Huang, H., Song, S., Liu, B., Zeng, Y.: Robust stereo visual SLAM for dynamic environments with moving object. IEEE Access **9**, 32310–32320 (2021)
9. Linus, N., Clemens, F., Bartsch, K., Rueckert, E.: ROMR: A ros-based open-source mobile robot. HardwareX, e00426 (2023). https://www.sciencedirect.com/science/article/pii/S2468067223000330

10. Mueller, A.: Modern robotics: mechanics, planning, and control [bookshelf]. IEEE Control Syst. Mag. **39**(6), 100–102 (2019)
11. Mur-Artal, R., Tardós, J.D.: ORB-SLAM2: an open-source SLAM system for monocular, stereo, and RGB-D cameras. IEEE Trans. Robot. **33**(5) (2017)
12. Placed, J.A., et al.: A survey on active simultaneous localization and mapping: state of the art and new frontiers (2022)
13. Quigley, M., et al.: ROS: An Open-Source Robot Operating System, vol. 3 (2009)
14. Rottmann, N., Studt, N., Ernst, F., Rueckert, E.: ROS-mobile: an Android application for the robot operating system. arXiv preprint arXiv:2011.02781 (2020)
15. Stachniss, C.: Simultaneous Localization and Mapping, pp. 1–29. Springer, Heidelberg (2016)
16. Todoran, H.G., Bader, M.: Extended Kalman Filter (EKF)-based local SLAM in dynamic environments: a framework. In: Borangiu, T. (ed.) Advances in Robot Design and Intelligent Control. AISC, vol. 371, pp. 459–469. Springer, Cham (2016). https://doi.org/10.1007/978-3-319-21290-6_46

# Probabilistic Fusion of Surface and Underwater Maps in a Shallow Water Environment

Hannan Ejaz Keen(✉) and Karsten Berns

RPTU Kaiserslautern-Landau, Kaiserslautern, Germany
{keen,berns}@informatik.uni-kl.de

**Abstract.** Global warming has increased the frequency of floods around the world. These floods destroy the environment and create navigational problems for rescue operations. This paper presents a novel probabilistic mapping technique that fuses the surface and underwater information using Octomaps to deliver the obstacle map of an unstructured area. An adaptive obstacle detection algorithm retrieves underwater information from a forward-looking high-frequency sonar, whereas a three-dimensional LiDAR maps the surface water environment. The mapping methodology is tested at several shallow water bodies, and promising results are provided.

**Keywords:** Underwater mapping · Surface water mapping · Sonar · LiDAR · Surface Water Vehicles · USV

## 1 Introduction

Floods are one of the significant effects of environmental destruction caused by global warming. The number of floods and the extent of their destruction increase each passing year. Floods destroy road networks and disrupt assistance via land, while aerial rescue has always been complicated and expensive. On the other hand, boats are better suited for flood rescue operations because they access the affected areas more quickly and do not require helipads. However, flood brings several unknown structures, such as fallen trees, broken fences, or other debris, which poses extreme navigational challenges for boats. Although these structures can be seen from the surface, their unknown underwater characteristics, such as depth and dimension, may damage the boats. Furthermore, other structures that are not even visible from the surface are equally dangerous. Contrarily, floating materials such as leaves and other vegetation do not exist under the water and should not be considered as obstacles. Hence, a fusion of surface water and underwater obstacle detection is crucial for collision-free uninterrupted long-term navigation.

Despite the availability of several sensors such as LiDAR, stereo cameras, or RADAR for detection applications, these sensors are not feasible for the underwater environment because flood water contains a high amount of sand particles


T. Petrič et al. (Eds.): RAAD 2023, MMS 135, pp. 195–202, 2023.
https://doi.org/10.1007/978-3-031-32606-6_23

which affects water density and turbidity in a flooded environment. Conversely, water is an efficient medium for sound transmission that travels many kilometers without decaying. Hence, this work uses a forward-looking high-frequency sonar better suited for underwater detection applications.

This paper presents a novel probabilistic model to generate an obstacle map that fuses the surface and underwater obstacles' information provided by LiDAR and sonar, respectively. The fusion process takes place using Octomaps that show a maximum drift of ±15 cm between the detected points of the same obstacles in both sensors.

The remaining paper is organized as follows. Related work is discussed in Sect. 2. Section 3 describes the methodology of the obstacle detection and mapping algorithms. The experiments and results are presented in Sect. 4. Finally, Sect. 5 concludes the paper.

## 2 Related Work

State-of-the-art maps such as google maps or open street maps help tremendously in navigation [7] but are only feasible when the change in environment is limited. Therefore such maps are unstable in a destructed region where the change is enormous. Hence, new ways to map the environment have been under consideration for a long time. In our previous works [6,14], this change in the environment within the agriculture domain is studied, and an algorithm for the generation of elevation maps in agriculture fields is presented.

Satellite imagery [1,11], and aerial imagery [2] do not differentiate between surface and underwater obstacles, hence infeasible for our application. Therefore, unmanned surface vehicles (USV) [3–13] are most beneficial in gathering information about surface water or underwater obstacles and for avoiding collisions. Furthermore, [10] presents a similar system as ours regarding obstacle detection, but with two key differences. 1) The detection of targets such as other ships is relatively easy as such a problem can be solved using many state-of-the-art object detection, and deep learning algorithms. In contrast, our application requires the detection of random destructive obstacles without the pre-knowledge of their shape or size, and it is challenging to train such a generalized network. Therefore, our system includes a LiDAR to get a better 3D understanding of the environment. 2) Our application also requires underwater obstacle knowledge for safe navigation, as the rescue boat could be at least 50 cm inside water which is not addressed in the cited work.

Raber et al. [9] have developed the small USV (sUSV) to monitor the Structure from Motion (SfM) of coral reef in a clean water body using a camera, which is not feasible for a flooded environment because the flood brings a high volume of sand particles. Therefore, we primarily focus on using sonar sensors as sound works better in most underwater conditions. Moreover, a complex system is presented in [12] to generate mosaics of underwater scenes using similar forward-looking multibeam sonar, but it does not consider the surface water environment. Furthermore, [5] provides a study using sonar to find sea-surface

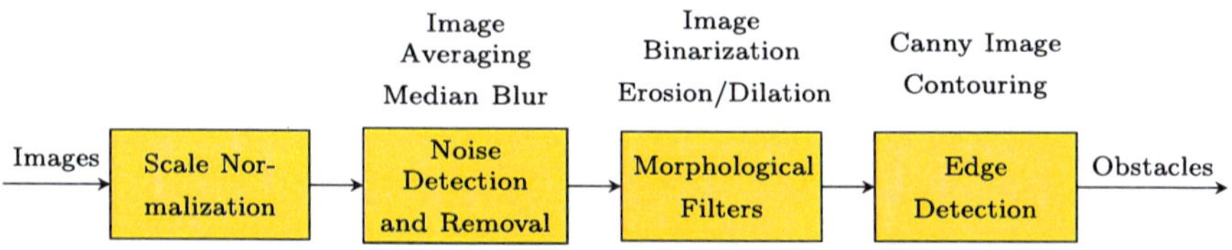


**Fig. 1.** Block Diagram - Underwater Obstacle Detection

obstacles by detecting the wakes of ships, but the system's reliability is not provided. Our system detects and fuses surface water and underwater obstacles to generate a long-term obstacle map using LiDAR and Sonar.

# 3 Methodology

## 3.1 Underwater Detection

The physical phenomenon of sonar is essential for computing the probability distribution of sonar data. To the extent of our knowledge, image processing and sonar modeling have been extensively studied, but a concrete relationship between the two still needs to be explored. Therefore, our system follows a model of uncertainty which is developed based on sound propagation in shallow water and the physical design of our sonar. Such a model helps to detect obstacles based on the probability distribution of each pixel in the sonar image as presented mathematically in Eq. 1.

$$P^S_{ptx} = \begin{cases} 1 - (\frac{0.4}{d_{water}}.d_{ptx}), & \text{if } d_{ptx} \leq \frac{d_{water}}{cos(90^\circ - vfov)} \\ 0.6, & \text{otherwise} \end{cases} \tag{1}$$

where $P^S_{ptx}$ is the probability of each sonar data point, whereas $d_{water}$ is the depth of water (via altimeter) and $d_{ptx}$ is the distance of each pixel in the Cartesian coordinate system from sonar's origin. The vertical field of view of sonar (in our case 20°) is denoted by $vfov$. The factor $\frac{d_{water}}{cos(90^\circ - vfov)}$ is the nearest point of the seabed at a specific height. Before this point, sonar receives no reflection from the bed and has very low noise. Therefore, the model is more confident (0.6) before that region.

The images with pixel-wise probability distributions are forwarded to the obstacle detection module, where they are initially scale-normalized to increase SNR. Sonar images often contain noise due to design limitations of sonar and undesired sound frequencies. The discussion on these noises is beyond the scope of the work presented in this paper. These noises are assumed to be independent across the incoming images, so they are removed by averaging the $N$ number of consecutive images. The parameter $N$ can be opted based on the acceptable trade-off between required frames per second and noise reduction. Furthermore, morphological operations such as dilation and erosion are performed to differentiate between obstacles' intensity and their shadows in the images. Only the

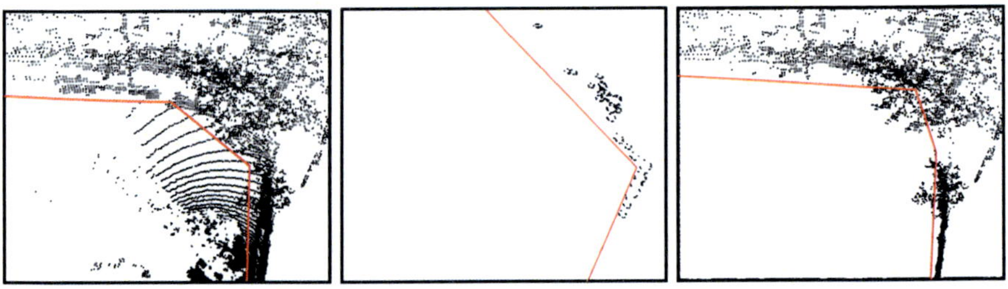

(a) LiDAR Scan (b) Sonar Scan (c) Fused LiDAR Scan

**Fig. 2.** LiDAR and Sonar scans at same timestamp. (a, b) sensor data c) shows the fusion of LiDAR and Sonar models to clear leaves. *The colored lines are only for visualization.*

intensities are considered for this study, and detected shadows are ignored. A canny filter further processes the image to provide the contour of each obstacle. The block diagram of the underwater obstacle detection algorithm is provided in Fig. 1.

### 3.2 Surface Water Detection

The LiDAR scanner under consideration generates beams within the near-infrared range, which is beneficial particularly for our surface water application. Since water has a high density, infrared rays absorb completely over its surface and provide a negligible amount of light back to the scanner. Therefore, the water's surface acts as a stealth body for LiDAR and makes it easy to detect obstacles on its surface.

Another challenge is to detect safe obstacles, such as leaves that float over the surface, and should be filtered from the LiDAR scan to increase the accuracy of the obstacle map. To tackle this problem, the z-axis information of the obstacle seems to be an effective parameter. Our system determines the obstacle's height from the water surface by LiDAR, while Sonar provides information about the depth of the obstacle under the water. The following equation computes the probability of each point being an obstacle.

$$P_{ptx} = \begin{cases} P^L_{ptx}, & \text{if } h^L \geq C_h \\ max(P^L_{ptx} \cdot \frac{h^L}{C_h}, P^S_{ptx} \cdot d^S), & \text{else if } C_h > h^L > 0 \\ P^S_{ptx}, & \text{otherwise} \end{cases} \tag{2}$$

where $P^S_{ptx}$ is the probability of a point in a sonar image, $P^L_{ptx}$ is the probability of a point in a LiDAR scan, $h^L$ is the height of a point above the surface of the water, and $d^S$ is the depth of point under the water. Finally, $C_h$ is the minimum height of the obstacle set to detect it as an obstacle without considering the underwater profile of the same obstacle. Figure 2 shows the filtered scan after dropping the vegetation points. Afterwards, this scan is forwarded to the mapping algorithm.

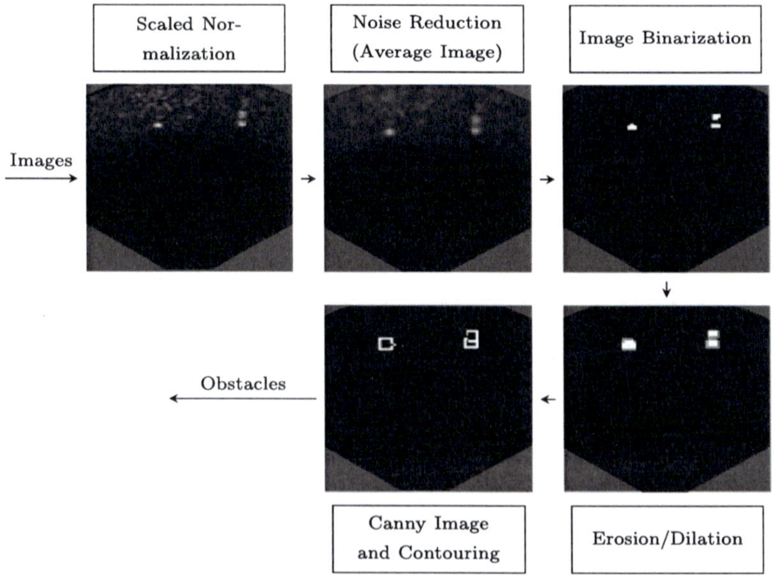


**Fig. 3.** Underwater Obstacle Detection

### 3.3 Fused Mapping

The preprocessing steps for the presented mapping methodology are derived from our previous study [6], with some tweaks and parameters related to surface water.

Every scan goes through a series of adaptive steps in which undesired data is filtered out. Steps include 1) *Box Filter:* where extra points associated with the water drone are removed from the scan. 2) *Outlier Removal:* where all outliers are filtered based on their K-nearest neighbors. 3) *Transformation:* Lastly, the data is transformed to the global coordinate system using the sensor to boat transformations. At this point, the transformed data received from the surface and underwater detection modules are sent to the obstacle map.

Additional challenges are associated with long-term obstacle mappings, such as memory and time constraints. We partially addressed this problem in our previous work using Octomaps [4]. From the perspective of the application, the entire shape of each obstacle is not required so that fewer essential points can be identified. Algorithm 1 transforms the 3D map into a 2D obstacle map based on the probability distribution of each scan. By comparing probability values on the z-axis, we keep only the points with the highest probability computed by Eq. 2.

## 4 Experimentation and Results

The methodology described in Sect. 3, has been tested on several water bodies. The entire experiment is conducted using Basilisk (USV) presented in our

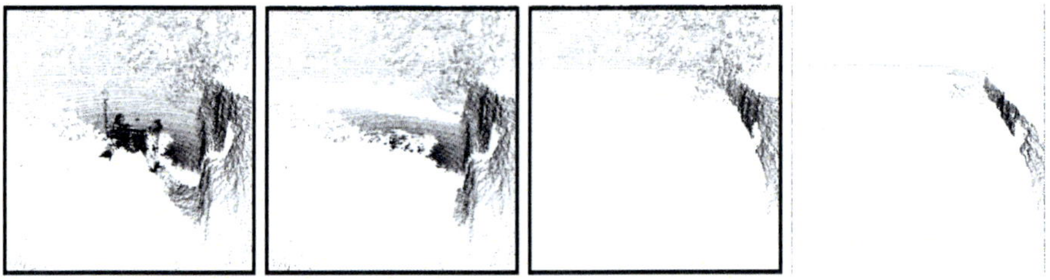


**Fig. 4.** Results for Mapping Methodology a) Raw LiDAR scan b) Box filter to clean the usv from the scan. c) Removal of floating non-destructive obstacles such as leaves by fusing LiDAR and Sonar scans d) Points filtration using of Algorithm 1

**Algorithm 1.** Point Reduction for Obstacle Map

**Input:** $S_i$ => Incoming scan.
$P_{ptx}$ => Probability Vector of points in scan.
**Require:** $S_f$ => Filtered scan with lesser points.
  **while** each point $k$ in $S_i$ **do**
    **while** each point $p$ in z-direction of $k$ **do**
      **if** $P_p \geq P_k$ **then** ▷ $P_p$ is probability of point p
        Delete $k$ in $S_i$
        $k \leftarrow p$
      **else**
        Delete $p$ in $S_i$
      **end if**
    **end while**
    $S_f \leftarrow k$
  **end while**

previous work [8]. It includes a commercial-grade multi-antenna GPS with an integrated IMU, a 90° vertical field of view (fov) with 128 lines of LiDAR. Additionally, a stereo camera and a forward-looking multibeam imaging sonar with a 120° horizontal field of view are used. In order to handle the large amount of data generated by the sensor system described above, an 8th Generation 6-Core i7 processor with 64GB of RAM and 16GB of Nvidia 2080 graphics memory is added.

The center axis between the rotors of the Basilisk is assumed to be the robot's body frame. The sonar receives the sound waves in 20° of vertical fov. Therefore, it is tilted 10° downwards to avoid backscattering from surface bubbles.

The Basilisk is driven at a maximum speed of 0.8 m/s and with a maximum curvature angle of 25°. Due to the unavailability of an actual flood scenario, the system is tested on two different shallow water sites, i.e., THW Testing Site at Hoya and Hamburg Port Testing Site, as shown in Fig. 5. Tests are made in November 2021 with an average 0–5 °C temperature and light rain.

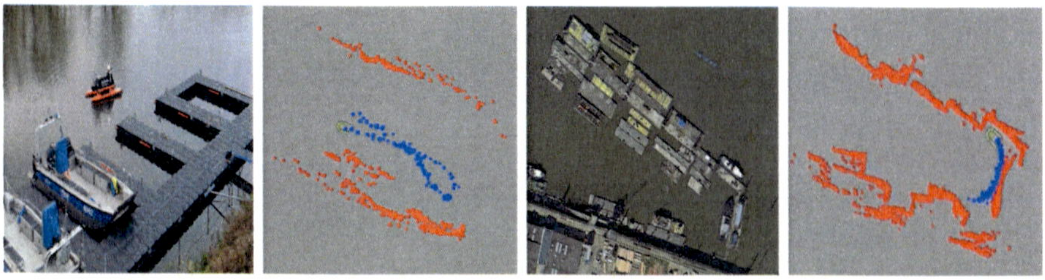

(a) Hoya (b) Obstacle Map (c) Hamburg Port (d) Obstacle Map

**Fig. 5.** Red points show the surface water map whereas blue points show underwater map. (a, c) depict the sites whereas (b, d) show the generation of map. The labeled arc area is the static ship standing at the site.

Figure 3 displays the underwater obstacle detection steps described in part Sect. 3.1. The parameters of each step are tailored to the specific testing site. The computation time of the underwater obstacle detection algorithm is primarily affected by the frequency of images under consideration in the image averaging step. The noise decreases with more images, but computation has been slower. Therefore, the selected value of N for experiments provides real-time computation and better results in terms of noise.

Additionally, Fig. 4 illustrates the steps of surface water filtration presented in Sect. 3.2. Finally, Fig. 5 illustrates the final maps created at the Hoya and Hamburg port testing sites in 2021. There is a maximum drift of ±15 cm between surface and underwater maps using Octomaps.

## 5 Conclusion

The primary problem of the destructed environment is the lack of information regarding changes to maps. There is thus a dire need for a mapping mechanism that caters to the problems posed by the different scenarios of surface and underwater obstacles. This paper presents a probabilistic model that fuses surface and underwater detected obstacles in a map using Octomaps. An obstacle's surface and underwater profile are used to compute the probability of detection. We plan to use a stereo camera to gather information about the texture of surface water environments. Moreover, the detection probability for obstacles should be considered based on their composition and roughness.

## References

1. Bioresita, F., Puissant, A., Stumpf, A., Malet, J.P.: A method for automatic and rapid mapping of water surfaces from sentinel-1 imagery. Remote Sens. **10**(2) (2018). https://doi.org/10.3390/rs10020217. https://www.mdpi.com/2072-4292/10/2/217

2. Fernandez-Diaz, J.C., et al.: Early results of simultaneous terrain and shallow water bathymetry mapping using a single-wavelength airborne lidar sensor. IEEE J. Sel. Top. Appl. Earth Obs. Remote Sens. **7**(2), 623–635 (2014). https://doi.org/10.1109/JSTARS.2013.2265255
3. Han, J., Cho, Y., Kim, J.: Coastal SLAM with marine radar for USV operation in GPS-restricted situations. IEEE J. Oceanic Eng. **44**(2), 300–309 (2019). https://doi.org/10.1109/JOE.2018.2883887
4. Hornung, A., Wurm, K., Bennewitz, M., Stachniss, C., Burgard, W.: OctoMap: an efficient probabilistic 3D mapping framework based on octrees. Auton. Robots **34** (2013). https://doi.org/10.1007/s10514-012-9321-0
5. Karoui, I., Quidu, I., Legris, M.: Automatic sea-surface obstacle detection and tracking in forward-looking sonar image sequences. IEEE Trans. Geosci. Remote Sens. **53**(8), 4661–4669 (2015). https://doi.org/10.1109/TGRS.2015.2405672
6. Keen, H.E., Berns, K.: Generation of elevation maps for planning and navigation of vehicles in rough natural terrain. In: Berns, K., Görges, D. (eds.) RAAD 2019. AISC, vol. 980, pp. 488–495. Springer, Cham (2020). https://doi.org/10.1007/978-3-030-19648-6_56
7. Keen, H.E., Jan, Q.H., Berns, K.: Drive on pedestrian walk. TUK campus dataset. In: 2021 IEEE/RSJ International Conference on Intelligent Robots and Systems (IROS), pp. 3822–3828 (2021). https://doi.org/10.1109/IROS51168.2021.9636848
8. Meckel, D., Keen, H., Heupel, C., Berns, K.: Transferring off-road control concepts to watercraft used in flooded areas. In: Berns, K., Dressler, K., Kalmar, R., Stephan, N., Teutsch, R., Thul, M. (eds.) ICVTS 2022, pp. 121–136. Springer, Cham (2023). https://doi.org/10.1007/978-3-658-40783-4_9
9. Raber, G.T., Schill, S.R.: Reef rover: a low-cost small autonomous unmanned surface vehicle (USV) for mapping and monitoring coral reefs. Drones **3**(2) (2019). https://doi.org/10.3390/drones3020038. https://www.mdpi.com/2504-446X/3/2/38
10. Sinisterra, A., Dhanak, M., Kouvaras, N.: A USV platform for surface autonomy. In: OCEANS 2017 - Anchorage, pp. 1–8 (2017)
11. Verpoorter, C., Kutser, T., Tranvik, L.: Automated mapping of water bodies using Landsat multispectral data. Limnol. Oceanogr. Methods **10** (2012). https://doi.org/10.4319/lom.2012.10.1037
12. Vilarnau, N.H.: Forward-looking sonar mosaicing for underwater environments (2014)
13. Watanabe, A., Kuri, M., Nagatani, K.: Field report: autonomous lake bed depth mapping by a portable semi-submersible USV at Mt. Zao Okama Crater lake. In: 2016 IEEE International Symposium on Safety, Security, and Rescue Robotics (SSRR), pp. 214–219 (2016). https://doi.org/10.1109/SSRR.2016.7784301
14. Zaheer, M.H., Mehdi, S.A., Keen, H.E., Berns, K.: Detection of fungus in gladiolus fields using a quadcopter. In: Zeghloul, S., Laribi, M.A., Sandoval, J. (eds.) RAAD 2021. MMS, vol. 102, pp. 127–134. Springer, Cham (2021). https://doi.org/10.1007/978-3-030-75259-0_14

# Effective Visual Content of eHMI for Autonomous Vehicles in Pedestrian Zones

Qazi Hamza Jan(✉), Priyanka Dahiwal, and Karsten Berns

Robotic Research Lab, Department of Computer Science,
RPTU Kaiserslautern-Landau, Gottlieb-Daimler-Str, Building-48,
67663 Kaiserslautern, Germany
{hamza,berns}@cs.uni-kl.de, pdahiwal@rhrk.uni-kl.de

**Abstract.** Comprehensive and lucid interaction is of utmost importance when driving Autonomous Vehicles (AVs) in pedestrian zones. It facilitates pedestrian safety by conveying the vehicle's intentions and engaging them for priority. The focus of this work is to analyze the effects of various exterior car displays on pedestrian behavior. The analysis is conducted on an autonomous bus used as an intra-university transport system, equipped with LED displays. The experiments are performed in two phases: first, highly rated interaction content is chosen based on questionnaires from the participants. Secondly, the interaction contents from phase 1 are shown in sequence. The results provide insight into pedestrian understanding based on the type and attributes of signals with various combinations.

**Keywords:** Human Machine Interaction · Pedestrian Zones · Interaction Systems · Autonomous Vehicles · LED Displays · eHMI

## 1 Introduction

Safe and smooth driving of AVs in pedestrian zones requires interaction with pedestrians [5]. Interaction provides pedestrians with a trustworthy and friendly perspective, reducing confusion when encountering an AV. Therefore, interacting modules are necessary for smooth and safe driving. Researchers have largely used eHMI [2,6,7] as a visual aid to convey messages. Numerous text messages [1,3] are used for conveying intention or commands for pedestrians.

To understand the efficacy of the interaction system, experiments were performed in this work using two basic components of interaction systems: display and sound. A thorough analysis was conducted to evaluate pedestrian understanding of interactions by showing different signals on the display, as can be seen in Fig. 1.

Plethora of content can be conveyed through such modules, including text, images, and animations, along with various attributes such as size, blinking speed, color, etc. Uniqueness in this work is refining the display content through


T. Petrič et al. (Eds.): RAAD 2023, MMS 135, pp. 203–210, 2023.
https://doi.org/10.1007/978-3-031-32606-6_24

**Fig. 1.** Experimental setup for static Testing Phase-1 (left) and phase-2 (right).

experiments where pedestrian opinions are integrated, and the system is continuously updated. The goal of this work is to narrow down various visual, such as animations, texts, and signs, for effective and straightforward interaction given the insufficient time during crossing an AV. This creates a concrete base for the system, reducing the chances of mishaps in the real environment.

## 2 System and Approach

The interaction system was designed for a driverless bus [4], intended to operate on the RPTU Kaiserslautern-Landau campus. The minibus can transport passengers from building to building, and the interacting modules were developed to convey information about the vehicle's state, the priority between the vehicle and pedestrians, and the passenger entry and exit states. The LED display is mounted at the front of the vehicle, as shown in Fig. 1, which ensures it is not easily overlooked when crossing the vehicle from the front.

Figure 2 illustrates the overall system design, from the input GUI to the output interacting modules. To aid understanding, the structure has been hierarchized into the Physical layer, Logical layer, and System interface, which are explained further in this section.

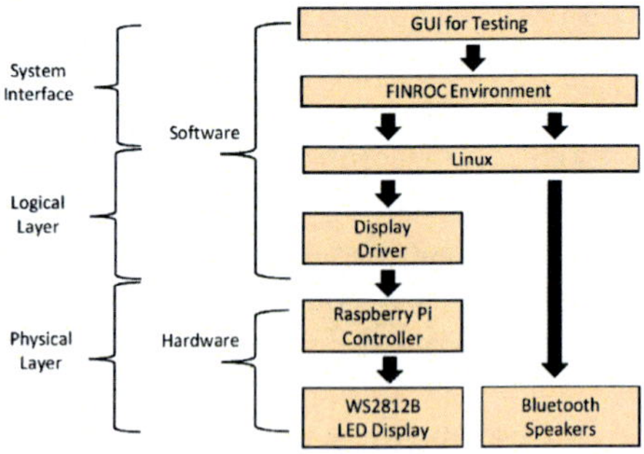


**Fig. 2.** System design.

### 2.1 Physical Layer

The physical layer of the system comprises all hardware components located at the front and back ends. This includes the Raspberry Pi 4, 4 x WS2812b Display Strip, and speakers. The Raspberry Pi 4 was chosen as the system controller due to its computing ability for real-time scenarios, low weight, and power consumption.

The WS2812b LED Display size is an $8 \times 16$ $(h \times l)$ matrix strip. It is possible to connect multiple strips in series to achieve the desired configuration. After evaluating the size of the text and the brightness of the LED strip, it was decided to use four strips arranged in a way that forms a display board of 16 by 64 LED matrix. This arrangement was considered readable from a distance of 10 m (a sufficient distance at walking speed). To swiftly view different types of signals such as texts or images, the code was designed to overlay an RGB image over the given display matrix. In this way, every text and image was first created in an image editor tool, and then the ".png" files were stored in the Raspberry Pi.

### 2.2 Logical Layer

The logical layer comprises the software part, which includes the display device driver and the Linux system connection. The Raspberry Pi controls the display through the display device drivers, which link the display to the Linux system.

The display device driver contains all information about the LED display, such as the number of rows, columns, LEDs, brightness control, color information, pin connection information, and code of color assigned to each pixel. It also contains information about the image folder and the process of reading and rendering images to the display. Through the Linux operating system, the hardware part of the system is connected to the FINROC architecture[1] (software used in the Robotic Research Lab).

### 2.3 System Interface Layer

This layer serves as a possibility to control the output on the interacting modules. A GUI was built in order to activate the required outputs.

## 3 Methodology

Static testing was executed, where the vehicle was parked in a controlled environment, and participants were asked to stand in front of the vehicle. Static testing supports lesser uncertainty in pedestrians' reactions. This testing gave us insight into the pedestrian mindset and their responses towards particular attributes of the interacting content. Static testing was conducted in two phases.

**Phase 1:** Static testing was conducted while the system delivered all primitive variations. It facilitated the decision on which signs have the most impact and are

[1] https://www.finroc.org/.

easy to understand. Also, suggestions throughout the experiments were collected, which were implemented in the second phase.

**Phase 2:** The directed content from phase 1 was shown in a sequence for phase 2. Participants were required to cross the vehicle with the sequence running. The purpose of this phase was to know the ease of understanding the sequence of interaction while driving in a real scenario. This is sometimes challenging due to the short interaction period. The interacting modules were also mounted on the target vehicle.

The participants were allowed to stand near the vehicle and were also asked to change positions according to certain test cases, such as standing parallel or perpendicular to the vehicle. Based on the test scene being shown, participants were asked questions about recognition of signs, visibility, and understandability. After explaining the test scene, their opinions were collected through a questionnaire.

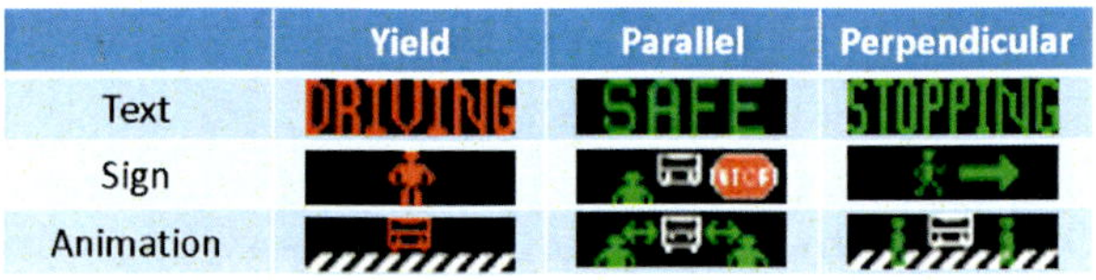


**Fig. 3.** Initial distribution of content.

To begin with the interaction content, primary categories were introduced based on the activities observed around a driving AV; yielding, parallel and perpendicular crossing. These activities were further divided into representations such as texts, signs, and animations. Finally, the representations were displayed in different color combinations and animation styles, resulting in over 200 possible combinations. Example of some primitive design are shown in Fig. 3.

## 3.1 Participants

In phase 1 of static testing, 14 participants aged 23 to 58 were included. Of these, 1 was a female participant and 13 were male participants. All participants were interacting for the first time.

In the second phase of static testing, data from 20 participants was collected. 14 participants were male and 6 were female with the same age range mentioned above. These participants included students from the university, research and development associates, and professors. Out of these 20 participants, 6 were also involved in the first phase of testing. Some of these participants were aware of the experiment as they had been involved in a similar kind of activity, but some of them were completely new to the experiment.

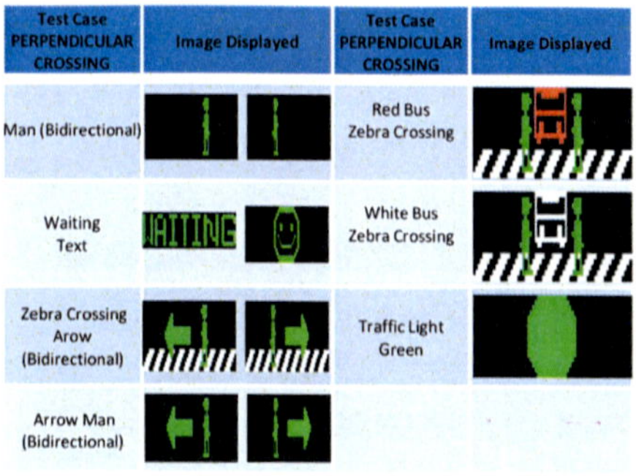


(a) Test cases used in Phase 1 showing various combinations for *Yielding and Perpendicular Crossing.*

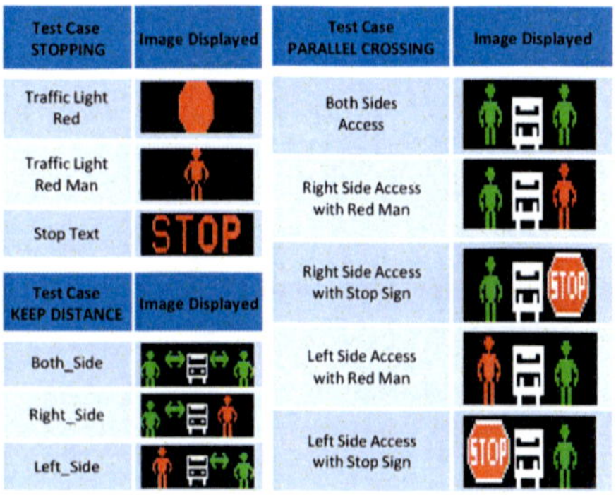


(b) Test cases used in Phase 1 showing various combinations for *Non-Yielding Perpendicular Crossing, Parallel Crossing* and *Keep Safe distance.*

**Fig. 4.** Test phase 1

## 3.2 Questionnaire

The questionnaire included questions aimed at gaining insight into pedestrian's understanding of the displayed content and the sequence of the content. Two different objectives were targeted during the two phases of static testing. In phase 1, the objective was to collect opinions from pedestrians and narrow down to valid content. In phase 2, the objective was to understand the sequence of the content. This can be seen for phase 1 testing in Figs. 4a, 4b, and 5. Figures 6a and 6b show test scenes with their respective sequences used in phase 2. One example is the yielding scenario in perpendicular crossing, which includes the following sequence: DRIVING - EYES (to check the environment) - STOPPING (with a countdown timer) - Perpendicular crossing test case (with Red Bus, ZC, Ani-

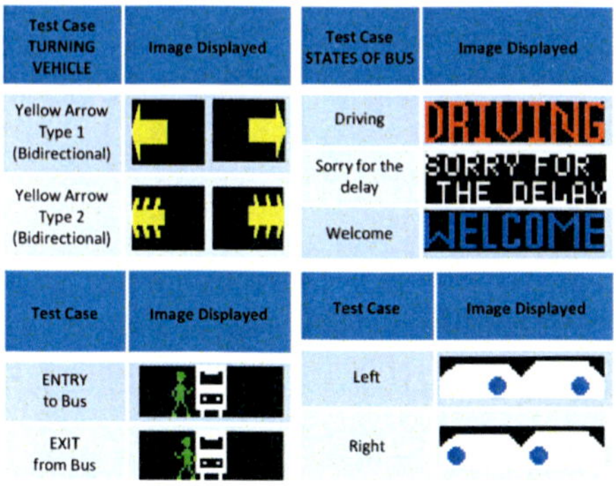


**Fig. 5.** Test cases used in Phase 1 showing two types of *Turning arrows, Entry and Exit* Cases for Bus, *States* of Bus and *Eyes*.

mation of walking man) - RESUMING (with a countdown timer) - DRIVING. With such sequences, questions were asked in a yes-no form regarding recognition, understandability, whether the participants would follow the sequence, and whether they would feel safe in natural circumstances.

| Test Scenes | Sequence |
|---|---|
| Yielding | DRIVING |
| | EYES |
| | HAPPY EMOTICON |
| | STOPPING 5...1 |
| | RED BUS |
| | RESUMING 5...1 |
| | DRIVING |

| Test Scenes | Sequence |
|---|---|
| Non-Yielding | DRIVING |
| | EYES |
| | STOP TEXT |
| | STOP MAN |
| | HAPPY EMOTICON |
| | DRIVING |

| Test Scenes | Sequence |
|---|---|
| Parallel Crossing | DRIVING |
| | EYES |
| | Happy Emoticon |
| | Parallel Crossing Both |
| | Driving |

| Test Scenes | Sequence |
|---|---|
| Vehicle Turning | DRIVING |
| | EYES |
| | VEHICLE TURNING LEFT and Right |
| | DRIVING |

(a) Test Scenario and Sequences used in Phase 2 for *Yielding*, *Non-Yielding*, *Parallel Crossing* and *Vehicle Turning*.

| Test Scenes | Sequence |
|---|---|
| KEEP SAFE DISTANCE BOTH, RIGHT, LEFT | DRIVING |
| | EYES |
| | KEEP SAFE DISTANCE |
| | DRIVING |

| Test Scenes | Sequence |
|---|---|
| ENTRY_EXIT | DRIVING |
| | STOPPING 5... |
| | EXIT FROM BUS |
| | WAITING |
| | ENTRY TO BUS |
| | RESUMING 5... |
| | DRIVING |

| Test Scenes | Sequence |
|---|---|
| SORRY_FOR_DELAY | Driving |
| | Stopping 5.... |
| | Sorry for delay |
| | Waiting |
| | Resuming 5... |
| | Driving |

(b) Test Scenario and Sequences used in Phase 2 for *Keep Safe Distance*, *Entry- Exit from Bus* and *Sorry-For-Delay*.

**Fig. 6.** Test phase 2

# 4 Results

During static testing phase 1, we established the interaction requirements. The readability of signs, text, or figures depended on the display size. The context of messages conveyed changed depending on the position of the bus or pedestrian. The display position could target different audiences in surrounding areas, and different messages needed to be conveyed simultaneously to successfully interact with them. Animations of text, signs, or figures were better understood than still ones. Colors like red were perceived as dangerous or alert, while green was considered safe. Zebra crossings played an important role in pedestrian decision-making. Blinking arrows were mostly associated with the bus intention. In static testing phase 2, some participants were also involved in phase 1, and they found the experiment much more understandable.

**Static Testing Phase 1:** Two subjective opinions, *Recognition* and *Preferences*, were reckoned for the displayed content. *Recognition* is how easy it is to recognize the content, and *preference* is which content is liked independent of recognition.

Based on the data collected through the questionnaire, bar plots are shown. In these plots, the X-axis refers to the test case while the Y-axis refers to the percentage of recognition or preference. Under perpendicular crossing, various combinations are tested, and their recognition and preferences are shown on the

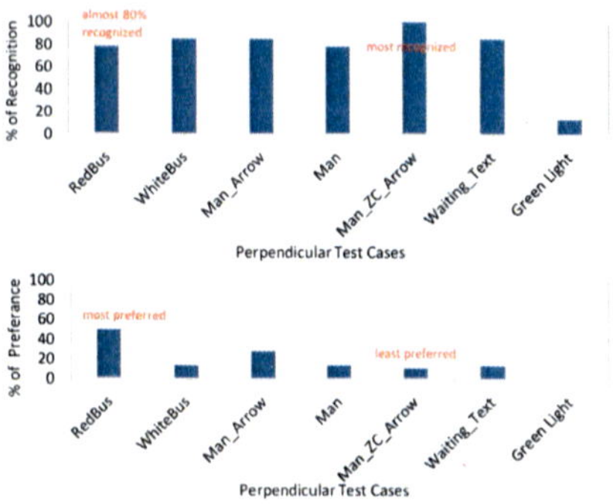


(a) Result of Phase 1 showing percentage of recognition (top) and preference (bottom) vs test cases in *Yielding Perpendicular Crossing.*

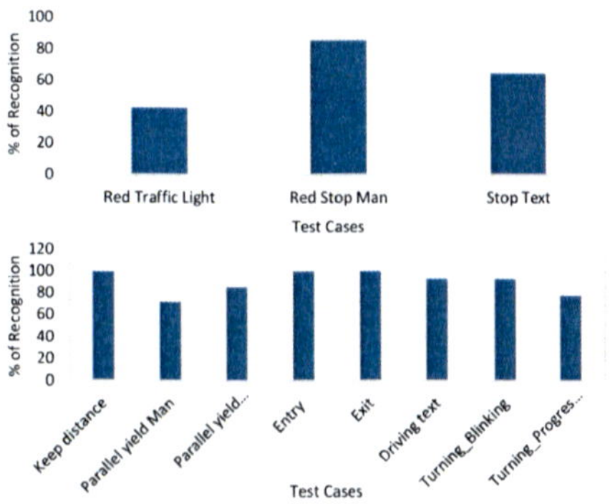


(b) Result of Phase 1 showing percentage of recognition (top) vs test cases (bottom) in *Non-Yielding Perpendicular Crossing* and other test cases.

**Fig. 7.** Test phase 1

top and bottom of Fig. 7a, respectively. The results from these plots imply that the *Man-Zebra Crossing-Arrow* combination is the most recognized but is the least preferred. On the other hand, the *Red Bus-Zebra Crossing-Man* combination is recognized by 80% of participants, and it is also the most preferred. Similarly, the bar plots shown in Fig. 7b (above plot) for non-yielding perpendicular crossing indicate that the *red stop man* is the most recognized sign for the same test case. Figure 7b (below plot) shows the plots for other cases, which indicate that *keep safe distance, entry case, and exit* case are perceived as expected. While some test cases are not understood by all participants, the majority of participants understood them.

**Static Testing Phase 2:** Figure 8 is a bar plots for phase 2, test sequences for *Yielding, Entry - Exit, Sorry for Delay,* were 100% recognized, understood,

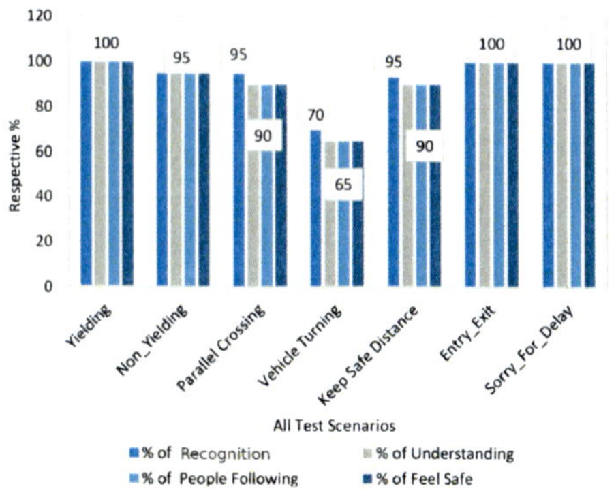


**Fig. 8.** Result of Phase 2 showing percentage of recognition, understanding, follow and feel safe vs Test Scenarios.

followed and feel safe while doing so. Two test scenes, which are *Parallel crossing* and *Keep safe Distance* perceived as similar. *Non-yielding* scenario, is well perceived by 95% of participants. *Vehicle turning* scenario is yet to improve such that its perception level should go above 90%, currently it is in the range of 60 to 70%.

## References


1. Bazilinskyy, P., Kooijman, L., Dodou, D., de Winter, J.C.: Coupled simulator for research on the interaction between pedestrians and (automated) vehicles. In: Driving Simulation Conference Europe, Antibes, France (2020)
2. Carmona, J., Guindel, C., Garcia, F., de la Escalera, A.: eHMI: review and guidelines for deployment on autonomous vehicles. Sensors **21**(9), 2912 (2021)
3. Deb, S., Rahman, M.M., Strawderman, L.J., Garrison, T.M.: Pedestrians' receptivity toward fully automated vehicles: research review and roadmap for future research. IEEE Trans. Hum.-Mach. Syst. **48**(3), 279–290 (2018)
4. Jan, Q.H., Berns, K.: Safety-configuration of autonomous bus in pedestrian zone. In: VEHITS, pp. 698–705 (2021)
5. Jan, Q.H., Klein, S., Berns, K.: Safe and efficient navigation of an autonomous shuttle in a pedestrian zone. In: Berns, K., Görges, D. (eds.) RAAD 2019. AISC, vol. 980, pp. 267–274. Springer, Cham (2020). https://doi.org/10.1007/978-3-030-19648-6_31
6. Lim, D., Kim, B.: UI design of eHMI of autonomous vehicles. Int. J. Hum.-Comput. Interact. **38**(18–20), 1944–1961 (2022)
7. Troel-Madec, M., et al.: eHMI positioning for autonomous vehicle/pedestrians interaction. In: Adjunct Proceedings of the 31st Conference on l'Interaction Homme-Machine, pp. 1–8 (2019)

# Medical Robotics

# A Novel Intraoperative Force Estimation Method via Electrical Bioimpedance Sensing

Zhuoqi Cheng[1(✉)] and Leonardo S. Mattos[2]

[1] Maersk Mc Kinney Moller Institute, University of Southern Denmark, Odense, Denmark
zch@mmmi.sdu.dk
[2] Department of Advanced Robotics, Istituto Italiano di Tecnologia, Genova, Italy
Leonardo.DeMattos@iit.it

**Abstract.** During minimally invasive robotic surgery (MIRS), intraoperative force sensing on the surgical tool plays an important role to mitigate the risk of inadvertent tissue damage. This study proposes a novel method for estimating the force exerted by the surgical tool tip based on the electrical bioimpedance (EBI) of the contacting tissue. When the surgical tool is pressing on the tissue, the tissue is deformed corresponding to the exerted force. Meanwhile, the measured EBI value changes accordingly since tissue deformations change the contact area between the tool and the tissue. An ex-vivo experimental study was performed and different machine learning methods were tested for correlating the exerted force and the EBI values. The experimental results show that a Feed-forward Multi-layer Neural Network (F-MNN) can provide good results in terms of efficiency and accuracy. The exerted force on the tissue can be accurately estimated with a median error of 0.072 N and with about 7 ms testing time. In addition, the proposed method has significant advantages over other techniques since it requires little hardware modification and allows fast and seamless integration with the existing surgical robotic system.

**Keywords:** Force sensing · electrical bioimpedance · robot-assisted surgery

## 1 Introduction

Minimally invasive robotic surgery (MIRS) has been demonstrated to significantly improve surgical outcomes, but different aspects of current systems can still be improved. One of the main issues is the perception feedback for surgeons. For most existing surgical robots, only visual information is provided. Haptic feedback is considered valuable for improving the surgical quality, but it is absent for most robotic systems. Without force feedback, surgeons were found to generally increase the average applied force on the tissue [17], tending


T. Petrič et al. (Eds.): RAAD 2023, MMS 135, pp. 213–220, 2023.
https://doi.org/10.1007/978-3-031-32606-6_25

to increase the risk of tissue damages. For example, as reported by Giannarou *et al.* [7], the allowed pressing force on brain tissue during a neurosurgery should be constrained to be less than 0.3 N. Failure to stay below this force threshold can result in an iatrogenic injury, which in turn carries a risk of disability or even death [9,14]. In addition, the magnitude of the exerted force on liver is required to be controlled. Liver tissue can be damaged if a stress over 162.5 kPa is generated by the exerted force [16].

For restoring the pushing force feedback capability in MIRS, the most common method has been to design integrated force sensors for the surgical tools. In the study of Arata *et al.* [1], a force/torque sensor was designed and mounted in the driving units of a slave robot arm. By this means, the forces and moments acting at the surgical instrument were sensed indirectly. In another study [12], Marcus *et al.* developed a handheld robot for assisting the surgeon in handling delicate tissue by constraining the applying force on the tissue during microsurgery. In addition, a miniaturized 6-DOF force/torque sensor was developed [10]. This sensor was attached to the distal end of a forceps instrument, enabling direct measurement of forces and moments acting on it. However, the above designs require complicated and costly hardware development. As an alternative, a novel forceps with controllable constant torque has been developed [5]. Also, other researchers have proposed to estimate the exerted force by observing the tissue deformation using laparoscopic vision [7].

In this study, a new force sensing method is proposed by measuring the electrical bioimpedance (EBI) of the contacting tissue during the tool-tissue interaction. The proposed method is verified based on a bipolar robotic surgical tool which is commonly used for electrification [8]. Such tools have a pair of electrically isolated jaws which can be used as bipolar electrodes for measuring the EBI of the tissue in contact [2]. When the forceps is pushing on the soft tissue, the tip of the forceps is immersed into it. This changes the contact area between the tool tip and the tissue, which consequently changes the electric field applied on the tissue under measurement, altering the measured EBI value. Meanwhile, a corresponding reacting force is generated. This study aims to model the relationship between the exerted force and the measured EBI value, which can allow estimating the pushing force based on EBI measurements. The proposed method can be easily integrated into the readily available surgical system, requiring only small modifications to existing instruments.

## 2 Methodology

This study uses a hepatic MIRS scenario as an example for illustrating the proposed force sensing method. The proposed method allows setting the maximum interaction force to be 1 N to avoid damaging the liver tissue [15,16].

As shown in Fig. 1(A), the EBI of the contacting tissue is measured continuously via two jaws of a bipolar forceps. An alternative voltage $U$ is applied between them and the reciprocal current $I$ is measured. According to [4], the current $I$ can be obtained by integrating the current density $J$ through the tissue range $r$.

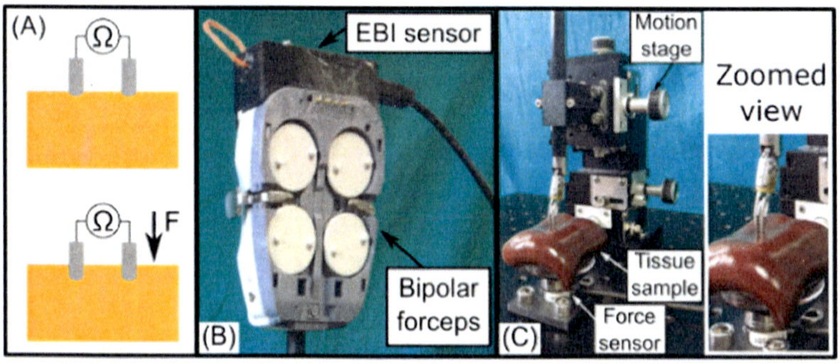


**Fig. 1.** (A) $|Z|$ is measured between two jaws of the forceps during its pressing on the soft tissue; (B) The EBI sensor prototype integrated to a da Vinci Maryland bipolar forceps; (C) The experimental setup for collecting both the pressing force and the EBI value simultaneously.

$$|Z| = \int_{\Omega} (\frac{J}{\hat{\sigma}}) d\Omega = \frac{1}{2\pi\hat{\sigma}} (\frac{1}{r_0} - \frac{1}{r}) \tag{1}$$

where $r_0$ is the equivalent radius of the electrode immersing in the tissue, and $\hat{\sigma}$ represents the complex conductivity of tissue. When the forceps is pressing deeper, $r_0$ increases gradually. In this case, the measured EBI value $|Z|$ decreases accordingly. Thus, the force estimation can be done by the correlation between the pressing force $F$ and the EBI value of the contacting tissue $|Z|$.

## 3 Experimental Setup

An ex-vivo experiment was conducted for measuring the exerted force and the corresponding EBI value simultaneously. The EBI sensor designed and presented in the previous study [2] was adopted. The EBI sensor integrates a micro-controller (Atmega328P, Atmel Inc., USA) and an impedance converter (AD5933, Analog Devices Inc., USA), allowing accurate EBI measurement at multiple frequencies. A prototype was made as shown in Fig. 1(B), which could be directly mounted on a da Vinci Maryland bipolar forceps (Intuitive Surgical Inc., USA). The measured EBI values were streamed in real-time to a PC via an USB connection. In this study, five excitation frequencies (20, 40, 60, 80 and 100 kHz) were chosen and applied for the EBI measurement. The frequencies within this range were found to provide both intracellular and extracellular electrical characteristics of the tissue under measurement [13].

Figure 1(C) shows the experimental setup for collecting both the pressing force and the EBI value simultaneously. To measure the downward force exerted from the top of the tissue by the forceps, a force sensor Nano17 (ATI Industrial Automation, Inc., USA) was used and placed beneath the tissue sample carrier. The Maryland bipolar forceps was fixed to a precision motion stage (Siskiyou Co., USA) with the EBI sensor connected to its poles. In this work, we upgraded the motion stage shown in Fig. 1(C) to a motorized version. The forceps was driven to move 8 mm downwards and then 8 mm upwards in a constant speed. Two

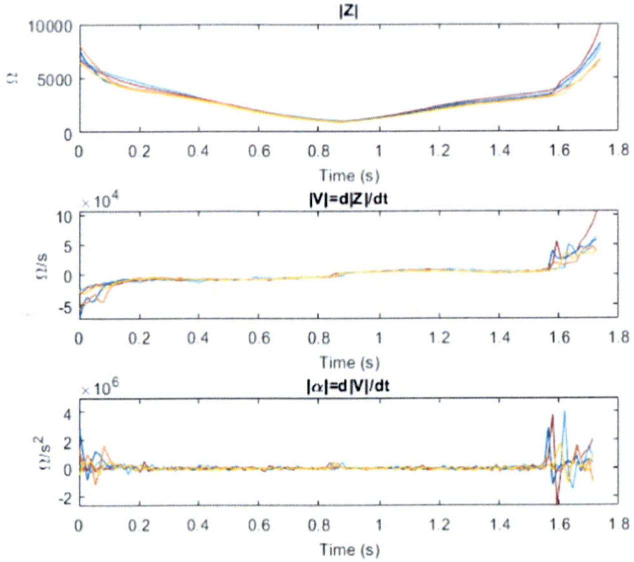


**Fig. 2.** Example of input EBI data: the top figure is the 5D EBI values $|Z|$, the middle figure is the 5D velocity of the EBI values $|V|$ and the bottom figure is the 5D acceleration of the EBI values $|\alpha|$. Different colours are used for representing different $|Z|$ components at 5 different excitation frequencies.

speed settings were tested, namely 0.1 and 0.05 mm/s. During the experiment, the pressing force in Z direction and the corresponding EBI values were recorded at the same time.

## 4 Data Analysis

The aim of this procedure is to construct a regression model $f \in \mathcal{F}$ parametrized in $\Theta \in \mathbb{R}^d$ such that $\hat{F} = f(\mathtt{X}; \Theta)$ is close to the real output $F$. The collected time-varying input-output pairs $\mathcal{P}_t = \{\mathtt{X}_t, F_t\}$ were divided into the training dataset $\mathcal{P}_t^A$ for establishing the prediction model $f$ and the testing datasets $\mathcal{P}_t^B$ for evaluating the performance of the model $f$. Considering that soft tissue is normally modelled as a spring-damper system, the velocity of the EBI values $|V|$, can play an important role for predicting $F$. In addition, the acceleration of the EBI values $|\alpha|$ could potentially improve the performance of the regression model, and thus it was considered for model comparison. In the following regression study, three different input components, i.e., 5D-$[|Z|]$, 10D-$[|Z|, |V|]$ or 15D-$[|Z|, |V|, |\alpha|]$, were evaluated in terms of accuracy and efficiency. The regression study was implemented in MATLAB 2018b. Figure 2 shows the three types of inputs computed on one set of the collected data. The EBI values $|Z|$ at 5 different excitation frequencies were plotted in different colours.

Artificial Neural Network (ANN) is selected to solve the regression problem, which does not require the mathematical model of the system and has the learning capacity to fit any complex function and non-linear relationships. The following factors which could limit the performance of the model were investigated in this study, namely, the amount of neurons in the hidden layer, the number of hidden layers, and the types of ANN and the dimension of inputs. To find the best model $f^*$, both of the Feed-forward Neural Network (F-NN) and Cascade-forward NN (C-NN) with the back-propagation learning strategy were implemented in this work. Meanwhile, the adopted NN models were constructed with both single layer (SNN) and multi-layer (MNN) for performance comparison. The Broyden-Fletcher-Goldfarb-Shanno algorithm [11] was used to calculate the maximum or minimum gradient during the network parameters optimization. To improve the effectiveness of the ANN models, we adopted a series of optimization processing, like setting 1.05 as the increase learning rate and 6 as the maximum validation failures.

The performance of the model $f$ was evaluated by measuring the Root Mean Square Error (RMSE) and the Pearson correlation coefficient (Corr) (Eq. 2).

$$\begin{aligned} RSME &= \sqrt{\frac{1}{N}\sum_{t=1}^{N}(\hat{F}_t - F_t)^2} \\ Corr &= \frac{1}{N-1}\sum_{t=1}^{N}\left(\frac{\hat{F}_t-\mu_{\hat{F}}}{\sigma_{\hat{F}}}\right)\left(\frac{F_t-\mu_F}{\sigma_F}\right) \end{aligned} \tag{2}$$

where $N$ demotes the length of the observed force sequences $F$ and the predicted sequences $\hat{F}$. $\mu_{\hat{F}}$ and $\sigma_{\hat{F}}$ are the average and standard deviation of $\hat{F}$, while $\mu_F$ and $\sigma_F$ represent the same values of $F$.

## 5 Results

The comparison experiments were implemented with the leave-one-out strategy. In total, 26 groups of datasets were collected, half of which were collect in a pushing speed of 0.1 mm/s and the others were in a speed of 0.05 mm/s. We use 25 datasets to build the ANN-based models and a randomly selected dataset (from those acquired using 0.05 mm/s) to evaluate and compare the performance among them. Figure 3 shows the RMSE and correlation coefficient of the four models including two MNN models (C-MNN and F-MNN) with two hidden layers and two SNN models (C-SNN and F-SNN). For the results showing in Fig. 3, both MNN models choose 20 and 10 neurons for each layer and both SNN models have 30 nodes in their hidden layer.

The comparison results of RMSE illustrate that the F-MNN is the best regression model with 10D inputs, which has the lowest RMSE (median value 0.0722 $N$, standard deviation 0.0078). Although all of the ANN models have similar results of Corr ($\approx$0.92) according to the tests on the 10D and 15D inputs datasets, the F-MNN model gets the most robust results with the lowest standard deviation (0.0116).

The first step of model construction proved that F-MNN with 10D inputs ($\mathtt{X} = [|Z|, |V|]$) is the best choice for mapping the EBI values to the pressing

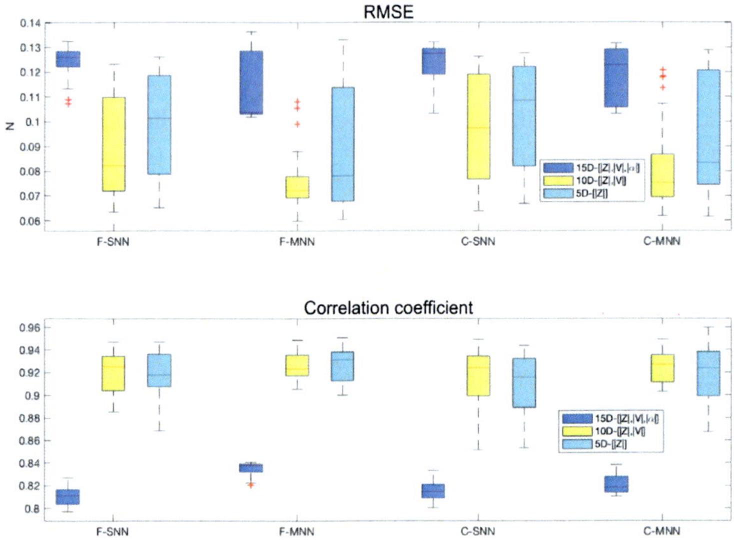


**Fig. 3.** The comparison results of RMSE and correlation coefficient among the four ANN-based models, i.e., F-SNN, F-MNN, C-SNN and C-MNN. For each model, it is tested by the three types of inputs with 5D (purple), 10D (yellow) and 15D (cyan). (Color figure online)

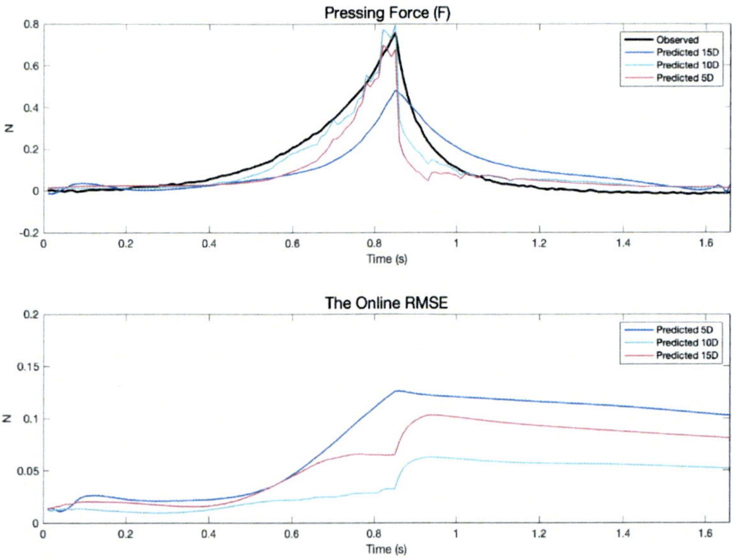


**Fig. 4.** An example of the comparison results of the predicted pressing forces and the online RMSE on the testing datasets with 5D, 10D and 15D inputs.

force within the sets of tested models. Furthermore, the number of neurons in the hidden layer were tested in order to optimise the performance of ANN-based model $f$. Even though the results were close to each other, the F-MNN model was found to be the best model for fast and robust computation with high accuracy.

Figure 4 shows the predicted pressing force $\hat{F}$ and the calculated RMSE based on the F-MNN model. The results demonstrate that the designed model can track the observed force with 10D inputs accurately. The RMSE is smaller than 0.02 N during the forceps pressing into the tissue and smaller than 0.06 N when it was pulled up.

## 6 Discussion and Conclusion

This study proposed a new method for estimating the pushing force of surgical tool on soft tissue, aiming to incorporate force feedback onto the RMIS system with minimal hardware modifications. When the forceps presses on the tissue, the indentation of the tissue changes the contact area between the forceps and the tissue, and this is reflected on the measured EBI value. Although the electrical conductivity of bio-tissue can be compression-dependent, the compression rate in this study is generally low (about 20–30%) which corresponds to a <8% increase of $|Z|$ according to the study of Dodde *et al.* [6]. Therefore, the change of contacting area between the electrodes and the tissue during the tool-tissue interaction is considered as the main reason.

The regression model shows that the 10D model has the highest fitting accuracy. This is reasonable since the mechanical property of soft tissue can be modelled as a spring-damper system. Nevertheless, the second derivative, corresponding to $|\alpha|$, is found not contributing to $F$ estimation. Instead, it introduces noise to the model. This may be because a constant speed of forceps movement was used in the experiment. To investigate this factor, more experiments with dynamic tool-tissue interaction will be planned in the future study.

The main limitation of the proposed method is that the trained regression model is tissue specific. Since different tissues have different electrical properties [3], it would be required to characterise each tissue of interest before implementing the proposed method. In this case, an automated tissue identification method by EBI sensing [2] can also be used. With this prior knowledge, the proposed force sensing method can be applied in a more general manner. Nevertheless, considering the current limitations regarding model dependency on tissue type and the low controllability of the contacting area between the electrode and the tissue, the proposed method may be more appropriate for surgical scenarios requiring less precision of force measurement.

## References

1. Arata, J., Mitsuishi, M., Warisawa, S., Tanaka, K., Yoshizawa, T., Hashizume, M.: Development of a dexterous minimally-invasive surgical system with augmented force feedback capability. In: 2005 IEEE/RSJ International Conference on Intelligent Robots and Systems, pp. 3207–3212. IEEE (2005)

2. Cheng, Z., et al.: Design and integration of electrical bio-impedance sensing in surgical robotic tools for tissue identification and display. Front. Robot. AI **6**, 55 (2019)
3. Cheng, Z., Dall'Alba, D., Schwaner, K.L., Fiorini, P., Savarimuthu, T.R.: Robot assisted electrical impedance scanning for tissue bioimpedance spectroscopy measurement. Measurement **195**, 111112 (2022)
4. Cheng, Z., Savarimuthu, T.R.: Monopolar, bipolar, tripolar, and tetrapolar configurations in robot assisted electrical impedance sensing. Biomed. Phys. Eng. Express **8**(5), 055014 (2022)
5. Cheng, Z., Savarimuthu, T.R., Foong, S., Tan, U.X.: Design of adjustable constant force/torque mechanisms for medical applications. J. Mech. Robot. **15**(2), 025001 (2023)
6. Dodde, R., Bull, J., Shih, A.: Bioimpedance of soft tissue under compression. Physiol. Meas. **33**(6), 1095 (2012)
7. Giannarou, S., Ye, M., Gras, G., Leibrandt, K., Marcus, H.J., Yang, G.Z.: Vision-based deformation recovery for intraoperative force estimation of tool-tissue interaction for neurosurgery. Int. J. Comput. Assist. Radiol. Surg. **11**(6), 929–936 (2016)
8. Giulianotti, P.C., Bianco, F.M., Daskalaki, D., Gonzalez-Ciccarelli, L.F., Kim, J., Benedetti, E.: Robotic liver surgery: technical aspects and review of the literature. Hepatobiliary Surg. Nutr. **5**(4), 311 (2016)
9. Gonenc, B., et al.: Preliminary evaluation of a micro-force sensing handheld robot for vitreoretinal surgery. In: 2012 IEEE/RSJ International Conference on Intelligent Robots and Systems, pp. 4125–4130. IEEE (2012)
10. Kim, U., Lee, D.H., Yoon, W.J., Hannaford, B., Choi, H.R.: Force sensor integrated surgical forceps for minimally invasive robotic surgery. IEEE Trans. Rob. **31**(5), 1214–1224 (2015)
11. Kou, C.X., Dai, Y.H.: A modified self-scaling memoryless Broyden-Fletcher-Goldfarb-Shanno method for unconstrained optimization. J. Optim. Theory Appl. **165**(1), 209–224 (2015)
12. Marcus, H.J., et al.: A "smart" force-limiting instrument for microsurgery: laboratory and in vivo validation. PloS One **11**(9), e0162232 (2016)
13. Martinsen, O.G., Grimnes, S.: Bioimpedance and Bioelectricity Basics. Academic Press (2011)
14. Payne, C.J., Latt, W.T., Yang, G.Z.: A new hand-held force-amplifying device for micromanipulation. In: 2012 IEEE International Conference on Robotics and Automation, pp. 1583–1588. IEEE (2012)
15. Rosen, J., Brown, J., De, S., Hannaford, B.: Macro and micro soft-tissue biomechanics and tissue damage: application in surgical robotics. In: Rosen, J., Hannaford, B., Satava, R. (eds.) Surgical Robotics, pp. 583–618. Springer, Boston (2011). https://doi.org/10.1007/978-1-4419-1126-1_24
16. Rosen, J., Brown, J.D., De, S., Sinanan, M., Hannaford, B.: Biomechanical properties of abdominal organs in vivo and postmortem under compression loads. J. Biomech. Eng. **130**(2), 021020 (2008)
17. Semere, W., Kitagawa, M., Okamura, A.M.: Teleoperation with sensor/actuator asymmetry: task performance with partial force feedback. In: Proceedings of the 12th International Symposium on Haptic Interfaces for Virtual Environment and Teleoperator Systems, HAPTICS 2004, pp. 121–127. IEEE (2004)

# Remote Center of Motion Based on Null Space and Virtual Joints in a Surgical Robotics Research Platform

Miha Deniša[1(✉)], Jan Kranjec[1,2], and Aleš Ude[1]

[1] Humanoid and Cognitive Robotics Lab., Department of Automatics, Biocybernetics, and Robotics, Jožef Stefan Institute, Ljubljana, Slovenia
{miha.denisa,ales.ude}@ijs.si

[2] Faculty of Electrical Engineering, University of Ljubljana, Ljubljana, Slovenia
jk6238@student.uni-lj.si

https://abr.ijs.si/

**Abstract.** We propose an approach for Remote Center of Motion (RCM) control in a Robot-assisted Minimally Invasive Surgery (RMIS) setting. We define Remote Center of Motion constraint as a primary task. Its control includes an implementation of a virtual joint. Task tracking of the surgical tool's end-effector is treated as a secondary task and controlled by exploiting the redundancy of our system. To evaluate the approach, a RMIS research platform including two 7-DOF robot arms and a surgical tool was used. The platform is based on a modular and open platform for surgical robotics research [8]. A pick and place task is learned through teleoperation, via kinesthetic guidance of one of the robot arms. The results show applicability of our proposed approach, as the RCM constraint is ensured with high precision, while there is no drastic rise in the tracking error of desired pose of the surgical tool's end-effector.

**Keywords:** Robot-assisted Minimal Invasive Surgery · Remote Center of Motion · Research Platform

## 1 Introduction

Robot-assisted Minimally Invasive Surgery (RMIS) has gained popularity in the past 20 years [10], with systems like the da Vinci (Intuitive, Inc.) being available and widely used worldwide. Robotic surgical systems typically have multiple arms with surgical tools and an endoscope, all teleoperated by surgeons using remote controls. Compared to classic endoscopic surgery, robotic surgical tools have articulated wrists, allowing the surgeon a larger range of motion and the

We would like to thank Bojan Nemec and Leon Žlajpah (JSI) for advice regarding task velocity kinematics, and Kim Lindberg Schwaner (SDU) for the help with implementation of the MOPS platform.


T. Petrič et al. (Eds.): RAAD 2023, MMS 135, pp. 221–228, 2023.
https://doi.org/10.1007/978-3-031-32606-6_26

system allows for a more relaxed and ergonomically correct working position. Experimental research on surgical robotics, especially introduction of (partial) autonomy, is carried out in simulation or on various dedicated research platforms. The da Vinci Research Kit (dVRK) platform [4] is most closely related to the most common surgical system, but its availability is limited. While the Raven-II research platform [2] is more accessible, it is difficult to use for autonomous tasks, as its tendon driven actuators introduce large uncertainties for state estimation. On the other hand, the MOPS research platform [7,8] is designed as a modular and open systems, and it is based on use of general purpose robot arms. With this in mind we set up a MOPS-based system using KUKA LWR robot arms and Xi surgical tools.

One of the main constraints of RMIS control, is ensuring the tool moves through the insertion point of the patient's body, which can be achieved via Remote Center of Motion (RCM) approaches. Several various approaches to ensure RCM have been researched: based on implicit active constrains [5], constrained kinematic controllers [1], using redundant 7-DOF robots [6], passive admittance controllers [3], etc. While the majority of RMIS platforms include control via RCM, we implemented and evaluated our own approach based on virtual joints and null space. We performed an initial evaluation on our platform with a pick and place task.

## 2 Research Platform

While our RMIS research platform is based on the MOPS platform [8], some differences need to be noted. The main distinctions in our implementation are the lack of cameras, use of 7-DOF Kuka LWR robots, and substitution of dedicated teleoperation devices with a compliant robot, which can be guided by a human in gravity compensation mode. The manipulator side of our surgical robot system, seen on the left of Fig. 1, consists of the 7 DOF Kuka arm with a surgical tool attached to its flange. The surgical tool has four actuators on top, which control the clamp at the end of the rod. This translates to 3 DOF of rotation with one additional DOF for opening/closing the clamp.

The operators side, seen on the right of Fig. 1, is represented by a compliant robot, which the user can move to execute the task. While substituting a teleoperation device for a robot has it drawbacks, e.g. higher price, several potential benefits can be determined. The robot can exert not only forces at its end-effector but also torques, which would benefit transmitting information to the user performing the task. In addition, the users side can be calibrated (moved) to represent the current surgical tool orientation and thus ease the teleoperation. In addition to changes in the hardware setup, certain adaptation on the software side were implemented.

Besides the needed changes to adapt the MOPS system to our equipment, our own robot control, including the null space based RCM, was implemented.

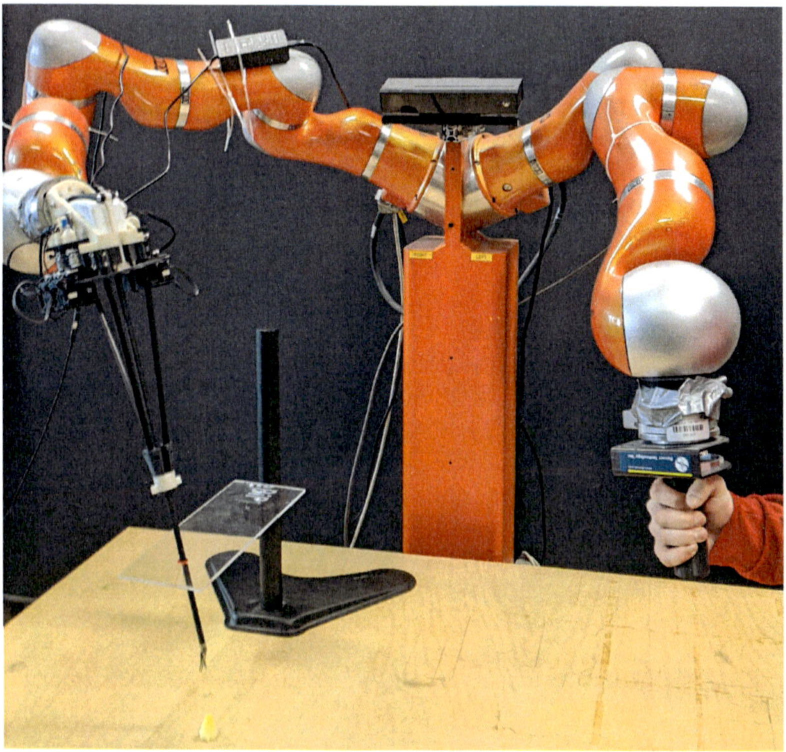

**Fig. 1.** RMIS research platform consists of two LWR Kuka robot arms. The operator side can be seen on the right, where a robot arm is used as a substitute for teleoperation device. Additional remote buttons are used. The manipulator side, seen on the left, comprises of a robot arm, MOPS based adapter for actuating the surgical tool, and the Xi tool itself.

## 3 Remote Center of Motion

As mentioned, one of the main constraints during RMIS is ensuring the tool's rod moves along the insertion point. We ensure this via RCM control and denote the insertion point $\boldsymbol{p}_{RCM}$ as the RCM point. We treat the RCM constraint as a primary task end enchance its control with a virtual prismatic joint. Tracking of the desired pose of the surgical tool's end-effector is treated as a secondary task. While it directly controls the 3 DOFs of the tool, it influences the robot arm joints only through its null space. We omit further details on determining the desired tool's end-effector based on information gained from the manipulator's side and refer the reader to [8].

First, let's only look at the primary task control, i.e. ensuring the tool moves through the RCM point. This means the $\boldsymbol{p}_{RCM}$ must always be somewhere on the tool's rod. For the first step, let's select an arbitrary point $\boldsymbol{p}_{rod}$ somewhere on the rod. A general task velocity kinematics equation [9] can be written as,

$$\dot{\boldsymbol{q}} = \boldsymbol{J}^{\dagger}\dot{\boldsymbol{x}}, \tag{1}$$

where $\dot{\boldsymbol{q}}$ is the control joint velocity vector, $\boldsymbol{J}^{\dagger}$ is a pseudo-inverse of the robot's Jacobian matrix, and $\dot{\boldsymbol{x}}$ is the time derivative of the desired end-effector position and orientation. To ensure $\boldsymbol{p}_{rod} = \boldsymbol{p}_{RCM}$ we adapt (1) as,

$$\dot{\boldsymbol{q}}_K = \boldsymbol{J}_K^{\dagger}\dot{\boldsymbol{x}}_A, \tag{2}$$

where $A$ denotes the primary task and $\dot{\boldsymbol{x}}_A$ is defined as the difference between the actual position on the rod and the desired one $\dot{\boldsymbol{x}}_A = \boldsymbol{p}_{rod} - \boldsymbol{p}_{RCM}$. The control joint velocity vector only includes the robot arm's 7 DOF $\dot{\boldsymbol{q}}_K = [\dot{q}_{K1}, \ldots, \dot{q}_{K7}]^T$ and $\boldsymbol{J}_K^{\dagger}$ is the pseudo-inverse of the Kuka arm Jacobian, as only the robot arm can effect the primary task. In addition we only use the top three rows of $\boldsymbol{J}_K$ as the primary task only includes positions.

While (2) ensures a point on the rod stays at $\boldsymbol{p}_{RCM}$, the rod cannot move along it as is the definition of the primary task. To enable this we introduce a virtual prismatic joint $q_{virt}$ at the start of the rod. If we adapt $\boldsymbol{J}_K^{\dagger}$ to also include the virtual joint, we can rewrite (2) as

$$\dot{\boldsymbol{q}}_A = \boldsymbol{J}_A^{\dagger}\dot{\boldsymbol{x}}_A, \tag{3}$$

where $\boldsymbol{J}_A$ is the first three rows of the adapted Kuka Jacobian matrix and $\dot{\boldsymbol{q}}_A = [\dot{q}_{K1}, \ldots, \dot{q}_{K7}, \dot{q}_{virt}]^T$. Control using (3) will now ensure the execution of the primary task.

For the secondary task, the clamp at the tip of the tool $\boldsymbol{x}_c$ should follow the desired pose $\boldsymbol{x}_d$ gained via human movement of the robot arm at the users side. If we just look at the secondary task, (1) can be rewritten as,

$$\dot{\boldsymbol{q}}_B = \boldsymbol{J}_B^{\dagger}\dot{\boldsymbol{x}}_B, \tag{4}$$

where $B$ denotes the secondary task and $\dot{\boldsymbol{x}}_B = \boldsymbol{x}_d - \boldsymbol{x}_c$ which now includes position and orientation. The control joint velocity vector now includes 7 robot arm joints and 3 joints for the surgical tool $\dot{\boldsymbol{q}}_B = [\dot{q}_{K1}, \ldots, \dot{q}_{K7}, \dot{q}_{X1}, \dot{q}_{X2}, \dot{q}_{X3}]^T$.

As the secondary task should not interfere with the primary one, we take advantage of primary task redundancy as the number of operational space variables needed is $m = 3$ and the number of available DOFs is $n = 7$. The primary task velocity kinematics (3) can now be rewritten to include the secondary task:

$$\dot{\boldsymbol{q}}_A = \boldsymbol{J}_A^{\dagger}\dot{\boldsymbol{x}}_A + (\boldsymbol{I} - \boldsymbol{J}_A^{\dagger}\boldsymbol{J}_A)\dot{\boldsymbol{q}}_{ns}. \tag{5}$$

The secondary task is included through the null-space joint velocity vector,

$$\dot{\boldsymbol{q}}_{ns} = [\dot{q}_{B1}, \ldots, \dot{q}_{B7}, 0]^T, \tag{6}$$

which only includes the first seven values of the secondary task's joint velocity vector, as the last three ones are $[q_{X1}, q_{X2}, q_{X3}]$ are independent from the primary task.

Our proposed approach for RCM control can now be written as,

$$\begin{aligned}\dot{\boldsymbol{q}}_A &= \boldsymbol{J}_A^{\dagger}\dot{\boldsymbol{x}}_A + (\boldsymbol{I} - \boldsymbol{J}_A^{\dagger}\boldsymbol{J}_A)\dot{\boldsymbol{q}}_{ns},\\ \dot{\boldsymbol{q}}_B &= \boldsymbol{J}_B^{\dagger}\dot{\boldsymbol{x}}_B,\\ \dot{\boldsymbol{q}}_{ns} &= [\dot{q}_{B1},\dots,\dot{q}_{B7},0]^T,\\ \dot{\boldsymbol{q}} &= [\dot{\boldsymbol{q}}_A,\dot{q}_{B8},\dots,\dot{q}_{B10}]^T,\end{aligned} \tag{7}$$

where $\dot{\boldsymbol{q}}$ is the main control joint velocity vector.

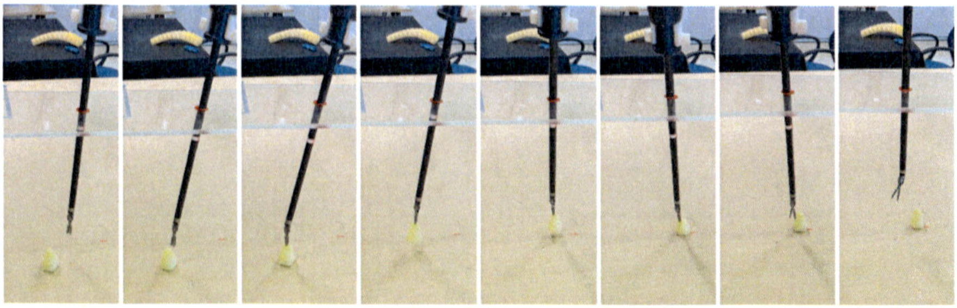

(a)

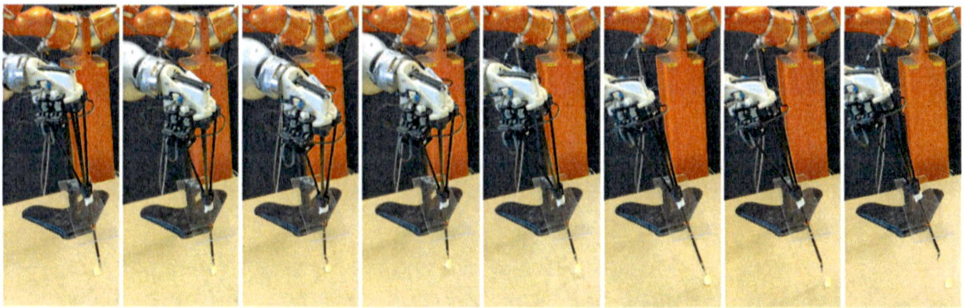

(b)

**Fig. 2.** Snapshot sequences of the surgical tool movement under RCM constraint. (a) Close-up of the tool's end-effector. The red circle on the acrylic glass marks the RCM point $\boldsymbol{p}_{RCM}$. We can observe the pick and place task is executed while the RCM constraint is preserved. (b) A wider view of the same execution also shows the robot arm movements. (Color figure online)

## 4 Evaluation

Our research platform was used to evaluate the proposed RCM approach. A pick and place task was demonstrated by guiding the compliant robot (see Fig. 1). The complaint robot movement in task space was recorded and used for later executions of the task. The demonstrated pick and place task was executed with

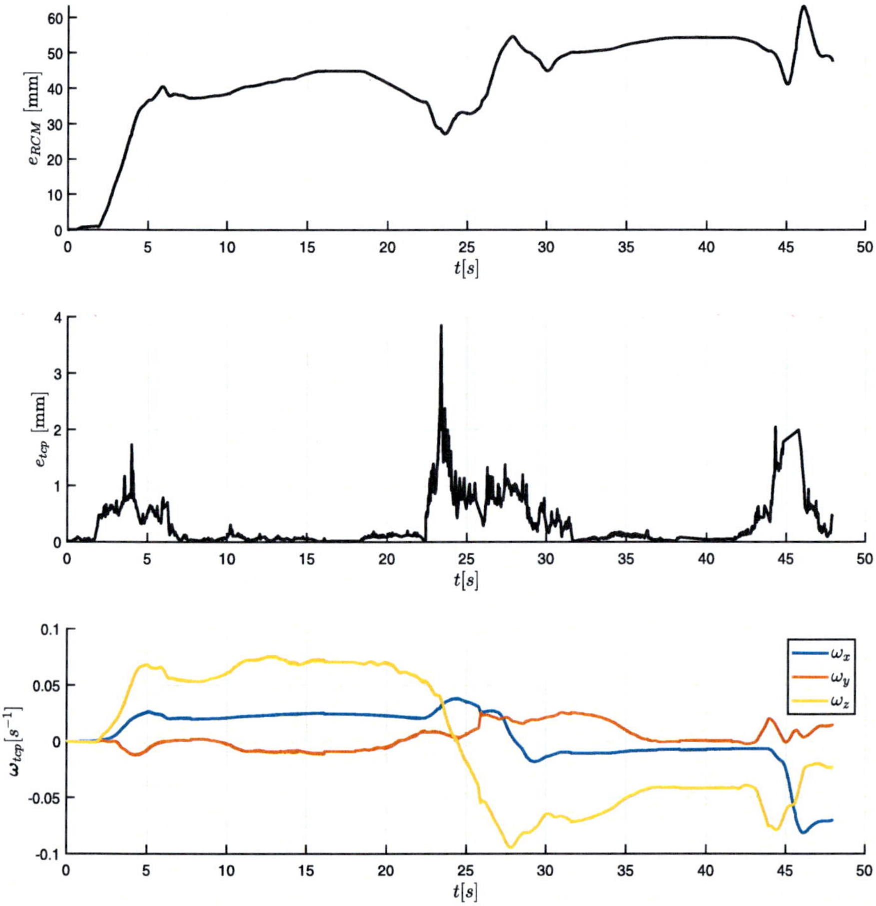


**Fig. 3.** Tracking errors with no RCM constraint. As expected, RCM error, shown on the top graph, rises at the start of the movement and remains high throughout the execution. Tracking error of the desired tool's end-effector, remains low most of the execution. While positional error $e_{tcp}$ is shown in the middle graph, orientational errors $\boldsymbol{\omega}_{tcp}$ can be seen in the bottom graph.

and without the RCM constraint (incl. the physical contraint). Standard task velocity kinematics (4) were utilized in the case when no RCM constraint was enforced. Execution of the task under the RCM constraint enforced by the proposed approach (7) can be seen in Fig. 2. The RCM point $\boldsymbol{p}_{RCM}$ was determined by moving the point on the tool, marked with a white ring, to the desired RCM point, marked with a red ring.

While both execution, with and without the RCM constraint, successfully executed the task, differences can be observed by comparing Fig. 3 and Fig. 4. Error in RCM $\boldsymbol{e}_{RCM} = \|\dot{\boldsymbol{x}}_A\| = \|\boldsymbol{p}_{rod} - \boldsymbol{p}_{RCM}\|$ is, as expected, high if no RCM constraint is used. We can observe that $\boldsymbol{e}_{RCM}$ is significantly smaller throughout

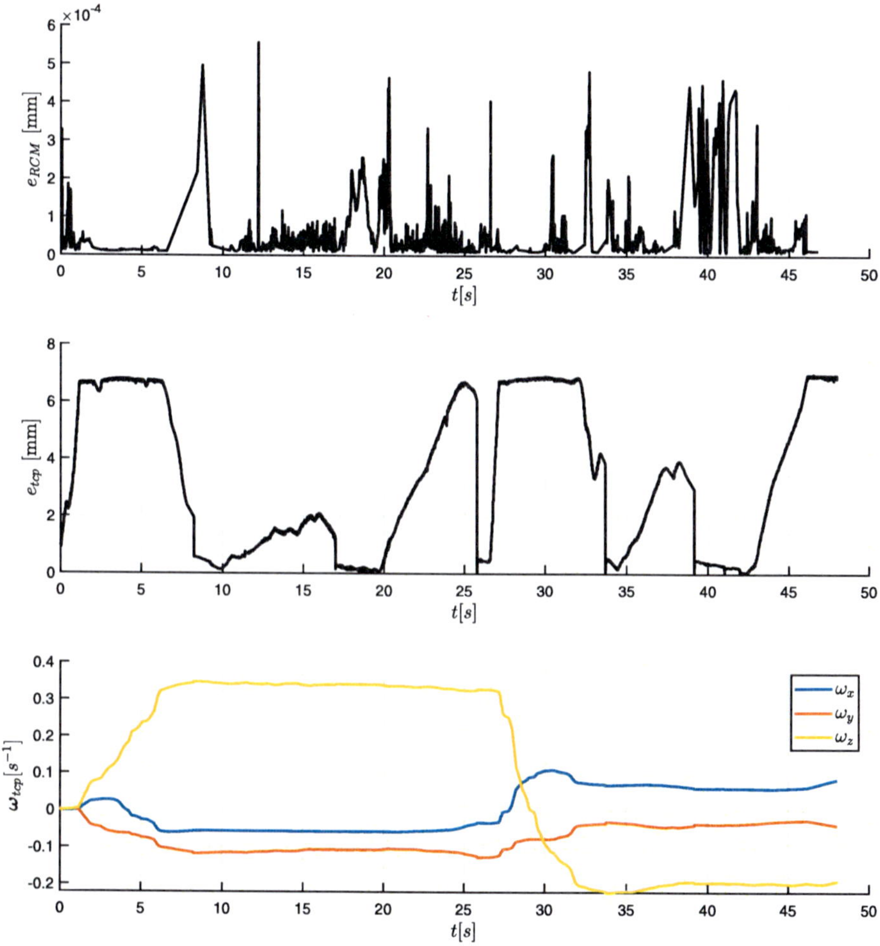


**Fig. 4.** Tracking errors with the RCM constraint. While RCM error $\boldsymbol{e}_{RCM}$ is significantly reduced, compared to the execution without the RCM constraint, a slight rise in tracking error of the desired tool's end-effector can be seen in the bottom two graphs.

the execution if our proposed RCM approach is used. We should note, that while value $\boldsymbol{p}_{RCM}$ remains constant throughout the execution, $\boldsymbol{p}_{rod}$ is calculated based on current robot arm joint angles $\boldsymbol{q}_K$ and direct kinematics. If we compare the middle two graphs, we can observe the tracking error $e_{tcp} = \|\dot{\boldsymbol{x}}_B\| = \|\boldsymbol{x}_d - \boldsymbol{x}_c\|$ is higher when RCM approach is used, as this task is now treated as a secondary task. This is to be expected as the secondary task is being enforced through null space. The mean value still remains relatively low at $\overline{e}_{tcp} \approx 3.4$ mm. Similar behaviour can be observed for orientational errors $\boldsymbol{\omega}_{tcp}$.

## 5 Conclusion

We proposed an approach for RCM in a RMIS setting based on a primary task with a virtual joint and a secondary task exploiting the redundancy of the system. The proposed approach for RCM was evaluated on our own implementation of the MOPS research platform [8], which includes two 7-DOF Kuka LWR arms. While one is used as teleoperation device, the other moves the surgical tool. The results show applicability of our approach, as we can ensure RCM at high precision. Future works include comparison to other RCM approaches, evaluation on a more surgically relevant task, and improving the secondary task tracking.

## References

1. Aghakhani, N., Geravand, M., Shahriari, N., Vendittelli, M., Oriolo, G.: Task control with remote center of motion constraint for minimally invasive robotic surgery. In: 2013 IEEE International Conference on Robotics and Automation, pp. 5807–5812. IEEE (2013)
2. Hannaford, B., et al.: Raven-ii: an open platform for surgical robotics research. IEEE Trans. Biomed. Eng. **60**(4), 954–959 (2012)
3. Kastritsi, T., Doulgeri, Z.: A passive admittance controller to enforce remote center of motion and tool spatial constraints with application in hands-on surgical procedures. Robot. Auton. Syst. **152**, 104073 (2022)
4. Kazanzides, P., Chen, Z., Deguet, A., Fischer, G.S., Taylor, R.H., DiMaio, S.P.: An open-source research kit for the da Vinci® surgical system. In: 2014 IEEE International Conference on Robotics and Automation (ICRA), pp. 6434–6439. IEEE, May 2014. https://doi.org/10.1109/ICRA.2014.6907809, http://ieeexplore.ieee.org/document/6907809/
5. Leibrandt, K., Marcus, H.J., Kwok, K.W., Yang, G.Z.: Implicit active constraints for a compliant surgical manipulator. In: 2014 IEEE International Conference on Robotics and Automation (ICRA), pp. 276–283. IEEE (2014)
6. Sandoval, J., Su, H., Vieyres, P., Poisson, G., Ferrigno, G., De Momi, E.: Collaborative framework for robot-assisted minimally invasive surgery using a 7-DoF anthropomorphic robot. Robot. Auton. Syst. **106**, 95–106 (2018)
7. Schwaner, K.L., Iturrate, I., Andersen, J.K., Jensen, P.T., Savarimuthu, T.R.: Autonomous bi-manual surgical suturing based on skills learned from demonstration. In: 2021 IEEE/RSJ International Conference on Intelligent Robots and Systems (IROS), pp. 4017–4024. IEEE (2021)
8. Schwaner, K.L., Iturrate, I., Andersen, J.K.H., Dam, C.R., Jensen, P.T., Savarimuthu, T.R.: MOPS: a modular and open platform for surgical robotics research. In: 2021 International Symposium on Medical Robotics (ISMR), pp. 1–8. IEEE (2021)
9. Siciliano, B., Sciavicco, L., Villani, L., Oriolo, G.: Robotics: Modelling, Planning and Control. Advanced Textbooks in Control and Signal Processing, Springer, London (2010). https://doi.org/10.1007/978-1-84628-642-1
10. Taylor, R.H., Menciassi, A., Fichtinger, G., Fiorini, P., Dario, P.: medical robotics and computer-integrated surgery. In: Siciliano, B., Khatib, O. (eds.) Springer Handbook of Robotics, pp. 1657–1684. Springer, Cham (2016). https://doi.org/10.1007/978-3-319-32552-1_63

# A Handheld Forceps-Like Tracker for Learning Robotic Surgical Tasks from Demonstration

Iñigo Iturrate(✉), Kim Lindberg Schwaner, Thiusius Rajeeth Savarimuthu, and Zhuoqi Cheng

The Maersk Mc-Kinney Moller Institute, University of Southern Denmark, Odense, Denmark
{inju,kils,trs,zch}@mmmi.sdu.dk

**Abstract.** This paper presents a novel approach to collecting demonstrations of surgical procedures that can be used to transfer skills from experts to robots using standard learning from demonstration. We use a handheld forceps-like magnetic tracker similar to the grasper of an Intuitive Surgical Inc. Large Needle Driver to record the demonstrations. Our approach does not depend on a surgical robot for the demonstration phase. We show that the resulting demonstrations can be executed on a surgical robot system by encoding them as Dynamic Movement Primitives. Our evaluation shows that users are quicker at demonstrating trajectories with the device rather than by teleoperating the robot, at the expense of slightly more noise due to hand tremors.

**Keywords:** robot-assisted minimally invasive surgery · autonomy in robotic surgery · learning from demonstration

## 1 Introduction

Robot-assisted Minimally Invasive Surgery (RMIS) presents multiple advantages for both patients and doctors compared to traditional laparoscopic surgery [2,5]. However, the workload experienced by surgeons is still high [8] and there is significant repetition in many surgical procedures, such as suturing or dissection. Consequently, full or partial automation of robotic surgery has a very high impact potential and is a topic of active research interest.

As, in the operating room, most surgical robots for MIS are teleoperated by surgeons, learning from demonstration (LfD) has been proposed as a way to transfer expert-like surgical skills to robots. The surgeon performs the operation as they normally would, using teleoperation, and signals from different encoders and sensors, e.g., the robot and endoscopic camera, are recorded. Then, either the trajectories are used to learn motion primitives directly [3,7,16] or features in the scene tracked using visual observation are used to plan motions [1,9,11], or to adapt learned motion primitives [13].

While the above approach allows skills to be learned from expert robot operators, it suffers from the limitation that expert data is hard to come by. Commercial systems used in hospitals do not allow reading robot sensor data. This


T. Petrič et al. (Eds.): RAAD 2023, MMS 135, pp. 229–236, 2023.
https://doi.org/10.1007/978-3-031-32606-6_27

requires surgeons to operate research systems, e.g., the daVinci Research Kit (dVRK) [6], which handle differently than the surgical systems used in hospitals [10]. For this reason, the amount of data available for LfD is very limited.

To address the above limitations, we propose a novel demonstration modality for surgical tasks based on a handheld forceps tracker through which the demonstrator can mimic the tool operation in an RMIS system. This allows medical personnel trained in surgical procedures such as suturing – but not qualified in the use of RMIS systems – to provide expert-like demonstrations of surgical procedures manually, while making the data collected easily transferable to a robotic system. It provides a simple way to collect large amounts of data that can be used to train essential robot skills. Thus, although the collected skills may not directly be able to complete a surgical operation on their own, they can form part of a larger skill library and can be scaffolded as the basis of more complex procedures. To the extent of our knowledge, this is the first method for demonstration of surgical motions that does not rely on teleoperation.

Our contributions are the following:

1. A framework for collection of surgical skill demonstrations that does not require access to a surgical robot, but easily transfers to one.
2. The design for a forceps-like tool to be used with the above methodology.
3. A comparison of our proposed demonstration collection framework with robot teleoperation, as is common in most RMIS systems.

## 2 Materials and Methods

### 2.1 Handheld Trajectory Tracker

The designed forceps-like tracker is shown in Fig. 1A-B. It is designed with jaws that are the same dimensions as those of a large needle driver from Intuitive Surgical Inc. The Trakstar Model 90 sensor (NDI DIGITAL INC., Canada) is used for the 6 Degree of Freedom (DOFs) trajectory tracking. This sensing system is documented with a measurement rate of 80 Hz, a spatial position error of 1.4 mm and an orientation error of 0.5°. Specifically, the Trakstar system includes a coil module, which is placed in the demonstration workspace, and an electromagnetic (EM) sensor (red tube in Fig. 1A) which is embedded inside the tracker. In order not to affect the sensed magnetic signal, the tracker is fabricated using PLA through 3D printing. A non-ferromagnetic metal screw is used as the jaws' pivot to provide sufficient tear strength during object grasping.

### 2.2 Robotic Platform

For both the teleoperated demonstrations, as well as all subsequent executions of the task either from teleoperated recordings or recordings using the proposed device, we make use of MOPS, a modular and open system for robotic surgery (see Fig. 1C for a partial view of the system or [12] for further details). This system consists of two Universal Robots UR5e collaborative robot manipulators mounted with Intuitive Surgical large needle drivers through custom adapters. Two 3D Systems Touch haptic devices are used for teleoperation.

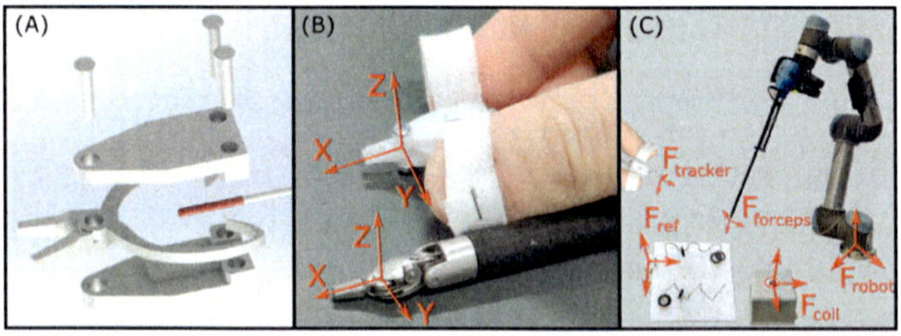


**Fig. 1.** (A) The exploded view of the forceps-like tracker. The red tube is the embedded EM sensor. (B) The jaws of the tracker are designed with the same dimension as the large needle driver's from Intuitive Surgical Inc. (C) For demonstration, the trajectory of the tracker $F_{\text{tracker}}$ is registered to the coordinate of the referenced sensor $F_{\text{ref}}$ via $F_{\text{coil}}$. In reproduction phase, the task execution requires the coordinate transformation between the robot base $F_{\text{robot}}$ and $F_{\text{coil}}$, considering that $F_{\text{forceps}}$ with respect to $F_{\text{robot}}$ is internally retrievable.

## 2.3 Proposed System Overview

As shown in Fig. 1C, the system calibration is done within five coordinate frames. We use $F_{\text{tracker}}$, $F_{\text{coil}}$, $F_{\text{ref}}$, $F_{\text{robot}}$, and $F_{\text{forceps}}$ to denote the coordinates of the tracker, the EM transmitter (coil), the referenced EM sensor attached to the task plate, the robot base and the robotic forceps end-effector respectively.

The EM sensor system streams the position and orientation information of both tracker and the reference sensor with respect to the EM transmitter. Through transformation among $F_{\text{tracker}}$, $F_{\text{coil}}$, and $F_{\text{ref}}$, the trajectory of the tracker can be registered to $F_{\text{ref}}$ during the demonstration phase. In the reproduction phase, the coordinate transformation between $F_{\text{robot}}$ and $F_{\text{ref}}$ can be achieved by controlling the tip of the robotic forceps to multiple known positions with respect to $F_{\text{ref}}$. Given the kinematic information of the robot, the tip position with respect to $F_{\text{robot}}$ can be retrieved.

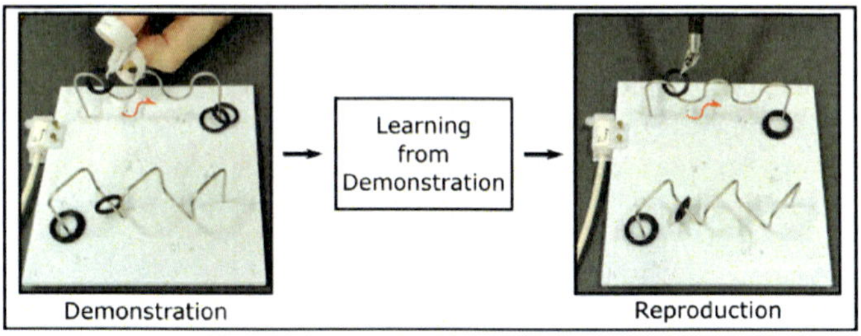


**Fig. 2.** Proposed system flow: A demonstration of a surgical skill is recorded by manually guiding the forceps-like tracker, encoded using any existing LfD, and executed on a real surgical robot. The mapping from demonstration to robot is simplified, as the tracking device is designed similarly to existing surgical tools.

### 2.4 Encoding Demonstrated Motions

Using Dynamic Movement Primitives (DMPs) [4], we encode demonstrated motions into adaptive robot skills as a dynamical system of the form:

$$\tau^2 \ddot{\mathbf{y}} = \alpha_y \big( \beta_y (\mathbf{g} - \mathbf{y}) + \tau \dot{\mathbf{y}} \big) + \mathbf{f}(x), \tag{1}$$

$$\tau \dot{x} = -\alpha_x x, . \tag{2}$$

Equation (1) is known as the *transformation system* and defines the policy's trajectory, from an initial state, $\mathbf{y}_0 \in \mathbb{R}^3$ to an attractor, $\mathbf{g} \in \mathbb{R}^3$. The constants, $\alpha_y \in \mathbb{R}^+$ and $\beta_y \in \mathbb{R}^+$, act as gains, and are typically chosen to achieve critical damping. The time constant, $\tau \in \mathbb{R}^+$ affects the duration of the movement. The system is modulated by a non-linear forcing term, $\mathbf{f}(x)$, given by Eq. (3), which can be learned to fit any demonstration.

The phase variable, $x$, controls the time evolution through Eq. (2), known as the *canonical system*. Notice that, as long as $\alpha_x > 0$, then $x \to 0$, which implies that the term $\mathbf{f}(x)$ will vanish, leaving us with second-order spring-damper dynamics that – under the assumption that $\alpha_y$ and $\beta_y$ are chosen for critical damping – guarantee the convergence of $\mathbf{y}$ to $\mathbf{g}$.

The forcing term

$$\mathbf{f}(x) = \frac{\sum_{i=1}^{N} \psi_i(x) \boldsymbol{w}_i}{\sum_{i=1}^{N} \psi_i(x)} x \left( \mathbf{g} - \mathbf{y}_0 \right), \tag{3}$$

$$\psi_i(x) = \exp\left( -h_i \left( x - c_i \right)^2 \right), \tag{4}$$

is a mixture of $N \in \mathbb{Z}^+$ Gaussian radial basis functions Eq. (4) parameterized by their centers $c_i$ and widths $h_i$, and weighted by $w_i \in \mathbb{R}^3$.

To learn $\mathbf{f}(x)$ from demonstration, we can formulate

$$\mathbf{f}_{\text{desired}} = \tau^2 \ddot{\mathbf{y}}_{\text{demo}} - \alpha_y \big( \beta_y \left( \mathbf{g} - \mathbf{y}_{\text{demo}} \right) - \tau \dot{\mathbf{y}}_{\text{demo}} \big), \tag{5}$$

where $\{\mathbf{y}_{\text{demo}}, \dot{\mathbf{y}}_{\text{demo}}, \ddot{\mathbf{y}}_{\text{demo}}\}$ are the positions, velocities and accelerations observed during a given demonstration. As Eq. (3) is linear in its parameters, the above problem amounts to a least-squares weighted linear regression.

For the rotational degrees of freedom, we adopt the formulation from [15], using quaternions to represent orientations in $S^3$.

## 3 Experimental Evaluation

We evaluated our system in two ways: First, we compared trajectories recorded using A) teleoperation of the surgical robot in a manner analogous to traditional RMIS, B) our proposed tool and methodology. Second, we encoded the demonstrated trajectories as DMPs to show that our method can be used to learn generalizable representations of surgical skills.

### 3.1 Comparison with Teleoperated Demonstrations

The first experiment was designed to evaluate the speed of demonstration using the designed tracking system. We used the one hand task of the SBT3 - wire chaser (3-Dmed, U.S., Fig. 2) as a benchmark. The task consisted in grasping the ring (13 mm diameter), moving it from the right end to the left along the wire track, and then back. This was repeated three times for each experimental condition 1) using the proposed tracking device and 2) by teleoperating the MOPS robot. The subjects were told to focus on speed and not accuracy. The results are shown in Table 1. The participants averaged 13.4 $\pm$ 2.9 s with the manual tracker and 39.9 $\pm$ 10 s using robot teleoperation. A Welch's t-test rejected the null hypothesis that the sample means were equal with $p < 0.01$.

**Table 1.** Comparison of the demonstration time for the tracking device (*Manual Tracker*) and robot teleoperation (*Robot*). The participants focused on speed.

| | **Demonstration Time [s]** | | | | | |
|---|---|---|---|---|---|---|
| | **Manual Tracker** | | | **Robot** | | |
| **Participant** | Trial 1 | Trial 2 | Trial 3 | Trial 1 | Trial 2 | Trial 3 |
| A | 19.8 | 16.0 | 12.7 | 62.0 | 38.2 | 35.2 |
| B | 12.2 | 10.8 | 10.4 | 35.5 | 32.1 | 50.0 |
| C | 12.9 | 12.9 | 12.6 | 40.5 | 31.2 | 34.5 |

The second task was the same, but the ring was pre-grasped at a fixed location, and all trajectory data was transformed to the center of the grasped ring. The users were asked to focus on accuracy and attempt to keep the ring centered on the track, avoiding contact between the ring and wire. To compare the trajectories, we defined a reference as the path formed by the wire frame, and calculated Fréchet distances [14], shown in Table 2, which averaged $0.0818 \pm 0.0015$ using the tracker, and $0.080 \pm 0.0006$ with robot teleoperation. A Welch's t-test

**Table 2.** Comparison of the Fréchet distance from the demonstration to the reference for the two experimental conditions: tracking device (*Manual Tracker*) and robot teleoperation (*Robot*). The participants focused on accuracy.

| | **Frchet Distance from Reference [m]** | | | | | |
|---|---|---|---|---|---|---|
| | **Manual Tracker** | | | **Robot** | | |
| **Participant** | Trial 1 | Trial 2 | Trial 3 | Trial 1 | Trial 2 | Trial 3 |
| A | 0.0810 | 0.0806 | 0.0814 | 0.0801 | 0.0803 | 0.0801 |
| B | 0.0830 | 0.0842 | 0.0820 | 0.0795 | 0.0810 | 0.0797 |
| C | 0.0835 | 0.0795 | 0.0812 | 0.0810 | 0.0794 | 0.0793 |

rejected the null hypothesis that the sample means for the two populations were equal with $p < 0.01$. Figure 3 shows all demonstration trajectories compared to the reference under both experimental conditions.

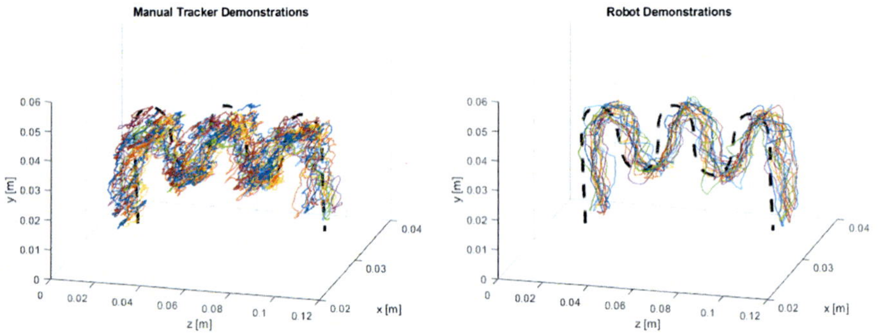


(a) Demonstrations using tracker device. (b) Demonstrations using teleoperation.

**Fig. 3.** Comparison of demonstration approaches. The reference trajectory is denoted by the thick dashed black line.

### 3.2 Integration into a Learning from Demonstration Framework

In the second experiment, we integrated our recording device into a full LfD workflow, as described in Sect. 2.3, with the objective of evaluating the resulting policies when executed autonomously on a robotic system. We used the demonstrations recorded using the manual tracker for the second task in Sect. 3.1. Figure 4a shows a DMP trained from demonstration recorded using the proposed tracking system. The number of Gaussian Basis Functions was set to $N = 50$ in order to filter the hand tremors present in the recording. Figure 4b shows the corresponding execution on a real robot.

## 4 Discussion

Our results show that the proposed tracker device is viable for manually recording demonstrations of surgical skills and using these to train robot policies. One of the main advantages of the tracker is the speed of demonstration, which was considerably faster than for robot teleoperation. We believe this is due to the lack of haptic feedback and of proprioception when teleoperating the robot.

As seen from Table 2 and Fig. 3, demonstrations performed with the manual tracker tend to show larger deviations from the reference trajectory than using robot teleoperation. We found that hand tremor had a significant effect on the recorded trajectories, and was influenced by the position of the arm and elbow of the demonstrator. An advantage of teleoperation in this regard is that haptic

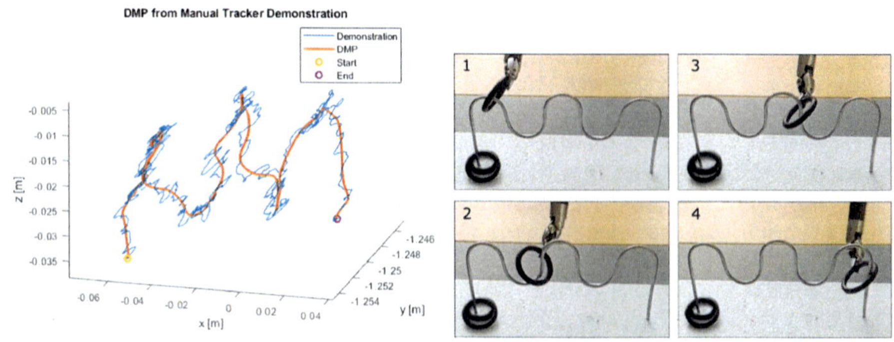


(a) DMP trained on a demonstration provided with the proposed manual tracker. (b) Autonomous execution on MOPS surgical robot platform.

**Fig. 4.** Execution of the DMP learned from the tracker demonstration.

control devices low-pass filter hand tremor. Note, however, that as seen in Fig. 4a, the additional noise from manual demonstrations can easily be handled by DMPs by selecting an appropriate number of basis functions.

A major advantage of our proposed tracking device is that it does not require a surgical robot to perform the demonstrations. As the device is handheld, the demonstrator's manual skills easily transfer over to the demonstration. These two factors mean that personnel with no experience in surgical robot teleoperation can still easily provide demonstration of skills that can subsequently be transferred to a surgical robotic system. Additionally, the system is easily portable and can be setup in various environments after a system calibration.

## 5 Conclusion

In this paper, we have presented a new tracker device that can be used to provide manual demonstrations of surgical robot skills without access to a surgical robot. Our experiments have shown that users can quickly record task demonstrations and benefit from their existing manual skills compared to demonstrations provided using teleoperation. However, the data is sensitive to hand tremors. We have shown that, despite this noise, the device can be integrated into an existing LfD framework, and a policy, for instance a DMP, can be trained from the recorded data and used to successfully execute the skill on a surgical robot.

## References

1. Berthet-Rayne, P., Power, M., King, H., Yang, G.Z.: Hubot: A three state human-robot collaborative framework for bimanual surgical tasks based on learned models. In: 2016 IEEE International Conference on Robotics and Automation (ICRA), pp. 715–722. IEEE (2016)

2. Dalager, T., Jensen, P.T., Eriksen, J.R., Jakobsen, H.L., Mogensen, O., Søgaard, K.: Surgeons' posture and muscle strain during laparoscopic and robotic surgery. Br. J. Surg. **107**(6), 756–766 (2020)
3. Deniša, M., Schwaner, K.L., Iturrate, I., Savarimuthu, T.R.: Semi-autonomous cooperative tasks in a multi-arm robotic surgical domain. In: 2021 20th International Conference on Advanced Robotics (ICAR), pp. 134–141. IEEE (2021)
4. Ijspeert, A.J., Nakanishi, J., Hoffmann, H., Pastor, P., Schaal, S.: Dynamical movement primitives: Learning attractor models for motor behaviors. Neural Comput. **25**(2), 328–373 (2013)
5. Jørgensen, S.L., et al.: Nationwide introduction of minimally invasive robotic surgery for early-stage endometrial cancer and its association with severe complications. JAMA Surg. **154**(6), 530 (2019)
6. Kazanzides, P., Chen, Z., Deguet, A., Fischer, G.S., Taylor, R.H., DiMaio, S.P.: An open-source research kit for the da Vinci® surgical system. In: 2014 IEEE International Conference on Robotics and Automation (ICRA), pp. 6434–6439. IEEE (2014)
7. Knoll, A., Mayer, H., Staub, C., Bauernschmitt, R.: Selective automation and skill transfer in medical robotics: A demonstration on surgical knot-tying. Int. J. Med. Robot. Comput. Assist. Surg. **8**(4), 384–397 (2012)
8. Lau, E., Alkhamesi, N.A., Schlachta, C.M.: Impact of robotic assistance on mental workload and cognitive performance of surgical trainees performing a complex minimally invasive suturing task. Surg. Endosc. **34**, 2551–2559 (2020)
9. Murali, A., et al.: Learning by observation for surgical subtasks: Multilateral cutting of 3D viscoelastic and 2D orthotropic tissue phantoms. In: 2015 IEEE International Conference on Robotics and Automation (ICRA), pp. 1202–1209. IEEE (2015)
10. Richter, F., Funk, E.K., Park, W.S., Orosco, R.K., Yip, M.C.: From bench to bedside: The first live robotic surgery on the dVRK to enable remote telesurgery with motion scaling. In: 2021 International Symposium on Medical Robotics (ISMR). IEEE (2021)
11. Schulman, J., Gupta, A., Venkatesan, S., Tayson-Frederick, M., Abbeel, P.: A case study of trajectory transfer through non-rigid registration for a simplified suturing scenario. In: 2013 IEEE/RSJ International Conference on Intelligent Robots and Systems, pp. 4111–4117. IEEE (2013)
12. Schwaner, K.L., Iturrate, I., Andersen, J.K.H., Dam, C.R., Jensen, P.T., Savarimuthu, T.R.: MOPS: A modular and open platform for surgical robotics research. In: 2021 International Symposium on Medical Robotics (ISMR). IEEE, Atlanta, GA, USA (2021)
13. Schwaner, K.L., Iturrate, I., Andersen, J.K.H., Jensen, P.T., Savarimuthu, T.R.: Autonomous bi-manual surgical suturing based on skills learned from demonstration. In: 2021 IEEE/RSJ International Conference on Intelligent Robots and Systems (IROS), pp. 4017–4024. IEEE, Prague, Czech Republic (2021)
14. Tao, Y., et al.: A comparative analysis of trajectory similarity measures. GISci. Remote Sens. **58**(5), 643–669 (2021)
15. Ude, A., Nemec, B., Petrić, T., Morimoto, J.: Orientation in cartesian space dynamic movement primitives. In: 2014 IEEE International Conference on Robotics and Automation (ICRA), pp. 2997–3004. IEEE (2014)
16. Van Den Berg, J., et al.: Superhuman performance of surgical tasks by robots using iterative learning from human-guided demonstrations. In: 2010 IEEE International Conference on Robotics and Automation, pp. 2074–2081. IEEE (2010)

# An Approach for Combining Transparency and Motion Assistance of a Lower Body Exoskeleton

Jakob Ziegler(✉), Bernhard Rameder, Hubert Gattringer, and Andreas Müller

Institute of Robotics, Johannes Kepler University Linz, Altenberger Str. 69, 4040 Linz, Austria
{jakob.ziegler,bernhard.rameder,hubert.gattringer,a.mueller}@jku.at
https://www.jku.at/en/institute-of-robotics

**Abstract.** In this paper, an approach for gait assistance with a lower body exoskeleton is described. Two concepts, transparency and motion assistance, are combined. The transparent mode, where the system is following the user's free motion with a minimum of perceived interaction forces, is realized by exploiting the gear backlash of the actuation units. During walking a superimposed assistance mode applies an additional torque guiding the legs to their estimated future position. The concept of adaptive oscillators is utilized to learn the quasi-periodic signals typical for locomotion. First experiments showed promising results.

**Keywords:** gait · exoskeleton · transparency · motion assistance

## 1 Introduction

In order to cope with the increasing demand for rehabilitation and movement assistance due to the global demographic change, interest in robotic solutions as supplement to classical methods and therapies is growing tremendously. Naturally, also wearable robotic devices, like exoskeletons, are getting more and more focused on. The most commonly known type of exoskeletons are probably systems assisting the human walking motion that are e.g. applied in the field of robotic gait rehabilitation. Due to the close interaction between the human body and the robotic system, an immense variety of research topics is related to wearable robotics in general and exoskeletons in particular, which renders a detailed and exhaustive overview of the international status of research difficult. However, robustly recognizing the user's intended motion and adapting to it in an intelligent way, as well as the accurate timing of provided assistance can be defined as crucial factors in assistive wearable robotics [5].

Considering human gait, the cyclic nature of locomotion patterns is used to learn the current movement and to anticipate the appropriate assistance. To this end, the methodology of adaptive oscillators have already been successfully


T. Petrič et al. (Eds.): RAAD 2023, MMS 135, pp. 237–244, 2023.
https://doi.org/10.1007/978-3-031-32606-6_28

applied in the past [1,3,10]. These publications all focused on assisting the flexion and extension of the hip joints during walking. In [1] adaptive oscillators are used to synchronize the assistive torque to the net muscle torque, which is assumed to be quasi-periodic and estimated with the measured hip angle and average values of the dynamic parameters of leg and hip joint of adult persons. The authors of [10] used adaptive oscillators paired with non-sinusoidal functions representing nominal joint angle trajectories based on experimental data to synchronize to the current gait phase of the user. The assistance torque is then calculated based on predefined torque patterns as a function of the gait phase. In [3], a modified version of a Hopf oscillator is utilized as adaptive oscillator, which is then coupled with a kernel-based non-linear filter to track and learn the periodic features of the hip angle trajectory. Assistance is realized by calculating the difference between the current position and the estimated future position of the joint based on the learned signal and a phase shift.

In line with above mentioned publications, this paper deals with a methodology towards the assistance of gait with a lower extremity exoskeleton. Adaptive oscillators are used to learn the current quasi-periodic movement, represented by the measured hip angle trajectories, and to anticipate the appropriate assistance. To support this well-known methodology and to increase the overall robustness, artificially generated gait patterns are used to tune the learning gains and to limit the parameters of the estimated signals. Foundation of the overall functionality of the robotic device is the implementation of transparency [4], where the system follows the user's free motion without hindering it significantly. As a novel approach, this is established in a simple but effective way by exploiting the gear backlash of the actuation units.

## 2 Setup

Nowadays several exoskeletons are commercially available. However, a commercial system is typically not suitable for research purposes, as the possibilities to investigate and adapt the implemented methods and algorithms are usually extremely limited. On this account, an exoskeleton prototype is currently being developed at the Institute of Robotics of the Johannes Kepler University Linz. It serves as a testing platform, where novel control strategies, motion assistance approaches and the like can be implemented and tested in a low-level manner with full access to the functionality of the system.

In the current status four actuation units are in operation, two positioned at the hip joints and two at the knee joints of the user. One actuation unit (*Exoskeleton Drive GEN.1*, Maxon, Switzerland) consists of a brushless DC motor with an internal incremental encoder, a control unit, a planetary gear head (gear ratio 448/3) and an absolute encoder at the gear output.

The modular conception of the system allows for an easy change between two (hip joints) or four (hip and knee joints) actively driven degrees of freedom. Figure 1a and 1b shows the current status of the exoskeleton prototype. For the measurements and experiments described in this paper the version with detached knee actuators was used.

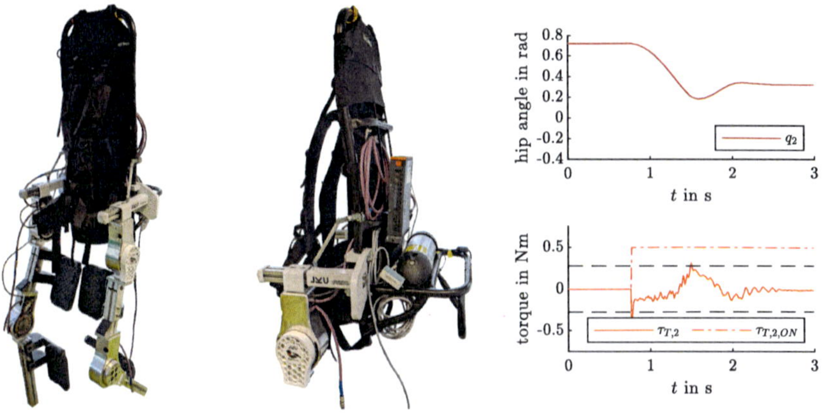


(a) hip and knee joints (b) hip joints (c) transparency mode check

**Fig. 1.** Current setup of the exoskeleton and show-case of transparency mode

## 3 Transparency

To ensure overall usability of a wearable robotic system it is highly desirable that even when no active assistance is required, the device should not hinder the user's free motion in any way. In consequence, unwanted and disturbing effects, for instance due to friction, inertia or gravity, have to be compensated in order to make the powered but otherwise not assisting device *transparent* to the user. As no force or torque sensors or the like are included in the current setup, this transparency has to be realized with the existing components. In a simple but surprisingly effective approach the gear backlash of the actuators is exploited. Inside this gear backlash, which has an angular range of about $\pm 0.5°$, the gear output side of the actuators can be moved without the motion being transferred to the electric motor on the gear input side. Both, the motor position $q_{M,j}$ and the actuator position $q_j$ on the gear output side of actuator $j$ are measured and hence the difference of these positions due to the gear backlash can be calculated. When the gear output is moving without the motion being transferred to the electric motor, this position difference can be used to control the system in a way so that it becomes transparent to the user. The according torque $\tau_{T,j}$ for the $j^{th}$ actuator is calculated as

$$\tau_{T,j} = K_T(q_j - q_{M,j}), \tag{1}$$

where $K_T$ is a tune-able virtual stiffness. Applying this torque establishes the desired behavior by significantly decreasing the otherwise quite noticeable and interfering effects, mainly caused by inertia. In addition to the restriction of $\tau_{T,j}$, due to the limited backlash, the resulting torque of the system was kept within bounds by choosing the virtual stiffnesses in an appropriate way.

The functionality of the transparent mode was demonstrated by applying the torque $\tau_{T,j}$ to the hip joint, which was initially deflected by a certain angle, causing it to behave like a pendulum. The joint quickly converged to its equilibrium (Fig. 1c), which is, due to an eccentrical center of mass, not the vertical. The strong damping of the oscillation is caused by the inertia and friction of the planetary gear.

# 4 Motion Assistance

Two critical factors for the support of movement with an actuated prosthetic or orthotic device are the robust and correct recognition of the user's motion intention, as well as the accurate timing of the provided assistance [5]. Thus, when the device is designed to assist a specific movement task, it has to adapt to the characteristics of the intended motion. Concerning human gait, the motion patterns are characterized by their (quasi-)periodic behavior.

## 4.1 Learning of Periodic Signals

One method to adapt to a signal with cyclic nature and to extract its relevant features is the use of so-called adaptive oscillators, the mathematical description of a non-linear dynamical system capable of learning periodic signals [2,6,9]. If the input signal has several frequency components, such an oscillator will generally adapt to the component closest to its initial intrinsic frequency. In [7] it is proposed to use a set of adaptive oscillators, coupled in parallel, to achieve sort of a dynamic Fourier series decomposition of any periodic input signal. Each of the individual oscillators will adapt to the frequency component closest to its intrinsic frequency. If thereby the individual intrinsic frequencies $\omega_i = i\omega$ of the $i = [1, \ldots, N]$ oscillators are defined as multiples of a fundamental frequency $\omega$. The update rules are given by

$$\dot{\phi}_i = i\omega + \kappa_\phi p \cos \phi_i \tag{2}$$

$$\dot{\omega} = \kappa_\omega p \cos \phi_1 \tag{3}$$

$$\dot{\alpha}_0 = \kappa_\alpha p \tag{4}$$

$$\dot{\alpha}_i = \kappa_\alpha p \sin \phi_i, \tag{5}$$

where $\phi_i$ is the phase and $\alpha_i$ is the amplitude of the $i^{th}$ oscillator and $\kappa_\phi$, $\kappa_\omega$ and $\kappa_\alpha$ are learning gains. The periodic perturbation $p = q_j - \hat{q}_j$ is defined as the difference between the measured position $q_j$ of the $j^{th}$ actuator and its estimation [9]

$$\hat{q}_j = \alpha_0 + \sum_{i=1}^{N} \alpha_i \sin \phi_i. \tag{6}$$

The choice of the learning gains is crucial for the overall behavior of this algorithm. A recommendation in this regard is proposed in [8], based on which the parameters were initially set. To simulate above methodology and to fine tune the parameters, artificially generated gait patterns were used. To this end we used the approach described in a previous publication [11] to synthesize hip angle trajectories according to a simulated person walking at variable gait speed.

### 4.2 Assistive Torque

Adaptive oscillators can be used to synchronize to a quasi-periodic input signal, which could be the angular position of joints or even electromyographic measurements of the leg muscles [1,3,10]. In the present case the joint angles of the hip actuators are used. Once the frequency, phase and amplitude components of the input signal and thus its periodic behavior is learned, it can be reproduced based on the current parameters. An estimate of the current joint angle $\hat{q}_j(t)$ and its future value $\hat{q}_j(t + \Delta t)$ according to a specified time horizon $\Delta t$ are therefore provided. Based thereon, an assistive torque $\tau_{A,j}$ can be calculated as

$$\tau_{A,j} = K_A(\hat{q}_j(t + \Delta t) - \hat{q}_j(t)), \tag{7}$$

where $K_A$ is a tune-able virtual stiffness [3]. Thus, the limbs of the user are smoothly assisted towards their estimated future position, while the system adapts to changes in the gait pattern in terms of frequency and shape, e.g. due to varying gait speed.

Following this strategy an additional torque of course can only be considered as *assistive* when the reproduced periodic signal resembles the measured signal and the discrepancy between estimated and actual movement is small. To ensure usability an additional weight

$$w_{A,j} = \frac{1}{2}\left(1 - \tanh\frac{p - p_{max}}{\epsilon}\right) \tag{8}$$

is defined, where $p(t)$ again is the signal difference between input and estimation, $p_{max}$ adjusts the allowed maximum of this difference and $\epsilon$ tunes the steepness of the tanh function. The final torque

$$\tau_j = \tau_{T,j} + w_{A,j}\tau_{A,j} \tag{9}$$

applied onto the $j^{th}$ actuator is then the combination of the torque components according to transparency and motion assistance.

## 5 Experimental Results

First experiments with the described exoskeleton prototype were successfully conducted, where the assistive torque was superimposed on the torque according to the presented transparent mode. These initial tests were performed by persons of our institute walking on level ground. Figure 2 shows a typical adaption process during gait, starting with the transition from stand to walk. The measured angles of the right ($q_1$) and left ($q_2$) hip joint are shown as solid lines, their estimations $\hat{q}_1$ and $\hat{q}_2$, respectively, as dashed lines. At about 0.5 s the adaption process is (re-)activated and the adaptive oscillator system is reset to defined initial conditions. As can be seen, the estimated signal of the left leg is close enough to the measured joint angle after one gait cycle and the assistive torque (bottom, solid red line) is smoothly switched on. In contrast, the learned signal of the right leg is not accurate enough during the first gait cycle, causing the assistive torque (bottom, solid blue line) to be switched off again, and the motion assistance starts with the third gait cycle. Activated assistance is shown as lines with dots, where +1 corresponds to the right hip joint and −1 to the left hip joint. Transparency is constantly activated. The according torques $\tau_{T,j}$ are shown as dashed lines. Due to safety reasons, the total desired torque is limited to ±0.275 Nm (black dashed lines). In Fig. 3 the parameters of the oscillator system of the left leg is shown for the same measurement. After the convergence phase during the first few gait cycles the periodic behavior of the motion signals is learned accurate enough for activating the assistance mode, while the system constantly adapts to the current hip angle trajectories.

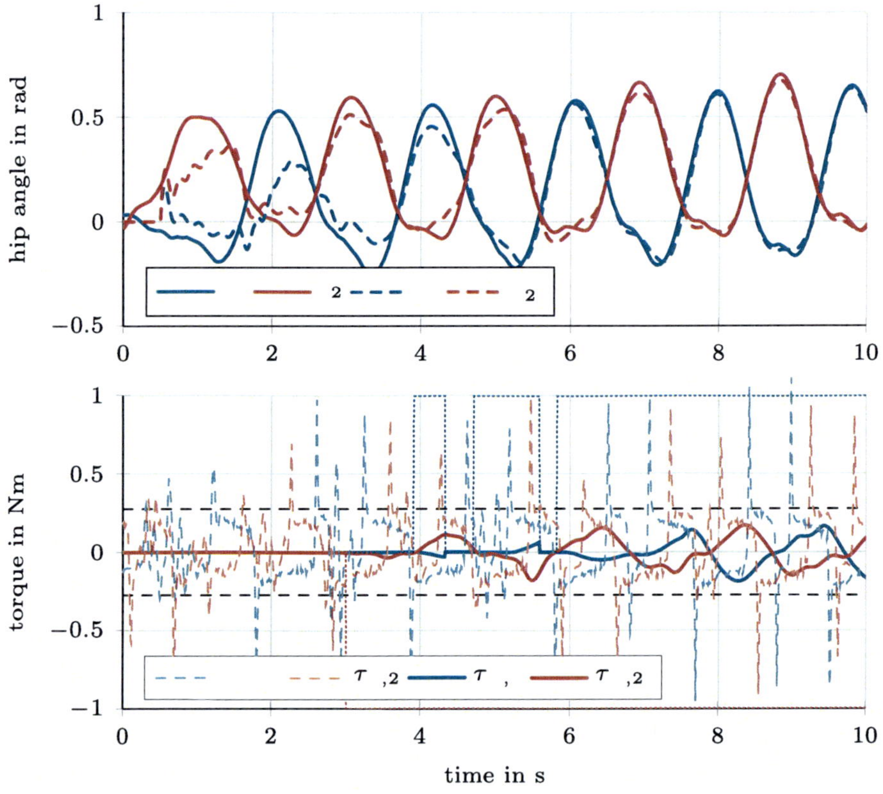


**Fig. 2.** Adaption process with measured (solid) and estimated (dashed) signals on the top and torques according to calculated assistance (solid) and transparency (dashed) on the bottom (activated assistance dotted)

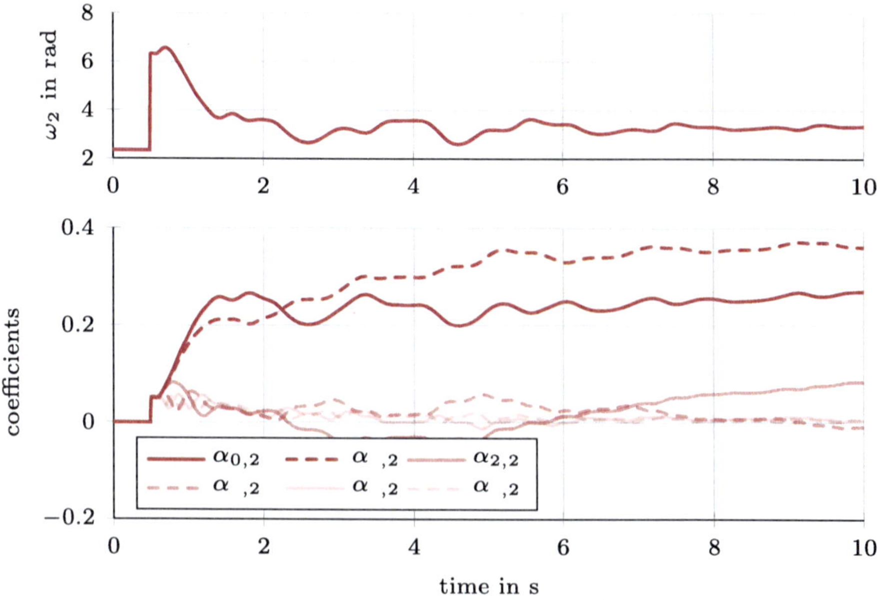


**Fig. 3.** Adaption of the angular frequency (top) and the coefficients (bottom)

## 6 Conclusion

This paper presents an approach to assist gait motion using a lower body exoskeleton, which consists of two main parts. On the one hand, a transparent mode is realized, where the device is not applying any assistance or resistance. Unwanted effects due to friction, inertia or gravity are compensated significantly, minimizing the perceived interaction forces between user and exoskeleton. To this end, the gear backlash of the actuators is exploited, representing a low-cost solution as no additional force or torque sensors, or similar, are necessary. On the other hand, an assistance mode is implemented to support the user while walking. The methodology of adaptive oscillators is utilized to learn the quasi-periodic motion of the legs during gait. Artificially generated gait patterns were used to tune the learning gains and to define bounds for the system parameters to increase robustness. Based on the learned motion an assistive torque is calculated and superimposed to the torque needed to achieve transparency.

The introduced methods were implemented and tested on an exoskeleton prototype. Initial experiments with persons starting to walk from stance showed promising results and the functionality of the approach was confirmed. Ascribable to a lack of usable standard tests, the effects of transparency and assistance were assessed just by subjective evaluation, supported by measurements of actuator positions and torques. Extensive experiments not only considering gait, but also stair ascent and descent and other cyclic movement tasks, as well as designing objective performance measures are in the focus of ongoing research.

**Acknowledgement.** This work has been supported by the "LCM - K2 Center for Symbiotic Mechatronics" within the framework of the Austrian COMET-K2 program.

## References

1. Aguirre-Ollinger, G.: Exoskeleton control for lower-extremity assistance based on adaptive frequency oscillators: adaptation of muscle activation and movement frequency. Proc. Inst. Mech. Eng. [H] **229**(1), 52–68 (2015)
2. Buchli, J., Righetti, L., Ijspeert, A.J.: Frequency analysis with coupled nonlinear oscillators. Phys. D Nonlinear Phenom. **237**(13), 1705–1718 (2008). https://doi.org/10.1016/j.physd.2008.01.014
3. Giovacchini, F., et al.: A light-weight active orthosis for hip movement assistance. Robot. Auton. Syst. **73**, 123–134 (2015). https://doi.org/10.1016/j.robot.2014.08.015
4. Just, F., et al.: Exoskeleton transparency: feed-forward compensation vs. disturbance observer. at - Automatisierungstechnik **66**(12), 1014–1026 (2018). https://doi.org/10.1515/auto-2018-0069
5. Koller, J.R., Jacobs, D.A., Ferris, D.P., Remy, C.D.: Learning to walk with an adaptive gain proportional myoelectric controller for a robotic ankle exoskeleton. J. Neuroeng. Rehabil. **12**(1), 97 (2015). https://doi.org/10.1186/s12984-015-0086-5
6. Righetti, L., Buchli, J., Ijspeert, A.J.: Dynamic Hebbian learning in adaptive frequency oscillators. Phys. D Nonlinear Phenom. **216**(2), 269–281 (2006). https://doi.org/10.1016/j.physd.2006.02.009
7. Righetti, L., Ijspeert, A.J.: Programmable central pattern generators: an application to biped locomotion control. In: Proceedings of the - IEEE International Conference on Robotics and Automation, vol. 2006, pp. 1585–1590. IEEE (2006). https://doi.org/10.1109/ROBOT.2006.1641933
8. Ronsse, R., De Rossi, S.M.M., Vitiello, N., Lenzi, T., Carrozza, M.C., Ijspeert, A.J.: Real-time estimate of velocity and acceleration of quasi-periodic signals using adaptive oscillators. IEEE Trans. Robot. **29**(3), 783–791 (2013). https://doi.org/10.1109/TRO.2013.2240173
9. Ronsse, R., et al.: Oscillator-based assistance of cyclical movements: model-based and model-free approaches. Med. Biol. Eng. Comput. **49**(10), 1173–1185 (2011). https://doi.org/10.1007/s11517-011-0816-1
10. Seo, K., Lee, J., Lee, Y., Ha, T., Shim, Y.: Fully autonomous hip exoskeleton saves metabolic cost of walking. In: 2016 IEEE International Conference on Robotics and Automation (ICRA), pp. 4628–4635. IEEE (2016). https://doi.org/10.1109/ICRA.2016.7487663
11. Ziegler, J., Gattringer, H., Müller, A.: Generation of parametric gait patterns. In: Altuzarra, O., Kecskemethy, A. (eds.) Advances in Robot Kinematics 2022. ARK 2022. Springer Proceedings in Advanced Robotics, vol. 24, pp. 375–382. Springer, Cham (2022). https://doi.org/10.1007/978-3-031-08140-8_41

# On the Stiffness Modelling of the ProHep-LCT Robotic Needle Insertion Instrument

Bogdan Gherman[1], Corina Radu[2], Andrei Caprariu[1], Nadim Al Hajjar[2], Calin Vaida[1], Andra Ciocan[2], Paul Tucan[1], Emil Mois[2], and Doina Pisla[1](✉)

[1] CESTER, Technical University of Cluj-Napoca, 400114 Cluj-Napoca, Romania
doina.pisla@mep.utcluj.ro

[2] Iuliu Hatieganu University of Medicine and Pharmacy, 400347 Cluj-Napoca, Romania

**Abstract.** Minimally invasive treatment of liver cancer through percutaneous approaches (interstitial brachytherapy, intra-tumoral chemotherapy) requires an accurate needle placement achieved through accurate insertion devices and robust control. This paper presents the stiffness modelling of a needle insertion device used within a robotic system for liver cancer treatment, able to insert up to 6 rigid brachytherapy needles using linear trajectories for improved accuracy. The device is highly versatile and has been designed to be used in conjunction with computer tomography or with an intra-operatory ultrasound, as an imaging control device. The developed model will be used for further stiffness analysis and optimization for increased needle placement accuracy within interstitial brachytherapy liver cancer treatment.

**Keywords:** Medical robot · needle insertion · stiffness model · stiffness analysis

## 1 Introduction

Liver cancer is one of the deadliest cancers nowadays [1], with a worldwide incidence of over one million cases projected at the level of 2025. Hepatocellular carcinoma (HCC) is by far the most common form, accounting for more than 90% of liver cancers. Although surgery and liver transplant are the most efficient treatment options [2], most patients (more than 75%) cannot benefit from this treatment, because of the metastases or because of the poor condition of the patient. Other therapeutic approaches include brachytherapy, percutaneous radiofrequency ablation (RFA), microwave ablation or percutaneous intra-tumoral chemotherapy [3]. All these techniques applied in a minimally invasive approach involve percutaneous needle insertion from the patient's skin up to a target point into the tumor within the patient's body [4]. The general accuracy requirement is that the needle must be placed within an error of about 1–2 mm from the target point, to avoid puncturing important vessels or other tissue structures. Depending on the actual approach, the required accuracy may vary, for example, in the case of interstitial brachytherapy, according to the final position of the needle regarding the target point into the tumor, the irradiation dose (or the irradiation time) can be modified to destroy the tumoral cells (if


T. Petrič et al. (Eds.): RAAD 2023, MMS 135, pp. 245–252, 2023.
https://doi.org/10.1007/978-3-031-32606-6_29

the surrounding healthy tissue is not severely affected), [4]. Uncontrolled needle deflection, especially for rigid needles, is one of the main causes of inaccuracy [5]. Manual insertion may be difficult, due to a lack of accurate control (and tremor) of the surgeon's hand, which is why a robotic solution is ultimately preferred. Although the insertion forces are small (usually around 20 N), the needle insertion instruments may have an inherent compliance, due to various reasons, such as tight space (which leads to small and thin elements) or weight constraints [6]. In the case of industrial serial robots, the static stiffness is within a range of $10^5$ N/m [7], which, of course changes according to the robot configuration. Several methods have been developed for modelling the robot stiffness, such as: Finite Element Analysis (FEA), Matrix Structure Analysis (MSA) and Virtual Joint Modelling (VJM) [8]. All these approaches present their own strengths and weaknesses and can be used in various stages of the development or the modelling of optimal trajectories.

This paper presents the stiffness modelling using the MSA approach of the needle insertion instrument of the ProHep-LCT parallel robot for HCC treatment using interstitial brachytherapy. The targeted outcome is to define the MSA-based representation of the device and further of the ProHep-LCT robotic system targeting its optimization with minimal costs for an improved needle placement accuracy. Section 2 presents the ProHep-LCT robotic system architecture for HCC treatment using percutaneous needle placement; Sect. 3 presents initial testing results using the needle insertion module; Sect. 4 presents the MSA stiffness modelling; Sect. 5 presents some conclusions and future development directions.

## 2 The ProHep-LCT Robotic System for HCC Treatment

The ProHep-LCT robotic system for HCC treatment (Fig. 1) consists of two identical parallel robots: one for guiding the Intra-operatory Ultrasound (IUS) probe and the other to guide the Needle Insertion Device (NID), [9–13].

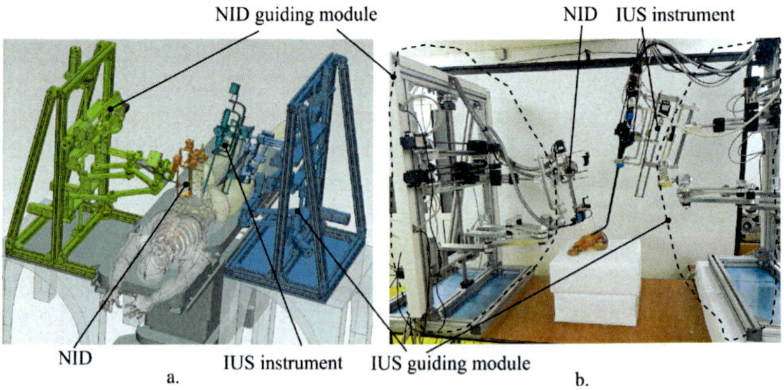


**Fig. 1.** The ProHep-LCT robotic system for HCC treatment: a. CAD model within medical environment; b. the experimental model in laboratory tests.

The IUS robot guides a commercially available ultrasound probe (Hitachi Arietta 70, [14]), used to provide the required data (the target points for the needles, into the tumors), as input into the control system of the NID robot. Each robot has an identical parallel architecture with 5 degrees of freedom (DoFs), 3 prismatic active joints and 2 revolute joints, guiding a mobile platform on which the two instruments (IUS and NID) are mounted. Each robot drives a guiding module. The IUS guiding module (which actuates the ultrasound probe) consists of 2 DOFs (insertion/retraction of the probe and its rotation) and another 2 DOFs which actuate the probe's levers (thus controlling the probe's tip). The NID module (Fig. 2) is based on a Gantry (Cartesian) mechanism with 3 + 1 DOFs (the needle gripper) and can manipulate and insert multiple needles (up to six in the current configuration).

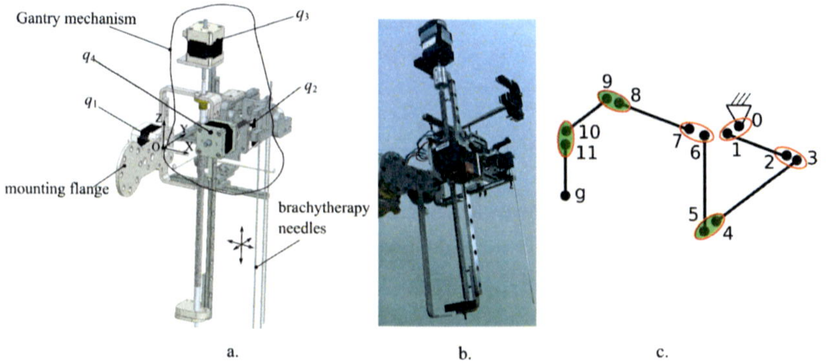


**Fig. 2.** The NID device for the insertion of multiple needles in interstitial brachytherapy procedures: a. CAD model; b. experimental model; c. the NID MSA-based representation

## 3 NID Experimental Stiffness Evaluation

The ProHep-LCT has been designed considering the constraints imposed by the medical environment and the medical protocol described in [15, 16], which considers that the data regarding the tumor position and the needle trajectory is monitored using the IUS. The NID has been developed considering its use in other HCC treatment approaches, like the one described in [15], where the needle trajectory is monitored using Computer Tomography (CT). This has led to a very compact architecture, to fit into the CT bore, together with the patient and CT couch. Unfortunately, this has also led to an unwanted relatively high level of compliance observed during validation tests. To study the behavior of the NID in real environment, laboratory tests using the layout presented in Fig. 3 have been performed to measure the stiffness of the NID in terms of displacements at the level of the end-effector (the needle gripper) using a dial gauge indicator. The obtained results are presented in Fig. 4. The displacement of the actuating element has been measured on two axes, corresponding to the OX and OY axis at various insertion speeds, namely 10 mm/s (which is the maximum insertion speed of the instrument), 5 mm/s and 0.5 mm/s. Five different measurements have been taken for every axis and every insertion speed, using various areas of the liver. On average, the higher the insertion speed, the larger is the measured displacement at the level of the end-effector of the NID, with the peaks varying between 0.5 and 0.72 mm on both axes (the peak values are presented in Fig. 4 a–f).

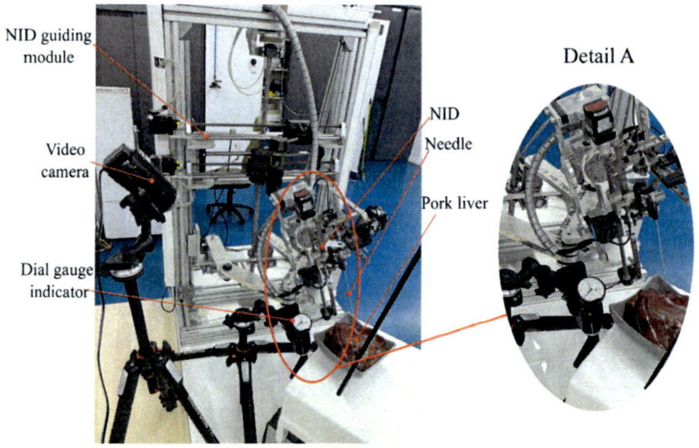


**Fig. 3.** Stiffness measurement layout in laboratory conditions using pork liver.

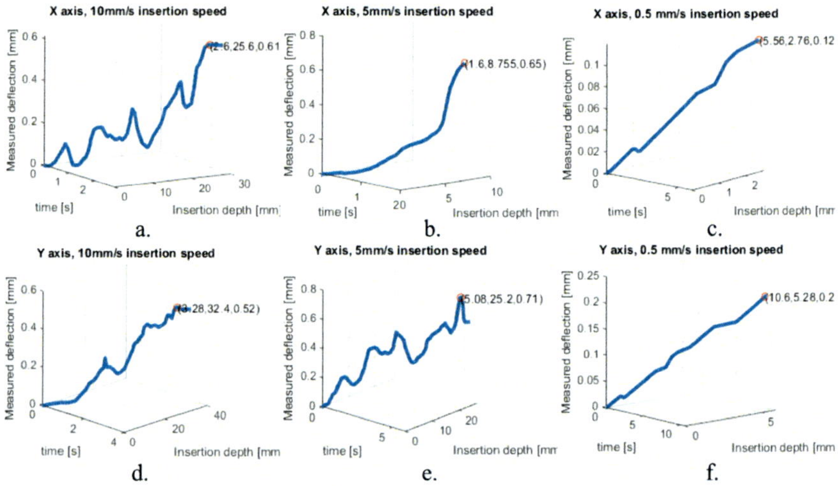


**Fig. 4.** Stiffness measurement test results.

At lower insertion speeds (0.5 mm/s), the measured displacements are rather low, but, as proved in [17], the lower the insertion speed, the higher the needle deflection and thus, ultimately reduces the placement accuracy. This is especially true in skin puncturing, which is known to be tougher than parenchyma. The measured insertion force in the parenchyma is of maximum 2 N, which leads to an overall stiffness of the NID of about $2.85 \times 10^3$ N/m, which is lower than the average of a serial robot. Of course, needle deflection is the major issue during needle insertion, causing a poor accuracy of the

NID, but a small improvement of the NID stiffness might lead to an improved placement accuracy.

## 4 Stiffness Modelling of the Needle Insertion Device

The NID stiffness has been modelled using Matrix Structure Analysis (MSA). Although this simplifies the robot structure, thus reducing the overall accuracy of the stiffness model, it provides the advantage of speed, especially for structural optimization, compared to FEA [18]. As Fig. 2a. Shows, the NID has a serial architecture and has been considered as a unique beam within the MSA algorithm. The MSA-based representation of the NID (with links and nodes) is presented in Fig. 2c. The NID links are modeled as elastic elements, while the joints <0,1>, <2,3> and <6,7> are modeled as passive rigid joints, and <4,5>, <8,9> and <10,11> are modeled as elastic prismatic active joints. According to the MSA method, the stiffness matrix $K$ transforms the displacement vector $\delta$ ($\left[\delta_{px}, \delta_{py}, \delta_{pz}, \delta_{ox}, \delta_{oy}, \delta_{oz}\right]^T$, $\delta_{px}$ – displacement along the OX axis; $\delta_{ox}$ – angular displacement around the OX axis) into the static wrench $W$ ($\left[F_x, F_y, F_z, T_x, T_y, T_z\right]^T$):

$$W = K \cdot \delta \tag{1}$$

The stiffness model in the case of the links (1,2), (3,4), (5,6), (7,8), (9,10), (11,g) is:

$$\begin{bmatrix} W_i \\ W_j \end{bmatrix} = \begin{bmatrix} K_{uu}^{ij} & K_{vu}^{ij} \\ K_{uv}^{ij} & K_{vv}^{ij} \end{bmatrix} \cdot \begin{bmatrix} \delta_i \\ \delta_j \end{bmatrix} \tag{2}$$

with $\delta_0 = 0_{6x1}$ and the matrices $K_{uu}^{ij}, K_{uv}^{ij}, K_{vu}^{ij}, K_{vv}^{ij}$ are expressed in the OXYZ coordinate system (Fig. 2a.) and represent the wrench reaction from one side ($u$ or $v$) of the link due to the same side displacement or the opposite ($u$ or $v$). The aggregated stiffness matrix $K_L$ which yields the links stiffness model is presented in Eqs. (3).

$$K_L = \begin{bmatrix} \begin{bmatrix} K_{uu}^{12} & K_{vu}^{12} \\ K_{uv}^{12} & K_{vv}^{12} \end{bmatrix} & & \\ & \ddots & \\ & & \begin{bmatrix} K_{uu}^{11g} & K_{vu}^{11g} \\ K_{uv}^{11g} & K_{vv}^{11g} \end{bmatrix} \end{bmatrix}_{72\times72}; \quad \begin{bmatrix} -I_{72x72} & K_L \end{bmatrix} \begin{bmatrix} W_1 \\ \vdots \\ W_g \\ \delta_1 \\ \vdots \\ \delta_g \end{bmatrix}_{144\times1} = 0_{72\times1} \tag{3}$$

In the case of the rigid joints < 0,1 >, < 2,3 > and < 6,7 >, the constraints are expressed as $\Omega_R$ (considering the NID in quasi-vertical position), while for the active prismatic joints <4,5>, <8,9> and <10,11>, these are expressed as $\Omega_P$, Eqs. (4), which leads to the constraint equations expressed in Eqs. (5) and (6). The boundary conditions are further integrated into the stiffness model. The actuators are placed on the links (3,4), (5,6) and (11,g), which add the constraints in Eqs. (7). Furthermore, considering that $\delta_0 = 0_{6x1}$ and that the external loading (the insertion force and the gripper weight) is $W_{ef}$, it leads to the constraints presented in Eqs. (8).

$$\Omega_R = \begin{bmatrix} 000001 \end{bmatrix};\ \Omega_P = \begin{bmatrix} 100000 \\ 010000 \\ 000000 \\ 000100 \\ 000010 \\ 000001 \end{bmatrix} \tag{4}$$

$$\begin{bmatrix} I_{6\times 6} - I_{6\times 6} \end{bmatrix} \cdot \begin{bmatrix} \delta_i \\ \delta_j \end{bmatrix} = 0_{6\times 1};\ \begin{bmatrix} \Omega_R\ \Omega_R \end{bmatrix} \cdot \begin{bmatrix} W_i \\ W_j \end{bmatrix} = 0_{1\times 1}\ (i = 0, 2, 6;\ j = 1, 3, 7) \tag{5}$$

$$[\Omega_P - \Omega_P] \cdot \begin{bmatrix} \delta_i \\ \delta_j \end{bmatrix} = 0_{6\times 1};\ \begin{bmatrix} \Omega_P\ \Omega_P \end{bmatrix} \cdot \begin{bmatrix} W_i \\ W_j \end{bmatrix} = 0_{6\times 1};\ \begin{bmatrix} \Omega_P & 0_{1\times 6} \\ 0_{1\times 6} & \Omega_P \end{bmatrix} \cdot \begin{bmatrix} W_i \\ W_j \end{bmatrix} = 0_{6\times 1}$$
$$(i = 4, 8, 10;\ j = 5, 9, 11) \tag{6}$$

$$\begin{bmatrix} I_{6\times 6}\ I_{6\times 6} \end{bmatrix}_{6\times 12} \cdot \begin{bmatrix} W_i \\ W_j \end{bmatrix}_{12\times 1} = \begin{bmatrix} W_{qk} \end{bmatrix}\ (i = 3, 5, 11;\ j = 4, 6, 12;\ k = 1, 2, 3) \tag{7}$$

$$W_g = W_{ef};\ \Omega_P \cdot \delta_1 = 0_{6\times 1};\ \Omega_R \cdot W_1 = 0_{1\times 1} \tag{8}$$

Assembling all constraints into the Cartesian stiffness matrix model, it yields an equation like the one presented in eq. (9), which after separating the nodes displacements into internal $\delta_{\text{int}}$ (EI) and external $\delta_{ext}$, , linked to the end-effector ($\delta_g$ – G) lead to the NID stiffness matrix. An example is given in Eqs. (11), for $q_1 = 125.5\,\text{mm}$; $q_2 = 133.1\,\text{mm}$; $q_3 = 88.9\,\text{mm}$ and $W_{ef} = \begin{bmatrix} 0030000 \end{bmatrix}^T$ [N].

$$A_{144\times 144} \cdot \begin{bmatrix} W_1 \\ \vdots \\ W_g \\ \delta_1 \\ \vdots \\ \delta_g \end{bmatrix}_{144\times 1} = \begin{bmatrix} 0_{138\times 1} \\ W_{ef} \end{bmatrix}_{144\times 1};\ A_{144\times 144} = \begin{bmatrix} -I_{72\times 72} & K_L \\ 0_{37\times 72} & \Omega_{rC} \\ \Omega_{rCP} & 0_{28x72} \\ -\Omega_{1rCP} & \Omega_{2rCP} \\ II & 0_{6\times 72} \end{bmatrix} \tag{9}$$

$$K_1 = \begin{bmatrix} K_L \\ \Omega_{rC} \\ 0_{28\times 72} \\ \Omega_{2rCP} \end{bmatrix};\ S_1 = \begin{bmatrix} -I_{72\times 72} \\ 0_{36\times 72} \\ \Omega_{rCP} \\ -\Omega_{1rCP} \end{bmatrix};\ K_{ef} = -\begin{bmatrix} GI_{6\times 72}\ 0_{6\times 72} \end{bmatrix} \cdot \begin{bmatrix} S_1 & K_1 \\ EII_{6\times 72} & 0_{6\times 72} \end{bmatrix} \cdot \begin{bmatrix} K_G \\ C_G \end{bmatrix} \tag{10}$$

$$K_{ef} = \begin{bmatrix} 1.25 \cdot 10^{3}0.17 \cdot 10^{3}0006.65 \cdot 10^{3} \\ 0.17 \cdot 10^{3}1.25 \cdot 10^{3}0000.17 \cdot 10^{3} \\ 003.12 \cdot 10^{3}-2.56 \cdot 10^{2}1.21 \cdot 10^{3}0 \\ 00-2.56 \cdot 10^{2}8.6 \cdot 10^{2}-1.92 \cdot 10^{2}0 \\ 001.21 \cdot 10^{3}-1.92 \cdot 10^{2}9.14 \cdot 10^{2}0 \\ 6.65 \cdot 10^{3}0.17 \cdot 10^{3}0003.8 \cdot 10^{2} \end{bmatrix} \quad (11)$$

The same active coordinates have been used in a FEA simulation (Eqs. 12) using Abaqus [19], producing similar results and thus validating the stiffness model. The developed stiffness model has the advantage of speed (a Matlab run for this system takes up to 0.009 ms), being of great help in further optimization. An increased section by 4mm (thus doubling the section area of the current experimental model) of the links, have reduced by ~50% the total deformation at the level of the end-effector.

$$K_{ef} = \begin{bmatrix} 1.67 \cdot 10^{3}0.23 \cdot 10^{3}0007.14 \cdot 10^{3} \\ 0.24 \cdot 10^{3}1.65 \cdot 10^{3}0000.28 \cdot 10^{3} \\ 003.88 \cdot 10^{3}-1.98 \cdot 10^{2}1.335 \cdot 10^{3}0 \\ 00-1.98 \cdot 10^{2}8.8 \cdot 10^{2}-1.86 \cdot 10^{2}0 \\ 001.335 \cdot 10^{3}-1.86 \cdot 10^{2}8.95 \cdot 10^{2}0 \\ 7.14 \cdot 10^{3}0.28 \cdot 10^{3}0004.06 \cdot 10^{2} \end{bmatrix} \quad (12)$$

## 5 Conclusions

The paper presents the stiffness modelling and analysis of a robotic instrument for needle insertion, designed to be used within the ProHep-LCT robotic system for HCC minimally invasive treatment. A set of initial measurements taken in laboratory conditions using a pork liver have indicated the deflection estimated at the level of the end-effector. The stiffness model has been achieved using the Matrix Structure Analysis, which can be used within simulations to study the influence of various dimensional parameters of the needle insertion device. Future work targets the extension of the current model for the entire ProHep-LCT robotic system.

**Acknowledgement.** This work was supported by a grant of the Ministry of Research, Innovation and Digitization, CNCS/CCCDI—UEFISCDI, project number PN-III-P2-2.1-PED-2021-2790 694PED—Enhance, within PNCDI III.

## References

1. Llovet, J.M., et al.: Hepatocellular carcinoma. Nat. Rev. Dis. Primers **7**, 6 (2021)
2. European Association for the Study of the Liver: EASL clinical practice guidelines: management of hepatocellular carcinoma. J. Hepatol. **69**, 182–236 (2018)
3. Inchingolo, R., et al.: Locoregional treatments for hepatocellular carcinoma: current evidence and future directions. World J. Gastroenterol. **25**, 4614–4628 (2019)

4. Chen, S., Wang, F., Lin, Y., Shi, Q., Wang, Y.: Ultrasound-guided needle insertion robotic system for percutaneous puncture. Int. J. Comput. Assist. Radiol. Surg. **16**(3), 475–484 (2021)
5. Li, M., Lei, Y., Xu, T.: Sensitivity of influential factors on needle insertion experiments: a quantitative analysis on phantom deformations and needle deflections. Chin. J. Mech. Eng. **33**(1), 1–11 (2020)
6. Guo, J. et al. The Design of Compact Robotic-Assisted Needle Position System with MPC-Based Remote Control. Complex. 5406084:1–5406084:13 (2020)
7. Xiong, G., Ding, Y., Zhu, L.: Stiffness-based pose optimization of an industrial robot for five-axis milling. Robot. Comput. Manuf. **55**, 19–28 (2018)
8. Wu, K., Li, J., Zhao, H., Zhong, Y.: Review of industrial robot stiffness identification and modelling. Appl. Sci. **12**, 8719 (2022)
9. Plitea, N., Pisla, D., Vaida, C., Gherman, B., Tucan, P.: PRoHep-LCT-Parallel robot for the minimally invasive treatment of hepatic carcinoma, patent pending A1017 (2018)
10. Vaida, C., et al.: A new robotic system for minimally invasive treatment of liver tumours. Proc. Rom. Acad., Series A. **21**(3), 263–271 (2020)
11. Gherman, B., Birlescu, I., Burz, A., Pisla, D.: Automated medical instrument for the insertion of brachytherapy needles on parallel trajectories, patent pending A00806 (2019)
12. Birlescu, I., et al.: Automated medical instrument for the manipulation of a laparoscopic ultasound probe, patent pending A00752 (2019)
13. Gherman, B., et al.: Singularities and workspace analysis for a parallel robot for minimally invasive surgery. In: International Conference Automation, Quality and Testing, Robotics (AQTR), pp. 1–6 (2010)
14. Hersant, J., et al.: Comparison between conventional duplex ultrasonography and the dual-gate Doppler mode for hemodynamic measurements of the carotid arteries. Ultrasonography **41**(2), 373–381 (2022)
15. Radu, C., et al.: Integration of real-time image fusion in the robotic-assisted treatment of hepatocellular carcinoma. Biology **9**, 397 (2020)
16. Gherman, B., et al.: Risk assessment-oriented design of a needle insertion robotic system for non-resectable liver tumors. Healthcare **10**, 389 (2022)
17. Tucan, P., et al.: Design and experimental setup of a robotic medical instrument for brachytherapy in non-resectable liver tumors. Cancers **14**, 5841 (2022)
18. Klimchik, A., Pashkevich, A., Chablat, D.: Fundamentals of manipulator stiffness modeling using matrix structural analysis. Mech. Mach. Theory **133**, 365–394 (2019)
19. Navidtehrani, Y., Betegón, C., Martínez-Pañeda, E.: A simple and robust Abaqus implementation of the phase field fracture method. App. in Eng. Sci. **6**, 1000650 (2021)

# Kinematics and Dynamics

# Optimizing Robot Positioning Accuracy with Kinematic Calibration and Deflection Estimation

Leon Žlajpah and Tadej Petrič(✉)

Department of Automatics, Biocybernetics and Robotics, Jožef Stefan Institute, Ljubljana, Slovenia
{leon.zlajpah,tadej.petric}@ijs.si

**Abstract.** To achieve higher positioning accuracy, it is common practice to calibrate the robot. An essential part of the calibration is the estimation of the kinematic parameters. Due to various nonlinear influences on the end-effector position accuracy, such as joint and link flexibility, standard methods of identifying kinematic parameters do not always give a satisfactory result. In this paper, we propose a strategy that considers deflection-dependent errors to improve the overall positioning accuracy of the robot. As joint/link deflections mainly depend on gravity, we include the compensation of gravity-induced errors in the estimation procedure. In the first step of the proposed strategy, we compute the joint position errors caused by gravity. In the next step, we apply an existing optimization method to estimate the kinematic parameters. We propose to use an optimization based on random configurations. Such an approach allows good calibration even when we want to calibrate a robot in a bounded workspace. Since calibration is generally time consuming, we investigated how the number of measured configurations influences the calibration. To evaluate the proposed method, we used a simulation of the collaborative robot Franka Emika Panda in MuJoCo.

**Keywords:** serial manipulators · calibration · error compensation

## 1 Introduction

There is considerable motivation and potential to increase the use of collaborative robots for different applications where positional accuracy is important. Compared to conventional industrial robots, they are more compliant and, therefore, more suitable for many tasks where the robot is in contact with the object or environment. However, for many applications it is still challenging for robots to achieve the required position accuracy. The position accuracy of the robot can be improved by calibrating the robot. Calibration is defined here in a broader sense as identifying robot parameters that affect the accuracy of determining the pose (position and orientation) of robot end-effector based on the given positions in the joints. Calibration is done in four steps: modeling, measurements, estimation, and compensation.

This work was supported by Slovenian Research Agency grant N2-0269 and P2-0076.


T. Petrič et al. (Eds.): RAAD 2023, MMS 135, pp. 255–263, 2023.
https://doi.org/10.1007/978-3-031-32606-6_30

The displacement between the desired pose and the actual pose of the end-effector can be caused by incorrect kinematic and/or non-kinematic parameters [11, 12]. Kinematic parameters describe the geometry of the robot structure and are used in the kinematic model. Usually, robot manufacturers provide nominal kinematic parameters, e.g., Denavit-Hartenberg (DH) parameters, which differ from the actual parameters due to deviations in link lengths, offset distances between joints, and misaligned joint axes.

The sources of non-kinematic errors are joint flexibility, link flexibility, gearbox errors, etc. To be able to compensate for these errors, corresponding models have to be developed. As the non-kinematic errors can be highly nonlinear, it is not easy to develop models including all error sources [1]. Therefore, the non-kinematic errors are in most cases not considered. Recently, some researchers have proposed calibration methods free of any modeling. In [16] integrated algorithm based on the standard DH model and artificial neural network (ANN) is proposed to calibrate the robot. Similarly, in [15] ANN is used together with geometric parameter identification to compensate pose errors. An improved robot accuracy utilizing a combination of geometric calibration and ANN is also reported in [3]. From the practical point of view, it is also important to know whether additional modeling beyond kinematic modeling will improve the accuracy of the robot. For most robots, link and joint flexibility contribute only a minor part of pose errors [2, 7]. Errors due to other non-kinematic sources are even smaller [2].

Which error sources have to be considered also depends on the accuracy of the measurement of the robot end-effector pose. For calibration, several measurement systems have been proposed, such as laser trackers [8, 14], conventional or optical coordinate measuring machines [10], optical tracking systems [17], or ball bars [9], which all have different repeatability and accuracy. To achieve good identification accuracy, it is important to select enough measurement poses. As a high number of measurement poses increases the computational complexity is good to optimize the number and location of poses [5]. A good solution is to use random robot configurations, especially whenever it is not possible to move the robot in the whole workspace.

We propose an approach to robot calibration that focuses on correcting the robot's kinematic model and compensates for errors due to joint and link deflections. For better transparency, we implemented it in a simulation environment. First, we define the kinematic model based on DH parameters and the linearized model for compensating gravity-induced errors. In the next section, we describe the estimation of model parameters. Then, we present our setup, which is realized in a simulation environment based on the general-purpose physics engine MuJoCo, and data generation. Finally, we present the results, emphasizing position accuracy improvement due to DH parameter correction and compensation of gravity-induced errors.

## 2 Robot Error Modelling

### 2.1 Kinematic Errors

To describe the robot kinematics we use Denavit-Hartenberg (DH) convention, where a set of four parameters $\mathrm{DH}_i = (a_i, \alpha_i, d_i, \theta_i)$ is used to define the homogenous matrix ${}^{i-1}\mathbf{T}_i$ describing the transformation between frames attached to links $i-1$ and $i$ connected by joint $i$. Assuming that joint $i$ is revolute and $q_i$ is a joint variable (position), ${}^{i-1}\mathbf{T}_i(q_i)$ equals

$$^{i-1}\mathbf{T}_i(q_i, \mathrm{DH}_i) = \mathrm{Rot}_{z_{i-1}}(\theta_i + q_i) \cdot \mathrm{Trans}_{z_{i-1}}(d_i) \cdot \mathrm{Trans}_{x_i}(r_i) \cdot \mathrm{Rot}_{x_i}(\alpha_i) \quad (1)$$

where $\mathrm{Trans}_{\boldsymbol{a}}(\cdot)$ and $\mathrm{Rot}_{\boldsymbol{a}}(\cdot)$ represent the translation and rotation along vector $\boldsymbol{a}$, respectively. Combining all transformations from the robot base to the robot end-effector, we obtain the pose (position and orientation) of the robot end-effector

$$^{0}\mathbf{T}_E(\boldsymbol{q}, \mathrm{DH}) = \left(\prod_{i=1}^{n} {}^{i-1}\mathbf{T}_i(q_i, \mathrm{DH}_i)\right) {}^{n}\mathbf{T}_E = \begin{bmatrix} \mathbf{R} & \boldsymbol{p} \\ \mathbf{0} & 1 \end{bmatrix}, \quad (2)$$

where ${}^{n}\mathbf{T}_E$ is the fixed transformation matrix between the robot flange and the end-effector, $\boldsymbol{p}$ and $\mathbf{R}$ are the position and orientation of the end-effector, and $n$ is the number of joints.

Errors in kinematic parameters cause differences between the actual and the nominal end-effector pose. Let $\mathrm{DH}_i = (a_i, \alpha_i, d_i, \theta_i)$ be the nominal parameters provided by the robot manufacturer and $\widehat{\mathrm{DH}}_i = (\hat{a}_i, \hat{\alpha}_i, \hat{d}_i, \hat{\theta}_i)$ the actual parameters. Then the position and orientation errors due to kinematic errors equal

$$\boldsymbol{e}_p = \boldsymbol{p}(\boldsymbol{q}, \widehat{\mathrm{DH}}) - \boldsymbol{p}(\boldsymbol{q}, \mathrm{DH}), \qquad \boldsymbol{e}_R = \log(\mathbf{R}(\boldsymbol{q}, \widehat{\mathrm{DH}})\mathbf{R}(\boldsymbol{q}, \mathrm{DH})^T). \quad (3)$$

In the proposed calibration procedure only the position $\boldsymbol{p}$ of the robot end-effector will be considered.

### 2.2 Non-kinematic Errors

The robot end-effector pose is also influenced by non-geometric error sources, which may be highly nonlinear and changing. According to robotics literature, e.g. [2,11,12], the majority of non-kinematic errors are contributed by compliance errors due to the flexibility of joints and links. As joint and link compliance can be predicted with the use of an appropriate model, the errors due to flexibility can be included in a model based calibration. As we are considering static calibration, we include errors due to gravity in the calibration. As proposed in [4] the relation between the gravity forces and joint deviations can be simplified to

$$\Delta \boldsymbol{q}_i = \mathbf{S}_i \boldsymbol{g}_i(\boldsymbol{q}), \qquad i = 1, \ldots, n, \quad (4)$$

where $\mathbf{S}_i$ are effective stiffness coefficients and $\boldsymbol{g}(\boldsymbol{q})$ are joint torques due to gravity. Assuming that $\Delta \boldsymbol{q}$ is small, the relation between the end-effector position and $\Delta \boldsymbol{q}$ is

$$\boldsymbol{p}_M = \boldsymbol{p}(\boldsymbol{q} + \Delta \boldsymbol{q}) \approx \boldsymbol{p}(\boldsymbol{q}) + \mathbf{J}(\boldsymbol{q})\Delta \boldsymbol{q} = \boldsymbol{p}(\boldsymbol{q}) + \mathbf{J}(\boldsymbol{q})\mathbf{S}\boldsymbol{g}(\boldsymbol{q}), \quad (5)$$

where $\boldsymbol{p}_M$ are actual (measured) positions of the end-effector. Knowing the static model of the robot, which also includes the payload, the joint torques $\boldsymbol{g}$ are known, and it is necessary only to estimate the stiffness coefficients $\mathbf{S}_i$.

## 3 Estimation

Combining kinematic and non-kinematic errors the position error equals

$$\boldsymbol{e}_p = \boldsymbol{p}(\boldsymbol{q} + \Delta \boldsymbol{q}, \widehat{\mathrm{DH}}) - \boldsymbol{p}(\boldsymbol{q}, \mathrm{DH}). \quad (6)$$

A robot manufacturer usually provides the nominal values of DH parameters or some other data, which can be used to calculate DH parameters. However, the actual DH parameters of a robot may differ from the nominal values due to manufacturing or assembly errors. Therefore, to reduce the positioning errors of the end-effector, it is necessary to identify the actual kinematic parameters of a robot. On the other hand, robots are in general assumed to be rigid. Hence, there should be no joint and link deflection. Therefore, in the calibration it is necessary to check this and to estimate the stiffness coefficients $\mathbf{S}_i$. A straightforward method for calculating DH parameters is a geometric approach based on known joint axes and linear algebra. The detailed procedure is given in [17]. To estimate the stiffness coefficients $\mathbf{S}$ we can solve (5) in the least square sense.

As an alternative, an optimization method can be used to estimate the kinematic and non-kinematic parameters [6, 13]. In this case we can define objective function as

$$\{\mathrm{DH}, \mathbf{S}\} = \arg\min_{\mathrm{DH},\mathbf{S}} \sum_{k=1}^{m} \left( \|\boldsymbol{p}(\boldsymbol{q}_k + \mathbf{S}\boldsymbol{g}(\boldsymbol{q}_k), \mathrm{DH}) - \boldsymbol{p}_{M,k}\| \right) , \tag{7}$$

where $m$ is the number of measurement samples, $\boldsymbol{p}_M$ and $\boldsymbol{q}_k$ are measured end-effector positions and joint positions for sample $k$, respectively. Equation 7 is used to find the best estimate of DH parameters and/or stiffness coefficients $\mathbf{S}$. By measuring the robot configurations, i.e., joint positions $\boldsymbol{q}$, the end-ffector positions can be calculated using the forward kinematics $\boldsymbol{p}(\boldsymbol{q}, \mathrm{DH})$. The error between the measured end-effector positions $\boldsymbol{p}_M$ and the calculated positions $\boldsymbol{p}$ is then used to define the optimization problem, which is then minimized to obtain the best estimates for the DH parameters and stiffness coefficients. This optimization approach was used by [6, 13] to estimate kinematic and non-kinematic parameters. The optimization process requires multiple samples of the robot configuration, where the number of samples is denoted by $m$. The objective function is then summed over all samples, and the DH parameters and stiffness coefficients $\mathbf{S}$ that minimize the objective function are selected.

## 4 Analysis in Simulation Environment

To analyze the calibration procedure we have used a simulation environment. Our simulation setup consists of general purpose physics engine MuJoCo where we have implemented a model of Franka Emika Panda robot. The robot model is controlled by a control system based on Matlab, which allows easy motion programming and acquisition of robot states.

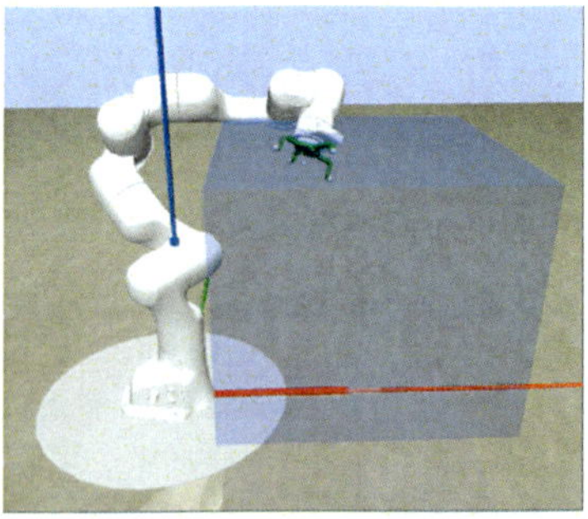

**Fig. 1.** Franka Emika Panda robot in MuJoCo simulation environment. Blue box shows the restricted robot workspace region. (Color figure online)

Besides the inaccurate kinematic parameters and robot compliance an important role in calibration has also measurement accuracy. Namely, to be able to compensate for errors due to specific sources, their contribution to the position error has to be larger than the measurement noise. All these can be easily defined in

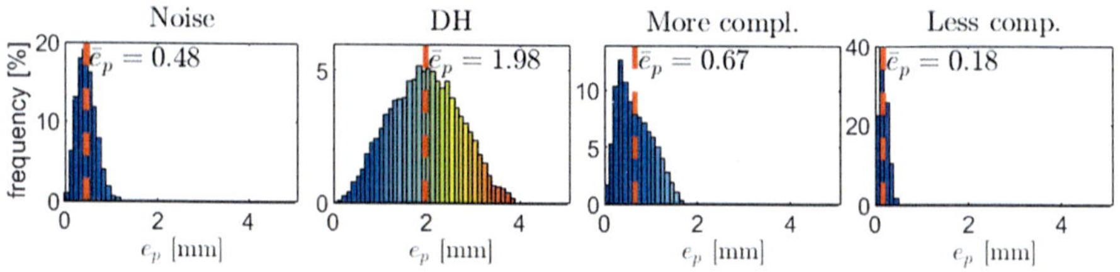


**Fig. 2.** Position error distribution for all error sources: measurement noise, inaccurate DH parameters and robot compliancy.

the simulation environment. In this work, we have assumed that the measurement accuracy equals to the accuracy of a motion capture system. Therefore, we have added to the robot end-effector position normally distributed random noise with a mean of 0mm and a standard deviation of 0.3 mm.

The independent position error distributions caused by measurement noise, incorrect DH parameters, and deflections due to more or less compliant robot are shown in Fig. 2. We can see that the initial DH parameters produce errors significantly larger than measurement noise. The compliances have been selected so that errors for more compliant robot are larger than the error due to measurement noise and lower for the less compliant robot.

For calibration we have used three types of motion patterns. The first type is a combination of independent joint motions to estimate DH parameters using a geometric approach (C trajectories). The second one was the motion between 10000 random configurations in the whole workspace of the robot (R configurations). The last one was the motion between 1000 random configurations in a smaller region (blue box in Fig. 1) simulating the case when obstacles prevent movement in the whole workspace (A configurations).

### 4.1 Results

To estimate DH parameters using geometric approach we have used C trajectory. We have captured 40–100 points for each joint, depending on the range of each joint. We have calculated DH parameters for four cases: a stiff and compliant robot without measurement noise and a stiff and compliant robot with present measurement noise. To verify the obtained position accuracy, we always used the maximal set of measured R configurations. Figure 3 shows the comparison of the position error distributions using calculated $\mathrm{DH}_G$ for different situations regarding the measurement noise and the compliance of the robot. Comparing these error distributions with the one in Fig. 2, we can see that the non-kinematic errors decrease the accuracy and have a similar effect on the estimation of DH parameters as the measurement noise.

Using the geometric approach and C trajectories we first calculate the kinematic parameters. After calculating $\mathrm{DH}_G$ for a compliant robot with measurement noise using the geometric approach (par. set: $\mathrm{DH}_G$) we can improve the accuracy by also compensating the non-kinematic errors, i.e., robot deflection. For that, we have used the optimization (7) on the same measurement data where only $\mathbf{S}$ has been optimized for a given $\mathrm{DH} = \mathrm{DH}_G$ (par. set: $\mathrm{DH}_G/\mathbf{S}_C$).

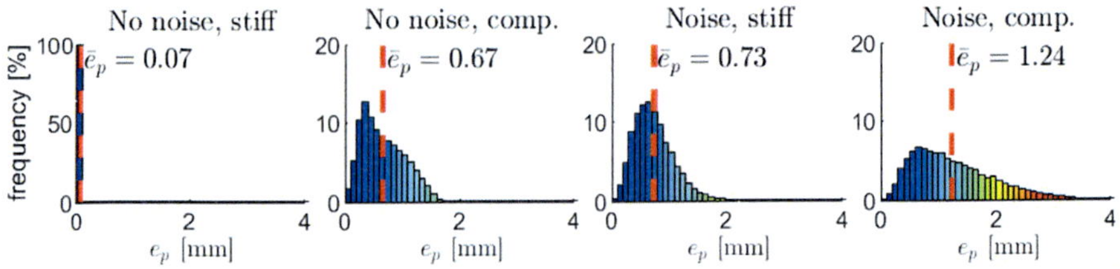


**Fig. 3.** Position error distribution for $DH_G$ parameters estimated using geometric approach under different conditions: a) no noise and stiff robot, b) no noise and compliant robot, c) noise and stiff robot, and d) noise and compliant robot.

Another possibility to estimate DH and **S** simultaneously is using the optimization (7). The optimization approach allows using any measured poses as long as we capture enough well conditioned poses. For comparison, we have calculated parameters using data obtained from C trajectory, random poses A, and a subset of 200 poses from random R poses. So, we have obtained three sets of parameters $\{DH_C, \mathbf{S}_C\}$, $\{DH_A, \mathbf{S}_A\}$ and $\{DH_R, \mathbf{S}_R\}$, respectively. To analyze the influence of the robot compliance, we have calculated the parameters for a more compliant robot and for a less compliant robot. As before, we have verified the obtained position accuracy on the complete set of R poses. The results are shown in Figs. 4 and 5. Our analysis shows that using optimization approach for parameter estimation results in better accuracy than geometric approach. Next, comparing $\{DH_C, \mathbf{S}_C\}$ with $\{DH_A, \mathbf{S}_A\}$ and $\{DH_R, \mathbf{S}_R\}$ we see that C type trajectories are not optimal for estimation of stiffness parameters. The reason is that deflection is not changing much between points on C trajectory. This can be seen especially in the case when the robot is less compliant and the error due to deflection is in the range of the measurement. Therefore, it is better to use random configurations for optimization. On the other hand, the difference between achieved accuracy for $\{DH_A, \mathbf{S}_A\}$ is not significantly worse than for $\{DH_R, \mathbf{S}_R\}$, which suggests that calibration can be successfully performed even with poses in a bounded workspace. So, the advantage of optimization approach is also that it can be used in situation where the robot is not allowed to move in the whole workspace.

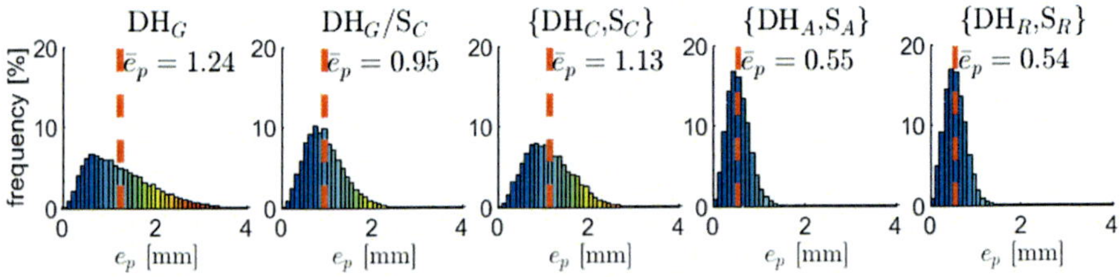


**Fig. 4.** Position error distribution for more compliant robot versus the estimated parameters using geometric approach and optimization based on different trajectories.

Finally, for efficient calibration it is necessary to measure a lot of different poses. It is quite clear that increasing the number of measurements does not reduce the error

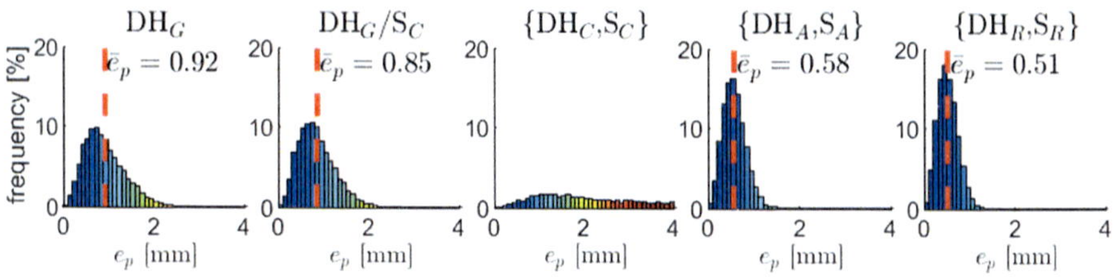


**Fig. 5.** Position error distribution for less compliant robot versus the estimated parameters using geometric approach and optimization based on different trajectories.

indefinitely. Our results show that the optimization approach gives better results than the geometric approach. So, we analyzed how many random poses have to be measured to calculate model parameters that give satisfactory final accuracy. From the R set of random poses we have prepared 10 training subsets of 50, 100, 200, 500 and 100 poses. Using these subsets, we have calculated DH$_R$ and $\mathbf{S}_R$ parameters and verified the achieved accuracy on the whole set of R poses. The position error statistics for mean and maximal errors are shown in Fig. 6. We can see that errors are decreasing with the increased number of samples in the training set and converges to the noise mean value. As there is no significant improvement for 200 or more samples, 200 measurement samples are enough for good calibration to be used for parameter calculation. From Figs. 4 and 5 it is obvious that random poses bounded to a subregion of robot workspace give slightly worse results, but using random poses in the non-constraint workspace is still the best option for good calibration.

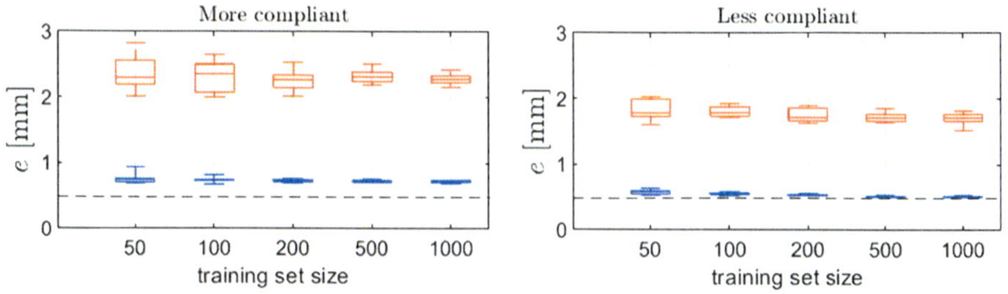


**Fig. 6.** Position error statistics versus number of samples in a training set for more and less compliant robot (blue: mean error, red: maximal error, dotted line: noise mean value).

## 5 Conclusion

In this paper, we showed how to improve the positioning accuracy of robots by highlighting the importance of both kinematic calibration and deflection estimation. Since the deflections depend mainly on gravity, we use a model-based calibration with kinematic and static models of the robot. We used a geometric approach and optimization to estimate the kinematic parameters. The analysis in the simulation environment showed that optimization gives better results than the geometric approach. Next, for optimization it is optimal to use random robot configurations. To make the calibration efficient, we shown that 200 random points is enough to achieve a position accuracy that is close to the measurement noise. Another advantage of the optimization approach is that we can calibrate the robot even when it is not possible to move in the whole workspace during the calibration measurement. In the future, we plan to verify the proposed approach on a real robot with measurement systems of different accuracy.

## References

1. Chen, X., Zhang, Q., Sun, Y.: Non-kinematic calibration of industrial robots using a rigid-flexible coupling error model and a full pose measurement method. Robot. Comput.-Integr. Manuf. **57**(November 2018), 46–58 (2019)
2. Elatta, A.Y., Gen, L.P., Zhi, F.L., Daoyuan, Y., Fei, L.: An overview of robot calibration. Inf. Technol. J. **3**(1), 74–78 (2004)
3. Gadringer, S., Gattringer, H., Müller, A., Naderer, R.: Robot calibration combining kinematic model and neural network for enhanced positioning and orientation accuracy. IFAC-PapersOnLine **53**(2), 8432–8437 (2020)
4. Joubair, A., Bonev, I.A.: Non-kinematic calibration of a six-axis serial robot using planar constraints. Precis. Eng. **40**, 325–333 (2015)
5. Klimchik, A., Caro, S., Pashkevich, A.: Optimal pose selection for calibration of planar anthropomorphic manipulators. Precis. Eng. **40**, 214–229 (2015)
6. Li, K.L., Yang, W.T., Chan, K.Y., Lin, P.C.: An optimization technique for identifying robot manipulator parameters under uncertainty. Springerplus **5**(1), 1–16 (2016). https://doi.org/10.1186/s40064-016-3417-5
7. Ma, L., Bazzoli, P., Sammons, P.M., Landers, R.G., Bristow, D.A.: Modeling and calibration of high-order joint-dependent kinematic errors for industrial robots. Robot. Comput.-Integr. Manuf. **50**(May 2017), 153–167 (2018)
8. Neubauer, M., Gattringer, H., Müller, A., Steinhauser, A., Höbarth, W.: A two-stage calibration method for industrial robots with joint and drive flexibilities. Mech. Sci. **6**(2), 191–201 (2015). https://doi.org/10.5194/ms-6-191-2015
9. Nubiola, A., Slamani, M., Bonev, I.A.: A new method for measuring a large set of poses with a single telescoping ballbar. Precis. Eng. **37**(2), 451–460 (2013). https://doi.org/10.1016/j.precisioneng.2012.12.001
10. Nubiola, A., Slamani, M., Joubair, A., Bonev, I.A.: Comparison of two calibration methods for a small industrial robot based on an optical CMM and a laser tracker. Robotica **32**(3), 447–466 (2014). https://doi.org/10.1017/S0263574713000714
11. Renders, J.M., Rossignol, E., Becquet, M., Hanus, R.: Kinematic calibration and geometrical parameter identification for robots. IEEE Trans. Robot. Autom. **7**(6), 721–732 (1991)
12. Roth, Z.S., Mooring, B.W., Ravani, B.: An overview of robot calibration. IEEE J. Robot. Autom. **3**(5), 377–385 (1987)
13. Švaco, M., Šekoranja, B., Šuligoj, F., Jerbić, B.: Calibration of an industrial robot using a stereo vision system. Procedia Eng. **69**, 459–463 (2014). https://doi.org/10.1016/j.proeng.2014.03.012
14. Vincze, M., Prenninger, J., Gander, H.: A laser tracking system to measure position and orientation of robot end effectors under motion. Int. J. Robot. Res. **13**(4), 305–314 (1994). https://doi.org/10.1177/027836499401300402

15. Wang, Z., Chen, Z., Wang, Y., Mao, C., Hang, Q.: A robot calibration method based on joint angle division and an artificial neural network. Math. Probl. Eng. **2019**, 12, 9293484 (2019). https://doi.org/10.1155/2019/9293484
16. Xu, W., Dongsheng, L., Mingming, W.: Complete calibration of industrial robot with limited parameters and neural network. In: IRIS 2016 - 2016 IEEE 4th International Symposium on Robotics and Intelligent Sensors: Empowering Robots with Smart Sensors, pp. 103–108 (2017)
17. Žlajpah, L., Petrič, T.: Kinematic calibration for collaborative robots on a mobile platform using motion capture system. Robot. Comput.-Integr. Manuf. **79**, 102446 (2023)

# Dynamically Isotropic Gough-Stewart Platform Design Using a Pair of Triangles

Yogesh Pratap Singh(✉) and Ashitava Ghosal

Indian Institute of Science, Bangalore 560012, India
{yogeshsingh,asitava}@iisc.ac.in

**Abstract.** An ideal dynamically isotropic Gough Stewart platform (GSP) has its first six modes of vibration as same, and this enables one to use effectively designed identical dampers to attenuate vibration from a source to a sensitive payload. A modified GSP (MGSP) capable of being dynamically isotropic is considered in this work. For a dynamically isotropic MGSP, the use of a force transformation matrix leads to a set of coupled transcendental equations in terms of design variables. It was observed that all the design variables for an MGSP were related by pairs of triangles. This work develops a general analytical closed-form solution for a dynamically isotropic MGSP using geometrical relations between these triangle pairs. The presented closed-form algebraic solutions can be used directly and supersede the need for any other complex algorithms. Additionally, the design variable in their explicit form offers straightforward solutions, flexible design and can ensure mechanical feasibility. The designs obtained in this work were validated by numerical simulations results done in ANSYS.

**Keywords:** Gough-Stewart platform · Dynamic isotropy · Natural frequency · Modified Gough-Stewart platform

## 1 Introduction

A Gough Stewart Platform (GSP) based isolator is proposed in the literature for micro-vibration control in spacecraft [1,2]. A dynamically isotropic GSP is of particular interest to this application as it has equal first six natural frequencies [2–4]. It is easy to tune a damper for passive vibration control, given all resonance peaks lying close to each other (ideally the same). This is also an effective vibration isolation condition as the region of isolation associated with any degree of freedom (DOF) is not affected by resonance peaks of the cross DOFs in the amplitude versus frequency curve [2]. Moreover, the use of a dynamically isotropic configuration converts a multi-input-multi-output (MIMO) system into several single-input-single-output (SISO) systems, and we can make use of a decoupled control strategy simplifying active vibration control [5]. In a non-isotropic design, the coupling among the six DOFs complicates the controller design leading to a reduction in control accuracy [5]. Apart from the


T. Petrič et al. (Eds.): RAAD 2023, MMS 135, pp. 264–272, 2023.
https://doi.org/10.1007/978-3-031-32606-6_31

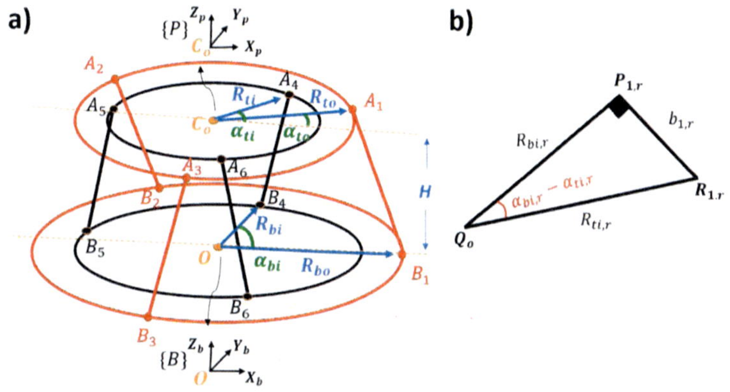


**Fig. 1. a)** Modified GSP **b)** Right angle triangle case in previous work [2].

vibration isolation standpoint, a dynamically isotropic configuration is singularity free. Dynamic isotropy also implies the maximization of the lowest natural frequency [3], which is a favorable criterion for stability.

A modified GSP/two-radii GSP (MGSP) was studied for dynamic isotropy [2,4–7] as a recourse to a traditional 6-6 GSP, which cannot be dynamically isotropic [3,4,7]. In such MGSPs, the anchor points are described on two radii instead of one radius in a traditional 6-6 GSP (refer to Fig. 1(a)), and this can be shown to achieve dynamic isotropy [2,4–7]. Afzali et al. [6] and others [4,5,7] presented different approaches to developing an analytical solution for an MGSP. In all these cases, arriving at a design is challenging as the presented solutions are in implicit form or coupled. Moreover, such approaches lacked addressing design flexibility and feasibility to satisfy space constraints or the intersection of legs in the 3D space.

In our previous work, we presented closed-form solutions in their explicit form by adopting a partial geometry-based approach [2]. A set of solutions was obtained by using geometrical relation in a right-angled triangle, resulting in valuable observations for a dynamically isotropic MGSP. This paper is based on the development of a general analytical closed-form solution using a purely geometric approach. This work concludes that the 3-D dynamically isotropic MGSP design problem can be simplified into sets of triangles in 2-D space related by certain geometrical relationships. All the design parameters of an MGSP can be deduced from these triangles, and this leads to an algebraic solution that can be used directly, superseding the need for any other complex algorithms to solve the set of transcendental dynamic equations for this problem.

## 2 Formulation

A modified two-radii Gough-Stewart Platform (MGSP) consist of a top mobile platform, a fixed base, and six prismatic actuated legs in between them as shown in Fig. 1(a). The anchor points are on two radii with variables $R_{bi}$ and $R_{bo}$ denoting the inner and outer radius of the base platform while $R_{ti}$ and $R_{to}$ representing the inner and outer radii of the mobile platform, respectively. There are two sets of three identical legs having rotational symmetry of 120°. The symbol $H$ denotes the height between the two platforms. The coordinates of a point on the base frame $\{B\}$ are represented by $\{x_b, y_b, z_b\}$ and on the moving frame $\{P\}$ are represented by $\{x_p, y_p, z_p\}$. The vector $\boldsymbol{OB_1}$ (magnitude equal to $R_{bo}$) is chosen along $\boldsymbol{X_b}$. The variable $\alpha_{bi}$, $\alpha_{to}$, $\alpha_{ti}$ denote angles made by vectors $\boldsymbol{OB_4}$, $\boldsymbol{C_oA_1}$, $\boldsymbol{C_oA_4}$ with $\boldsymbol{X_b}$, respectively. Each leg is connected to the mobile platform through a spherical joint and to the base platform through a spherical or universal joint.

We made use of the force transformation matrix ( $[\mathbf{B}]$ ) for an MGSP [1,2]. With all legs assumed to have an axial stiffness of $k$ in their joint space, the stiffness matrix $[\mathbf{K}_T]$ in the task space [1,2] is given by

$$[\mathbf{K}_T] = k[\mathbf{B}][\mathbf{B}]^T \tag{1}$$

The force transformation matrix ($[\mathbf{B}]$) for MGSP is given by

$$[\mathbf{B}]_{6\times 6} = \begin{bmatrix} \boldsymbol{s_1} & \cdots & \boldsymbol{s_6} \\ ({}^B[\mathbf{R}]_P({}^P\boldsymbol{p_1})) \times \boldsymbol{s_1} & \dots & ({}^B[\mathbf{R}]_P({}^P\boldsymbol{p_6})) \times \boldsymbol{s_6} \end{bmatrix} \tag{2}$$

where $\boldsymbol{s_j} = \dfrac{{}^B\boldsymbol{t} + {}^B[\mathbf{R}]_P\, {}^P\boldsymbol{p_j} - {}^B\boldsymbol{b_j}}{l_j}$, $j = 1, \dots, 6$. The vector ${}^B\boldsymbol{t}$ is directed from centre of the base platform to the centre of the mobile platform, $\boldsymbol{S}_j$ $(= l_j\boldsymbol{s}_j)$ is a vector along the respective leg of an MGSP with length $l_j$ while ${}^P\boldsymbol{p_j}$ and ${}^B\boldsymbol{b_j}$ are the coordinates of an anchor point on the mobile and base platform expressed in their respective frames. All the formulations are considered at the neutral position of the platform with ${}^B[\mathbf{R}]_P = [\mathbf{I}]$ and ${}^B\boldsymbol{t} = [\ 0\ 0\ H\ ]^T$. This is a fair assumption for precise control applications like micro-vibration isolation that require small motion. The mass/inertia of the legs can be neglected owing to their small values. With proper choice of the coordinate system, the mass matrix can be written as $[\mathbf{M}] = \text{diag}[\, m_p,\ m_p,\ m_p,\ I_{xx},\ I_{yy},\ I_{zz}]$, where $m_p$ denote the payloads' mass (including mobile platform) and $I_{xx}$, $I_{yy}$, and $I_{zz}$ denotes its principal moment of inertia along each direction with respect to its centre of mass (COM). For a neutral configuration, $[\mathbf{B}]$ is a constant matrix and will have an infinite number of possible configurations [7] as attachment points on the mobile platform are not uniquely determined. Using Eqs. (1) and (2), we can write the natural frequency matrix in task space [2,4,7] as:

$$[\mathbf{G}] = [\mathbf{M}]^{-1}[\mathbf{K_T}] = [\mathbf{M}]^{-1}k[\mathbf{B}][\mathbf{B}]^T = \begin{bmatrix} [\mathbf{P}] & [\mathbf{T}] \\ [\mathbf{T}]^T & [\mathbf{U}] \end{bmatrix} \tag{3}$$

with $[\mathbf{P}] = \text{diag}(\lambda_1, \lambda_2, \lambda_3)$, $[\mathbf{U}] = \text{diag}(\lambda_4, \lambda_5, \lambda_6)$ and $[\mathbf{T}] = \begin{bmatrix} \mu_{11} & -\mu_{12} & 0 \\ \mu_{12} & \mu_{11} & 0 \\ 0 & 0 & \mu_{33} \end{bmatrix}$

The expressions for $\lambda_i$ and $\mu_{ij}$ are given in appendix. The leg length ratio, $a$, relates the lengths of two sets of legs by

$$l_2 = al_1 \tag{4}$$

where $l_1 = |\boldsymbol{S_1}| = |\boldsymbol{S_2}| = |\boldsymbol{S_3}|$ and $l_2 = |\boldsymbol{S_4}| = |\boldsymbol{S_5}| = |\boldsymbol{S_6}|$ (see appendix for details).

For the dynamic isotropy at the neutral position, all six eigenvalues of the natural frequency matrix in Eq. (3) must be equal. To obtain the design parameters, we have to solve a set of coupled transcendental equations generated from conditions of dynamic isotropy given by

$$\lambda_1 = \lambda_2 = \lambda_3 = \lambda_4 = \lambda_5 = \lambda_6 = \omega^2 \text{ and } \mu_{11} = \mu_{12} = \mu_{33} = 0 \tag{5}$$

where, $\omega$ is the natural frequency of the MGSP and $\lambda_1$ to $\lambda_6$ are the eigenvalues of the matrix [**G**].

# 3 Design of Dynamically Isotropic MGSP

The unknowns (design parameters) for an MGSP are $R_{bo}$, $R_{to}$, $R_{bi}$, $R_{ti}$, $H$, $a$, $\alpha_{to}$, and $(\alpha_{bi} - \alpha_{ti})$. Simplifications of dynamic isotropy conditions in Eq. (5) give rise to several useful geometrical observations, which can be seen in Table 1 and Fig. 2(a). If any set of geometries satisfies conditions in Table 1, the design parameters obtained will represent a dynamically isotropic configuration. It is to be noted that $K(= I_{xx}/I_{zz})$ and $Q(= I_{xx}/m_p)$ are payload properties that are known to us. The necessary condition of $\lambda_4 = \lambda_5$ is practically valid for symmetrical payloads ($I_{xx} = I_{yy}$). The challenge to obtain the sides for any general triangles in Fig. 2(a) is overcome by obtaining sides of the corresponding right angle triangle $\Delta Q_o P_{1,r} R_1$ in Fig. 2(b) (denoted by subscript '$r$' for right angle case). We make use of analytical results for a right-angled triangle (see Fig. 1(b)) obtained in our previous work [2]. The relation in Fig. 1(b) was used to initiate the solution, and all other variables were obtained analytically (using Eq. (5)) [2] as:

$$R_{ti,r} = \sqrt{QC_2}, R_{to,r} = \frac{\sqrt{QC_1}}{a}, H_r = \sqrt{\frac{Q(KC_1C_2 - a^2)}{KC_1^2}},$$
$$\theta = \arctan\left(\frac{a}{\sqrt{KC_1C_2 - a^2}}\right) \text{ and } (\alpha_{bi,r} - \alpha_{ti,r}) = 90^\circ - \theta \tag{6}$$

where the leg length ratio $a$ can be taken as input (free variable) with $K$ and $Q$ as known.

A general analytical solution in an explicit form is tedious due to coupling among variables. However, the study of the specific cases for $\Delta Q_o P_{1,r} R_1$ (Right-angle case) led to several useful observations and geometrical interpretations. The geometries, i.e., $\Delta Q_o P_{1,r} R_1$ and $\Delta Q_o P_{2,r} R_2$ shown in Fig. 2(b) can be constructed from the known results in Eq. (6) ( for other design variables in $\Delta Q_o P_{2,r} R_2$ for right angle triangle case, substitute $f = 1$ in Table 2). Now, these geometries are perturbed to generate the new geometries, i.e., $\Delta Q_o P_1 R_1$ and $\Delta Q_o P_2 R_2$ as shown in Fig. 2(b). The perturbations done along line $\overline{P_{1,r} R_1}$

**Table 1.** Geometrical interpretations from dynamic isotropy conditions

| Case | Condition used | Simplified Equation | Geometrical Interpretation in Fig. 2(a) |
|---|---|---|---|
| 1 | $\lambda_1 = \lambda_2$ and Eq. (4) | ${b_1}^2 = R_{ti}^2 + R_{bi}^2 - 2R_{ti}R_{bi}\cos(\alpha_{bi} - \alpha_{ti})$ | $\Delta Q_o P_1 R_1$ :triangle with sides $R_{bi}$, $R_{ti}$ and $b_1$ (cosine rule) |
| 2 | $\lambda_1 = \lambda_2$ and Eq. (4) | ${b_2}^2 = R_{to}^2 + R_{bo}^2 - 2R_{to}R_{bo}\cos(\alpha_{to})$ | $\Delta Q_o P_2 R_2$ :triangle with sides $R_{bo}$, $R_{to}$ and $b_2$ |
| 3 | $\mu_{11} = 0$ | $R_{ti}R_{bi}\sin(\alpha_{bi} - \alpha_{ti}) = a^2 R_{to}R_{bo}\sin(\alpha_{to})$ | Area of $\Delta Q_o P_1 R_1 = a^2 \times$ (Area of $\Delta Q_o P_2 R_2$ ) |
| 4 | $\mu_{12} = 0$ | $R_{ti}(R_{ti} - R_{bi}\cos(\alpha_{bi} - \alpha_{ti})) = a^2 R_{to}(R_{bo}\cos(\alpha_{to}) - R_{to})$ | $(\overline{Q_o R_1})P_{g1} = a^2\,(\overline{Q_o R_2})P_{g2}$ |
| 5 | $\lambda_4 = \lambda_3$ and $\lambda_4 = \lambda_6$ | $\dfrac{a^2 R_{to}^2}{a^2+1} + \dfrac{R_{ti}^2}{a^2+1} = 2Q$ | $\overline{P_3 R_3}$ is a constant magnitude line treating $R_{to}$ and $R_{ti}$ as perpendicular basis. |
| 6 | $\mu_{12} = 0$ and $\mu_{11} = 0$ | $\tan\theta_1 = \tan\theta_2$ | $\angle Q_o R_1 P_1 = 180° - \angle Q_o R_2 P_2 \implies \theta_1 = \theta_2 = \theta$ |

where $b_1 = H\sqrt{C_1}$, $b_2 = \dfrac{H\sqrt{C_2}}{a}$, $C_1 = \left(\dfrac{3a^2+1}{2}\right)$, $C_2 = \left(\dfrac{a^2+3}{2}\right)$

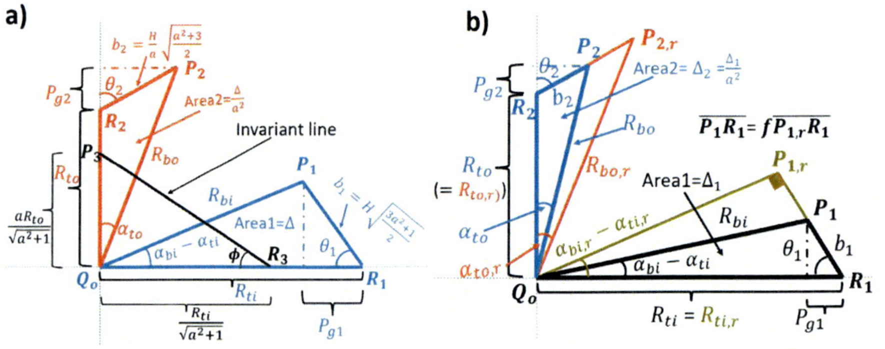


**Fig. 2.** **a)** Geometrical interpretation of design variables **b)** General triangle from a right-angled triangle.

will keep the leg length ratio $a$ constant. This means such perturbations along line $\overline{P_{1,r}R_1}$ to generate $\overline{P_1R_1}$(hence $\Delta Q_oP_1R_1$) also keeps few design variables such as $R_{ti}(= R_{ti,r}), R_{to}(= R_{to,r})$ and $\theta_1\backslash\theta_2 = \theta$ the same (see Eq. (6)). Different $a$ values generate new sets of triangles corresponding to their respective right-angled triangle (see Fig. 3). The new height $H$ for the general case will now be given as $H = fH_r$ (see Eq. (6)) because $\overline{P_1R_1} = f\ \overline{P_{1,r}R_1}$ (their lengths are directly proportional to $H$, see Fig. 2(b)), where $f$ is a ratio (free variable) whose value could be less than 1, equal to 1 (right-angle case) or greater than 1 (extrapolation of line) depending on the design values we want for an MGSP.

Interestingly, the new triangles due to scaling ($\Delta Q_oP_1R_1$ and $\Delta Q_oP_2R_2$) satisfy all the isotropic/geometric conditions in Table 1 and represents a dynamically isotropic design. The design variables for the general case ($R_{bi}$, $\alpha_{bi}$ −

**Table 2.** General solution for design variables in their explicit form

| | | | |
|---|---|---|---|
| Variable | $R_{ti}(=\overline{Q_oR_1})$ | $R_{to}(=\overline{Q_oR_2})$ | $H(\propto \overline{P_1R_1}$ or $\propto \overline{P_2R_2})$ |
| Solution | Same as Eq. (6) | Same as Eq. (6) | $=fH_r$(see Eq. (6)) |
| Variable | $R_{bi}(=\overline{Q_oP_1})$ | $R_{bo}(=\overline{Q_oP_2})$ | |
| Solution | See Eq. (7) | $\sqrt{\frac{QC_1}{a^2}+\frac{fQ(KC_1C_2-a^2)(2C_1+fC_2)}{KC_1^2a^2}}$ | |
| Variable | $\alpha_{bi}-\alpha_{ti}$ $(=\angle R_1Q_oP_1)$ | | $\alpha_{to}$ $(=\angle R_2Q_oP_2)$ |
| Solution | $\arctan\left(\frac{fa\sqrt{(KC_1C_2-a^2)}}{KC_1C_2-f(KC_1C_2-a^2)}\right)$ | | $\arctan\left(\frac{b_2\sin(\theta_2)}{R_{to}+b_2\cos(\theta_2)}\right)$ |

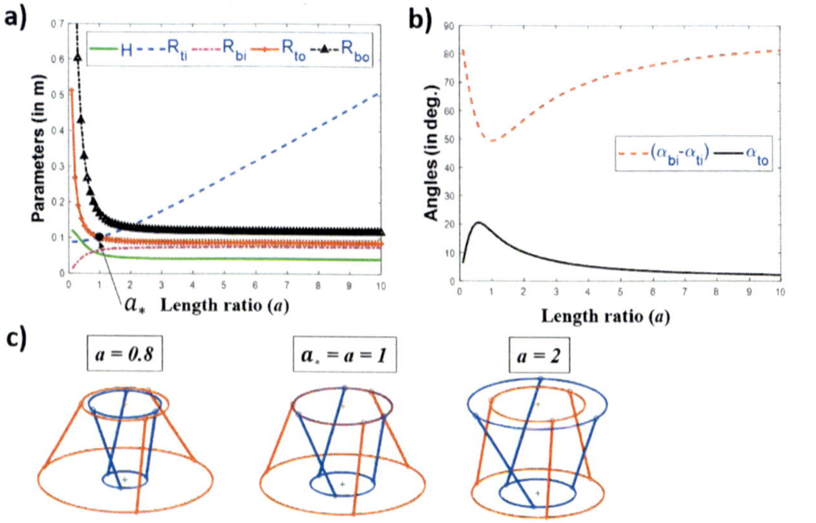


**Fig. 3.** **a)** & **b)** Variation of parameters for $f = 1$, **c)** Configuration transition.

$\alpha_{ti}$, $R_{bo}$, $\alpha_{to}$) which are different from the right-angled triangle case can be computed analytically in terms of ratios $f$, $a$ and payload properties ($K$, $Q$). This can be done by making use of geometrical relations in Fig. 2(b) using known variables $(R_{ti}, R_{to}, \theta_1\backslash\theta_2)$ and the new value of $H$. For, eg., $R_{bi}$ can be obtained as $\sqrt{(R_{ti}-b_1\cos\theta_1)^2+(b_1\sin\theta_1)^2}$, which gives :

$$R_{bi}=\sqrt{\frac{QKC_1C_2+fQ(KC_1C_2-a^2)(f-2)}{KC_1}} \tag{7}$$

Similar to $R_{bi}$, general analytical solutions to all the variables can be obtained using geometry and are listed in Table 2. The result obtained in Table 2 can be directly used to design a dynamically isotropic MGSP.

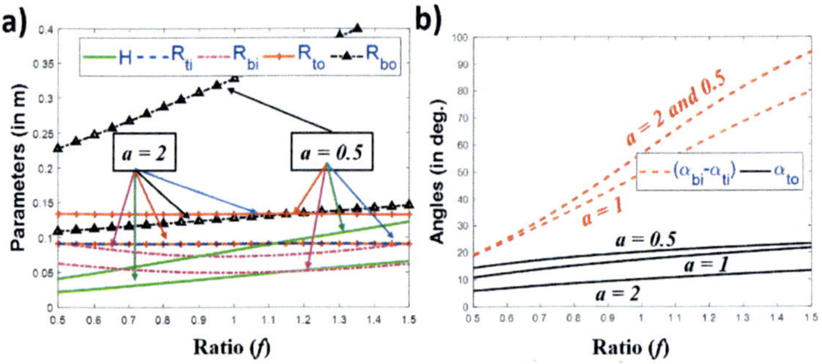


**Fig. 4.** **a)** Variation of geometrical parameters for $a = 0.5, 2$ **b)** Angle variation.

## 4 Observation

**1) Parameter variation**: The variation of design parameters for right-angled triangle case ($f = 1$) with respect to leg length ratio $a$ can be seen in Fig. 3(a) and (b) for a typical payload of $K = 0.590887$, $Q = 5.089 \times 10^{-3} m^2$, and $m_p = 5$ Kg. Each $a$ corresponds to a different right-angled triangle case, which served as the base/initial formulation in Fig. 2(b). An interesting observation from the expressions in Table 2 and Fig. 3(a) can be seen at $a = 1$. At this point, $R_{to} = R_{ti}$ or the two radii on the top platform become equal (irrespective of $f$). This is also a point of configuration transition ($a_*$) where there is a transition from outer-outer (inner-inner) type leg connections to outer-inner (cross leg) type legs, as shown in Fig. 3(c). Choosing a design with $a < 1$ avoids interference of legs in the 3D space, and in this case, the legs will never intersect, ensuring feasibility. For $a > 1$, such interference needs to be investigated. The parameter variation at $a = 2,\ 0.5$ with different scaling values $f$ is shown in Fig. 4. The value of $R_{ti}$ and $R_{to}$ can be seen interchanged for $a = 2$ and $a = 0.5$ due to the reciprocal property between $a$ and its inverse. These graphs can help us to design MGSP within our design constraints using free variables $a$ and $f$.
**2) Dynamically isotropic natural frequency**: On substituting closed form expression for all design variables in Eq. (3), we obtain an equal value of all six natural frequencies ($\omega$) of the platform as

$$\lambda_1 = \lambda_2 = \lambda_3 = \lambda_4 = \lambda_5 = \lambda_6 = \omega^2 \text{ and } \omega = \sqrt{2k/m_p} \tag{8}$$

**3) Invariant line**: It can be observed from Case 5 (Table 1) and Fig. 2(a) that there is an invariant line $\overline{P_3 R_3}$ whose length remains constant ($= \sqrt{2Q}$) for any value of $a$ (any triangle). However angle $\phi$ ($= \arctan(\sqrt{C_1/C_2})$) varies with $a$.
**4) Traditional GSP**: The known fact that a traditional GSP ($R_{bi} = R_{bo}$ and $R_{to} = R_{ti}$) fails to give dynamic isotropy can also be verified from this approach. For this, $\Delta Q_o P_1 R_1$ and $\Delta Q_o P_2 R_2$ must be congruent, implying the angle $\theta_2$ to be obtuse and the condition $\tan\theta_1 = \tan\theta_2$ thus cannot be satisfied.

## 5 Validation

Simulations were performed in ANSYS for the above-mentioned designs treating mobile and base platforms as rigid bodies (shell 181 element) and legs as ideal springs (link 180 elements). In each of the cases, the analytical results closely match the simulation results. For, e.g., assuming $a = 2$ and $f = 0.75$, $m_p = 5$ Kg, $k = 10^5$ N/m, $K = 0.590887$, and $Q = 5.089 \times 10^{-3} m^2$ (approximate values for a hardware under design and fabrication), the first six natural frequencies were obtained using the analytical result in Eq. (8) ($= \sqrt{2k/m_p}$) as 31.83 Hz. The first six natural frequency obtained with simulation are 31.68, 31.70, 31.76, 31.82, 31,82, and 31.85 Hz, which closely matches the analytical results, validating our design.

## 6 Conclusion

A pair of triangles were used to arrive at a general design of a dynamically isotropic MGSP. Using this geometry-based approach, a closed-form solution in its explicit form was derived, providing feasibility and flexibility. Such designs with the first six natural frequencies equal can ease passive and active micro-vibration isolation.

## Appendix

The value of all $\lambda's$ and $\mu's$ in Eq. (3) are given as:

$$\lambda_1 = \lambda_2 = \frac{3k\,\left(l_1^2\Psi_1 + l_2^2\Psi_2\right)}{\left(2m_p l_1^2 l_2^2\right)},\ \lambda_3 = \frac{3kH^2\left(l_1^2 + l_2^2\right)}{\left(m_p l_1^2 l_2^2\right)},\ \lambda_4 = \frac{3k\Psi_5 H^2}{\left(2I_{xx} l_1^2 l_2^2\right)},$$

$$\lambda_5 = \lambda_4 \frac{(I_{xx})}{(I_{yy})},\ \lambda_6 = \frac{3k\left(\Psi_6^2 l_1^2 + \Psi_7^2 l_2^2\right)}{\left(I_{zz} l_1^2 l_2^2\right)},\ \mu_{11} = \frac{-3kH\left(-\Psi_6 l_1^2 + \Psi_7 l_2^2\right)}{\left(2l_1^2 l_2^2\right)},$$

$$\mu_{33} = -2\mu_{11},\ \mu_{12} = \frac{3kH\left(\Psi_3 l_1^2 + \Psi_4 l_2^2\right)}{\left(2l_1^2 l_2^2\right)}$$

where,

$$\Psi_1 = R_{ti}^2 + R_{bi}^2 - 2R_{ti}R_{bi}\cos(\alpha_{bi} - \alpha_{ti}),\ \Psi_2 = R_{to}^2 + R_{bo}^2 - 2R_{to}R_{bo}\cos(\alpha_{to}),$$

$$\Psi_3 = R_{ti}^2 - R_{ti}R_{bi}\cos(\alpha_{bi} - \alpha_{ti}),\ \Psi_4 = R_{to}^2 - R_{to}R_{bo}\cos(\alpha_{to}),\ \Psi_5 = R_{ti}^2 l_1^2 + R_{to}^2 l_2^2$$

$$\Psi_6 = R_{ti}R_{bi}\sin(\alpha_{bi} - \alpha_{ti}),\ \Psi_7 = R_{to}R_{bo}\sin(\alpha_{to})$$

$$l_1 = \left(\sqrt{R_{to}^2 + R_{bo}^2 - 2R_{to}R_{bo}\cos(\alpha_{to}) + H^2}\right)$$

$$l_2 = \left(\sqrt{R_{ti}^2 + R_{bi}^2 - 2R_{ti}R_{bi}\cos(\alpha_{bi} - \alpha_{ti}) + H^2}\right)$$

## References


1. Hanieh, A.A.: Active isolation and damping of vibrations via Stewart platform. PhD Thesis, ULB ASL (2003)
2. Singh, Y.P., Ahmad, N., Ghosal, A.: Design of dynamically isotropic modified Gough- Stewart platform using a geometry-based approach. In: Khang, N.V., Hoang, N.Q., Ceccarelli, M. (eds.) ASIAN MMS 2021. MMS, vol. 113, pp. 258–268. Springer, Cham (2022). https://doi.org/10.1007/978-3-030-91892-7_24

3. Afzali-Far, B., Lidström, P.: Analytical index of dynamic isotropy and its application to hexapods. Precis. Eng. **52**, 242–248 (2018)
4. Tong, Z., He, J., Jiang, H., Duan, G.: Optimal design of a class of generalized symmetric Gough-Stewart parallel manipulators with dynamic isotropy and singularity-free workspace. Robotica **30**(2), 305–314 (2012)
5. Yun, H., Liu, L., Li, Q., Li, W., Tang, L.: Development of an isotropic Stewart platform for telescope secondary mirror. Mech. Syst. Sig. Process. **127**, 328–344 (2019)
6. Afzali-Far, B., Lidström, P.: A class of generalized Gough-Stewart platforms used for effectively obtaining dynamic isotropy – an analytical study. In: MATEC Web of Conferences, vol. 35, p. 02002:P.1–P.5 (2015)
7. Jiang, H.Z., Feng He, J., Tong, Z.Z., Wang, W.: Dynamic isotropic design for modified Gough-Stewart platforms lying on a pair of circular hyperboloids. Mech. Mach. Theory **46**(9), 1301–1315 (2011)

# Inverse and Direct Kinematics of a Modified Stewart-Gough Platform

Martin Bem(✉), Simon Reberšek, Igor Kovač, and Aleš Ude

Department of Automatics, Biocybernetics, and Robotics, Jožef Stefan Institute, Jamova cesta 39, 1000 Ljubljana, Slovenia
martin.bem@ijs.si
https://abr.ijs.si

**Abstract.** This paper presents the development and testing of a numerical method for the direct and inverse kinematic calculation of a passive flexible Stewart-Gough platform with special axis-offset universal joints. The approach models each leg using DH parameters with 6 partially redundant degrees of freedom. The DH parameter leg description is used to form a system of 6 homogeneous closed-loop transformation chains, each yielding 12 equations for a total of 72 equations. The system of equations is not independent and is solved using the Newton-Raphson method. Depending on which derivatives we use for the Jacobians, we can also use this model to calculate the calibration parameters of the mechanism. The approach was tested using the CAD model of the platform and proved to be submicrometer accurate for direct and inverse kinematics.

**Keywords:** Direct kinematics · inverse kinematics · calibration · parallel platform

## 1 Introduction

Competition and pricing pressures in industry demand high productivity and fast time-to-market strategies [1]. The rise of diversity and model variety coupled with short product life cycles require changes of established production paradigms. Industrial robots can be programmed to perform any movement inside their working space and are therefore flexible. Robotic periphery is however often totally application specific and has to be redesigned, remanufactured and redeployed each time the workpiece changes. To mitigate this a passive flexible Stewart - Gough platform with 6 degrees of freedom called the Hexapod was developed and tested [2,3]. The initial version of mechanism has no actuators or position measurement systems and is moved kinesthetically or by robot. When the correct pose is reached the position is held in place by special hydraulic brakes in each Hexapod leg. The mechanism proved to be mechanically stiff, affordable and reliable however determining whether a certain pose is within its working space remained an open problem. Furthermore, a Hexapod equipped


T. Petrič et al. (Eds.): RAAD 2023, MMS 135, pp. 273–280, 2023.
https://doi.org/10.1007/978-3-031-32606-6_32

with linear encoders in each leg was developed. The encoder data is to be used to determine the position and orientation of the top plate in order to better control production process parameters. What complicates the situation is the fact that the Hexapod uses proprietary patented Cardan joints with non intersecting axis to increase mechanical stiffness while locked. In order to address all of the above-mentioned challenges a novel numeric approach described below was developed. It enables the calculation of direct and inverse Hexapod kinematics as well as Hexapod calibration.

## 2 Stewart - Gough Platform Direct and Inverse Kinematics Calculation

Inverse kinematics of a parallel mechanism is relatively simple to solve [5] and has been used in the past to compute an optimal arrangement of hexapods for a given set of workpieces [9]. A Stewart-Gough platform with arbitrary top and bottom plate geometries is schematically shown in Fig. 1.

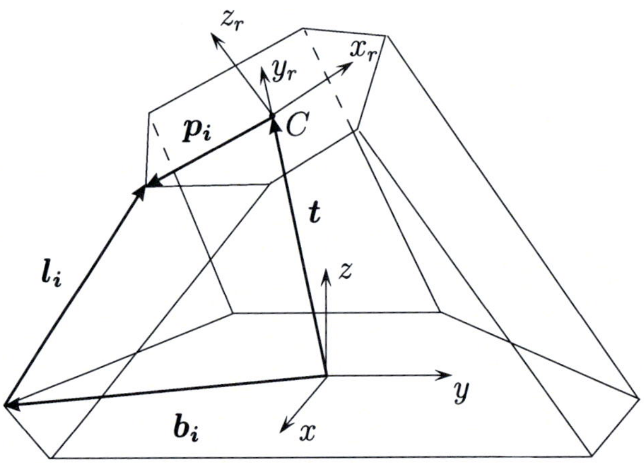


**Fig. 1.** Stewart-Gough platform schematic view

The leg length can be calculated by using the inverse kinematic equations described in [4]. Equation (1) describes the closed chain of vectors shown in Fig. 1,

$$\mathbf{G}_i(x) = \boldsymbol{l}_i = -\boldsymbol{b}_i + \boldsymbol{t} + \mathbf{R}\boldsymbol{p}_i,\ i = 1, 2, \ldots, 6, \tag{1}$$

where

$$\mathbf{R} = \begin{bmatrix} \cos\beta\cos\gamma & \cos\gamma\sin\alpha\sin\beta - \cos\alpha\sin\gamma & \gamma\sin\alpha\sin\gamma + \cos\alpha\cos\gamma\sin\beta \\ \cos\beta\sin\gamma & \cos\alpha\cos\gamma + \sin\alpha\sin\beta\sin\gamma & \cos\alpha\sin\beta\sin\gamma - \cos\gamma\sin\alpha \\ -\sin\beta & \cos\beta\sin\alpha & \cos\alpha\cos\beta \end{bmatrix} \tag{2}$$

and $x$, $y$, $z$ are the top plate position offsets, $\alpha$ is the roll (rotation about $x_r$), $\beta$ the pitch (rotation about the $y_r$), $\gamma$ the yaw (rotation about $z_r$), and $\boldsymbol{x} = [x, y, z, \alpha, \beta, \gamma]$ is the top plate pose. Leg lengths can then be calculated by Eq. (3)

$$l_i = \sqrt{\boldsymbol{l}_i * \boldsymbol{l}_i}. \tag{3}$$

The calculation of direct kinematics is a much harder issue. The direct kinematics determine the position and orientation of the top plate from the known leg lengths. The main problem is that for a certain leg length configuration there are several possible valid top plates poses. A possible solution to remove ambiguity is to use additional sensors in the parallel platform construction. This approach is unusable for us because low price is very important for our product. The other is to use numerical approaches that find a valid solution near a previously known solution [6], and [7]. Newton-Raphson method is used in our approach as in [8]. This iterative method is described by the following equation

$$\boldsymbol{x}_{(j+1)} = \boldsymbol{x}_j - \mathbf{J}^{-1}(\mathbf{G}(\boldsymbol{x}_j) - \boldsymbol{q}^m) = \boldsymbol{x}_j - \mathbf{J}^{-1}\psi\boldsymbol{x}_j, \tag{4}$$

where $\boldsymbol{x}_{j+1} = [x_{j+1}, y_{j+1}, z_{j+1}, \alpha_{j+1}, \beta_{j+1}, \gamma_{j+1}]$ is the next step pose approximation, $\boldsymbol{x}_j = [x_j, y_j, z_j, \alpha_j, \beta_j, \gamma_j]$ is the current pose approximation, $\boldsymbol{q}^m = [l_1^m, ..., l_6^m]$ is the measured leg length vector, $\mathbf{G}(x)$ is the inverse kinematic function and $\mathbf{J}$ the Jacobian given by the followin equation

$$\mathbf{J} = \left[\frac{\partial\psi(\boldsymbol{x}_j)}{\partial(\boldsymbol{x})}\right]^{-1}. \tag{5}$$

Thus the valid top plate poses are those that meet the condition $\psi(\boldsymbol{x}) = 0$.

This approach, however, does not help us solve the direct kinematic problem for our proposed Hexapod. The main issue is that as mentioned before, the Hexapod uses 12 special Cardan joints were the two perpendicular axis of rotation do not intersect. Therefore, we cannot use vectors in our chain because the modified Cardan joints are not symmetric. In our approach, the homogeneous transformation is used to form closed loops. The inverse kinematic function in Eq. (1) has to be therefore modified for our mechanism.

## 3 Hexapod with Modified Cardan Joints Direct and Inverse Kinematics Calculation

As mentioned before the offset in Cardan joint axis requires extension in Eq. 1. For our mechanism, a new closed loop homogeneous transformation chain has to be determined. Hexapod marked with coordinate systems and corresponding transformations between each pair of coordinate systems is shown in Fig. 2. The transformation loop is composed of two parts as shown in Eq. (6)

$$\mathbf{A}_i - \mathbf{B}_i = 0,\ i = 1, 2, \ldots, 6. \tag{6}$$

The homogeneous transformation $\mathbf{A}_i$ is defined by Eq. (7)

$$\mathbf{A}_i = \mathbf{T}_{top} \cdot \mathbf{T}_{tp,i}, \ i = 1, 2, \ldots, 6. \tag{7}$$

The transformation chain $\mathbf{B}_i$ is defined by Eq. (8)

$$\mathbf{B}_i = \mathbf{T}_{bp,i} \cdot \mathbf{T}_{DH}, \ i = 1, 2, \ldots, 6, \tag{8}$$

where $\mathbf{T}_{bp,i}$ represents the fixed transformation from the center of the bottom plate to the bottom of the i-th leg, $\mathbf{T}_{tp,i}$ represents the fixed transformation from the center of the top plate to the top of the i-th leg, $\mathbf{T}_{top}$ represents the transformation between the centers of bottom and top plate and the $\mathbf{T}_{DH}$ describes the homogeneous transformation of i-th leg using Denavit-Hartenberg standard notation. Since all the legs are identical we need to determine the DH parameters only for one leg. From CAD model we need to create schematic and position the coordinate systems according to DH convention which is shown in Fig. 2 (Table 1).

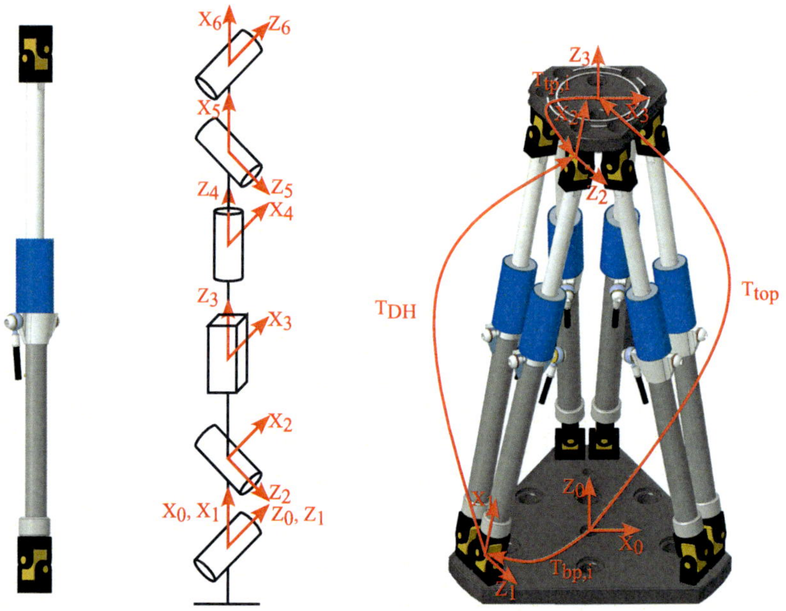


**Fig. 2.** Stewart-Gough platform schematic view

Putting equations together, we get a system of equations

$$\mathbf{T}_{top} \cdot \mathbf{T}_{tp,i} - \mathbf{T}_{bp,i} \cdot \mathbf{T}_{DH} = 0. \tag{9}$$

**Table 1.** Hexapod DH parameters

| | $\theta_i$ | $\alpha_i$ | $a_1$ | $d_i$ |
|---|---|---|---|---|
| 1 | $\theta_1$ | 0 | 0 | 0 |
| 2 | $\theta_2 + \pi/2$ | $+\pi/2$ | $7mm$ | 0 |
| 3 | 0 | $+\pi/2$ | 0 | 0 |
| 4 | $\theta_4$ | 0 | 0 | $l_3$ |
| 5 | $\theta_5 + \pi/2$ | $+\pi/2$ | 0 | 0 |
| 6 | $\theta_6 + \pi/2$ | $-\pi/2$ | $7mm$ | 0 |

Since $\mathbf{T}_{top} \cdot \mathbf{T}_{tp,i}$, $\mathbf{T}_{bp,i} \cdot \mathbf{T}_{DH}$ are all homogeneous transformations, we can form a system of equations for each leg as described in Eq. (10)

$$\mathbf{G}(x) = \begin{bmatrix} f_{i,1} & f_{i,2} & f_{i,3} & f_{i,4} \\ f_{i,5} & f_{i,6} & f_{i,7} & f_{i,8} \\ f_{i,9} & f_{i,10} & f_{i,11} & f_{i,12} \\ 0 & 0 & 0 & 1 \end{bmatrix} = \begin{bmatrix} 0\,0\,0\,0 \\ 0\,0\,0\,0 \\ 0\,0\,0\,0 \\ 0\,0\,0\,1 \end{bmatrix}, \quad i = 1, 2, \ldots, 6. \tag{10}$$

It is obvious that for a valid pose the conditions given in Eq. (11) must be met for each Hexapod leg

$$f_{i,1}, f_{i,2}, ..., f_{i,12} = 0, \ i = 1, 2, \ldots, 6. \tag{11}$$

For all six Hexapod legs this gives us 72 equations that need to be solved using a modified approach described in 2. To find the solutions we need to find the roots of all of the equations. The error of each configuration is therefore directly given by evaluating $\mathbf{G}(x_j)$. In our case the aforementioned equation uses a pseudo inverse Jacobian matrix since the equations in the system are not independent therefore the Jacobian matrix is not square. The iterative Newton-Raphson equation in our case can be written as Eq. (12)

$$\boldsymbol{x}_{j+1} = \boldsymbol{x}_j - \mathbf{J}^{\#}\mathbf{G}(\boldsymbol{x}_j). \tag{12}$$

Our system modeled the Hexapod with this method has 36 partially redundant degrees of freedom. Each leg has 6 degrees of freedom (4 cardan joint angles $\delta_i, \epsilon_i, \zeta_i, \eta_i$, leg length $l_i$, and leg axial rotation $\theta_i$) and the top plate has 6 degrees of freedom $x, y, z, \alpha, \beta, \gamma$. It is also worth mentioning that depending on which degrees of freedom we take as constants we can calculate both direct and inverse kinematics as well as calibrate the Hexapod.

## 4 Direct Hexapod Kinematics

To iteratively calculate direct kinematics using this approach we use Eq. (9) where the vector $\boldsymbol{x}$ is composed by the 6 top plate degrees of freedom and the 30 hexapod leg degrees of freedom

$$\boldsymbol{x} = [x, y, z, \alpha, \beta, \gamma, \delta_1, ..., \delta_6, \epsilon_1, ..., \epsilon_6, \zeta_1, ..., \zeta_6, \eta_1, ..., \eta_6, \theta_1, ..., \theta_6], \tag{13}$$

while the $\boldsymbol{l}_i$ vector is obtained from the measurements. The Jacobian is calculated as specified in Eq. (14)

$$\mathbf{J} = \begin{bmatrix} \frac{\partial f_{1,1}}{\partial x} & \frac{\partial f_{1,1}}{\partial y} & \cdots & \frac{\partial f_{1,1}}{\partial \theta_6} \\ \frac{\partial f_{1,2}}{\partial x} & \frac{\partial f_{1,2}}{\partial y} & \cdots & \frac{\partial f_{1,2}}{\partial \theta_6} \\ \vdots & \vdots & \ddots & \vdots \\ \frac{\partial f_{2,1}}{\partial x} & \frac{\partial f_{2,1}}{\partial y} & \cdots & \frac{\partial f_{2,1}}{\partial \theta_6} \\ \vdots & \vdots & \ddots & \vdots \\ \frac{\partial f_{6,12}}{\partial x} & \frac{\partial f_{6,12}}{\partial y} & \cdots & \frac{\partial f_{6,12}}{\partial \theta_6} \end{bmatrix}. \tag{14}$$

To calculate the final results, we iteratively apply Eq. (12) until the error given by the equation below

$$\boldsymbol{x}_j - \mathbf{J}^{\#}\mathbf{G}(\boldsymbol{x}_j) < err\_goal \tag{15}$$

gets smaller than the goal error *err_goal*. For the first step of approximations we use a known valid initial state of the hexapod $x_0$ that was measured from CAD model.

## 5 Inverse Hexapod Kinematics

To iteratively calculate inverse kinematics using this approach, we need to solve Eq. (9) where the unknown variable vector $\boldsymbol{x}$ is composed of 6 unknown leg lengths and 30 hexapod leg degrees of freedom

$$\boldsymbol{x} = [l_1, \ldots, l_6, \delta_1, \ldots, \delta_6, \epsilon_1, \ldots, \epsilon_6, \zeta_1, \ldots, \zeta_6, \eta_1, \ldots, \eta_6, \theta_1, \ldots, \theta_6]. \tag{16}$$

The top plate pose $(x, y, z, \alpha, \beta, \gamma)$ is obtained from the measurements. The Jacobian is calculated as follows

$$J = \begin{bmatrix} \frac{\partial f_{1,1}}{\partial l_1} & \frac{\partial f_{1,1}}{\partial l_2} & \cdots & \frac{\partial f_{1,1}}{\partial \theta_6} \\ \frac{\partial f_{1,2}}{\partial l_1} & \frac{\partial f_{1,2}}{\partial l_2} & \cdots & \frac{\partial f_{1,2}}{\partial \theta_6} \\ \vdots & \vdots & \ddots & \vdots \\ \frac{\partial f_{2,1}}{\partial l_1} & \frac{\partial f_{2,1}}{\partial l_2} & \cdots & \frac{\partial f_{2,1}}{\partial \theta_6} \\ \vdots & \vdots & \ddots & \vdots \\ \frac{\partial f_{6,12}}{\partial l_1} & \frac{\partial f_{6,12}}{\partial l_2} & \cdots & \frac{\partial f_{6,12}}{\partial \theta_6} \end{bmatrix}. \tag{17}$$

To calculate the final results, we iteratively apply Eq. (12) until the error

$$\boldsymbol{x}_j - \mathbf{J}^{\#}\mathbf{G}(\boldsymbol{x}_j) < err\_goal \tag{18}$$

gets smaller than the goal error *err_goal*. For the first step of approximation, we use a known valid initial state of the hexapod legs $\boldsymbol{l}_0 = [l_1^0, l_2^0, \ldots, l_6^0]$ that was obtained from the CAD model.

## 6 Hexapod Calibration

To calibrate the hexapod using this approach, we need to solve Eq. (9) where the unknown variable vector $x$ is composed of the constants of matrix $\mathbf{G}(x)$ from Eq. (10), which we want to calibrate

$$\boldsymbol{x} = [const_1, const_2, \ldots, const_n], \tag{19}$$

The Jacobian in this case is a $72 \cdot m \times n$ matrix where m is the number of points in which we calibrate the hexapod and n is the number of constants to be calculated in this way. In each calibration point i, we measure the top plate pose $[x_i^m, y_i^m, z_i^m, \alpha_i^m, \beta_i^m, \gamma_i^m], i = 1, 2, \ldots, m$ as well as the hexapod leg length vector $[l_{i,1}^m, l_{i,2}^m, \ldots, l_{i,6}^m], i = 1, 2, \ldots, m$. Using Eq. (12), we can then iteratively calculate the values of all constants to an arbitrary precision.

## 7 Results

The method described above was implemented to test the inverse and direct kinematics so far. The convergence is very good but that is to be expected as the Newton-Raphson method has quadratic convergence. The method usually takes 3–4 iterative steps to reach sub micron precision depending on how far the current pose is from the initial approximation. The method was tested against a cad model and was shown to be correct. The current algorithm implementation written in Python takes about 0.02–0.04 s for each iteration for a total calculation time of about 0.06–0.16 s. We can observe that the algorithm is quite slow. The reason for is two-fold. Firstly, this approach is quite computationally expensive. The second is the sub optimal software implementation. For the future we have some software improvements that should significantly expedite the calculation process.

## 8 Conclusion

In the paper we have shown that inverse and direct kinematics can be numerically calculated to an arbitrary precision. The approach takes 3–4 iterations to find a valid solution. Using the current implementation this takes 0.06–0.16 s for each inverse or direct kinematic calculation. The results were validated against the cad model and were shown to be correct. These calculations are very important for us in order to better integrate our Hexapod technology in future projects. In the future we plan to implement the calibration using the method describe above. Currently the solution was implemented as a custom library in RoboDK. RoboDK is used for visualizing the hexapod, its workspace assessment, robotic cell design and offline programming. The data from this algorithm will be used in the future to better design robotic cells and to be used with the special position feedback enabled hexapod.

## References


1. Brogårdh, T.: Present and future robot control development—An industrial perspective. Ann. Rev. Control **31**(1), 69–79 (2007)
2. Bem, M., et al.: Reconfigurable fixture evaluation for use in automotive light assembly. 18th International Conference on Advanced Robotics (ICAR), Hong Kong, pp. 61-67 (2017)
3. T. Gašpar, et al.: Smart hardware integration with advanced robot programming technologies for efficient reconfiguration of robot workcells. In: Robotics and Computer-Integrated Manufacturing, vol. 66, pp. 1-17 (2020). art. 101979
4. Jamšek, M.: Razvoj krmilnika paralelne robotske platforme. Magistrsko delo, Fakulteta za Strojništvo (2017)
5. Lenarčič, J., Bajd, T.: Robotski Mehanizmi. Založba FE in FRI, Ljubljana (2003)
6. Tsai, M.-S., Shiau, T.-N., Tsai, Y.-J., Chang, T.-H.: Direct kinematic analysis of a 3-PRS parallel mechanism. Mech. Mach. Theory **38**(1), 71–83 (2003)
7. Gan, D., Dias, J., Seneviratne, L.: Unified kinematics and optimal design of a 3rRPS metamorphic parallel mechanism with a reconfigurable revolute joint. In: Mechanism and Machine Theory, vol. 96, part 2, pp. 239-254 (2016)
8. Li, Y., Xu, Q.: Kinematics and inverse dynamics analysis for a general 3-PRS spatial parallel mechanism. Robotica **23**(2), 219–229 (2005)
9. Gašpar, T., Kovač, I., Ude, A.: Optimal layout and reconfiguration of a fixturing system constructed from passive Stewart platforms. J. Manuf. Syst. **60**, 226–238 (2021)

# UR10e Robot Drift Compensation for Precision Measurement Applications

Michal Vocetka[1](✉), Dominik Heczko[1], Jan Babjak[1], Zdenko Bobovský[1], Václav Krys[1], Roman Ružarovský[2], and Robert Bočák[2]

[1] Department of Robotics, Faculty of Mechanical Engineering, VSB - Technical University of Ostrava, 17. Listopadu 2172/15, 708 00 Ostrava-Poruba, Czech Republic
michal.vocetka@vsb.cz

[2] Institute of Production Technologies, Faculty of Materials Science and Technology in Trnava, Slovak University of Technology in Bratislava, Ulica Jána Bottu č. 2781/25, 917 24, Trnava, Slovakia

**Abstract.** This paper deals with the possibility of low-cost on-line compensation for thermally induced error. The aim is to compensate the drift of the UR10e collaborative robot to the extent that the robot operates with the repeatability guaranteed by the manufacturer for the thermal steady state. Thus, the authors have developed a compensation system that predicts drift and, by applying it in the robot control, can reduce the difference in positioning between the cold and warm robot. A measurement nest equipped with confocal sensor and a pair of laser profilers was used to measure repeatability. The robot was also equipped with sensors for temperature detection. The tests were carried out under different conditions and ISO 9283:1998 standard was also considered. By applying the proposed methodology, drift can be compensated with a high degree of efficiency. In some measurements the drift was compensated completely, usually the compensation efficiency was around 90%. Thanks to the compensation system, it is not necessary to pre-warm the robot and, above all, it is possible to guarantee a predetermined level of repeatability regardless of the actual robot or ambient temperature.

**Keywords:** Collaborative robot · Repeatability · Drift · Robot warm-up · Robot compensation · Thermally induced error

## 1 Introduction

UR collaborative robots are a common cobots in university and industrial environments. Although URs were originally conceived for use in an application without major demands on stiffness, accuracy, and other parameters, due to their good availability, users often deploy these cobots in applications for which these were not primarily intended. Accuracy and repeatability can be improved in a variety of options. Common approaches include refining the kinematic model of the robot, [1, 2] based on diagnostics, usually performed with the use of a laser tracker. These procedures are already commonly used in industry [3] and are usually performed to improve the absolute accuracy.


T. Petrič et al. (Eds.): RAAD 2023, MMS 135, pp. 281–288, 2023.
https://doi.org/10.1007/978-3-031-32606-6_33

In applications such as welding or precision assembly, high repeatability may be required. This paper [4] presents methodology for robot positioning accuracy measurements based on vision technique. High-speed cameras have also been used in research investigating the effect of direction of approach to achievable repeatability [5]. An interesting comparison of a portable photogrammetry system with a laser tracker is described in [6]. The suitability of these systems for robot calibration is also discussed.

The accuracy of the robot can also be detected continuously, so the current position information can be used for on-line robot guidance. It is commonly used in industry in combination with a laser tracker [7], however, this is an extremely expensive solution. Based on a mathematical approach using a Jacobian matrix, identification of an industrial manipulator's base and end-effector frames in a workplace can be performed by [8], where authors used ARUCO [9] tags and RGB cameras to measure the position of robot coordinate systems.

Due to its motion, the robot heats up - which leads to thermal deformation that causes a slight change compared to the kinematic model. This phenomenon is commonly known as position drift. In the papers [10, 11], the authors analyze the effect of the error caused by the temperature change and suggest the potential possibility of increasing the reliability of their process, by selecting the position within the workspace that has the lowest sensitivity to errors attributed to temperature changes.

The authors of this paper also dealt with the possibilities of increasing the repeatability of the robot and especially with the on-line robot drift compensation. Research [12] proposed a methodology to compensate drift with a high degree of efficiency (70–100%). The advantage is that there is no need for permanent installation of an expensive measuring equipment. Only the temperature detection wiring harness remains permanently on the robot after the precision measurements have been taken.

This paper presents an application of the previously proposed and tested system, on a collaborative robot UR10e, which is installed in a measurement application. In this application, there is a need to measure the position of an automotive headlight in the robot's gripper. As known, the repeatable accuracy of this robot is $\pm 0.05$ mm according to the manufacturer datasheet [13] however, during the process, the robot structure heats up which inevitably leads to deformation of the structure due to thermal expansion. The authors believe that for the lightweight tubular structure of the UR robot, the system could achieve similar quality results as in the case of earlier tests carried out on the ABB IRB 1200 industrial robot. Compensation should therefore be possible with a success rate of at least 80%, while the low stiffness of the UR robots should not be a hindrance during the measurements as there is sufficient settling time.

## 2 Methodology

The methodology and equipment for drift measurement and compensation have been developed as part of previous research and are in detail described in [12]. The proposed methodology consists of the following steps:

1) Thermal diagnostics of the arm to determine a position for temperature sensors.
2) Integration of metrology equipment in the robot workspace, interconnection of this equipment and the temperature sensors with the control application in a PC or PLC.

3) Running a reference task using a target point in which repeatability is to be measured.
4) Evaluation of the results and determination of the dependence of the drift in the direction of all TCP axis on the actual temperature
5) Modification of compensation equations.
6) Verification of the defined compensation. Starting the robot's task with simultaneous measurement of repeatability in the compensated target point.

The first step is the thermal diagnosis of the robot. According to the assumption, the primary source of the heat are joint gearboxes (see Fig. 1 A), where the second joint of the UR10e robot can be seen. However, after an extended period, there is significant temperature increase of the structure in the base and joint 4 to 6 areas. For the lower and upper arms, a gradual heating towards the center of gravity of the arms is evident.

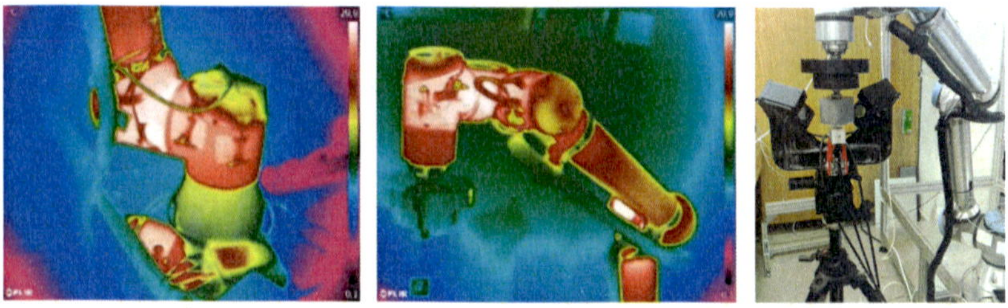

A B C

**Fig. 1.** Thermal diagnostics of the UR10e cobot. (A): Robot base and second axis. (B): Robot upper arm and tool. (C) UR10e with the test load inside the measuring nest

Robots in general are equipped with a temperature detection, however, this is a protective drive temperature detection only. The drive reaches the operating temperature in a short time and therefore this information is not useful in a drift compensation, since drift is caused by deformation of the structure, and it is the temperature of the structure that needs to be detected in real time.

Four Dallas DS18B20 [14] temperature sensors have been mounted on each axis. In total, twenty-four sensors on the robot and one that measures the ambient temperature were placed and sealed with aluminum tape. The data collection from these sensors is performed via a 1-Wire Bus by a microprocessor system built on the ARM STM32 platform, which checks the integrity of the data and then passes them on. At the time the measurements were realized, it was not possible to use a high-speed DIC cameras or laser tracker. As an alternative, a measurement nest based on the Keyence CL-P015/L015 confocal sensor [15] and a pair of Keyence LJ-X8080 laser profilers [16] was used. The sensor holder was created on an industrial 3D printer made of PC-ABS material, and our previous research [17] was used to correctly determine the angles that the laser profilers should take with the detected surface to increase the reliability of polished material scanning. The robot did not hold the car headlight during the tests, a test load was mounted instead. The entire measurement process is controlled by a measurement application tailor-made for this purpose and is operated on an Intel NUC PC.

## 3 Data Acquisition – Robot Drift Diagnostics

To determine to what extent drift affects the achievable repeatability of the UR10e robot, a verification cycle was defined in the robot. The robot cyclically repeated a defined trajectory and, after its execution, moved to the measurement nest. After the five seconds required for the oscillations to settle, the robot's repeatability was measured by the control application. This process was carried out over a period of eight hours, whereby the robot could be considered thermally stable after two and a half hours.

The graph in Fig. 2 shows the robot temperature difference between the cold and its operating temperature state. There is up to 12 °C difference, causing position drift, in this case by up to 0.28 mm. However, the manufacturer guarantees a value of ±0.05 mm for the temperature steady state. It is possible to preheat the robot, but such a process would require about three hours of time and even then, it would not be possible to guarantee a value of ±0.05 mm, since the robot temperature is not only affected by its own operation, but also by the ambient temperature, which is variable over the time. From the diagnostics results of which are shown in the graph (Fig. 2), can be seen that the drift affects the achieved repeatability in fundamental way.

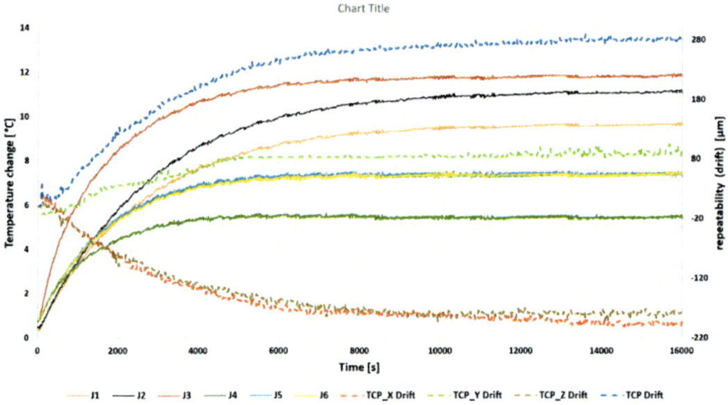


**Fig. 2.** UR10e drift and joint temperature measurement.

The next step is to determine the mathematical dependence of the drift with respect to the current temperature of the robot and its environment. The input variables in an Eq. (1) are the individual temperatures of the robot structure and the ambient temperature. The result is predicted TCP positioning error in the direction of a particular TCP axis at the current temperature. The prediction is based on the principle of linear regression.

After exporting the generated regression model from Regression Learner to Matlab, the coefficients for the calculation can be extracted. The compensation equation is formed by the sum of the coefficients and the current temperature, defined in the parallel robot task. This task is programmed to recompute the drift compensation at a frequency of 10 Hz based on all input temperatures measured on the robot arm. For a detailed description, see [12].

The resulting equation is then:

$$n_Z = [\text{trend} \cdot (I + A \cdot T_{25} + nJ_1 \cdot J_1 + nJ_2 \cdot J_2 + nJ_3 \cdot J_3 + nJ_4 \cdot J_4 + nJ_5 \cdot J_5 + nJ_6 \cdot J_6) + (\text{drift})/(\text{unit})] \tag{1}$$

The result of such an equation is the inverse value of the predicted drift for a specific temperature of the robot structure. In a robot task, with a frequency of 10 Hz, a compensation value is applied to compensate the predicted error. This will compensate the error to ideally "zero level", i.e., the state where the robot is thermally stable (see Fig. 3).

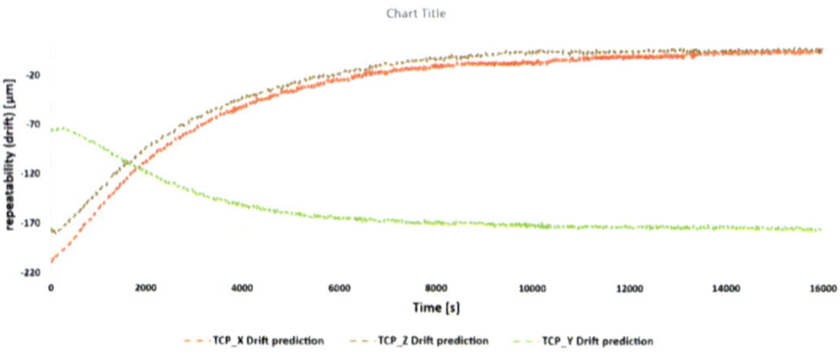


**Fig. 3.** Prediction of drift compensation of individual TCP axes

## 4 Results and Discussion

The use of the compensation procedure described above leads to a substantial mitigation of the drift effect which is evident from the waveforms in the graph (see Fig. 4). Without drift compensation, the repeatability ranges from 0 to 280 µm (as shown by the TCP_Drift waveform at Fig. 2). However, after activating drift compensation based on the actual robot temperature, the repeatability ranged from only 48 µm (the value is shown by the blue waveform at Fig. 4). In this particular measurement, the negative influence of drift was compensated to such an extent that the resulting repeatability variance is within the range declared by the manufacturer for the temperature steady state. One can therefore speak of a full compensation of the drift. The drift compensation in this case ensures the manufacturer's declared repeatability for any temperature, without the need to preheat the manipulator before starting its task.

Also, in another measurement at position C2 with a TCP velocity of 300 mm/s, an improvement was also achieved. The results are described in Fig. 5 below. This graph shows a comparison of the drift progress without and with its compensation. Without compensation, the robot drifts up to 350 µm. However, when the compensation is active, the total drift is only 71 µm. If the manufacturer's guaranteed value of 50 µm is subtracted, an improvement of 93% is achieved. This graph also shows the real value to which the system is compensated. Regardless of the actual robot or the ambient temperature that affects the cooling rate, the structure is compensated to the maximum operating temperature that is considered during operation. A prerequisite for correct operation is that the target was defined at the time the robot was warmed up to the operating temperature.

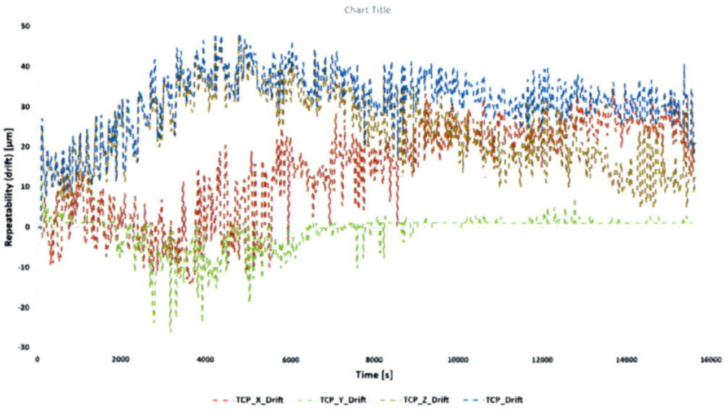


**Fig. 4.** Drift measurement with compensation applied

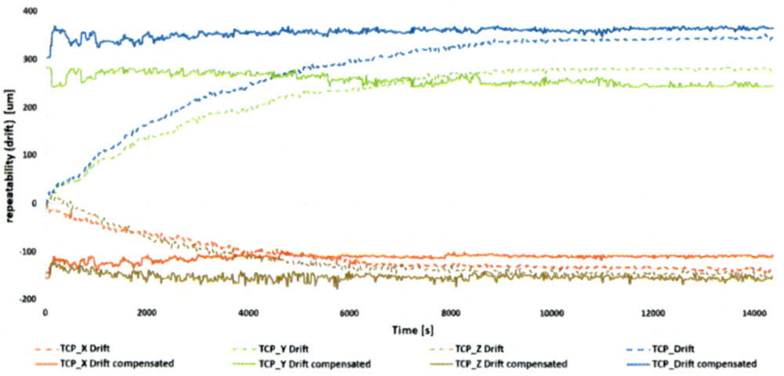


**Fig. 5.** Drift measurement with and without the compensation applied

# 5 Conclusion

After performance of all the measurements, it can be claimed that the collaborative robot UR10e is affected by drift and by applying its prediction, this effect can be substantially compensated. In some instances, full compensation was achieved, i.e., the robot positioned with repeatability up to 50 μm during the whole verification task which lasted eight hours. To demonstrate the ability to compensate for real processes, several cycles were carried out in which the robot was cycled and stopped. Despite these disturbances, the compensation system calculated a valid compensation parameter for a specific temperature of the structure that brought the robot closer to the target position. In general, it can be claimed that the compensation efficiency ranges from 70 to 95% depending on the ambient conditions.

The advantage of this solution when applied in a real industrial task is that it is not necessary to extend the robot or workstation with specialized metrology equipment, which is usually very expensive. However, it has to be mentioned that the repeatability is increased only in specific targets, not within a larger volume or in the entire working envelope of the robot. Authors are of the opinion that in a real task, a robot can usually be considered a single-purpose machine as is integrated in a specific application, equipped with a specific tool and performing a defined repetitive task. If a superior repeatability is required, there is certainly no need to increase it for all targets (the "standard repeatability" is usually sufficient). The specific positions for which the standard repeatability is insufficient are usually a few.

With a specific target compensation, better results will be achieved than by application of any general prediction. It should also be noted that the defined compensation will only be valid for a certain period of time. The compensation system needs to be updated. The robot like any machine is subject to wear and tear due to its own operation. This influence on repeatability cannot be predicted. The situation can be compared to the calibration of measuring tools according to ISO9001, which is carried out periodically by companies that have this standard established.

It can be assumed that the compensation equations defined for a particular target will also be applicable to surrounding targets if these are oriented in a similar way (the configuration of the arm joints will not differ substantially). Therefore, the next step of the research seems to be the definition of a region in which one defined set of compensation equations can be applied.

**Acknowledgement.** This article has been elaborated with the support of the Research Centre of Advanced Mechatronic Systems project, reg. no. CZ.02.1.01/0.0/0.0/16_019/0000867 in the frame of the Operational Program Research, Development and Education and by specific research project SP2023/060, financed by the state budget of the Czech Republic.

## References

1. Nubiola, A., Slamani, M., Joubair, A., Bonev, I.A.: Comparison of two calibration methods for a small industrial robot based on an optical CMM and a laser tracker. Robotica **32**(3), 447–466 (2014). https://doi.org/10.1017/S0263574713000714
2. Nubiola, A., Bonev, I.A.: Absolute calibration of an ABB IRB 1600 robot using a laser tracker. Rob. Computer-Integrated Manuf. **29**(1), 236–245 (2013). https://doi.org/10.1016/j.rcim.2012.06.004
3. Wu, Y., Klimchik, A., Caro, S., Furet, B., Pashkevich, A.: Geometric calibration of industrial robots using enhanced partial pose measurements and design of experiments. Robot. Computer-Integrated Manuf. **35**, 151–168 (2015). https://doi.org/10.1016/j.rcim.2015.03.007
4. Józwik, J., Ostrowski, D., Jarosz, P., Mika, D.: Industrial robot repeatability testing with high speed camera phantom V2511. Adv. Sci. Technol. Res. J. **10**(32), 86–96 (2016). https://doi.org/10.12913/22998624/65136
5. Vocetka, M., Huňady, R., Hagara, M., Bobovský, Z., Kot, T., Krys, V.: Influence of the approach direction on the repeatability of an industrial robot. Appl. Sci. **10**(23), 8714 (2020). https://doi.org/10.3390/app10238714

6. Filion, A., Joubair, A., Tahan, A.S., Bonev, I.A.: Robot calibration using a portable photogrammetry system. Robot. Computer-Integrated Manuf. **49**, 77–87 (2018). https://doi.org/10.1016/j.rcim.2017.05.004
7. Moeller, C., et al.: Real time pose control of an industrial robotic system for machining of large scale components in aerospace industry using laser tracker system. SAE Int. J. Aerosp. **10**(2), 100–108 (2017). https://doi.org/10.4271/2017-01-2165
8. Huczala, D., Oščádal, P., Spurný, T., Vysocký, A., Vocetka, M., Bobovský, Z.: Camera-based method for identification of the layout of a robotic workcell. Appl. Sci. **10**(21), 7679 (2020). https://doi.org/10.3390/app10217679
9. Oščádal, P., et al.: Improved pose estimation of aruco tags using a novel 3D placement strategy. Sensors **20**(17), 4825 (2020). https://doi.org/10.3390/s20174825
10. Kluz, R., Kubit, A., Trzepiecinski, T.: Investigations of temperature-induced errors in positioning of an industrial robot arm. J. Mech. Sci. Technol. **32**(11), 5421–5432 (2018). https://doi.org/10.17531/ein.2018.4.1
11. Kluz, R., Kubit, A., Sęp, J., Trzepieciński, T.: Effect of temperature variation on repeatability positioning of a robot when assembling parts with cylindrical surfaces. Eksploatacja i Niezawodność – Maintenance Reliab. **20**(4), 503–513 (2018). https://doi.org/10.17531/ein.2018.4.1
12. Vocetka, M., et al.: Influence of drift on robot repeatability and its compensation. Appl. Sci. **11**(22), 10813 (2021). https://doi.org/10.3390/app112210813
13. UR10e Specification [online]. Energivej 25 5260, Odense, Denmark: Universal Robots A/S [cit. 2023–01–10]. Available from: https://www.universal-robots.com/media/1807466/ur10e-rgb-fact-sheet-landscape-a4-125-kg.pdf (2023)
14. Programmable Resolution 1-Wire Digital Thermometer: DS18B20 [online]. [cit. 2022–05–02]. Available from: https://www.maximintegrated.com/en/products/sensors/DS18B20.html
15. Konfokální laserový měřicí systém řady CL-3000: Hlava senzoru CL-L015 (15 mm, čtyřbodový typ) [online]. Bedrijvenlaan 5, 2800 Mechelen, Belgie: KEYENCE [cit. 2022-05-02]. Available from: https://www.keyence.eu/cscz/products/measure/laser-1d/cl-3000/models/cl-l015/
16. 2D/3D Laser Profiler: LJ-X8000 series [online]. Avebury House, 219–225 Avebury Boulevard, Milton Keynes MK9 1AU, U.K.: Keyence [cit. 2021-10-04]. Available from: https://www.keyence.co.uk/products/measure/laser-2d/lj-x8000/models/lj-x8080/ (2021)
17. Heczko, D., Oščádal, P., Kot, T., Huczala, D., Semjon, J., Bobovský, Z.: Increasing the reliability of data collection of laser line triangulation sensor by proper placement of the sensor. Sensors **21**(8), 2890 (2021). https://doi.org/10.3390/s21082890

# Robot's Cartesian Stiffness Adjustment Through the Stiffness Ellipsoid Shaping

Branko Lukić(✉), Nikola Knežević, and Kosta Jovanović

School of Electrical Engineering, University of Belgrade, 11120 Belgrade, Serbia
{branko,knezevic,kostaj}@etf.rs

**Abstract.** One of the key properties that define the behaviour of the compliant robot with its environment is end-effector (EE) Cartesian stiffness. Typically, EE stiffness is represented as a stiffness matrix whose design can be impossible due to the lack of degrees of freedom to adjust all the elements within the stiffness matrix. Therefore, a tradeoff between matrix elements must be made. This paper proposed an approach for offline shaping of the stiffness matrix using stiffness ellipsoids as an alternative, where stiffness is shaped by adjusting the ellipsoid orientation and axis magnitudes. Shaping of the ellipsoid requires fewer parameters that need to be adjusted relative to the stiffness matrix. A criteria function for shaping the ellipsoid stiffness has been proposed. Optimal values of joint positions and stiffness were computed using the algorithm based on SLSQP (Sequential Least SQuare Programming). Further analysis is conducted on parts of the trajectory that cannot satisfy EE ellipsoid shaping criteria. This is overcome by starching the critical parts of the trajectory by dividing it into smaller increments. Results show improvement in ellipsoid shaping and reduced orientation error. The algorithm is tested in a simulation environment on the KUKALWR model.

**Keywords:** Stiffness ellipsoid · Cartesian stiffness · Joint stiffness

## 1 Introduction

In safe physical human-robot interaction (pHRI) or interaction of a robot with the environment in general [1], one of the most important roles is played by the robot's compliance. In addition to active compliance [2–4], which is realized through the control loop, there is also passive compliance, which is realized through elastic elements built into the transmission system of the robot [5, 6]. The end-effector (EE) stiffness matrix is a function of the joint positions (see (6)). This means that for a kinematically redundant robot with variable stiffness actuators (VSA) [7, 8], the Cartesian stiffness of the EE can be shaped by changing the configuration in null-space and/or by changing the stiffness of the joints. In general, the EE Cartesian stiffness matrix is a $6 \times 6$ symmetric matrix, resulting in 21 variables. Control of the EE Cartesian stiffness matrix and positions requires a robot with at least 14 DOFs that are VSA driven, not considering the physical limitations that exist and which makes the real number of required DOFs much larger and more complex to design and control.


T. Petrič et al. (Eds.): RAAD 2023, MMS 135, pp. 289–296, 2023.
https://doi.org/10.1007/978-3-031-32606-6_34

The EE stiffness matrix has a non-linear dependence on the robot configuration via the Jacobian matrix. Thus, for a typical robot with 6–7 DOFs, it is impossible to find an analytical set of joint positions and stiffness that will satisfy the desired EE position and stiffness matrix. Thus, optimization techniques using null space are necessary [9, 10]. A null-space motion optimization approach can lead to a locally optimal solution. When computing the optimal joint configuration, the null-space gradient-based optimization will ensure a smooth joint trajectory compared to nonlinear optimization methods that seek an optimal global solution that may differ significantly from the robot's current pose [11]. Shaping the EE stiffness can be divided into configuration-dependent stiffness (CDS) and common mode stiffness (CMS) [12]. The CDS originates from the redundant configuration of the robot and is used to direct the stiffness ellipsoid, while the CMS affects the magnitude (volume) of the ellipsoid. The CMS is an additional parameter that can be included in the stiffness ellipsoid design.

The focus of this work is on VSA-driven robots, which have two control variables: the position and stiffness of the actuator. The mathematical model of the LWR Kuka robot [13] has been used in simulations as a platform to emulate VSA-driven robot behavior. The approach for EE cartesian stiffness shaping proposed in this paper is based on the "offline" shaping of the stiffness ellipsoid in the phase of planning the trajectory and stiffness of the EE of the robot, whose shaping indirectly adjusts the EE stiffness matrix. Shaping is conducted in nonlinear optimization for defined criteria function in reduced joint position and joint stiffness space. Reduced space is computed for each iteration based on the maximally allowed variation within one iteration and the robot's physical limits. The analysis is conducted on parts of the trajectory that cannot satisfy EE ellipsoid shaping criteria. This problem is overcome by stretching the critical parts of the trajectory over time (the 3D space trajectory stays intact) by dividing it into smaller increments. Trajectory stretching is achieved using Radial Basis Functions (RBF) to code initial trajectory and interpolated trajectory [14], but it can be done with other interpolation methods.

## 2 EE Cartesian Stiffness Modelling

Joint stiffness matrix $K_j$ [7, 8] is defined as

$$K_j = -\frac{\partial \tau}{\partial q}, \tag{1}$$

where $\tau$ is the joint torque vector and $q$ is the joint position vector. The relation between joints torque vector ($\tau$) and Cartesian wrench ($F$) applied on the robot's EE is mapped with Jacobian matrix $J(q)$ as

$$\tau = J(q)^T F, \tag{2}$$

while wrench $F$ is proportional to the EE stiffness matrix $K_C$ and EE displacement $\Delta x$ which is a result of applying wrench $F$ to the robot's EE

$$F = K_C \Delta x. \tag{3}$$

Substituting (2) and (3) into (1) yields to

$$K_j = -\frac{\partial\left(J(q)^T F\right)}{\partial q} = -\frac{\partial\left(J(q)^T K_C \Delta x\right)}{\partial q} = J(q)^T K_C J(q) - \frac{\partial J(q)^T}{\partial q} K_C \Delta x, \tag{4}$$

where $\Delta x$ is the difference between the desired position ($x_d$) and achieved position ($x$). Thus, $-\partial \Delta x / \partial q = J(q)$. The $K_C$ is $m \times m$ symmetric matrix, $J(q)$ is $m \times n$ matrix, $q$ is $n$-dimension joint position vector and, $K_j$ is $n \times n$ diagonal matrix. Parameters $m$ and $n$ are the size of the workspace and joint space, respectively. When there is no displacement in equilibrium ($\Delta x = 0$) expression in (4) becomes

$$K_j = J(q)^T K_C J(q). \tag{5}$$

From (5) the EE Cartesian stiffness matrix expression is obtained as

$$K_C = \left(J(q) K_j^{-1} J(q)^T\right)^{-1}. \tag{6}$$

Equations (6) indicates that EE stiffness can be adjusted by:

1) by null space reconfiguration of the kinematically redundant robot,
2) variation of joint stiffness of VSA-driven robots,
3) simultaneously applying methods from 1) and 2).

All the listed methods can be applied to VSA-driven robots, while only 1) can be applied to robots driven by constant stiffness actuators.

A common approximation for the EE stiffness matrix is to divide it into four submatrices ($K_T$ and $K_R$ are the translational and rotational stiffness, while $K_{RT}$ and $K_{TR}$ are the coupling stiffness matrices). Only the submatrices on the main diagonal (translational and rotational stiffness) are of interest, while those on the secondary diagonals are ignored [8]

$$K_C = \begin{bmatrix} K_T & K_{RT} \\ K_{TR} & K_R \end{bmatrix}, \tag{7}$$

$$\hat{K}_C = \begin{bmatrix} K_T & 0 \\ 0 & K_R \end{bmatrix}. \tag{8}$$

It should be mentioned that $\hat{K}_C$ is an approximation of $K_C$ that is valid only locally.

## 3 EE Stiffness Ellipsoid Shaping

The EE Cartesian stiffness can be observed numerically through values in the stiffness matrix $K_C$ or through graphical representation with a stiffness ellipsoid for 3D space or stiffness ellipse for 2D space. For 2D space translator stiffness $K_T$ is a $2 \times 2$ matrix, while for 3D space is $3 \times 3$ matrix. Stiffness ellipsoids are a graphical approximation of how much EE achieve stiffness and in which direction. The stiffness ellipsoid is defined by axis magnitude ($\lambda_1$, $\lambda_2$ and $\lambda_3$) and orientation vector ($\vec{u}$). Orientation is always along the longest principal axis. Axis magnitudes and orientation vectors are obtained from $K_T$ with *Eigen Value Decomposition*

$$K_T = U\lambda U^T. \tag{9}$$

Elements of diagonal matrix $\lambda$ are magnitudes of the ellipsoid axis, while columns of orthogonal matrix $U$ present eigenvectors that define directions of the principal axis.

This paper proposes an offline approach to shaping the robot's EE Cartesian stiffness by applying optimization algorithms. An optimization algorithm is intended for planning the stiffness phase when the EE task trajectory is known in advance. The proposed criteria combine the shape and orientation error of the ellipsoid. Orientation error angle $\alpha_{error}$ is the angle between desired ellipsoid orientation vector ($\vec{u}$) and achieved ellipsoid orientation vector ($\vec{u}_a$). The shaping of the stiffness ellipsoid is achieved by a combination of null-space and joint stiffness variation. The optimization algorithm was applied to the simulation model of the KUKA LWR robot on the example of planar motion in the "*xy*" plane when the EE moves along a curved path.

The SLSQP (Sequential Least Square Programming) based optimization algorithm was used, and it was implemented in MATLAB [9, 15]. SLSQP is an algorithm that optimizes the values of nonlinear criteria functions with linear and nonlinear constraints. The limitations that exist in the proposed optimization algorithms refer to the achievable values of the position of the robot's joints and the defined range of stiffness that the robot can achieve. Non-linear constraints that must be satisfied are related to kinematics. The optimization algorithm must always satisfy the desired position of the EE. For the stiffness ellipsoid to be completely defined, it is necessary to meet the following requirements:

1) desired orientation of the ellipsoid,
2) the ratio of the axes of the ellipsoid,
3) and the magnitudes of the axes of the ellipsoid.

When the first two conditions are met, the third is achieved by applying CMS to the shaped ellipse to scale its magnitude. To optimize the first two conditions simultaneously, a criteria function $f$ that combines the orientation error and the ratio of the axes of the ellipsoid is proposed as

$$f = (1 + |\alpha_{error}|)\left(1 + \left|r - \frac{r_{max}}{r_{min}}\right|\right), \tag{10}$$

where $\alpha_{error}$ is ellipsoid orientation error in degrees, $r$ is the desired ratio of the ellipsoid axes, $r_{max}$ and $r_{min}$ are the longest and second longest axes of the ellipsoid.

In each iteration, the positions and stiffnesses of the joints are computed to minimize the value of the function $f$ as

$$\begin{bmatrix} q[p] \\ k_j[p] \end{bmatrix} = \underset{\substack{q \in [q_{\min}, q_{\max}] \\ k_j \in [k_{j\,\min}, k_{j\,\max}]}}{argmin} f(q, k_j), \tag{11}$$

where $p$ is the iteration number, and $q_{\min}$, $q_{\max}$, $k_{j\,\min}$ and $k_{j\,\max}$ are the minimum and maximum values of the positions and stiffness of the joints. The range of values of the variables in the optimization algorithm can be further narrowed, thereby reducing the search space and speeding up the search algorithm in each iteration. The new limit values of the variables ($q^*_{min}$, $q^*_{max}$, $k^*_{j\,min}$ and $k^*_{j\,max}$) are calculated in each iteration taking into account the maximum change speeds that the robot can achieve within one iteration (for positions joints -$\Delta q$, for joint stiffness -$\Delta k_j$) and physically feasible values as

$$q_{min\_iter}[p] = \max(q_{min}, q[p-1] - \Delta q), \tag{12}$$

$$q_{max\_iter}[p] = \min(q_{max}, q[p-1] + \Delta q), \tag{13}$$

$$k_{j\,min\_iter}[p] = \max(k_{j\,min}, k_j[p-1] - \Delta k_j), \tag{14}$$

$$k_{j\,max\_iter}[p] = \min(k_{j\,max}, k_j[p-1] + \Delta k_j). \tag{15}$$

The EE trajectory can be described parametrically. A suitable parametrization method to approximate the path is coding using RBF based on Gaussian kernel functions. The RBF-coded path allows the interpolation of data between two known values. This property is used to stretch trajectory in critical parts and to allow the algorithm to compute an optimal set of joint positions and stiffness with more slowly changing references. This interpolation is not a straight line, but it keeps some original path curvature. Thus, interpolated points will have a slower change of ellipsoid orientation between iterations. The trajectory stretching is activated only when the computed ellipsoid orientation error is above the defined threshold. The threshold presents an orientation angle error that will be tolerated.

## 4 Simulation Results

The optimization requirement for ellipsoid orientation error is set to 0 ($\alpha_{error} = 0°$), and the axis ratio of the ellipsoid is $r = 3$. Maximal joint position and stiffness change per iteration for all joints are set to 10° and 100Nm/rad, respectively.

The algorithm is tested for an example of curvilinear movement shown in Fig. 1.

Simulation results are shown in Fig. 2. Joint stiffness is shown in Fig. 2a, joint position in Fig. 2b, criteria function value $f$ in Fig. 2c and orientation error angle is shown in Fig. 2d. In most parts of the trajectory, desired EE stiffness ellipsoid shape is achieved. Critical parts are the initial position and parts of the trajectory with the biggest maneuver.

The trajectory stretching strategy is applied for $\alpha\_error \geq 10$ degrees. For the same initial values, the stretched trajectory is depicted in Fig. 3. Trajectory in 3D space is the same as in the previous case. The difference is in the number of samples in regions with higher orientation errors. The number of samples is increased, resulting in slower changes of trajectory in Cartesian space in critical parts.

Simulation results are shown in Fig. 4. Joint stiffness is shown in Fig. 4a, joint position in Fig. 4b, criteria function value f in Fig. 4c and orientation error angle $\alpha\_error$

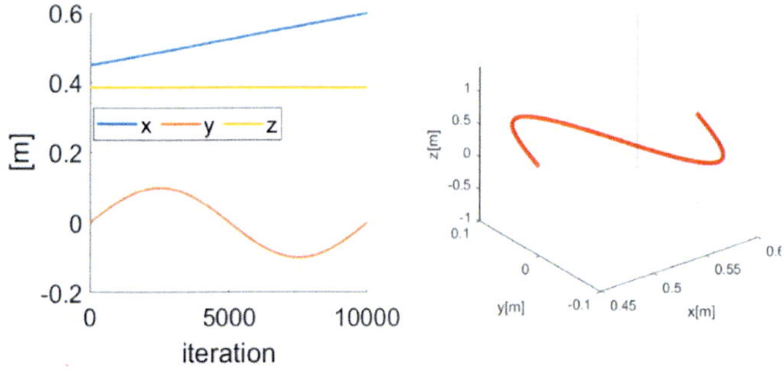


**Fig. 1.** EE trajectory. Left: *x*, *y* and *z* axis over iteration; right: EE trajectory in 3D space.

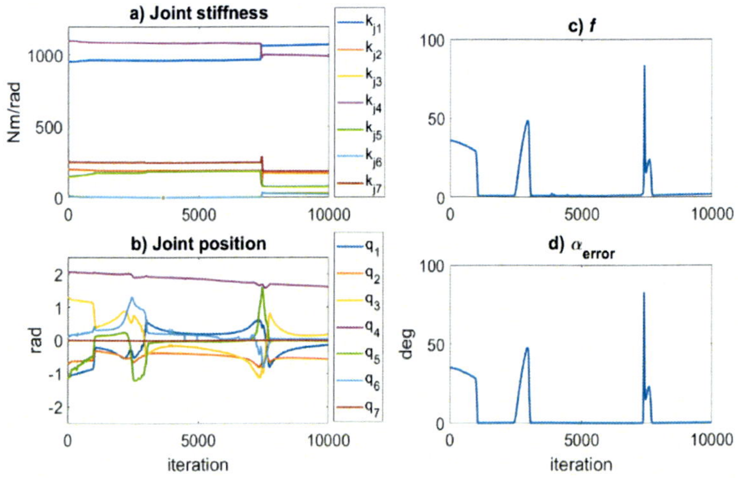


**Fig. 2.** Simulation results: a) Joint stiffness; b) Joint positions; c) criteria function; d) orientation angle error.

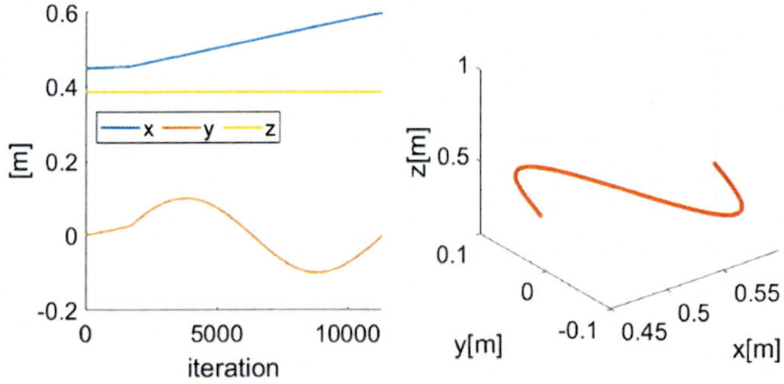


**Fig. 3.** Stretched trajectory. Left: *x*, *y* and *z* trajectory over iteration; right: trajectory in 3D space.

is shown in Fig. 4d. In most of the trajectories desired EE stiffness ellipsoid shape is achieved. Critical parts with the biggest maneuver are overcome, but the first parts of the trajectory have still some relatively high orientation errors. This is due to the applied optimization algorithm, which is gradient-based and therefore sensitive to the initial values. The criteria function at the high maneuver trajectory part has two small picks, demonstrating that this part of the trajectory is still the most demanding to overcome.

The illustration of ellipsoids along the path is depicted in Fig. 5.

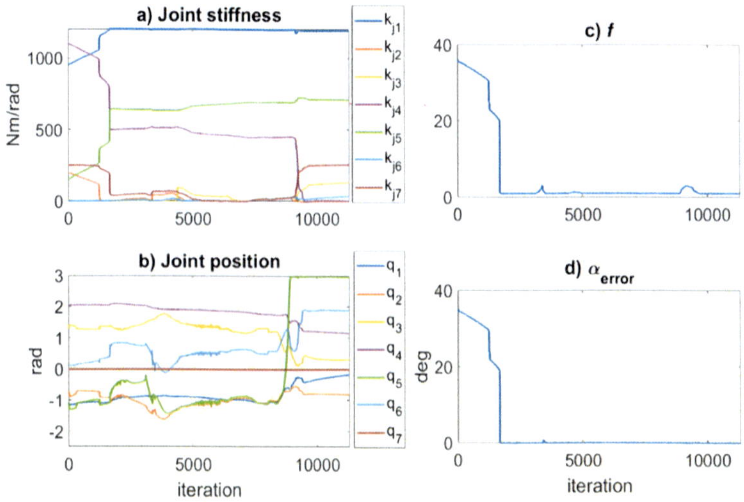


**Fig. 4.** Simulation results for stretch trajectory: a) Joint stiffness; b) Joint positions; c) criteria function; d) orientation angle error.

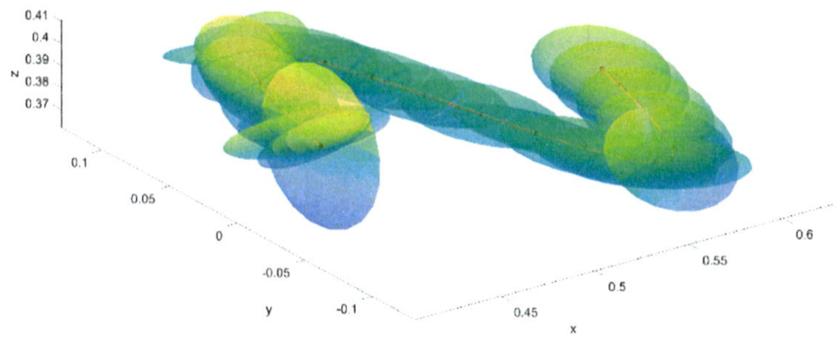


**Fig. 5.** The illustration of stiffness ellipsoids along the path.

## 5 Conclusion and Discussion

EE stiffness shaping approach that uses the graphical representation of the stiffness through a stiffness ellipsoid for the KUKA LWR robot model is successfully implemented. The obtained results generally meet the set requirements well. Stretching trajectory improved results in comparison with optimization with the original trajectory.

Results show that the achievable performances are limited by the choice of part of the workspace, as well as the shape of the trajectory. Optimization along an entire path is a computationally demanding job, and it is not easy to implement it to work in real-time, but it is possible to use it in the planning phase of stiffness and trajectory of the joints and the EE. The optimization algorithm is gradient-based and therefore sensitive to the initial values. Choosing optimal initial joint stiffness and position as well as exploiting the other optimization methods for obtaining better results is future work.

## References

1. De Santis, A., et al.: An atlas of physical human–robot interaction. Mech. Mach. Theory **43**(3), 253–270 (2008)
2. Hogan, N.: Impedance control: an approach to manipulation: part II-implementation. J. Dyn. Syst. Meas. Contr. **107**(1), 8–16 (1985)
3. Lukić, A., et al.: UKA LWR robot cartesian stiffness control based on kinematic redundancy. In: The 28th International Conference on Robotics in Alpe-Adria-Danube Region, RAAD 2019. Kaiserslautern, Germany (2019)
4. Dai, Y., et al.: A review of end-effector research based on compliance control. Machines **10**(2), 100 (2022)
5. Van Ham, R., et al.: Compliant actuator designs. IEEE Robot. Autom. Mag. **16**(3), 81–94 (2009)
6. Grioli, G., et al.: Variable stiffness actuators: The user's point of view. The Int. J. Robo. Res. **34**(6), 727–743 (2015)
7. Petit, F., Albu-Schäffer, A.: Cartesian impedance control for a variable stiffness robot arm. In: IEEE/RSJ International Conference on Intelligent Robots and Systems. San Francisco, CA, USA (2011)
8. Albu-Schäffer, A., et al.: Soft robotics: what Cartesian stiffness can obtain with passively compliant, uncoupled joints?. In: IEEE/RSJ International Conference on Intelligent Robots and Systems (IROS). Sendai, Japan (2004)
9. Knežević, N., et al.: End-effector Cartesian stiffness shaping - sequential least squares programming approach. Serbian J. Elec. Eng. 18(1), 1–14 (2021)
10. Lukić, A., et al.: Online cartesian compliance shaping of redundant robots in assembly tasks. Machines **11**(1), 35 (2023)
11. Knežević, N., Lukić, B., Jovanović, K.: Feedforward Control Approaches to Bidirectional Antagonistic Actuators Based on Learning, u Advances in Service and Industrial Robotics, pp. 337–345. Springer, Cham (2020)
12. Ajoudani, A., et al.: TeleImpedance: exploring the role of common-mode and configuration-dependant stiffness. In: 2012 12th IEEE-RAS International Conference on Humanoid Robots (Humanoids 2012). Osaka, Japan (2012)
13. Gaz, A., Flacco, F., De Luca, A.: Identifying the dynamic model used by the KUKA LWR: A reverse engineering approach. In: 2014 IEEE International Conference on Robotics and Automation (ICRA). Hong Kong, China (2014)
14. Žlajpah, L., Petrič, T.: Generation of Smooth Cartesian Paths Using Radial Basis Functions. In: Zeghloul, S., Laribi, M.A., Sandoval Arevalo, J.S. (eds.) RAAD 2020. MMS, vol. 84, pp. 171–180. Springer, Cham (2020). https://doi.org/10.1007/978-3-030-48989-2_19
15. Fu, Z., Liu, G., Guo, L.: Sequential Quadratic Programming Method for Nonlinear Least Squares Estimation and Its Application. Mathematical problems in engineering, vol. 2019 (2019)

# Analysis of Higher- Order Kinematics on Multibody Systems with Nilpotent Algebra

Daniel Condurache(✉)

Technical University of Iasi, D. Mangeron Street No.59, 700050 Iasi, Romania
daniel.condurache@tuiasi.ro

**Abstract.** This paper proposes a novel analysis method for higher-order acceleration vector fields in the case of rigid body motion and multibody systems, using the property of nilpotent algebra. The equations that allow the determination of higher-order accelerations field are given for the spatial serial kinematic chains. The results are in closed form and coordinate-free. As a case, the velocities, accelerations, jerks, and jounces fields are offered for a general kinematic spatial lower-pair kinematic chain.

**Keywords:** nilpotent algebra · higher-order kinematics · Lie group SE (3)

## 1 Introduction

The higher-precision robotic systems, new space docking procedures, artificial vision systems, and molecular dynamics require procedures to calculate higher-order accelerations. The key to the proposed approach starts with the property of rigid body displacements group of forming a Lie group, accompanied by its Lie algebra. A previous result offers an isomorphic representation of the Lie group, SE (3), with the group of the orthogonal dual tensors and Lie algebra, se (3), of the Lie algebra of dual vectors. The results obtained using dual algebras completely solve the problem of finding the field of higher-order accelerations utilizing a set of results obtained by the previous papers [1–3]. Moreover, the results can be extended for the nilpotent commutative algebra [4]. This paper proposes a novel computing method for studying the higher-order acceleration fields for rigid body motion and serial kinematic chains, using the calculus with the dual and the nilpotent algebras.

## 2 Higher-Order Kinematics of Rigid Body

Let be a rigid motion given by a curve in Lie group of the rigid body displacements SE (3) given by the homogenous matrix $\mathbf{g} = \begin{bmatrix} \mathbf{R} & \boldsymbol{\rho} \\ 0 & 1 \end{bmatrix}$ where $\mathbf{R} \in S\mathbb{O}_3$ is a proper orthogonal tensor [2, 3, 5], $\mathbf{R} = \mathbf{R}(t)$, and $\boldsymbol{\rho} = \boldsymbol{\rho}(t)$ a vector functions of a time variable, n-time differentiable. As shown in [1–3], the $n$-th order acceleration of a point of the rigid body given by the position vector $\mathbf{r}$ can be computed with the following equation [3]:

$$\mathbf{a}_{\mathbf{r}}^{[n]} = \mathbf{a}_n + \Phi_n \mathbf{r}; \, n \in \mathbb{N} \quad (1)$$


T. Petrič et al. (Eds.): RAAD 2023, MMS 135, pp. 297–305, 2023.
https://doi.org/10.1007/978-3-031-32606-6_35

where the invariants tensor $\Phi_n$ and the vector $\mathbf{a}_n$ is given by the below equations:

$$\Phi_n = \mathbf{R}^{(n)}\mathbf{R}^T \tag{2}$$

$$\mathbf{a}_n = \boldsymbol{\rho}^{(n)} - \Phi_n\boldsymbol{\rho} \tag{3}$$

Tensor $\Phi_\text{n}$ and vector $\mathbf{a}_\text{n}$ generalize the notions of velocity/acceleration tensor respectively velocity/acceleration invariant. They are fundamental in the study of the vector field of the $\text{n}^{\text{th}}$ order accelerations. The recursive formulas for computing $\Phi_\text{n}$ and $\mathbf{a}_\text{n}$ are [3]:

$$\begin{cases} \Phi_{n+1} = \dot{\Phi}_n + \Phi_n\Phi_1 \\ \mathbf{a}_{n+1} = \dot{\mathbf{a}}_n + \Phi_n\mathbf{a}_1 \end{cases}, n \geq 1,$$
$$\Phi_1 = \tilde{\boldsymbol{\omega}}, \mathbf{a}_1 = \mathbf{v}_Q - \tilde{\boldsymbol{\omega}}\boldsymbol{\rho}_Q \triangleq \mathbf{v} \tag{4}$$

The pair of vectors $(\boldsymbol{\omega}, \mathbf{v})$ is also known as *the spatial twist of rigid body.*

In [1–3], an iterative procedure is described to determine the instantaneous tensor $\Phi_n$, and vector $\mathbf{a}_n$ using the time derivative of spatial twist of rigid body motion, [1, 3]:

**Theorem 1.** *There is a unique polynomial with the coefficients in the non-commutative ring of Euclidian second order tensors* $\mathbf{L}(\mathbf{V}_3, \mathbf{V}_3)$ *such that the vector respectively the tensor invariants of the* $n^{th}$ *order accelerations will be written as:*

$$\begin{matrix} \mathbf{a}_n = \mathbf{P}_n(\mathbf{v}) \\ \Phi_n = \mathbf{P}_n(\tilde{\boldsymbol{\omega}}) \end{matrix}, n \in \mathbb{N}^* \tag{5}$$

*where* $\mathbf{P}_n$ *fulfills the relationship of recurrence:*

$$\begin{cases} \mathbf{P}_{n+1} = \mathbf{D}\mathbf{P}_n + \mathbf{P}_n(\tilde{\boldsymbol{\omega}}), n \in \mathbb{N}^* \\ \mathbf{P}_1 = \mathbf{I} \end{cases} \tag{6}$$

*with* $\mathrm{D} = \frac{d}{dt}$ *the derivative operator with respect to time.*

For $\text{n} = \overline{1, 4}$, it follows:

- *the velocity vector field invariants:*

$$\begin{cases} \mathbf{a}_1 = \mathbf{v} \\ \Phi_1 = \tilde{\boldsymbol{\omega}} \end{cases}$$
$$\mathbf{a}_{\boldsymbol{\rho}}^{[1]} = \mathbf{v} + \tilde{\boldsymbol{\omega}}\boldsymbol{\rho} \tag{7}$$

- *the acceleration vector field invariants:*

$$\begin{cases} \mathbf{a}_2 = \dot{\mathbf{v}} + \tilde{\boldsymbol{\omega}}\mathbf{v} \\ \Phi_2 = \dot{\tilde{\boldsymbol{\omega}}} + \tilde{\boldsymbol{\omega}}^2 \end{cases}$$
$$\mathbf{a}_{\boldsymbol{\rho}}^{[2]} = \dot{\mathbf{v}} + \tilde{\boldsymbol{\omega}}\mathbf{v} + (\dot{\tilde{\boldsymbol{\omega}}} + \tilde{\boldsymbol{\omega}}^2)\boldsymbol{\rho} \tag{8}$$

- *the jerk vector field invariants:*

$$\begin{cases} \mathbf{a}_3 = \ddot{\mathbf{v}} + \tilde{\boldsymbol{\omega}}\mathbf{v} + 2\dot{\tilde{\boldsymbol{\omega}}}\mathbf{v} + \tilde{\boldsymbol{\omega}}^2\mathbf{v} \\ \boldsymbol{\Phi}_3 = \ddot{\tilde{\boldsymbol{\omega}}} + \tilde{\boldsymbol{\omega}}\dot{\tilde{\boldsymbol{\omega}}} + 2\dot{\tilde{\boldsymbol{\omega}}}\tilde{\boldsymbol{\omega}} + \tilde{\boldsymbol{\omega}}^3 \end{cases}$$
$$\mathbf{a}_{\boldsymbol{\rho}}^{[3]} = \ddot{\mathbf{v}} + \tilde{\boldsymbol{\omega}}\dot{\mathbf{v}} + 2\dot{\tilde{\boldsymbol{\omega}}}\mathbf{v} + \tilde{\boldsymbol{\omega}}^2\mathbf{v} + (\ddot{\tilde{\boldsymbol{\omega}}} + \tilde{\boldsymbol{\omega}}\dot{\tilde{\boldsymbol{\omega}}} + 2\dot{\tilde{\boldsymbol{\omega}}}\tilde{\boldsymbol{\omega}} + \tilde{\boldsymbol{\omega}}^3)\boldsymbol{\rho} \tag{9}$$

The higher-order time differentiation of spatial twist of rigid body motion solves completely the problem of determining the field of the higher- order acceleration of rigid motion. Next, we present a new non-iterative procedure that permits the determination of the higher-order accelerations field using nilpotent algebra.

## 3 Mathematical Preliminaries of Nilpotent Algebra

Let be $\mathbb{R}$ the set of real numbers and $n \in \mathbb{N}$ a natural number. We will introduce the nilpotent algebra by: $\hat{\mathbb{R}} = \mathbb{R} + \varepsilon\mathbb{R} + \ldots \varepsilon^n\mathbb{R}$; $\varepsilon \neq 0, \varepsilon^{\mathrm{n}+1} = 0$. For n = 1, on obtain dual numbers algebra [5].

Two generic elements $\hat{x}, \hat{y} \in \hat{\mathbb{R}}$ will be written as following:

$$\hat{x} = x + x_1\varepsilon + \ldots + x_n\varepsilon^n;\ x_k \in \mathbb{R}, k = \overline{1, n} \tag{10}$$

$$\hat{y} = y + y_1\varepsilon + \ldots + y_n\varepsilon^n;\ y_k \in \mathbb{R}, k = \overline{1, n} \tag{11}$$

We will denote by $x = Re\hat{x}$ and $Mu\hat{x} = \sum_{k=1}^{n} \underline{x}_k\varepsilon^k$ the real parts and respectively, multidual parts of the multidual number $\hat{x}$. Also, we will denote by $\underline{x}_k = \frac{d\hat{x}}{d\varepsilon^k}$; $k = \overline{1, n}$, the $k$ order component of the hypercomplex number $\hat{x}$.

We will define the operations of additions and, respectively, multiplication of two hypercomplex numbers by:

$$\hat{x} + \hat{y} = \sum\nolimits_{k=0}^{n} (x_k + y_k)\varepsilon^k \tag{12}$$

$$\hat{x}\hat{y} = \sum\nolimits_{k=0}^{n} \left( \sum\nolimits_{p=0}^{k} x_p y_{p-k} \right) \varepsilon^k \tag{13}$$

In Eq. (12) and Eq. (13) we will denote by $\varepsilon^0 = 1, x_0 = x, y_0 = y$. The set $\hat{\mathbb{R}}$ with the addition operation (12) and multiplication operation (13) is a commutative ring with unit. An element $\hat{x} \in \hat{\mathbb{R}}$ is invertible if and only if $Re\hat{x} \neq 0$. The zero divisor from the ring $\hat{\mathbb{R}}$ is characterized by $Re\hat{x} = 0$. An element from $\hat{\mathbb{R}}$ is either invertible or zero divisor [6]. $\hat{\mathbb{R}}$ has a structure of (n + 1) − dimensional associative, commutative, and unitary generalized Clifford Algebra, over the field of real number $\mathbb{R}$.

Let be $f : \mathbb{I} \subseteq \mathbb{R} \to \mathbb{R}, f = f(x)$ a n-times differentiable real function. We will define the hypercomplex function of hypercomplex variable $f : \hat{\mathbb{I}} \subseteq \hat{\mathbb{R}} \to \hat{\mathbb{R}}, f = f(\hat{x})$ by equation:

$$f(\hat{x}) = f(x) + \sum\nolimits_{k=1}^{n} \frac{\Delta\hat{x}^k}{k!} f^{(k)}(x) \tag{14}$$

where we denoted by $\Delta\hat{x}$ : $\Delta\hat{x} = \hat{x} - x = \sum_{k=1}^{n} x_k \varepsilon^k$, and $\Delta\hat{x}^p = 0; p \geq n + 1$.

Using the Eq. (14), we will define the following hypercomplex functions:

$$\sin\hat{x} = \sin x + \sum_{k=1}^{n} \frac{\Delta\hat{x}^k}{k!} \sin\left(x + k\frac{\pi}{2}\right) \tag{15}$$

$$\cos\hat{x} = \cos x + \sum_{k=1}^{n} \frac{\Delta\hat{x}^k}{k!} \cos\left(x + k\frac{\pi}{2}\right) \tag{16}$$

If $Re\hat{x} = 0$ result:

$$\sin\Delta\hat{x} = \sum_{k=1}^{n} \frac{\Delta\hat{x}^k}{k!} \sin\left(k\frac{\pi}{2}\right) \tag{17}$$

$$\cos\Delta\hat{x} = 1 + \sum_{k=1}^{n} \frac{\Delta\hat{x}^k}{k!} \cos\left(k\frac{\pi}{2}\right) \tag{18}$$

### 3.1 Dual Vectors in Nilpotent Algebra

Let be $\underline{\mathbb{V}}_3 = \mathbb{V}_3 + \varepsilon_0 \mathbb{V}_3$; $\varepsilon_0 \neq 0$, $\varepsilon_0^2 = 0$, the set of dual vectors from three-dimensional Euclidean space. We will introduce the set of dual vectors in nilpotent algebra by: $\hat{\mathbb{V}}_3 = \underline{\mathbb{V}}_3 + \varepsilon\underline{\mathbb{V}}_3 + \ldots + \varepsilon^n\underline{\mathbb{V}}_3$; $\varepsilon \neq 0$, $\varepsilon^{n+1} = 0$ (for n = 1, obtain hyper-dual vectors [4]).

A generic vector from $\hat{\mathbb{V}}_3$ will be written as below:

$$\begin{aligned} \hat{\mathbf{a}} &= \underline{\mathbf{a}} + \underline{\mathbf{a}}_1\varepsilon + \ldots + \underline{\mathbf{a}}_n\varepsilon^n;\ \underline{\mathbf{a}},\ \underline{\mathbf{a}}_k \in \underline{\mathbb{V}}_3 \\ \hat{\mathbf{b}} &= \underline{\mathbf{b}} + \underline{\mathbf{b}}_1\varepsilon + \ldots + \underline{\mathbf{b}}_n\varepsilon^n;\ \underline{\mathbf{b}},\ \underline{\mathbf{b}}_k \in \underline{\mathbb{V}}_3 \end{aligned} \tag{19}$$

We will define the scalar product respectively the cross product of two vectors from $\hat{\mathbb{V}}_3$ by:

$$\hat{\mathbf{a}} \cdot \hat{\mathbf{b}} = \sum_{k=0}^{n} \sum_{p=0}^{k} \left(\underline{\mathbf{a}}_p \cdot \underline{\mathbf{b}}_{k-p}\right)\varepsilon^k \tag{20}$$

$$\hat{\mathbf{a}} \times \hat{\mathbf{b}} = \sum_{k=0}^{n} \sum_{p=0}^{k} \left(\underline{\mathbf{a}}_p \times \underline{\mathbf{b}}_{k-p}\right)\varepsilon^k \tag{21}$$

The triple vector product of three vectors $\hat{\mathbf{a}}$, $\hat{\mathbf{b}}$, $\hat{\mathbf{c}}$ is defined by $\langle\hat{\mathbf{a}}, \hat{\mathbf{b}}, \hat{\mathbf{c}}\rangle = \hat{\mathbf{a}} \cdot \left(\hat{\mathbf{b}} \times \hat{\mathbf{c}}\right)$.

All the three vectors $\hat{\mathbf{e}}_1, \hat{\mathbf{e}}_2, \hat{\mathbf{e}}_3$ represents a basis in the free module $\hat{\mathbb{V}}_3$ if and only if $Re\left(\hat{\mathbf{e}}_1, \hat{\mathbf{e}}_2, \hat{\mathbf{e}}_3\right) \neq 0$.

### 3.2 Euclidean Tensors in Nilpotent Algebra

An $\hat{\mathbb{R}}$ - linear application of $\hat{\mathbf{V}}_3$ into $\hat{\mathbf{V}}_3$ is called a Euclidean tensor:

$$\mathbf{T}\left(\hat{\lambda}_1\hat{\mathbf{v}}_1 + \hat{\lambda}_2\hat{\mathbf{v}}_2\right) = \hat{\lambda}_1\mathbf{T}\left(\hat{\mathbf{v}}_1\right) + \hat{\lambda}_2\mathbf{T}\left(\hat{\mathbf{v}}_2\right);\ \hat{\lambda}_1, \hat{\lambda}_2 \in \hat{\mathbf{R}}, \forall\hat{\mathbf{v}}_1, \hat{\mathbf{v}}_2 \in \hat{\mathbf{V}}_3. \tag{22}$$

Let $\mathbf{L}\left(\hat{\mathbf{V}}_3, \hat{\mathbf{V}}_3\right)$ be the set of tensors, then any $\hat{\mathbf{T}} \in \mathbf{L}\left(\hat{\mathbf{V}}_3, \hat{\mathbf{V}}_3\right)$, the transposed tensor denoted by $\hat{\mathbf{T}}^T$ is defined by $\hat{\mathbf{v}}_1 \cdot \left(\hat{\mathbf{T}}\hat{\mathbf{v}}_2\right) = \hat{\mathbf{v}}_2 \cdot \left(\hat{\mathbf{T}}^T\hat{\mathbf{v}}_1\right); \forall \hat{\mathbf{v}}_1, \hat{\mathbf{v}}_2 \in \hat{\mathbf{V}}_3$.

While $\forall \hat{\mathbf{v}}_1, \hat{\mathbf{v}}_2, \hat{\mathbf{v}}_3 \in \hat{\mathbf{V}}_3, Re(\langle \hat{\mathbf{v}}_1, \hat{\mathbf{v}}_2, \hat{\mathbf{v}}_3\rangle) \neq 0$, the determinant is:

$$\left\langle \hat{\mathbf{T}}\hat{\mathbf{v}}_1, \hat{\mathbf{T}}\hat{\mathbf{v}}_2, \hat{\mathbf{T}}\hat{\mathbf{v}}_3 \right\rangle = \det\hat{\mathbf{T}}\left(\hat{\mathbf{v}}_1, \hat{\mathbf{v}}_2, \hat{\mathbf{v}}_3\right). \tag{23}$$

For any vector $\hat{\mathbf{a}} \in \hat{\mathbf{V}}_3$, the associated skew-symmetric tensor will be denoted by $\hat{\mathbf{a}}\times$ and will be defined $(\hat{\mathbf{a}}\times)\hat{\mathbf{b}} = \hat{\mathbf{a}} \times \hat{\mathbf{b}}, \forall \hat{\mathbf{b}} \in \hat{\mathbf{V}}_3$.

The previous definition can be directly transposed into the following result: for any skew-symmetric tensor $\hat{\mathbf{A}} \in \mathbf{L}\left(\hat{\mathbf{V}}_3, \hat{\mathbf{V}}_3\right), \hat{\mathbf{A}} = -\hat{\mathbf{A}}^T$, a uniquely defined vector $\hat{\mathbf{a}} = vect\left(\hat{\mathbf{A}}\right), \hat{\mathbf{a}} \in \hat{\mathbf{V}}_3$ exists in order to have $\hat{\mathbf{A}}\hat{\mathbf{b}} = \hat{\mathbf{a}} \times \hat{\mathbf{b}}, \forall \hat{\mathbf{b}} \in \hat{\mathbf{V}}_3$. The set of skew-symmetric tensors is structured as a free $\hat{\mathbb{R}}$ - module of rank 3 and is isomorphic with $\hat{\mathbf{V}}_3$.

### 3.3 Differential Transform with Nilpotent Algebra

In this section will introduce a transformation that associates a hypercomplex function of a real variable to a real function of a real variable. This transformation permits a simultaneous determination of vector fields of the higher – order acceleration for a given rigid body motion.

So, being $f : \mathbb{I} \subseteq \mathbb{R} \to \underline{\mathbb{R}}, f = f(t)$, a dual function of real variable, $n$-th diferentiable, $n \in \mathbb{N}$. To this function, it will be associated the hypercomplex function of real variable $\breve{f}$ given by the following equation

$$\breve{f} = f + \varepsilon f + \cdots + \frac{\varepsilon^n}{n!} f^{(n)} = e^{\varepsilon \mathrm{D}} f \tag{24}$$

where $e^{\varepsilon \mathrm{D}} = 1 + \varepsilon \mathrm{D} + \cdots + \frac{\varepsilon^n}{n!} \mathrm{D}^n$with $\mathrm{D} = \frac{d}{dt}$the derivative operator with respect to time. The properties of this transformation are given by the following theorem:

**Theorem 2.** *Let $f$ and $g$ two real functions of class $C^n(\mathbb{I})$.*

$$\breve{\overline{f+g}} = \breve{f} + \breve{g}, \tag{25}$$

$$\breve{\overline{fg}} = \breve{f}\,\breve{g}, \tag{26}$$

$$\breve{\overline{\lambda f}} = \lambda \breve{f}, \forall \lambda \in \mathbb{R} \tag{27}$$

$$\breve{\overline{f(\alpha)}} = f\left(\breve{\alpha}\right), \alpha \in C^n(\mathbb{I}), \tag{28}$$

$$\dot{\breve{f}} = \breve{\dot{f}} \tag{29}$$

So, for $\underline{\mathbf{R}} = \underline{\mathbf{R}}(t) = (\mathbf{I} + \varepsilon_0 \boldsymbol{\rho}\times)\mathbf{R} = \mathbf{I} + (\sin\underline{\alpha})(\underline{\mathbf{u}}\times) + \left(1 - \cos\underline{\alpha}\right)(\underline{\mathbf{u}}\times)^2 = \exp \underline{\alpha}\,\underline{\mathbf{u}}\times\big) \in \underline{SO}_3, \forall t \in \mathbb{I} \subseteq \mathbb{R}$, the orthogonal dual tensor which corresponds by isomorphism to homogenous matrix $\mathbf{g} = \begin{bmatrix} \mathbf{R} & \boldsymbol{\rho} \\ 0 & 1 \end{bmatrix}$, can be defined de tensor in nilpotent algebra by differential transform given by **Theorem 2**:

$$\breve{\underline{\mathbf{R}}} = e^{\varepsilon \mathrm{D}}\underline{\mathbf{R}} = \underline{\mathbf{R}} + \dot{\underline{\mathbf{R}}}\varepsilon + \cdots + \frac{\underline{\mathbf{R}}^{(n)}}{n!}\varepsilon^n. \tag{30}$$

**Theorem 3.** *For any* $\underline{\mathbf{R}} = \underline{\mathbf{R}}(t) \in \underline{SO}_3$, *a n-th order differentiable function, the below relation takes place:*

$$\breve{\underline{\mathbf{R}}}\breve{\underline{\mathbf{R}}}^T = \breve{\underline{\mathbf{R}}}^T\breve{\underline{\mathbf{R}}} = \hat{\mathbf{I}} \tag{31}$$

$$\det\breve{\underline{\mathbf{R}}} = 1 \tag{32}$$

*If* $\underline{\mathbf{R}} = \underline{\mathbf{R}}\ \underline{\alpha}, \underline{\mathbf{u}}\big)$ *where the dual angle* $\underline{\alpha}$ *and the unit dual vector* $\underline{\mathbf{u}}$ *are the natural invariants of the orthogonal dual tensor* $\underline{\mathbf{R}}$, *takes place a Rodrigues-like formula:*

$$\breve{\underline{\mathbf{R}}} = \underline{\mathbf{I}} + \sin(\breve{\underline{\alpha}})\breve{\underline{\mathbf{u}}} \times +(1 - \cos\breve{\underline{\alpha}})(\breve{\underline{\mathbf{u}}}\times)^2 = \exp(\breve{\underline{\alpha}}\breve{\underline{\mathbf{u}}}\times) \tag{33}$$

*where:*

$$\breve{\underline{\alpha}} = \underline{\alpha} + \dot{\underline{\alpha}}\varepsilon + \ldots + \frac{\underline{\alpha}^{(n)}}{n!}\varepsilon^n, \tag{34}$$

$$\breve{\underline{\mathbf{u}}} = \underline{\mathbf{u}} + \dot{\underline{\mathbf{u}}}\varepsilon + \ldots + \frac{\underline{\mathbf{u}}^{(n)}}{n!}\varepsilon^n. \tag{35}$$

**Proof.** The proof of this theorem uses Rodrigues' formula for the tensors in $\underline{SO}_3$, $\underline{\mathbf{R}}(\underline{\alpha}, \underline{\mathbf{u}}) = \underline{\mathbf{I}} + (\sin\underline{\alpha})(\underline{\mathbf{u}}\times) + \ 1 - \cos\underline{\alpha}\big)(\underline{\mathbf{u}}\times)^2 = \exp\ \underline{\alpha}\underline{\mathbf{u}}\times\big)$, and the results of **Theorem 2**.

Using the previous theorem, if $\underline{\mathbf{R}} = (\mathbf{I} + \varepsilon_0 \boldsymbol{\rho}\times)\mathbf{R}$, the unique decomposition results:

$$\breve{\underline{\mathbf{R}}} = \left(\mathbf{I} + \varepsilon_0 \breve{\boldsymbol{\rho}}\times\right)\breve{\mathbf{R}};\ \varepsilon_0 \neq 0, \varepsilon_0^2 = 0 \tag{36}$$

After some algebra result the equations:

$$\breve{\boldsymbol{\Phi}} = \breve{\mathbf{R}}\mathbf{R}^T = \mathbf{I} + \boldsymbol{\Phi}_1\varepsilon + \ldots + \frac{\boldsymbol{\Phi}_n}{n!}\varepsilon^n \tag{37}$$

$$\breve{\mathbf{a}} = \breve{\mathbf{r}} - \breve{\mathbf{R}}\mathbf{R}^T\boldsymbol{\rho} = \mathbf{a}_1\varepsilon + \frac{\mathbf{a}_2}{2}\varepsilon^2 + \ldots + \frac{\mathbf{a}_n}{n!}\varepsilon^n \tag{38}$$

Equations (37) and (36) results from Eqs. (2), (3), and (30).

The orthogonal dual tensor $\breve{\underline{\Psi}} = \breve{\underline{\mathbf{R}}}\underline{\mathbf{R}}^T = \left(\mathbf{I} + \varepsilon_0\breve{\mathbf{a}}\times\right)\breve{\boldsymbol{\Phi}}$ encapsulated all the information about the higher-order acceleration field of order $k = \overline{1, n}$.

If $\underline{\mathbf{R}} \in \underline{S\mathbb{O}}_3$ models a helical motion around a fixed axis, then $\breve{\underline{\mathbf{u}}} = \underline{\mathbf{u}}$ and the Eq. (33) become:

$$\breve{\underline{\mathbf{R}}} = \underline{\mathbf{I}} + \sin\breve{\underline{\alpha}}(\underline{\mathbf{u}}\times) + \left(1 - \cos\breve{\underline{\alpha}}\right)(\underline{\mathbf{u}}\times)^2 = \exp(\breve{\underline{\alpha}}\underline{\mathbf{u}}\times) \tag{39}$$

In this case, from Eqs. (37), (38) and (39) and **Theorem 3**, after some algebra, results:

$$\begin{cases} \breve{\mathbf{\Phi}} = \mathbf{I} + (\sin\Delta\breve{\alpha})\mathbf{u}\times + (1 - \cos\Delta\breve{\alpha})(\mathbf{u}\times)^2 \\ \breve{\mathbf{a}} = \Delta\breve{\mathbf{p}} - (\sin\Delta\breve{\alpha})\mathbf{u}\times\boldsymbol{\rho} - (1 - \cos\Delta\breve{\alpha})\mathbf{u}\times(\mathbf{u}\times\boldsymbol{\rho}) \end{cases} \tag{40}$$

where with $\Delta\breve{\alpha}$ was denoted the multidual part of $\breve{\alpha}$. The mapping's $\sin\Delta\hat{\alpha}$ and $\cos\Delta\hat{\alpha}$ are polynomial and not transcendent, see Eq. 17 and Eq. 18, considering that $\Delta^p = 0; p \geq n+1$.

In this case ($\breve{\underline{\mathbf{u}}} = \underline{\mathbf{u}}$) orthogonal tensor $\breve{\underline{\mathbf{\Psi}}} = \breve{\underline{\mathbf{R}}}\underline{\mathbf{R}}^T$ are:

$$\breve{\underline{\mathbf{\Psi}}} = \breve{\underline{\mathbf{R}}}\underline{\mathbf{R}}^T = \exp\left(\breve{\underline{\alpha}}\underline{\mathbf{u}}\times\right)\exp(-\underline{\alpha}\underline{\mathbf{u}}\times) = \exp\left(\Delta\breve{\underline{\alpha}}\underline{\mathbf{u}}\times\right) \tag{41}$$

With Rodrigues's type formula, about Eq. (41) we obtain:

$$\breve{\underline{\mathbf{\Psi}}} = \underline{\mathbf{I}} + \sin\Delta\breve{\underline{\alpha}}(\underline{\mathbf{u}}\times) + \left(1 - \cos\Delta\breve{\underline{\alpha}}\right)(\underline{\mathbf{u}}\times)^2 \tag{42}$$

## 4 Higher-Order Kinematics of Multibody Systems

Let $C_k, k = \overline{0, m}$ a spatial kinematic chain of with m + 1 rigid bodies. The relative motion of the rigid body $C_k$ with respect to reference frame attached to $C_{k-1}$ is parameterization by the orthogonal dual tensor ${}^{k-1}\underline{\mathbf{R}}_k \in \underline{\mathbf{SO}}_3^{\mathbb{R}}$, $k = \overline{1, m}$. The relative motion of the terminal body $C_m$ with respect to reference frame attached to body frame of $C_0$ are described by the orthogonal dual tensor [4,7] by equation:

$$\underline{\mathbf{R}}_m = {}^0\underline{\mathbf{R}}_1{}^1\underline{\mathbf{R}}_2 \ldots {}^{m-1}\underline{\mathbf{R}}_m \tag{43}$$

$$\underline{\mathbf{R}}_m = exp(\underline{\alpha}_1{}^0\underline{\mathbf{u}}_1\times)exp(\underline{\alpha}_2{}^1\underline{\mathbf{u}}_2\times)\ldots exp(\underline{\alpha}_m{}^{m-1}\underline{\mathbf{u}}_m\times) \tag{44}$$

If unit dual vectors ${}^{k-1}\underline{\mathbf{u}}_k = \text{const}$, $k = \overline{1, m}$, the spatial kinematic chain is denoting general mC manipulator. The dual angles $\underline{\alpha}_p, p = \overline{1, m}$:

$$\underline{\alpha}_p = \alpha_p(t) + \varepsilon_0 d_p(t), t \in \mathbb{I} \subseteq \mathbb{R} \tag{45}$$

parametrized the relative motion of the rigid body $C_p$ with respect to body frame of $C_{p-1}$. The link between rigid bodies $C_p$ and $C_{p-1}$ is a general cylindrical joint C. In specific case on obtain revolute R, prismatic P, or helical H joints. The time functions

$\underline{\alpha}_p, p = \overline{1, m}$, in Eq. (44) are assumed to be $n$ times differentiable, $\forall t \in \mathbb{I} \subseteq \mathbb{R}$. Using differential transform on **Theorem 2** the following theorem can be proved:

**Theorem 4** *The vector fields of higher-order acceleration on the terminal body of general mC manipulator given by the kinematic mapping (44) result simultaneously from the hypercomplex orthogonal tensor*:

$$\breve{\underline{\boldsymbol{\Psi}}}_m = \exp\left[\underline{\mathbf{u}}_1 \times \Delta\breve{\underline{\alpha}}_1\right]\exp\left[\underline{\mathbf{u}}_2 \times \Delta\breve{\underline{\alpha}}_2\right]\ldots\exp\left[\underline{\mathbf{u}}_m \times \Delta\breve{\underline{\alpha}}_m\right] \tag{46}$$

*where* $\underline{\mathbf{u}}_1 = {}^0\underline{\mathbf{u}}_1$ *and:*

$$\underline{\mathbf{u}}_k = {}^0\underline{\mathbf{R}}_1 {}^1\underline{\mathbf{R}}_2 \ldots {}^{k-2}\underline{\mathbf{R}}_{k-1} {}^{k-1}\underline{\mathbf{u}}_k, k = \overline{2, m} \tag{47}$$

*is dual unit vector corresponding for instantaneous screw joint* $k$, *and* $\Delta(\widehat{\underline{\alpha}}_k)$ *denotes the multidual part of the time variable functions* $\breve{\underline{\alpha}}_k, k = \overline{1, m}$.

*Remark* The higher-order acceleration field of terminal body expressed in the body frame attached to $C_m$ are expressed by the hypercomplex orthogonal tensor:

$$\breve{\underline{\boldsymbol{\Psi}}}_m^B = \underline{\mathbf{R}}^T \breve{\underline{\boldsymbol{\Psi}}}_m \underline{\mathbf{R}} = \underline{\mathbf{R}}^T \breve{\underline{\mathbf{R}}} \tag{48}$$

By Eq. (43) and Eq. (44), after some algebra, on obtain:

$$\breve{\underline{\boldsymbol{\Psi}}}_m^B = \exp\left[\underline{\mathbf{v}}_1 \times \Delta\breve{\underline{\alpha}}_1\right]\exp\left[\underline{\mathbf{v}}_2 \times \Delta\breve{\underline{\alpha}}_2\right]\ldots\exp\left[\underline{\mathbf{v}}_m \times \Delta\breve{\underline{\alpha}}_m\right], \tag{49}$$

where unit dual vector $\underline{\mathbf{v}}_k, k = \overline{1, m}$ are:

$$\underline{\mathbf{v}}_k = [{}^{k-1}\underline{\mathbf{R}}_k {}^{k+1}\underline{\mathbf{R}}_{k+2} \ldots {}^{m-1}\underline{\mathbf{R}}_m]^{\mathrm{T}k-1}\underline{\mathbf{u}}_k \tag{50}$$

## 5 Conclusions

A general method for studying the vector field of arbitrary higher-order accelerations is described. It is proved that all information regarding the properties of the distribution of higher-order accelerations is encapsulated in the specified hypercomplex orthogonal tensor. Furthermore, higher-order kinematics properties of lower-pair serial chains with nilpotent algebra are proposed. The results interest the theoretical kinematics, higher-order kinematics analysis of a serial manipulator [6], control theory, and multibody kinematics.

## References


1. Condurache, D.: Higher-order relative kinematics of rigid body, and multibody systems. A novel approach with real and dual Lie algebras. Mech. Mach. Theory **176**, 104999 (2022)
2. Condurache, D.: Multidual algebra and higher-order kinematics. In: Pisla, D., Corves, B., Vaida, C. (eds.) EuCoMeS 2020. MMS, vol. 89, pp. 48–55. Springer, Cham (2020). https://doi.org/10.1007/978-3-030-55061-5_7

3. Condurache, D.: Higher-order kinematics of lower-pair chains with hyper-multidual algebra, ASME. In: 46th Mechanisms and Robotics Conference, Vol 7, August (2022)
4. Cohen, A., Shoham, M.: Application of hyper-dual numbers to multibody kinematics. ASME. J. Mechanisms Robotics **8**(1), 011015 (February 2016)
5. Messelmi, F.: Multidual numbers and their multidual functions. Electron. J. Math. Anal. Appl. **3**, 154–172 (2015)
6. Müller, A.: An overview of formulae for the higher-order kinematics of lower-pair chains with applications. In: Robotics and mechanism theory, Mechanism and Machine Theory, vol. 142 (2019)

# Motion Planning and Control

# An Experimental Setup to Test Time-Jerk Optimal Trajectories for Robotic Manipulators

Federico Lozer[1], Lorenzo Scalera[1], Paolo Boscariol[2], and Alessandro Gasparetto[1(✉)]

[1] University of Udine, 33100 Udine, Italy
{federico.lozer,lorenzo.scalera,alessandro.gasparetto}@uniud.it
[2] University of Padova, 36100 Vicenza, Italy
paolo.boscariol@unipd.it

**Abstract.** An experimental setup to test optimal time-jerk trajectories for robotic manipulators is presented in this paper. The setup is used, in this work, to test the execution of smooth motion profiles passing through a sequence of via-points, designed by means of the optimization of a mixed time-jerk cost function. Experimental tests are performed on a Franka Emika robot with seven degrees of freedom equipped with accelerometers to measure the motion-induced oscillations of the end-effector. The experimental results show a good agreement with the numerical tests and demonstrate the feasibility of the approach chosen for optimizing smooth trajectories for robotic manipulators.

**Keywords:** robotics · trajectory planning · vibrations · splines · jerk · Franka Emika

## 1 Introduction

Trajectory planning is a fundamental problem in robotics. Its purpose is to correlate a geometric path with a time sequence of position, velocity and acceleration to be executed by a robotic manipulator. Several approaches for trajectory planning have been investigated in the literature with the aim of enhancing productivity or other performance figures. One possible choice, which has been largely investigated in recent years, is to optimize the motion law with respect to the energy consumption of a robotic system [1].

Another interesting field of application of trajectory planning is the reduction of motion-induced oscillations. Indeed, robotic applications usually require smooth and jerk-limited trajectories during the prescribed task [2], to extend actuators lifetime, prevent mechanical damages and improve motion accuracy.

Splines are often chosen for representing an effective basis for the planning of robot trajectories with high order of continuity, sporting at the same time flexibility and a concise mathematical representation. In [3], Cook and Ho elaborated


T. Petrič et al. (Eds.): RAAD 2023, MMS 135, pp. 309–316, 2023.
https://doi.org/10.1007/978-3-031-32606-6_36

a trajectory planning method based on the "434" polynomial spline approach achieving acceleration continuity. That method was extended by Boscariol et al. [4], who presented the smoother "545" and the "5455" polynomial methods, which provide continuity up to jerk.

Smoothness, or lack thereof, is commonly regarded as the main cause of motion-induced oscillations, so its minimization is often the main goal of the motion planning. For example in [5], Fang et al. proposed a method based on S-curve point-to-point trajectories to solve a minimum-jerk optimization problem. In [6], Dai et al. experimented a genetic algorithm for minimum-jerk trajectories calculated in the configuration space of a UR3 robot. In [7], Wu et al. solved a constrained optimization problem for the minimum-jerk trajectories calculated with NURBS curves. In that case, jerk is reduced by using a cost function that is equal to the sum of the integrals of squared jerk of each robot joint.

The minimum-time calculation is also a focal point of trajectory planning, since a fast trajectory can increase the amount of tasks performed over a fixed amount of time. In this context, Abu-Dakka et al. proposed a method based on the PPGA1/PPGA2 genetic algorithm to calculate minimum-time trajectories using cubic splines [8]. Furthermore, in [9], Palleschi et al. used minimax method to calculate minimum-time $6^{th}$-order B-spline trajectories.

Most other works, like [10] by Gasparetto et al., investigated the optimization of a cost function that considers contributions which are proportional to both time and squared jerk, due to the need to have at the same time a fast and a smooth trajectory, proposing a mixed cost function that sets a trade-off between those two terms. Furthermore, Zanotto et al. in [11] have shown, through extensive experiments, the effectiveness of that approach by means of a comparison with the outcomes of a standard spline algorithm with imposed timing. The same optimum problem was tested on a SCARA robot by Huang et al. [12], solving it with a genetic algorithm (`NSGA-II`) and comparing their results to those of Gasparetto. However, to the best of our knowledge, that method has not been yet tested on a manipulator with 6 or 7 degrees of freedom (DOFs).

In this work, we present an experimental setup to test time-jerk optimal trajectories for robotic manipulators. We implement an optimization method for the planning of smooth trajectories based on a cost function composed of a term proportional to the total execution time, and one term proportional to the squared jerk along the trajectory. The setup comprises a Franka Emika robot with 7 DOFs equipped with three accelerometers to measure the motion-induced oscillations of the end-effector. The experimental results show a good agreement with the numerical tests and demonstrate the feasibility of the approach in optimizing smooth trajectories for robotic manipulators.

The paper is organized as follows: in Sect. 2 the trajectory planning approach is presented. Section 3 describes experimental setup, whereas Sect. 4 reports the experimental results. Finally, the conclusions are given in Sect. 5.

## 2 Time-Jerk Optimal Trajectories

The planning of time-jerk optimal trajectories aims at defining motion laws as smooth as possible and, at the same time, as fast as possible. Limiting jerk values produces smooth movements but leads to a slow trajectory. On the other hand, decreasing the duration of the trajectory produces higher levels of jerk. The key point is to find a suitable trade-off between a smooth and a fast trajectory.

The proposed algorithm is based on a parametrization of the motion profile by means of a set of time intervals $h_i$: each one of them representing the duration of the $i$-th segment of the trajectory, which connects the via-point $i$ to the via-point $i+1$. The set of $h_i$ with $1 \leq i \leq w_p$ (number of via points), i.e., the vector $\mathbf{h}$, is the decision variable vector, whose values are set by means of the minimization of the cost function in Eq. (1). The latter weighs both the sum of $h_i$, which equals the total duration $T$, and the sum of the $N$ values of the integral of joint jerks, being $N$ the number of robot joints. The trade off between the two elements of the cost function is set by $\alpha$, which is assumed to vary between 0 and 1.

$$\min_{\mathbf{h}} \alpha \sum_{i=1}^{w_p-1} h_i + (1-\alpha) \sum_{j=1}^{N} \int_0^T (\dddot{q}_j(t))^2 dt \tag{1}$$

The feasibility of the resulting trajectory $(\boldsymbol{q}(t), \ddot{\boldsymbol{q}}(t), \dddot{\boldsymbol{q}}(t))$ is ensured by the following constraints:

$$\begin{cases} \boldsymbol{q}_{min} < \boldsymbol{q} < \boldsymbol{q}_{max} & \boldsymbol{q}_{min}, \boldsymbol{q}_{max} \text{ lower and upper position bounds} \\ |\dot{\boldsymbol{q}}| < \dot{\boldsymbol{q}}_{max} & \dot{\boldsymbol{q}}_{max} \text{ velocity bound} \\ |\ddot{\boldsymbol{q}}| < \ddot{\boldsymbol{q}}_{max} & \ddot{\boldsymbol{q}}_{max} \text{ acceleration bound} \\ |\dddot{\boldsymbol{q}}| < \dddot{\boldsymbol{q}}_{max} & \dddot{\boldsymbol{q}}_{max} \text{ jerk bound} \\ |\boldsymbol{\tau}| < \boldsymbol{\tau}_{max} & \boldsymbol{\tau}_{max} \text{ torque bound} \\ |\dot{\boldsymbol{\tau}}| < \dot{\boldsymbol{\tau}}_{max} & \dot{\boldsymbol{\tau}}_{max} \text{ torque derivative bound} \end{cases} \tag{2}$$

The constrains in Eq. (2) include kinematic quantities, as well as joint torques, the latter being computed on the basis of the inverse dynamic model of the manipulator. In particular, the torques $\boldsymbol{\tau}$ are computed as:

$$\boldsymbol{\tau} = \boldsymbol{M}(\boldsymbol{q})\ddot{\boldsymbol{q}} + \boldsymbol{C}(\boldsymbol{q}, \dot{\boldsymbol{q}})\dot{\boldsymbol{q}} + \boldsymbol{F}_v\dot{\boldsymbol{q}} + \boldsymbol{f}_c\text{sign}(\dot{\boldsymbol{q}}) + \boldsymbol{g}(\boldsymbol{q}) \tag{3}$$

where $\boldsymbol{M}(\boldsymbol{q})$ is the mass matrix of the robot, and $\boldsymbol{C}(\boldsymbol{q}, \dot{\boldsymbol{q}})\dot{\boldsymbol{q}}$ considers the Coriolis and centrifugal terms. $\boldsymbol{F}_v$ and $\boldsymbol{f}_c$ are the viscous and Coulomb friction terms, respectively, whereas $\boldsymbol{g}(\boldsymbol{q})$ accounts for gravity. The optimization algorithm has been tested with a "434" spline [3,4], in which a $3^{rd}$ order polynomial function is used to describe all the segments between the via-points, with the exception of the first and the last one, where a $4^{th}$ polynomial is adopted, but the method can be extended to more elaborated spline functions as well. The trajectory between two adjacent via-points $P_k$ and $P_{k+1}$ with $2 \leq k \leq N-2$ can be expressed with the following function of time:

$$F_k(t) = B_{k,1} + B_{k,2}t + B_{k,3}t^2 + B_{k,4}t^3 \tag{4}$$

where $B_{k,n}$ are the polynomial coefficients of the polynomial equation. Instead, the trajectory between the first two via-points $P_1$ and $P_2$ and the last two way points $P_{N-1}$ and $P_N$ (i.e., $k = 1, N-1$) is given by:

$$F_k(t) = B_{k,1} + B_{k,2}t + B_{k,3}t^2 + B_{k,4}t^3 + B_{k,5}t^4 \tag{5}$$

The polynomial coefficients $B_{k,i}$ are computed to ensure the continuity of positions, speeds and accelerations. Jerk continuity is not however achieved, as jerk is either linearly changing or constant during each segment [3].

## 3 Experimental Setup

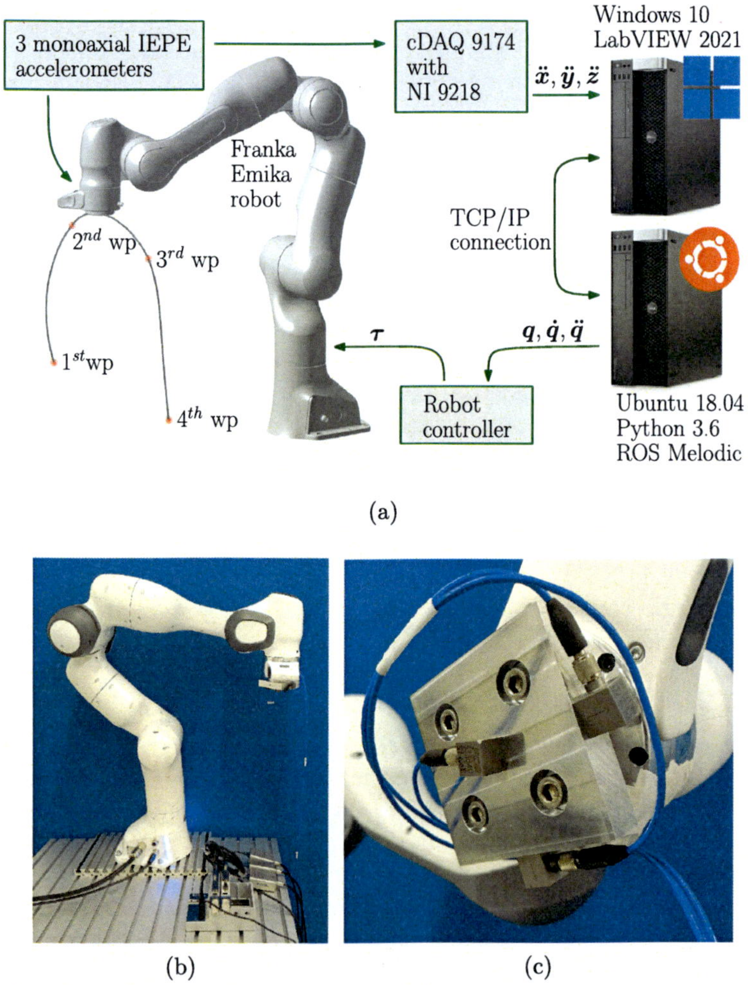


**Fig. 1.** (a) Overview of the experimental setup; (b) Franka Emika robot and data acquisition system; (c) aluminium flange and accelerometers on the end-effector.

The optimization problem in Eq. (1) with (2) is solved by a sequential quadratic programming algorithm using the `fmincon` Matlab function. To evaluate the method, 100 solutions have been run by varying $\alpha$ between 0 and 1 in equal steps. Each resulting trajectory has then been executed three times each on a Franka Emika robot with 7 DOFs (Fig. 1). The end-effector of the manipulator is equipped with a custom aluminium flange on which three monoaxial piezoelectric accelerometers are rigidly attached to. Data from accelerometers is acquired with a sampling frequency of 10 kHz by means of two a National Instruments NI9218 C Series modules, and a cDAQ 9174 chassis. Before post processing, raw acceleration data are filtered in Matlab with a moving-average filter that averages data over 5 samples.

The tests are carried out using two computers, one with Windows 10 operating system and one with Ubuntu 18.04: the two can exchange data using the TCP/IP protocol (Fig. 1). The workstation running Windows 10 is used for generating trajectories, for the data acquisition in LabVIEW 2021 and for the data analysis and post-processing in Matlab. The workstation running Ubuntu is used to control the robot by means of ROS Melodic Morenia and Python 3.6. The trajectories are executed on the Franka arm with a Python program that builds a message for the robot controller, specifying the sequence of joint position, velocity and acceleration which are sampled at 100 Hz.

**Table 1.** Results of Test 1 for some values of $\alpha$.

| $\alpha$ | $T$ $[s]$ | $J$ $[rad^2/s^5]$ | $\ddot{x}_{rms}$ $[m/s^2]$ | $\ddot{y}_{rms}$ $[m/s^2]$ | $\ddot{z}_{rms}$ $[m/s^2]$ | $\ddot{X}_{rms}$ $[m/s^2]$ |
|---|---|---|---|---|---|---|
| 0.20 | 12.07 | 0.60 | 0.23 | 0.28 | 0.18 | 0.40 |
| 0.40 | 10.25 | 1.37 | 0.27 | 0.32 | 0.20 | 0.46 |
| 0.60 | 8.95 | 2.69 | 0.31 | 0.36 | 0.21 | 0.52 |
| 0.80 | 7.60 | 6.09 | 0.34 | 0.43 | 0.25 | 0.56 |
| 1.00 | 2.17 | $3.51 \cdot 10^3$ | 1.14 | 1.76 | 1.19 | 2.40 |

## 4 Experimental Results

Figure 2 shows an example of trajectory, which has been evaluated for $\alpha = 0.5$. The figure reports the joint positions, velocities, and absolute acceleration over time. The overall acceleration is computed by the vector sum of the acceleration signals measured by each mono-axial accelerometer. Vertical solid lines in Fig. 2 indicate the passing on via-points.

Examples of measured data obtained by varying the weight $\alpha$ are reported in Table 1. In particular, total time $T$, sum of integral of squared joint jerks $J$, root mean square Cartesian accelerations ($\ddot{x}_{rms}$,$\ddot{y}_{rms}$,$\ddot{z}_{rms}$) and absolute root mean square acceleration $\ddot{X}_{rms}$ are reported for some values of $\alpha$. The data in Table 1 clearly shows that the relative weight $\alpha$ allows the user to set the desired trade-off between the total execution time and the acceleration measured on the end-effector of the robot.

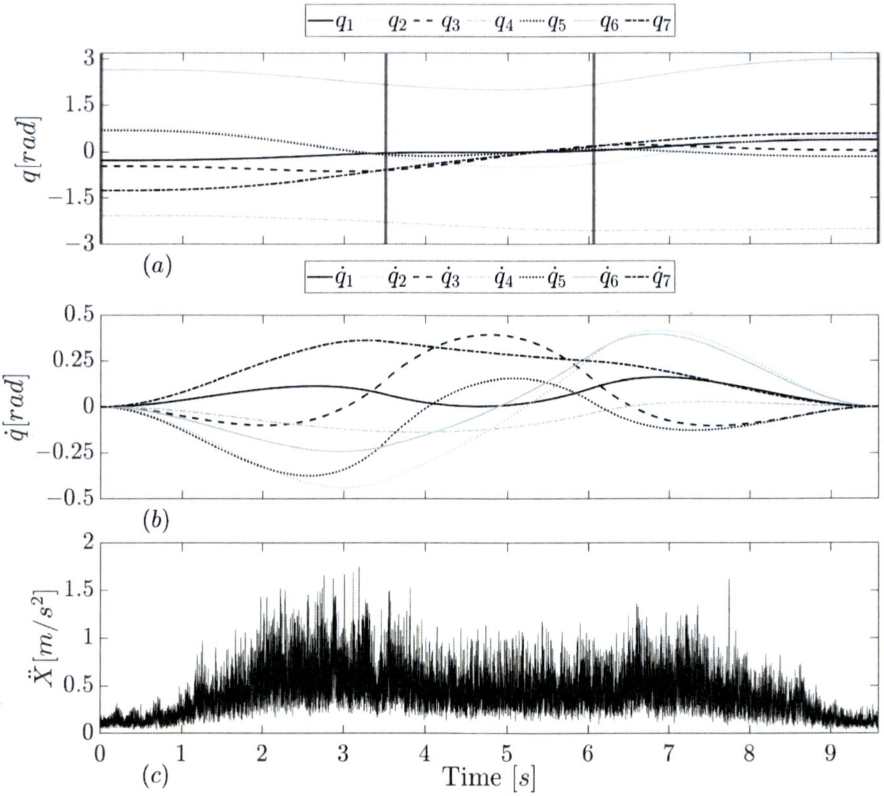


**Fig. 2.** Example trajectory: joint position, velocity and absolute acceleration signal for $\alpha = 0.5$. Vertical solid lines indicate target via-points.

Figure 3(a) shows the trend of $T$ and $J$ as a function of $\alpha$. Since $\alpha$ is increasing, $T$ is decreasing, presenting an inflection point approximately for $\alpha = 0.5$. On the contrary, $J$ is increasing over $\alpha$ with a parabolic trend. In fact, starting from low values of $\alpha$ and sweeping to higher values, the algorithm moves from jerk minimization to minimum time trajectory optimization, as in Eq. 1. Figure 3(b) shows $J$ as a function of $T$. As previously mentioned, these two quantities are inversely proportional to each other.

Finally, Fig. 3(c) shows $\ddot{X}_{rms}$ as a function of the total execution time $T$ for the three repeated tests. The resulting RMS accelerations are not only inversely proportional to the total execution time, but they are also closely related to the chosen metric, i.e. the integral of squared jerk $J$. Furthermore, the results of the three repeated tests reported in Fig. 3(c) show very close results throughout the tree repetitions of the same test, testifying a high repeatability of the experimental tests and of the acceleration measurements.

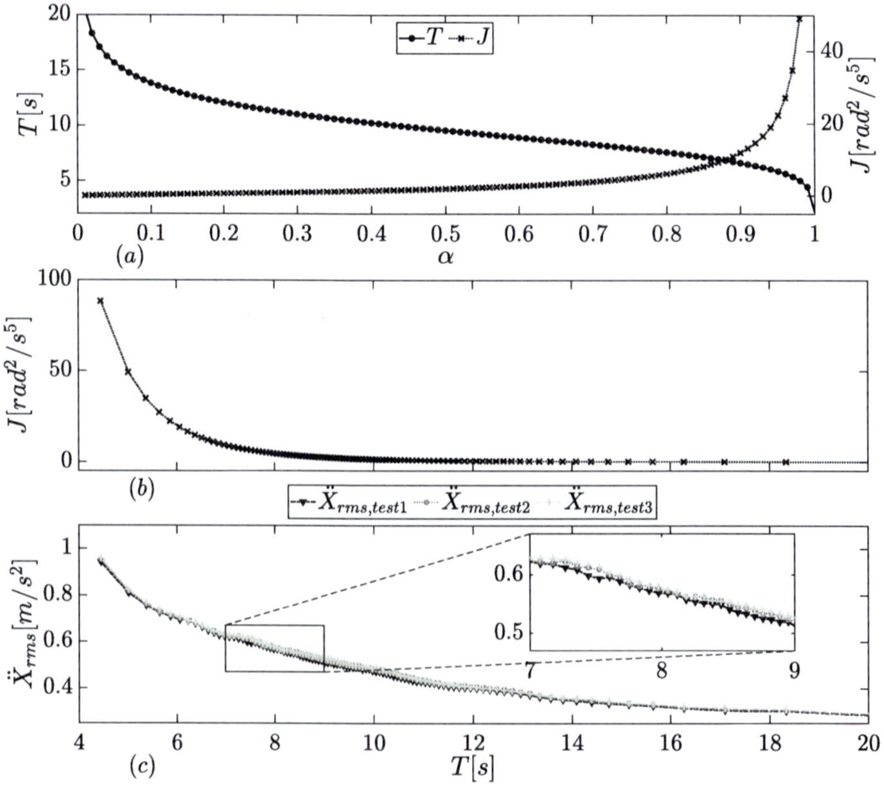


**Fig. 3.** (a) $T$ and $J$ vs. $\alpha$; (b) $J$ vs. $T$; (c) $\ddot{X}_{rms}$ vs. $T$.

## 5 Conclusions

In this paper an experimental setup to test time-jerk optimal trajectories for robotic manipulators has been presented. The experimental setup comprises a Franka Emika robot with 7 DOFs equipped with three accelerometers which measure the oscillations of the end-effector. The setup has been used for testing the smooth trajectories which were defined upon a "434" planning algorithm, by means of an optimization problem which minimizes a mixed cost function which weighs the total execution time as well as the integral of squared joint jerk. The experimental results shown a good agreement with the numerical results, demonstrating the feasibility of the proposed approach in optimizing smooth trajectories for robotic manipulators as well as proving the accuracy and the repeatability of the proposed experimental setup. Future works will include the exploitation of redundancy in the planning of time-jerk optimal trajectories, and the real-time computation of minimum-jerk trajectories for collaborative robotics, according to the approach presented in [13].

**Acknowledgments.** This research was supported by the Laboratory for Big Data, IoT, Cyber Security (LABIC) funded by Friuli Venezia Giulia, and the Laboratory for Artificial Intelligence for Human-Robot Collaboration (AI4HRC) funded by Fondazione Friuli. We thank LAMA FVG for the fabrication of the aluminium flange.

## References

1. Carabin, G., Scalera, L.: On the trajectory planning for energy efficiency in industrial robotic systems. Robotics **9**(4), 89 (2020)
2. Trigatti, G., Boscariol, P., Scalera, L., Pillan, D., Gasparetto, A.: A look-ahead trajectory planning algorithm for spray painting robots with non-spherical wrists. In: Gasparetto, A., Ceccarelli, M. (eds.) MEDER 2018. MMS, vol. 66, pp. 235–242. Springer, Cham (2019). https://doi.org/10.1007/978-3-030-00365-4_28
3. Cook, C.C., Ho, C.Y.: The Application of Spline Functions to Trajectory Generation for Computer-Controlled Manipulators, pp. 101–110. Springer, Cham (1984)
4. Boscariol, P., Gasparetto, A., Vidoni, R.: Planning continuous-jerk trajectories for industrial manipulators. In: Engineering Systems Design and Analysis, vol. 44861, pp. 127–136 (2012)
5. Fang, Y., Jie, H., Liu, W., Shao, Q., Qi, J., Peng, Y.: Smooth and time-optimal s-curve trajectory planning for automated robots and machines. Mech. Mach. Theory **137**, 127–153 (2019)
6. Dai, C., Lefebvre, S., Yu, K.M., Geraedts, J.M., Wang, C.C.: Planning jerk-optimized trajectory with discrete time constraints for redundant robots. IEEE Trans. Autom. Sci. Eng. **17**(4), 1711–1724 (2020)
7. Wu, G., Zhang, S.: Real-time jerk-minimization trajectory planning of robotic arm based on polynomial curve optimization. Proc. Inst. Mech. Eng. Part C: J. Mech. Eng. Sci. **236**, 10852–10864 (2022)
8. Abu-Dakka, F.J., Assad, I.F., Alkhdour, R.M., Abderahim, M.: Statistical evaluation of an evolutionary algorithm for minimum time trajectory planning problem for industrial robots. Int. J. Adv. Manuf. Technol. **89**(1), 389–406 (2017)
9. Palleschi, A., Garabini, M., Caporale, D., Pallottino, L.: Time-optimal path tracking for jerk controlled robots. IEEE Rob. Autom. Lett. **4**(4), 3932–3939 (2019)
10. Gasparetto, A., Zanotto, V.: A technique for time-jerk optimal planning of robot trajectories. Rob. Compt.-Int. Manuf. **24**(3), 415–426 (2008)
11. Zanotto, V., Gasparetto, A., Lanzutti, A., Boscariol, P., Vidoni, R.: Experimental validation of minimum time-jerk algorithms for industrial robots. J. Intell. Rob. Syst. **64**(2), 197–219 (2011)
12. Huang, J., Pengfei, H., Kaiyuan, W., Zeng, M.: Optimal time-jerk trajectory planning for industrial robots. Mech. Mach. Theory **121**, 530–544 (2018)
13. Scalera, L., Giusti, A., Vidoni, R., Gasparetto, A.: Enhancing fluency and productivity in human-robot collaboration through online scaling of dynamic safety zones. Int. J. Adv. Manuf. Technol. **121**(9), 6783–6798 (2022)

# A Force-Based Formation Synthesis Approach for the Cooperative Transportation of Objects

Mario Rosenfelder(✉), Henrik Ebel, and Peter Eberhard

Institute of Engineering and Computational Mechanics, University of Stuttgart, Stuttgart, Germany
{mario.rosenfelder,henrik.ebel,peter.eberhard}@itm.uni-stuttgart.de

**Abstract.** Increasing demands in automation entail that some tasks either cannot be accomplished by a single robotic manipulator or it is economically not meaningful to indiscriminately increase the robot's size or sophistication. At the same time, ongoing technological progress has paved the way for an increased usage of cost-effective robotic agents. Hence, the question whether the cooperation of multiple simple robotic agents is expedient to accomplish automation tasks is raised. Possible benefits are manifold, e.g., an increased versatility, resilience regarding failure, and the ability to solve more complex and larger tasks. A meaningful benchmark problem in distributed robotics is the cooperative object transportation where mobile robots organize around the object to move it by pushing. Thus, this paper deals with the question of how to allocate robotic agents around the object. Novelly, the problem is treated using second-order dynamics, explicitly including and limiting the exerted force of the agents. We derive a description of the manipulation space in terms of a zonotope which is useful for robotic manipulation beyond the scope of this benchmark problem. The proposed scheme's versatility and functionality is demonstrated by experimental results.

**Keywords:** Cooperative transportation · Force control · Grasp map

## 1 Introduction

Robotic automation has tremendously impacted several branches of industry over the last few decades, where the driving forces behind productivity increases have mainly been stand-alone robotic manipulators. To deal with the expanding set of automation tasks, cutting edge technologies have been utilized to push the bounds of these manipulators' possibilities yielding high-end, complex systems, often tailored for one specific task. However, e.g., in logistics, utilizing

This work is supported by the Deutsche Forschungsgemeinschaft (DFG, German Research Foundation) under Grant 433183605 and through Germany's Excellence Strategy (Project PN4-4 Theoretical Guarantees for Predictive Control in Adaptive Multi-Agent Scenarios) under Grant EXC2075-390740016.


T. Petrič et al. (Eds.): RAAD 2023, MMS 135, pp. 317–324, 2023.
https://doi.org/10.1007/978-3-031-32606-6_37

a cooperating group of simple robots rather than an extremely sophisticated individual robot might be beneficial in view of several criteria. While the advantages of a networked robotic system have been known for a long time, progress in cost-efficient processing units has boosted the interest in cooperative robotics recently; not only in scientific research but also in industry.

A meaningful benchmark problem for distributed robotics is the cooperative object transportation, as it incorporates several aspects which are crucial for networked systems, such as inter-robot communication, decentralized control, and self-coordination. While many previous solution approaches show some limitations , see, e.g., the overview given in [9], the scheme proposed in [2,3] stands out with its versatility regarding the object to be transported, the number of participating robots, and its autonomous reallocation of the agents. However, the scheme rests upon a geometric interpretation of the transportation problem and does not explicitly consider the exerted forces for organization and motion planning. The naturally applied, spatially distributed forces are nevertheless one major advantage over a single robot, enabling the group to manipulate large or heavy objects while maintaining preciseness.

To this end, the position-based scheme [2] has recently been extended to explicitly measure and govern the interacting forces between the agents and the object [7], although in simulations only. As a natural continuation, the present paper refines this force-based, i.e., based on second-order differential equations, consideration. In particular, the manipulation space spanned by the robotic agents is formulated in a novel way. The obtained closed set has the crucial advantage that it intrinsically takes the agents' physical limitations into account. This precise formulation is generic and may be used in distributed robotics beyond the scope of the given benchmark problem. Furthermore, constituting the main part of the present paper, this description is used to formulate an optimization-based formation synthesis problem allocating the robots around the object. Moreover, to the best of the authors' knowledge, it is the first time that the task is accomplished also in hardware experiments by explicitly measuring and governing the exerted forces. Further, coping with the benchmark problem by means of this second-order consideration is a step towards more generic problems in robotics such as general manipulation tasks since robotics in nature is about automating mechanical work such that the exerted forces represent the centerpiece.

This contribution is structured as follows. The problem setup and the considered mobile robots are introduced in Sect. 2. In Sect. 3, the formation synthesis is introduced for the force-based consideration of the manipulation task constituting the main contribution. The functionality of the proposed scheme is demonstrated in Sect. 4 before Sect. 5 gives a summary and an outlook.

## 2 Problem Description

As in our previous work [2,3], the object to be manipulated shall be any rigid body of polygonal, possibly non-convex, shape and not subject to any kinematic constraint. Further, the employed mobile robots shall be omnidirectional with a circular footprint of radius $r_\mathrm{r} \in \mathbb{R}_{>0}$. Crucially, each robot is equipped with one unidirectional force-sensing unit measuring the contact force between the object and the robot along the sensor's direction.

It is evident that the proposed scheme is a non-prehensile approach, i.e., the robots do not grasp the object such that transportation relies only on pushing forces. Thus, to transport the introduced broad class of objects along an arbitrary path, reorganization of the mobile robots is indispensable. This reorganization task, as well as all other control decisions, is solved in a rigorously distributed fashion avoiding a single link of failure but also requiring explicit communication among the agents. Particularly, each robot is capable of sending messages to and retrieving messages from other robots using a shared wireless network.

In contrast to the previous, geometrically-inspired work, the current manipulation objective shall not be quantified in terms of the object's current translational and rotational error retrieved from the path parametrization, but rather by a desired planar pushing force $\boldsymbol{f}_{\mathrm{t,des}} \in \mathbb{R}^2$ and a torque $\tau_{\mathrm{des}} \in \mathbb{R}$, which is subsumed to the desired generalized force $\boldsymbol{f}_\mathrm{d}^\top := \begin{bmatrix} \boldsymbol{f}_{\mathrm{t,des}}^\top & \tau_{\mathrm{des}} \end{bmatrix} \in \mathbb{R}^3$ exerted on the object. Notably, the more general formulation in terms of a desired force is not restricted to the considered planar transportation task but can in fact be transferred to an arbitrary, distributed manipulation of a rigid body in the space. Note that the manipulation objectives are usually described in the object-fixed frame of reference $\mathcal{K}_\mathrm{S}$ which, notation-wise, is denoted in the form ${}^\mathrm{S}\boldsymbol{f}_\mathrm{d}$ .

It is worth stressing that the underlying hybrid force-position-controller is not within the scope of this paper. The interested reader is referred to [7, Sect. IV] for an earlier, although conceptually similar, controller design in that regard. Instead, the focus of the present paper is to provide an experimentally validated blueprint on how $N \in \mathbb{Z}_{\geq 2}$ agents can cooperatively choose a favorable formation to manipulate an object based on a force-based, i.e., second-order consideration.

## 3 Force-Based Formation Synthesis

To push and rotate the object as desired, it is crucial to deploy the mobile robots around the object such that they are capable of cooperatively exerting ${}^\mathrm{S}\boldsymbol{f}_\mathrm{d}$. In a first step, it is thus essential to describe all feasible positions around the object in a mathematically sound manner. As in [2,7], the robots' center will be placed with a distance $d \in \mathbb{R}_{>r_\mathrm{r}}$ to the object edge such that the force sensor is just in contact with the corresponding edge, see Fig. 1. Moreover, the obtained dilated edges are clipped by a safety margin at pointed and concave corners yielding the closed set $\mathcal{I}_\mathrm{e} \subset \mathbb{R}_{\geq 0}$ of 1D-positions around the object. The cut-off at concave corners, which lets each robot have only one contact point with the object, can be seen as one significant drawback compared to the position-based scheme [2].

Finally, the formation synthesis problem consists of assigning $N$ positions $w_i \in \mathcal{I}_\mathrm{e}$, $i \in \{1, \ldots, N\}$, to execute the current objective ${}^\mathrm{S}\boldsymbol{f}_\mathrm{d}$. Summarizing the individual 1D-positions yields the configuration vector $\boldsymbol{w} := \begin{bmatrix} w_1 \cdots w_N \end{bmatrix}^\top \in \mathbb{R}_{\geq 0}^N$, which is used to investigate how the object can be manipulated by the

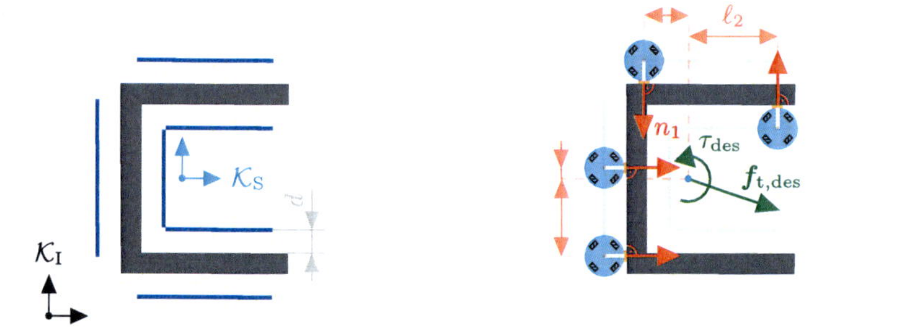


(a) edges of the dilated object (b) contact normals and lever arms

**Fig. 1.** Illustration of a polygonal object (dark gray), its dilated edges (dark blue), and an exemplary placement of the mobile robots around the object.

chosen formation. Particularly, for the contact points following from $\boldsymbol{w}$, it is vital to quantify the edge's normalized inner normal vectors ${}^{\mathrm{S}}\boldsymbol{n}_i$ and the resulting lever arms $\ell_i$ w.r.t. the object's center of mass by means of the matrix ${}^{\mathrm{S}}\boldsymbol{Z}(\boldsymbol{w}) \in \mathbb{R}^{2\times N}$ and the vector $\boldsymbol{\nu}(\boldsymbol{w}) \in \mathbb{R}^N$, respectively, see Fig. 1(b). These two quantities uniquely describe the force transmission properties of the configuration. Encapsulating both quantities forms the so-called grasp map $\boldsymbol{G}(\boldsymbol{w}) = \left[{}^{\mathrm{S}}\boldsymbol{Z}(\boldsymbol{w})^\top \; \boldsymbol{\nu}(\boldsymbol{w})\right]^\top \in \mathbb{R}^{3\times N}$ which is widely used in robotics manipulation and models of grasp analysis, see, e.g., [6]. Beyond that, the grasp map has also been used to characterize possible configurations for the non-prehensile transportation task [1]. However, to the best of the authors' knowledge, its connection to the spanned manipulation space and explicitly using it to derive favorable formations from a descriptive point of view is novel.

Each column $\boldsymbol{g}_i$ of the grasp map describes the manipulation characteristics of the related $i$th agent in the three-dimensional manipulation space, where the first two dimensions correspond to the translational force vector in the plane and the third dimension deals with the torque around the positive $z$-axis. Consequently, the manipulation space of the configuration $\boldsymbol{w}$ follows from the Minkowski sum of all individual line segments $\bar{f}_i\,\boldsymbol{g}_i$, where the maximum available pushing force $\bar{f}_i \in \mathbb{R}_{>0}$ explicitly takes robot $i$'s limitations into consideration. In this context, with the generators $\bar{f}_i\,\boldsymbol{g}_i$, the resulting manipulation space $\mathcal{G}(\boldsymbol{w}) \subset \mathbb{R}^3$ is a zonotope [10, Ch. 7.3], i.e., a convex, centrally-symmetric polytope. Figure 2(a) exemplarily shows the constructed zonotope $\mathcal{G}$ for the configuration $\boldsymbol{w}$ displayed in Fig. 1(b). In contrast to the purely geometrical considerations based on the generated open cones [3], this constitutes a more precise description of the configuration's actual manipulation potential. For instance, it takes into account the increasing pushing capability if the inner normal vectors of multiple robots have shares in a common direction which suits the force-based consideration. From another point of view, the obtained zonohedral manipulation space, meaning a three-dimensional zonotopic set, can be interpreted as a projection of an $N$-dimensional hypercube $\mathcal{F}$, where edge $i$ is of length $\bar{f}_i$, to the three-dimensional manipulation space using $\boldsymbol{G}(\boldsymbol{w})$ as the projection matrix.

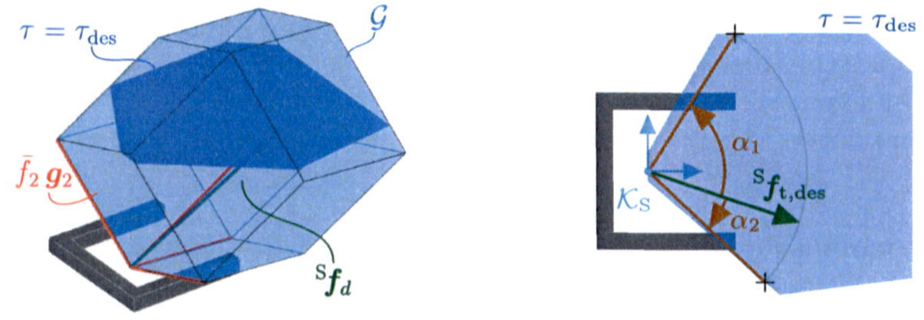


(a) zonohedral manipulation space (b) two-dimensional projection at $\tau_{\text{des}}$

**Fig. 2.** Obtained manipulation space $\mathcal{G}$ and translational robustness measure.

The previous, mechanically motivated zonohedral manipulation space can be utilized to formulate an optimization problem to deal with the formation synthesis. Optimization-based schemes have proven their value in previous works [2,3] since they can be formulated in a descriptive manner and are rather intuitive to tune. The main requirement for a valid configuration is that it accounts for the current manipulation objective. This requires that the objective is contained in the zonohedron, ${}^{\text{S}}\boldsymbol{f}_{\text{d}} \in \mathcal{G}(\boldsymbol{w})$, meaning that there exists a set of pushing forces $\tilde{\boldsymbol{f}}_{\text{p}} := \begin{bmatrix} f_1 \cdots f_N \end{bmatrix}^\top$ exerted by the robots such that the resulting force on the object is equal to the desired force, i.e., $\boldsymbol{G}(\boldsymbol{w})\tilde{\boldsymbol{f}}_{\text{p}} = {}^{\text{S}}\boldsymbol{f}_{\text{d}}$. On top of this, the zonohedron's shape and the location of ${}^{\text{S}}\boldsymbol{f}_{\text{d}}$ in $\mathcal{G}(\boldsymbol{w})$ allow to formulate descriptive robustness measures. To obtain a translational robustness measure, the zonohedron's projection at $\tau = \tau_{\text{des}}$ is investigated, see Fig. 2. Particularly, the intersection of the resulting convex polygon and a circle with radius $|{}^{\text{S}}\boldsymbol{f}_{\text{t,des}}|$ around $\mathcal{K}_{\text{R}}$'s origin is considered. The resulting robustness measure is formulated in terms of the minimal angle of the intersection points w.r.t. the direction of the desired force, i.e., $J_{\text{t}} = \min\{|\alpha_1|, |\alpha_2|\}$. Analogously, for rotational robustness purposes, the minimal distance from ${}^{\text{S}}\boldsymbol{f}_{\text{d}}$ to one of the facets in the manipulation space's $\tau$-direction is taken into account by the measure $J_{\text{r}}$. To formulate the formation synthesis problem as an optimization problem, necessary requirements for valid formations are enforced through constraints while preferable properties are represented through the cost function. This leads to

$$\underset{\boldsymbol{w}\in\mathbb{R}^N_{\geq 0}}{\text{minimize}} \quad -c_1 J_{\text{t}}(\boldsymbol{w}) - c_2 J_{\text{r}}(\boldsymbol{w}) \tag{1a}$$

$$\text{subject to} \quad \exists \tilde{\boldsymbol{f}}_{\text{p}} \in \mathcal{F} \text{ satisfying } \boldsymbol{G}(\boldsymbol{w})\tilde{\boldsymbol{f}}_{\text{p}} = {}^{\text{S}}\boldsymbol{f}_{\text{d}} \tag{1b}$$

$$\text{frD}\left({}^{\text{S}}\boldsymbol{E}(w_i), {}^{\text{S}}\boldsymbol{E}(w_j)\right) \geq d_{\min} \quad \forall i \neq j \tag{1c}$$

with $c_1, c_2 \in \mathbb{R}_{>0}$ and $d_{\min} \in \mathbb{R}_{>2r_{\text{r}}}$. Condition (1c) additionally ensures that the free Euclidean distance (frD) between the robots is large enough in order to avoid collisions. Further, the edges' lengths $\bar{f}_i$ of the hypercube $\mathcal{F}$ take the robot's pushing limits into account. Although the derivation of this optimization problem stems from a descriptive mechanical point of view, the problem is fairly intricate since it is non-convex and the objective function is a discontinuous function of the optimization variable $\boldsymbol{w}$ due to the polygonal object shape. Thus, a distributed augmented Lagrangian particle swarm optimization algorithm developed in [2] is utilized to solve (1). Once a feasible solution of (1) is found, the robots self-reliantly allot themselves to the positions around the object using the negotiation scheme described in [2, Sect. 5.2.2]. Moreover, the time-critical task of steering the robots to their desired position as well as controlling the exertion of force on the object is handled by a decoupled second program employed for each robot. The conceptual strategy of the hybrid controller governing the robots' relative postures w.r.t. the object and the desired forces to be exerted can be found in [7, Sect. IV].

## 4 Experimental Results

To substantiate the functionality and the versatility of the scheme, it is useful to investigate some case studies. This contribution refrains from investigating simulations, see [7] for simulation results, but rather looks at hardware experiments. This implies that the considered scenarios are limited w.r.t. the number of participating robots, the object scale, and the path length due to physical constraints of the available laboratory equipment. However, the experiments establish that the proposed scheme is conceptually suitable for the given transportation task.

The omnidirectional character of the robots is obtained by four circularly-arranged Mecanum wheels. Further, each robot is equipped with a self-designed force-sensing unit consisting of a compact compression load cell of type TE Connectivity FX29 as well as two leaf-springs to filter the normal force and to obtain a preload to avoid chattering, see [7, Fig. 1] for a photography. The sensor's data is processed onboard within the low-level hybrid controller running 100 Hz on a BeagleBone Blue board. This board is also concerned with the communication in the network using the LCM library [5]. Crucially, for the asynchronously running formation agents, the control logic (10 Hz), and the onboard low-level controller, individual program instances run for each robot. Apart from communication, these programs run fully separated from the other programs and robots. The first two sub-programs are implemented in C++ and run on external hardware to keep the mobile robots lightweight and cost-efficient.

In the first experiment, four robots shall transport a ⊔-shaped object along cornered line segments. The desired force ${}^{S}\boldsymbol{f}_{d}$ to transport it along the path is chosen to be proportional to the current pose deviation of the object retrieved from the path parametrization. The obtained performance is shown in Fig. 3(a) at three time instants, where the desired path is depicted as a black dashed line and the actual position of the object's center of mass is portrayed by the orange line. The robots' orientations and their exerted forces are indicated by white and red rectangles, respectively. Evidently, the formation synthesis sub-program utilizes two robots to push in the same direction in order to exert enough force to accomplish the given task. This behavior is intrinsically covered by the manipulation space's description as a closed set in the form of a zonohedron which highlights the advantages compared to purely geometrical formulations. Figure 4 displays the desired and actually exerted forces for two robots in an exemplary time interval, where the overlayed experimental photographs indicate the corresponding robots and the configuration. As given there, governing the desired forces is a

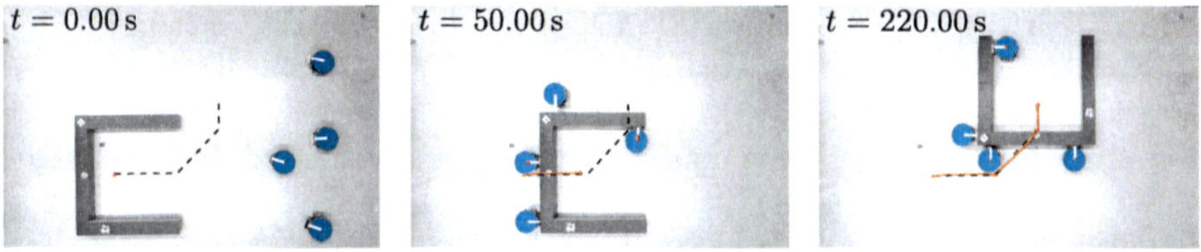


(a) transportation of ⊔-shaped object along a cornered path

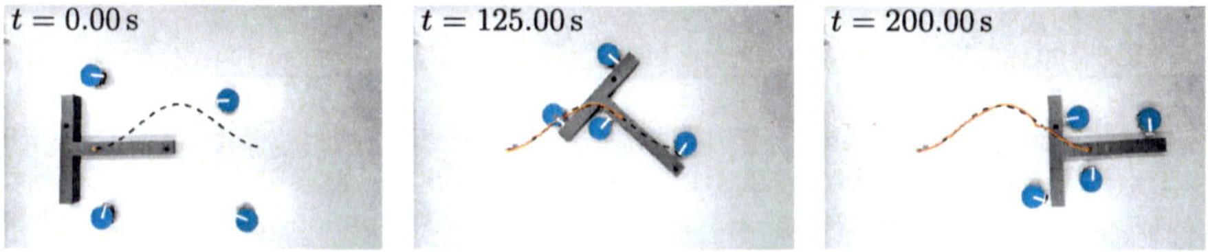


(b) transportation of a T-shaped object along a sinusoidal path

**Fig. 3.** Two experimental results for the proposed force-based scheme.

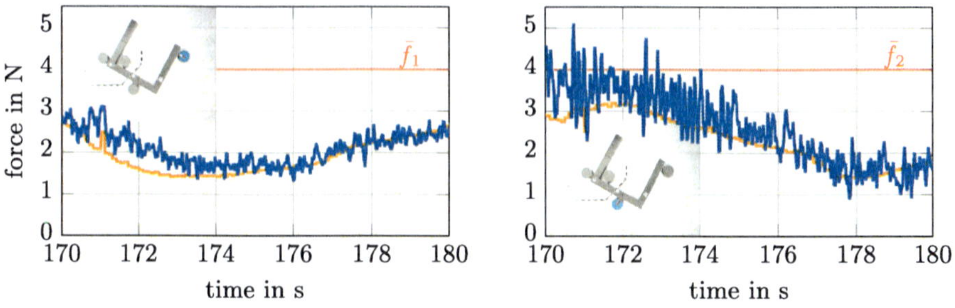


**Fig. 4.** Desired (——) and actually exerted (——) forces for two robots during an exemplary time interval for the ⊔-shaped object, see. Fig 3(a).

challenging task for the wheeled mobile robots which mainly stems from changing frictional properties and heavy slippage of the Mecanum wheels. Further, note that Fig. 2(a) visualizes the resulting zonohedron for the considered experiment at $t = 50$ s, see also Fig. 1(b) and Fig. 3(a) for the corresponding configuration.

A qualitatively equivalent behavior can be observed for the second experiment considering a T-shaped object that shall be pushed along a trigonometric function's segment, see Fig. 3(b). Crucially, the provided scheme is able to precisely transport both objects along the desired path. We refer to [8] for the experiments' video recordings in addition to the stills provided in Fig. 3.

## 5 Conclusions

The presented results indicate that solving tasks in distributed (mobile) robotics by explicitly taking into account and controlling the interacting forces is within the realm of possibility. The description of the manipulation space using zonotopes fits the actual manipulation possibilities of the agents such that their limits are intrinsically respected. Further, through the force measurements, a direct feedback from the object is obtained which is not given for purely position-based approaches. Although the proposed scheme is only applied to the planar transportation task of a rigid body, it is in our opinion universal and can easily be transferred to more general manipulation tasks in robotics, e.g., for manipulating flexible bodies or employing heterogeneous robotic swarms which has already been dealt with for position-level formation finding [4].

## References

1. Bertoncelli, F., Ruggiero, F., Sabattini, L.: Characterization of grasp configurations for multi-robot object pushing. In: Proceedings of the 2021 IEEE International Symposium on Multi-Robot and Multi-Agent Systems, pp. 38–46. Cambridge, UK (2021)
2. Ebel, H.: Distributed control and organization of communicating mobile robots: design, simulation, and experimentation. Dissertation, Schriften aus dem Institut für Technische und Numerische Mechanik der Universität Stuttgart, vol. 69, Shaker Verlag (2021). https://www.itm.uni-stuttgart.de/institut/team/ebel/Ebel2021.pdf
3. Ebel, H., Eberhard, P.: Non-prehensile cooperative object transportation with omnidirectional mobile robots: organization, control, simulation, and experimentation. In: Proceedings of the 2021 IEEE International Symposium on Multi-Robot and Multi-Agent Systems, pp. 1–10. Cambridge, UK (2021)
4. Ebel, H., Fahse, D.N., Rosenfelder, M., Eberhard, P.: Finding formations for the non-prehensile object transportation with differentially-driven mobile robots. In: Kecskeméthy, A., Parenti-Castelli, V. (eds.) ROMANSY 24 - Robot Design, Dynamics and Control ROMANSY 2022. CISM International Centre for Mechanical Sciences, vol. 606, pp. 163–170. Springer, Cham (2022). https://doi.org/10.1007/978-3-031-06409-8_17
5. Huang, A., Olson, E., Moore, D.: LCM: lightweight communications and marshalling. In: Proceedings of the 2010 IEEE/RSJ International Conference on Intelligent Robots and Systems, pp. 4057–4062. Taipei (2010)
6. Prattichizzo, D., Trinkle, J.C.: Grasping. In: Siciliano, B., Khatib, O. (eds.) Springer Handbook of Robotics, pp. 671–700. Springer, Berlin, Germany (2008)
7. Rosenfelder, M., Ebel, H., Eberhard, P.: A force-based control approach for the non-prehensile cooperative transportation of objects using omnidirectional mobile robots. In: Proceedings of the 2022 IEEE Conference on Control Technology and Applications (CCTA), pp. 349–356. Trieste, Italy (2022)
8. Rosenfelder, M., Ebel, H., Eberhard, P.: Experiment videos of the force-based non-prehensile cooperative transportation of objects with mobile robots (2023). https://doi.org/10.18419/darus-3331
9. Tuci, E., Alkilabi, M.H.M., Akanyeti, O.: Cooperative object transport in multi-robot systems: a review of the state-of-the-art. Front. Robot. AI **5**, 59 (2018)
10. Ziegler, G.M.: Lectures on Polytopes, vol. 152. Springer, New York, USA (2012)

# Center of Mass Wrench Constraints in Presence of Spatially Distributed Contacts

Milutin Nikolić[1(✉)], Branislav Borovac[1], Mirko Raković[1], and Milica Žigić[2]

[1] Faculty of Technical Sciences, University of Novi Sad, Novi Sad, Serbia
{milutinn,borovac,rakovicm}@uns.ac.rs
[2] Faculty of Sciences, University of Novi Sad, Novi Sad, Serbia
milica.zigic@dmi.uns.ac.rs

**Abstract.** Motion planning starts by planning the trajectory of the center of mass. With the increasing capabilities of humanoid robots, the case when contacts are spatially distributed should be considered. This paper shows the existence of contact configurations in which any acceleration of the center of mass is feasible. The procedure for identifying such configurations is presented. On the other hand, for the configurations in which the constraint on center of mass movement exists, it will be shown how to find that linear constraint, which defines the space of feasible motion. The proposed algorithm has low complexity. Additionally, it will be shown that the whole procedure needs to be run only once when the contact configuration changes.

**Keywords:** contact stability · motion constraints · centroidal dynamics

## 1 Introduction

The main indicator of dynamical balance during bipedal walking, zero momen point ($ZMP$) was introduced by M. Vukobratović and his closest associates [12,13]. The $ZMP$ notion considers only the existence of full contact between the feet and the ground. It is assumed that the friction is high and that sliding does not occur. When the $ZMP$ is inside the convex hull of the support area (but not on its edges), the humanoid is dynamically balanced and the robot will not start falling by rotating about the edge of the support area. The $ZMP$ notion in its original form can be used only for robots walking on flat horizontal surfaces.

Several authors have tried to generalize the $ZMP$ notion. In [8], authors have taken into account limited friction, and have shown that wrench acting on the foot is in convex cone, that can be computed from convex hull of a set


T. Petrič et al. (Eds.): RAAD 2023, MMS 135, pp. 325–333, 2023.
https://doi.org/10.1007/978-3-031-32606-6_38

of points in 5D space. Harada et al. [6] proposed generalized $ZMP$ ($GZMP$) which is applicable for the cases when a robot is touching the environment with its hand. In order for the motion to be feasible, the $GZMP$ needs to fall within the region on the ground surface, defined on the basis of the 3D convex hull of all the contact points and mass of the robot. The two major assumptions used: the humanoid robot is standing/walking on a *flat* floor and friction forces are *small enough*, make this approach less general. The same authors analyzed the influence of additional contact (i.e., grasping the environment) on the position of $ZMP$ [5], and exploited that contact in order to enable the robot to climb up the high step by holding the handlebar. In [1] authors have redefined ZMP as either a line containing all possible ZMPs or ZMP angle, representing the angle between ZMP line passing through center of mass (CoM) and vertical line.

In [10], the authors have addressed the issue of robot control in the presence of multiple contacts by employing a virtual linkage model. Based on it, internal and contact forces required to maintain stability of all contacts are calculated. In [11] feasible solution of wrench (FSW) is introduced. Gravitational and inertial forces have to be counter-balanced by contact forces so that their sum has to be in the space of feasible contact wrenches. Also, the authors have addressed the issue of low friction contact by introducing another criterion which can determine whether planned motion is feasible. In [2], the authors have employed the double descriptor method in order to calculate the gravito-inertial wrench cone. Although most general compared to all previous $ZMP$ generalizations, the algorithm has high computational complexity.

In this paper we will shown how to efficiently calculate CoM wrench constraint. Hence, the bounds of centroidal motion are also defined. The computation time is reduced by two orders of magnitude compared to the state of the art methods. This could be very beneficial for optimization-based control strategies [7,9], which are inherently very computationally expensive. This paper will also prove that contact configurations that enable arbitrary acceleration of CoM exist: the procedure on how to identify them will be presented as well.

## 2 Motion Constraints

Let's consider the system shown in Fig. 1a. The robot is in contact with the environment with both of its feet and with one of its hands. Since the motion planning is mainly based on motion of the CoM, it is of highest importance to find what kind of motions of CoM does current contact configuration permit. Nature of the contacts induces constraints on contact forces, while their spatial disposition defines the space of feasible total wrench. For the system from the Fig. 1a acceleration of center of mass and rate of change of angular momentum can be calculated using:

$$\begin{bmatrix} m\mathbf{a} \\ \dot{\mathbf{L}} \end{bmatrix} = \begin{bmatrix} \mathbf{F} \\ \mathbf{M} \end{bmatrix} + \begin{bmatrix} m\mathbf{g} \\ \mathbf{0} \end{bmatrix} \quad (1)$$

where $m$ is the mass of the system, $\mathbf{a}$ is the acceleration of the CoM, $\mathbf{L}$ is angular momentum, while $\mathbf{g}$ is gravitational acceleration. The total force acting on the system is denoted by $\mathbf{F}$, while the total torque that contact forces create about the CoM is denoted by $\mathbf{M}$. In order to ensure that the intended CoM motion (defined by acceleration $\mathbf{a}$ and rate of change of angular momentum $\dot{\mathbf{L}}$) is feasible there must exist a total wrench (which contains total force and total torque) such that (1) is fulfilled. Now the question of motion feasibility turns into a question of the existence of appropriate total forces and torques.

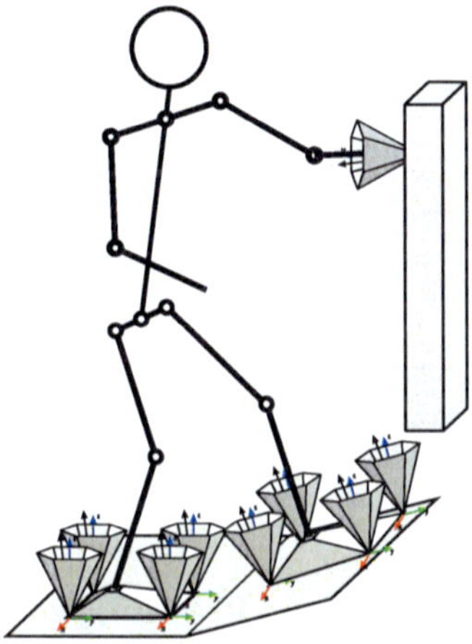

(a) A system with multiple contacts.

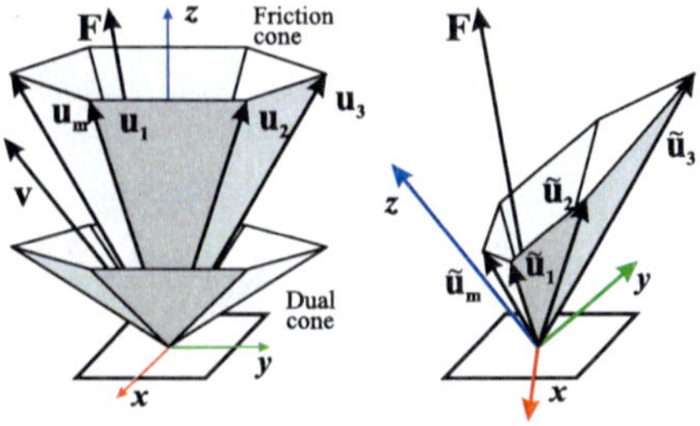


(b) Friction cone with its dual is shown on the left. Modified friction cone is shown on the right.

**Fig. 1.** Point contact and system with multiple spatially distributed contacts

As previously stated, the nature of the contact introduces the constraints on contact forces. In this paper only unilateral point contacts with Columb friction will be considered. Surface contacts can be understood as multiple point contacts placed at the corners of the perimeter of the contact surface [3]. The contact force (denoted by $\mathbf{F}^C$) at the point contact is bounded, so it needs to be inside the corresponding friction cone which can be approximated by a $m$-sided pyramid (Fig. 1b). Representation of friction cone in the span form is:

$$\mathbf{F}^C = \alpha_1 \mathbf{u}_1 + \alpha_2 \mathbf{u}_2 + \cdots + \alpha_m \mathbf{u}_m = \mathbf{U}\boldsymbol{\alpha}, \quad \alpha_i \geq 0 \tag{2}$$

where $\mathbf{U} = [\mathbf{u}_1\ \mathbf{u}_2\ \ldots\ \mathbf{u}_m]$ and $\boldsymbol{\alpha} = [\alpha_1\ \alpha_2\ \ldots\ \alpha_m]^T$.

For the case when multiple contacts are present (Fig. 1a) total wrench acting on the robot's CoM has the form:

$$\begin{bmatrix} \mathbf{F} \\ \mathbf{M} \end{bmatrix} = \sum_{i=1}^{n} \begin{bmatrix} \mathbf{R}_i \mathbf{F}_i^C \\ [\mathbf{r}_i]_\times \mathbf{R}_i \mathbf{F}_i^C \end{bmatrix} = \begin{bmatrix} \mathbf{R}_1 \mathbf{U}_1 & \ldots & \mathbf{R}_n \mathbf{U}_n \\ [\mathbf{r}_1]_\times \mathbf{R}_1 \mathbf{U}_1 & \ldots & [\mathbf{r}_n]_\times \mathbf{R}_n \mathbf{U}_n \end{bmatrix} \begin{bmatrix} \boldsymbol{\alpha}_1 \\ \vdots \\ \boldsymbol{\alpha}_n \end{bmatrix} = \begin{bmatrix} \boldsymbol{\Upsilon} \\ \boldsymbol{\Gamma} \end{bmatrix} \mathbf{A}$$

where $n$ is the number of separate contact points, $\mathbf{R}_i$ is the rotation matrix and $\mathbf{U}_i$ is the generating matrix for the $i$-th contact. Vectors $\mathbf{r}_1$ to $\mathbf{r}_n$ represent positions of contact points relative to CoM of the system. $[\bullet]_\times$ is a skew-symmetric cross product operator matrix. Matrix $\boldsymbol{\Upsilon}$ represents the matrix where all generating matrices are stacked horizontally, $\boldsymbol{\Gamma}$ represents the matrix where all cross products of position vectors and generating matrices are stacked horizontally and $\mathbf{A}$ represents the vector of $\boldsymbol{\alpha}_i$-s for all $n$ contacts. In the rest of the paper $i$-th columns of matrices $\boldsymbol{\Upsilon}$ and $\boldsymbol{\Gamma}$ will be denoted as $\boldsymbol{\upsilon}_i$ and $\boldsymbol{\gamma}_i$ respectively.

The key problem here is to find the space of feasible forces acting on CoM $\mathbf{F}$ for the given contact configuration. Space of forces acting on CoM directly influences the space of feasible accelerations of CoM.

From Farkas-Minkowski's theorem in form taken from [4] applied to case (3) follows:

**Corollary 1.** *For the robot system in contact with the environment, with a matrix of generating vectors $\boldsymbol{\Upsilon}$, the total force acting on robot $\mathbf{F}$ is feasible if and only if:*

- *For every vector $\mathbf{v} \in \mathbb{R}^3$ such that $\mathbf{v}^{\mathrm{T}}\boldsymbol{\Upsilon} \geq 0$ it holds that $\mathbf{v}^{\mathrm{T}}\mathbf{F} \geq 0$*

Analysis of this corollary it will be split in two cases.

## 2.1 First Case

We will consider the singular case where only zero-vector ($\mathbf{v} = \mathbf{0}$) satisfies $\mathbf{v}^{\mathrm{T}}\boldsymbol{\Upsilon} \geq 0$. It's trivial to see that if $\mathbf{v} = \mathbf{0}$, then $\mathbf{v}^{\mathrm{T}}\mathbf{F} \geq 0$ for every $\mathbf{F}$. This condition does not depend on total force at all, but only on matrix $\boldsymbol{\Upsilon}$, which describes contact configuration. In other words, *there are some contact configurations in which arbitrary total force can be applied.*

If for a vector $\mathbf{v}$ holds that $\mathbf{v}^{\mathrm{T}}\mathbf{R}_i\mathbf{U}_i \geq 0$ it means that vector $\mathbf{v}$ must lie in the dual cone[1] of the friction cone of $i$-th contact. After looking at the structure of matrix $\boldsymbol{\Upsilon}$ (3), together with corollary 1 we can conclude that when the intersection of dual cones of all contacts is $\{\mathbf{0}\}$, the arbitrary desired total force exists, hence arbitrary acceleration of the CoM. An illustration of such a case is given in Fig. 2a, where it can be clearly noted that there is no vector $\mathbf{v} \neq \mathbf{0}$ which would lie in all the friction cone duals.

## 2.2 Second Case

The second case considered will be the case where $\mathbf{v} \neq \mathbf{0}$ with the property $\mathbf{v}^{\mathrm{T}}\boldsymbol{\Upsilon} \geq 0$ exists. In this case it is possible to calculate the rotation matrix $\mathbf{R}_{\mathbf{v}}$, which rotates the global coordinate frame so that the direction of a new $z$-axis aligns with the vector $\mathbf{v}$. An example of such a rotation is depicted in Fig. 1b. After rotation the total wrench acting on the body would be:

$$\begin{bmatrix} \mathbf{F}^{\mathbf{v}} \\ \mathbf{M}^{\mathbf{v}} \end{bmatrix} = \begin{bmatrix} \mathbf{R}_{\mathbf{v}}\boldsymbol{\Upsilon} \\ \mathbf{R}_{\mathbf{v}}\boldsymbol{\Gamma} \end{bmatrix} \mathbf{A} \tag{3}$$

[1] Dual cone of cone $C$ in three dimensional space is defined as $C^* = \{\mathbf{y} \in \mathbb{R}^3 : \mathbf{y}^T\mathbf{x} \geq 0, \forall \mathbf{x} \in C\}$.

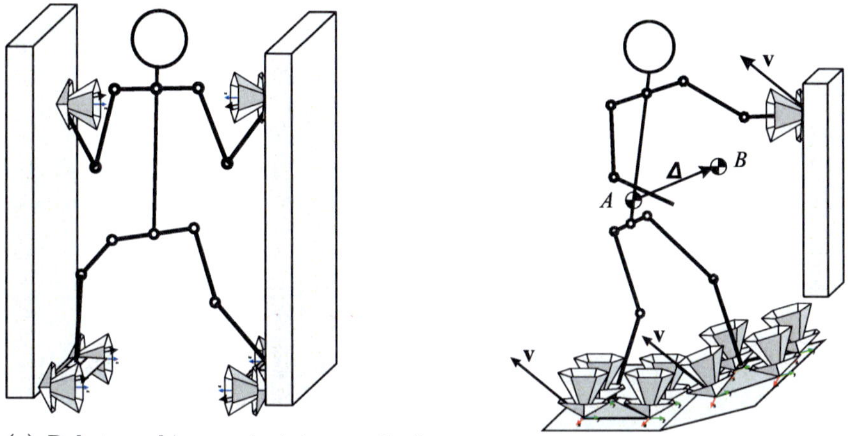


(a) Robot pushing against two vertical walls. Duals of all six friction cones intersect only at **0**, thus any acceleration of CoM is feasible

(b) Case where duals of friction cones intersect. Vector **v** that lies in duals of all friction cones is depicted.

**Fig. 2.** Two cases of a system with multiple spatial contacts

Since $z$-axis is aligned with $\mathbf{v}$ and $\mathbf{v}^{\mathrm{T}}\boldsymbol{\Upsilon} \geq 0$, all elements in the third row of matrix $\mathbf{R_v}\boldsymbol{\Upsilon}$ are greater than zero. Thus, columns of matrices $\mathbf{R_v}\boldsymbol{\Upsilon}$ and $\mathbf{R_v}\boldsymbol{\Gamma}$ would be normalized in order to obtain modified generating matrices:

$$\tilde{\boldsymbol{\upsilon}}_i = \frac{|\mathbf{v}|}{\mathbf{v} \bullet \boldsymbol{\upsilon}_i}\mathbf{R_v}\boldsymbol{\upsilon}_i, \quad \tilde{\boldsymbol{\Upsilon}} = [\tilde{\boldsymbol{\upsilon}}_i], \; i = 1 \ldots n \times m \tag{4}$$

$$\tilde{\boldsymbol{\gamma}}_i = \frac{|\mathbf{v}|}{\mathbf{v} \bullet \boldsymbol{\upsilon}_i}\mathbf{R_v}\boldsymbol{\gamma}_i, \quad \tilde{\boldsymbol{\Gamma}} = [\tilde{\boldsymbol{\gamma}}_i], \; i = 1 \ldots n \times m \tag{5}$$

The friction cone and its generating vectors after modification are depicted in Fig. 1b. It is clear that both the original friction cone and the modified friction cone generate the same space, but now, friction cones of all contacts would be expressed in the same coordinate frame. As a result this, Eq. (3) would assume the following form:

$$\begin{bmatrix} \mathbf{F^v} \\ \mathbf{M^v} \end{bmatrix} = \begin{bmatrix} \tilde{\boldsymbol{\Upsilon}} \\ \tilde{\boldsymbol{\Gamma}} \end{bmatrix} \tilde{\mathbf{A}} = \mathbf{P}\tilde{\mathbf{A}} \tag{6}$$

After modification, the third row of the matrix $\tilde{\boldsymbol{\Upsilon}}$ would be a row of ones. As a result we have reduced spatial case to planar case, and now we can compute wrench constraint by computing the convex hull of special 5D points by following the procedure described in [8]. The constraint can be written as:

$$\mathbf{W} \begin{bmatrix} \mathbf{F} \\ \mathbf{M} \end{bmatrix} \geq 0 \tag{7}$$

where $\mathbf{W}$ represents wrench constraint matrix about CoM - *WCM*. It is shown here how constraint on total wrench acting on the body can be constructed when duals of all friction cones intersect.

### 2.3 Shifted WCM

As stated in the previous paragraph, WCM is calculated for the CoM of the system. Each time the CoM shifts the WCM should be recalculated. Fortunately, we will shown that it is possible also to shift WCM.

If the WCM is calculated for point $A$ (current position of CoM) and the CoM shifts by vector $\boldsymbol{\Delta}$ to some point $B$ (Fig. 2b), the wrench acting on the new position of CoM can be calculated for point A:

$$\begin{bmatrix} \mathbf{F}^A \\ \mathbf{M}^A \end{bmatrix} = \begin{bmatrix} \mathbf{I}_{3\times3} & \mathbf{0}_{3\times3} \\ [\boldsymbol{\Delta}]_\times & \mathbf{I}_{3\times3} \end{bmatrix} \begin{bmatrix} \mathbf{F}^B \\ \mathbf{M}^B \end{bmatrix}. \tag{8}$$

Since WCM is calculated for point A, wrench calculated for point A must fulfill inequality (7). So after combining that (7) with (8) it is easy to obtain:

$$\mathbf{W}^{\boldsymbol{\Delta}} = \mathbf{W} \begin{bmatrix} \mathbf{I}_{3\times3} & \mathbf{0}_{3\times3} \\ [\boldsymbol{\Delta}]_\times & \mathbf{I}_{3\times3} \end{bmatrix} \tag{9}$$

$$\mathbf{W}^{\boldsymbol{\Delta}} \begin{bmatrix} \mathbf{F}^B \\ \mathbf{M}^B \end{bmatrix} \geq 0 \tag{10}$$

where $\mathbf{W}^{\boldsymbol{\Delta}}$ represents a shifted WCM. This shows, that WCM needs to be computed only once when the contact configuration changes. As CoM moves, as long as the contact configuration remains the same, the shifted WCM can be calculated by (9). It gives the same constraint (10) on the total wrench as WCM (7), but it's much faster to compute.

## 3 Simulation Results

In this section, the main results will be illustrated by two simulated scenarios. In the first case, the robot has to climb between two vertical walls as illustrated in Fig. 2a. In the second simulated case, the robot has to move sideways on the vertical wall while hanging by its hands at the top. The joint torques are calculated using a generalized task prioritization framework. More details about the framework can be found in [9]. In both cases, all friction cones are approximated by 4-sided pyramids and $\mu = 0.8$.

In the first simulated case only the hands and the edges of the feet are in contact with the wall. Most of the time only hands or only feet are in contact with the wall, except the short transition time between those two contact configurations. The illustrations of one and a half cycle of simulated motion are shown in the top row of Fig. 3. The robot first repositions its feet and after that repositions its hands. According to the first case (Sect. 2.1), we can see that any total force $\mathbf{F}$ is feasible for contact configurations that appear in this scenario, implying that arbitrary acceleration of the CoM is also feasible. As a result, the robot is able to climb between the vertical walls. Only thing that could hinder the robot from climbing are limited joint torques since internal load needs to be introduced in order to maintain all contacts stable.

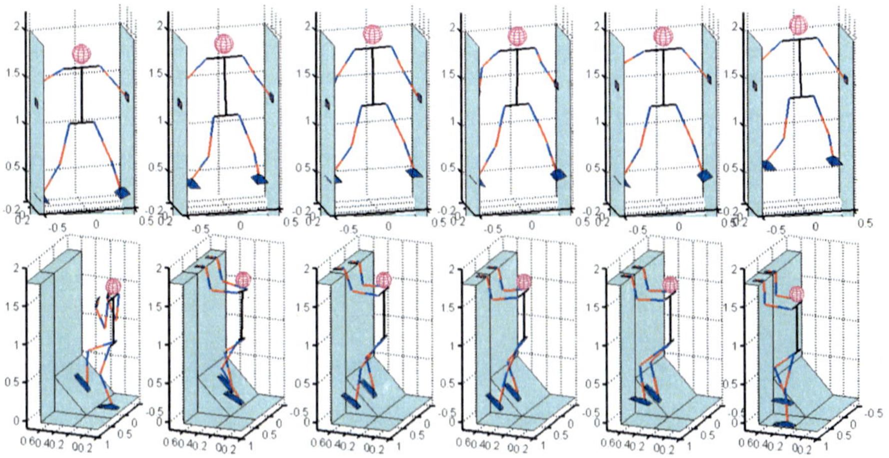

**Fig. 3.** Top row: A Robot climbing between the vertical walls. Bottom row: Robot moves sideways while holding on the top of the wall

In the second case considered, the contact configuration is changed substantially as the motion progresses. Hands can be only in contact with the horizontal top surface of the wall while the robot's feet can be in contact with the ground surface, with the surface angled at 45° or with a vertical wall, but only using front edge of the foot. In Fig. 3 a total of 6 phases of the motion are shown. After establishing the contact between the hands and the top of the wall, in each phase one of the limbs is repositioned, while others remain stationary. The sequence of repositioning is: left foot, left hand, right hand and right foot. Each phase has a different spatial contact configuration. In all phases of the motion, the duals of the friction cones intersect and the WCM has to be calculated. The $WCM$ computation was implemented in Python together with Numpy and CVXOPT packages, and it took on average $0.8ms$, $1.6ms$ and $2.7ms$ when 4, 8, 12 friction cones are present. In if the number of sides in the friction pyramids increases, the computation times would go up to 1.4 ms, 3.4 ms and 5.3 for 6 sided and 2.3 ms, 4.9 ms and 8.0 ms for 8 sided pyramids. WCM needs to be computed only once when the configuration changes and the shifted WCM could be used afterward. Hence, the procedure is highly time efficient and usable in real-time applications. During the simulation the WCM constraint was used to check the feasibility of the planned motion. At this stage the motion and contact sequence were planned manually.

## 4 Conclusions and Open Questions

In this paper an efficient procedure is presented for checking if current contact configuration allows realization of planned motion. It is shown that contact configurations which allow arbitrary acceleration of the CoM exist. In addition, the geometrical interpretation of this condition is given, stating that when all friction cones intersect only at **0** arbitrary acceleration of CoM is feasible. Such configurations enable unbounded movement of CoM in any direction. In the simulation examples, it is illustrated how such kind of configuration could be identified and exploited for climbing between vertical walls. In all phases of the motion, the duals of the friction cones intersect and the WCM has to be calculated.

For the case when acceleration of the CoM is bounded, the procedure is presented for calculating a wrench constraint matrix. Instead of recalculating WCM each time the CoM moves, the procedure for shifting WCM is created which enables for very expedient computation times. Actually, the computational times of WCM dropped almost 100 times. This result is significant, since it will enable this constraint to be used for the real-time control of humanoid robots

## References


1. Brecelj, T., Petrič, T.: Zero moment line - universal stability parameter for multi-contact systems in three dimensions. Sensors **22**(15) (2022). https://doi.org/10.3390/s22155656, https://www.mdpi.com/1424-8220/22/15/5656
2. Caron, S., Pham, Q.C., Nakamura, Y.: Leveraging cone double description for multi-contact stability of humanoids with applications to statics and dynamics. In: Robotics: Science and System (2015)
3. Caron, S., Pham, Q.C., Nakamura, Y.: Stability of surface contacts for humanoid robots: Closed-form formulae of the contact wrench for rectangular support areas. In: 2015 IEEE International Conference on Robotics and Automation (ICRA). IEEE (2015)
4. Gale, D., Kuhn, H.W., Tucker, A.W.: Activity analysis of production and allocation, vol. 13, chap. Linear programming and the theory of games, pp. 317–335 (1951)
5. Harada, K., et al.: Dynamical balance of a humanoid robot grasping an environment. In: 2004 IEEE/RSJ International Conference on Intelligent Robots and Systems (IROS 2004), vol. 2, pp. 1167–1173. IEEE (2004)
6. Harada, K., Kajita, S., Kaneko, K., Hirukawa, H.: Dynamics and balance of a humanoid robot during manipulation tasks. IEEE Trans. Robot. **22**(3), 568–575 (2006)
7. Kuindersma, S., Permenter, F., Tedrake, R.: An efficiently solvable quadratic program for stabilizing dynamic locomotion. In: 2014 IEEE International Conference on Robotics and Automation (ICRA), pp. 2589–2594 (2014). https://doi.org/10.1109/ICRA.2014.6907230
8. Nikolić, M., Borovac, B., Raković, M.: Dynamic balance preservation and prevention of sliding for humanoid robots in the presence of multiple spatial contacts. Multibody Syst. Dyn. **42**(2), 197–218 (2017). https://doi.org/10.1007/s11044-017-9572-9
9. Nikolić, M., Borovac, B., Raković, M., Savić, S.: A further generalization of task-oriented control through tasks prioritization. Int. J. Humanoid Rob. **10**(03) (2013)
10. Sentis, L., Park, J., Khatib, O.: Compliant control of multicontact and center-of-mass behaviors in humanoid robots. IEEE Trans. Robotics **26**(3), 483–501 (2010)

11. Takao, S., Yokokohji, Y., Yoshikawa, T.: FSW (feasible solution of wrench) for multi-legged robots. In: 2003 IEEE International Conference on Robotics and Automation, ICRA'03, vol. 3, pp. 3815–3820. IEEE (2003)
12. Vukobratović, M., Borovac, B.: Zero-moment point-thirty five years of its life. Int. J. Humanoid Rob. **1**(01), 157–173 (2004)
13. Vukobratović, M., Juricic, D.: Contribution to the synthesis of biped gait. IEEE Trans. Biomed. Eng. **1**, 1–6 (1969)

# Modification and Optimization of the Trajectory of an Industrial Robot to Scan a 3D-Surface for Quality Inspection

Atae Jafari-Tabrizi[1(✉)], Dieter P. Gruber[1], and Andrej Gams[2]

[1] Polymer Competence Center Leoben GmbH, Roseggerstraße 12, 8700 Leoben, Austria
{atae.jafari,dieter.gruber}@pccl.at

[2] Humanoid and Cognitive Robotics Lab, Department of Automatics, Biocybernetics and Robotics, Jožef Stefan Institute, Jamova cesta 39, 1000 Ljubljana, Slovenia
andrej.gams@ijs.si

**Abstract.** An industrial robot equipped with vision sensors can be used to address the unavoidable uncertainties occurring in the process of optical quality inspection. One of these uncertainties is the randomness in the positioning of the workpiece under inspection. It might not always be possible to accurately position the workpiece. In such cases, an industrial robot being solely able to follow a programmed trajectory will not be helpful. In this paper a solution is proposed, with which a robot can modify and optimize its scan trajectory with respect to the actual position of the workpiece, while the only sensor used was the camera mounted on the end effector of the robot, which is also used for the inspection. The metric used by the optimization algorithm acquired by the camera was the focus measure. With this assumption, the case where using additional sensors is not possible has been addressed. The proof-of-concept of this solution have been shown by simulations and also in a real-world experimental environment. In the simulations, the Nelder-Mead algorithm was used to simultaneously optimize the translational and rotational variations in the robotic trajectory. In the experiments, a simple line search method was used to address only the translational variation in the workpiece's position.

**Keywords:** Industrial robots · Automatic quality inspection · Machine vision

## 1 Introduction

### 1.1 Motivation and Problem Definition

Machine vision and deep-learning-based quality inspection techniques are being more and more implemented in different manufacturing industries [10]. Complicated shapes, different surface conditions and also cycle-time requirements


T. Petrič et al. (Eds.): RAAD 2023, MMS 135, pp. 334–341, 2023.
https://doi.org/10.1007/978-3-031-32606-6_39

necessitate employment of industrial robots for handling either workpieces under inspection or the sensors used for the purpose of data acquisition from the surfaces. Standard way of utilizing industrial robots is to program them for a specific task, such that they can follow certain trajectories depending on the commands that they receive. However, the robots are not flexible enough in case of an uncertainty that might occur during the process. Such a challenge can happen when the workpiece under inspection is not positioned accurately, for example, because of inaccurate grasping. Assuming that the machine vision camera is installed on the end-effector of the robot, in such a case, the robot cannot simply scan the surface using the pre-programmed trajectory. A possible solution would be to detect the position of the workpiece and adjust the trajectory accordingly. An additional challenge appears when it is not possible to install additional sensors which can help localize the workpiece. The solution for such a configuration is to use the camera on the robot to localize the workpiece.

### 1.2 State of the Art

In [12], the authors propose an automatic path generation method which utilizes CAD-information for an additive manufacturing robot. In [1], an automated offline programming (OP) method for robotic trajectories has been introduced, in which 3D vision helps an OP platform to locate the workpiece in the environment. For assembly operations, a learning by demonstration method was proposed [2]. While the demonstrator performs the tasks, a motion sensor digitizes it and uses the data to train the model. In [8], an automatic trajectory generation for industrial welding applications, based on 3D point cloud of the surface has been introduced. As a feasibility study, the proposed method has been validated by the authors by performing experimental work. For robotic grinding of weld surfaces, another point-cloud-based automatic trajectory generation method has been introduced by the authors in [3]. For robot-based optical inspection purposes, in [6] a CAD-based robot motion generation strategy has been proposed. Also a learning-based method has been introduced in order to optimize the scanning speed of the surface. In [11], a CAD-based algorithm has been developed for car painting. The algorithm is able to generate separate trajectories for each of the robotic manipulators to perform the painting task.

## 2 Proposed Approach

In this work the focus was on localizing the workpiece under inspection in order to inspect its 3D-surface. It was assumed that a scan trajectory already exists to scan the surface of the workpiece. However, the position of the workpiece was not known by the robot. Therefore, the aim was to determine the position of the workpiece, and adjust the trajectory accordingly. This was achieved only with the inspection camera which was mounted on the end effector of the robot. An optimization algorithm was used to evaluate the feedback obtained from the camera. The feedback from the camera (i.e., the acquired images) was quantified using the metric called focus measure. In order to demonstrate the feasibility of the proposed approach, it was tested both in simulation and in real-world experiments.

### 2.1 Defining the Environment

As mentioned above, in this work it was assumed that a scan trajectory for the 3D-surface already exists. However, it is not always possible for the workpieces to be accurately positioned (e.g., using a fixture). The reason for this might be that the workpieces are located inside boxes, which are carried by a conveyor belt inside the inspection cell.

Given that an original coordinate system exists (shown in Fig. 1a), with respect to which the robot's trajectory is programmed (depicted as $O(x, y, z)$), the position of the randomly located workpiece is denoted with coordinates $O'(x', y', z')$ (shown in Fig. 1b).

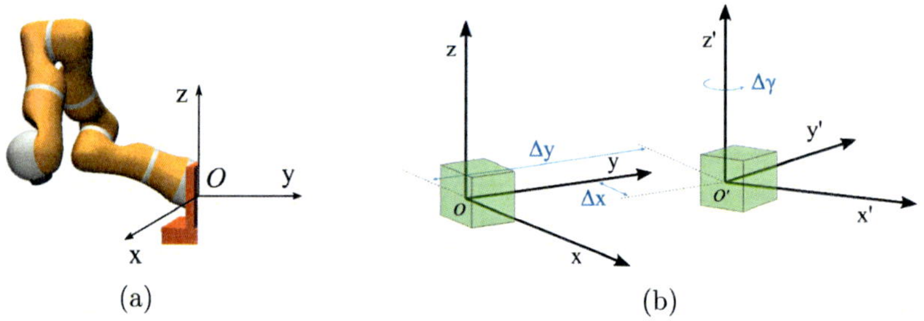


**Fig. 1.** (a) Coordinate system of the robot, $O(x, y, z)$, (b) Possible positional variations of the workpiece with respect to the robot coordinate system.

As it can be seen in Fig. 1b, assuming that the workpiece is located on a flat surface, the workpiece's position can vary in terms of translation in $x$- and $y$-directions ($\Delta x$ and $\Delta y$ respectively), and in terms of rotation around the $z$-axis ($\Delta\gamma$).

### 2.2 Focus Measure

Focus measure (FM) is an image processing metric for quantifying the image quality. Higher FM value means the features in an image are sharper. The behaviour of the FM metric chosen for this work in the vicinity of the camera's optimal distance from the surface (or the object) can be shown in Fig. 2. The positive integer denoting the FM reaches a maximum at the optimal distance, while it decreases equally as the camera moves away from the surface or towards it.

In this work, different FM operators available in the literature have been tested. The sensitivity of the FM outcome to the changes, and repeatability of

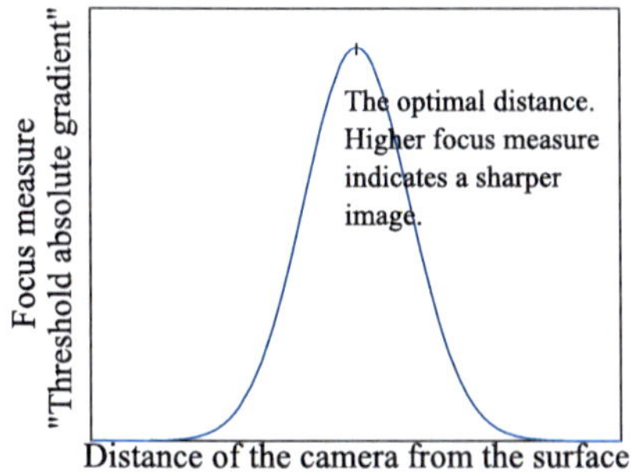


**Fig. 2.** The change of the focus measure "threshold absolute gradient" depending on the distance of the camera from the surface.

the operator were the criteria in choosing the most suitable FM operator for this work. Based on these criteria the operator called "threshold absolute gradient" has been selected to be used. This operator is based on image differentiation, where the calculation in horizontal direction for an image of $M \times N$ pixels is given as follows [9]

$$F_{\text{horizontal}} = \sum_{i=0}^{M-1} \sum_{j=0}^{N-2} |g(i, j+1) - g(i, j)|, \quad \text{while } |g(i, j+1) - g(i, j)| \geq v. \tag{1}$$

Here $g(i, j)$ is the grey level intensity at the pixel $(i, j)$, and $v$ is the threshold. Equation (1) is extended to calculate the FM both in horizontal and vertical directions, as mentioned in [9].

### 2.3 Objective Function for Optimizing the Robot Trajectory

Considering that higher FM values indicate higher image quality, the objective function used during the simulations and the experiments was the summation of the FM integer values.

$$\text{Objective} = \sum_{i=1}^{K} F_i. \tag{2}$$

Here $K$ is the total number of images taken during a single scan episode, and $F_i$ is the FM value of the area of interest in image $i$. In an experimental environment, during each scan episode, images would be acquired from the surface, and the FM values would be calculated for each image taken. The summation of the FM of the images taken can give an indication of overall scan quality. The objective function is maximum only in the ideal case when the camera is able to acquire the sharpest images during the whole scan episode.

## 3 Simulations

The code for simulations was written in Matlab. The simulations of the motion of the robot were done in MuJoCo HAPTIX [5] environment. For the sake of simplicity, and with the knowledge that the results can be extended further, in the simulations it was assumed that only deviations $\Delta y$ and $\Delta\gamma$ are present (see Fig. 1). Therefore the problem can be modelled as a 2-dimensional optimization problem, where the cost function is the summation of the FM values over the whole scan trajectory.

The Nelder-Mead [7] search was used to find the optimal adjustments in the trajectory. It is a direct search method (i.e., does not require derivative information of the objective function). And it was chosen as it requires less function evaluations in the initial iterations until it reaches an acceptable level of convergence. A detailed explanation of the Nelder-Mead method can be found in [4].

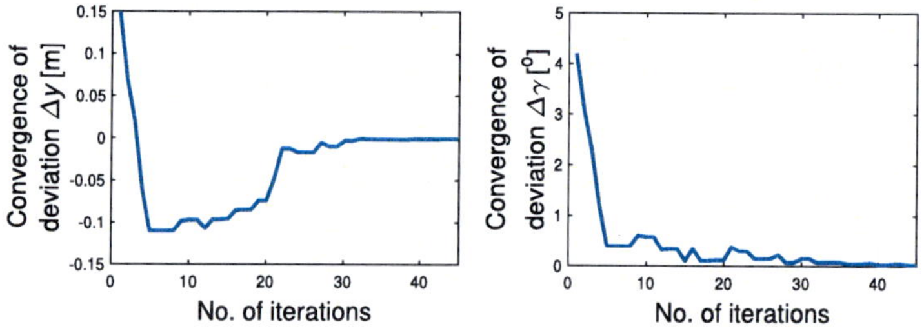


**Fig. 3.** Convergence of the $\Delta y$ and $\Delta\gamma$ values during the iterations

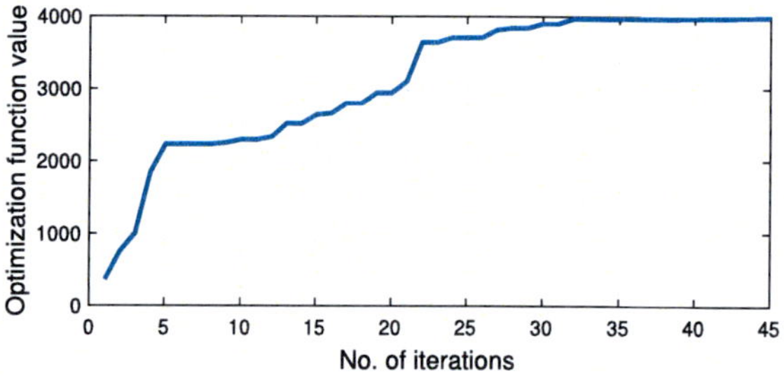


**Fig. 4.** The value of the optimization function during the iterations.

The results of the simulation are shown in Figs. 3 and 4. The initial deviations were randomly chosen as: 0.2 [m] in the $y$-axis ($\Delta y$), and 5° around the $z$-axis ($\Delta\gamma$). The correct positioning of the workpiece would require both of the variables to be zero. As it can be seen in Fig. 3, both $\Delta y$ and $\Delta\gamma$ start to converge rapidly already in the initial iterations. At around 30th iteration both parameters have reached to their respective expected values. Also the optimization function already reaches a value close to the optimum before 25th iteration (Fig. 4).

## 4 Experimental Work

For the experiments, a Kuka LWR-4 collaborative robot has been used. The camera was a black and white Basler acA1300-60gm, with a resolution of 1.3 MP, and a frame rate of 60 fps, with a lens which had a fixed focal length of 8.5 mm, and an aperture of f/1.3. The light source was a ring light, HPR2-100SW by CCS. The test surface under inspection can be seen in Fig. 5. In order to provide enough features in the image, a text was attached on the surface (see Fig. 5b). During the experiments it was assumed that the workpiece is deviated only in $y$-direction; that is, there is one unknown parameter (i.e., $\Delta y$). The most important reason for limiting the uncertainty to only one dimension was limitations in time and computational resources. This way, a simple unidimensional line search method could be applied to this problem. This search method is suitable for this work, as there existed only one optimal solution. The initial search step was chosen arbitrarily with trial and error. At every change in the direction of the search, the step would decrease. The termination criteria was based on comparing the difference between last two values with a threshold value. In the experiments, it has been seen that using the linear search method the robot is able to optimally modify the trajectory to scan the whole surface (Fig. 6).

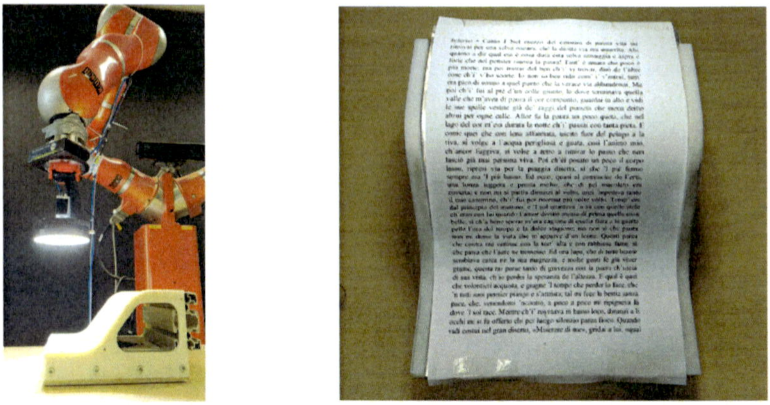

(a) (b)

**Fig. 5.** An overview of the experimental setup used. In 5a the camera and the light source attached to the end effector of the Kuka LWR along with the 3D-object under inspection can be seen. In 5b the text attached to the surface of the object can be seen.

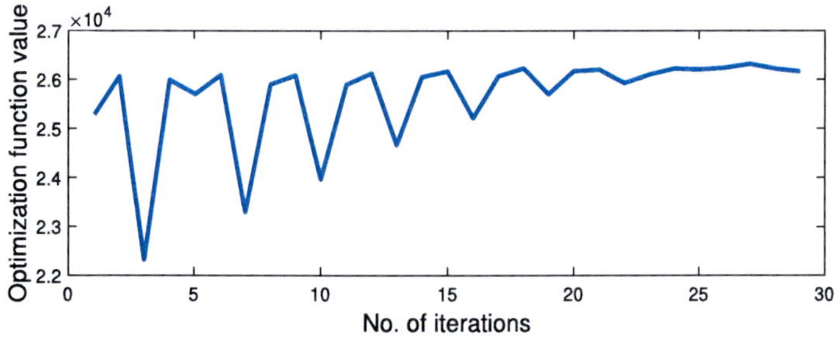


**Fig. 6.** The convergence to the optimum result using simple line search in the real-world experiments.

## 5 Discussion

Employing additional sensors could have simplified the approach. However, in this work it has been assumed that the only sensor available is the camera. In some applications using laser is not preferable, or is not possible. Therefore, the contribution of this work is to show the feasibility of such a solution. For the FM calculation to work properly, the surface of the workpiece must have patterns and features. This is necessary, as the FM function makes use of pixel value differences in an image to quantify the sharpness. This can impose some limitations on the application fields of this work. The Nelder-Mead search method, with its fast convergence to the vicinity of the optimum (as shown in Sect. 3) has proven to be a suitable search method for this work. An absolute optimal solution for this work, with noisy environment and solution is not necessary. The most important reason for this is that small positional deviations are still acceptable as long as they are within the depth of field of the camera, to guarantee that the acquired images are sharp.

## 6 Conclusion

In this work a method has been proposed to modify and optimize the scan trajectory of a robotic arm, which is used for vision-based quality inspection of a 3D-surface. The need for such an optimization arises when the position of the workpiece under inspection is not always fixed, and it can be different for each inspection episode. We assumed that additional sensors for localization of the workpiece are not available. The camera mounted on the end-effector of the robot was used as the only sensor to localize the surface. The proposed solution has been tested with both simulations and experimental work. As a proof-of-concept, this work has shown that the approach has the capacity to be further developed, and can be useful for real-world challenges. The future work will include extending the simulations and the experimental work to include uncertainties in the position and orientation of the 3D-surface in all possible directions in the working space.

**Acknowledgements.** The authors would like to thank Simon Reberšek for his technical support.

## References


1. Bedaka, A.K., Vidal, J., Lin, C.Y.: Automatic robot path integration using three-dimensional vision and offline programming. Int. J. Adv. Manuf. Technol. **102**(5–8), 1935–1950 (2019)
2. Duque, D.A., Prieto, F.A., Hoyos, J.G.: Trajectory generation for robotic assembly operations using learning by demonstration. Robot. Comput.-Integr. Manuf. **57**, 292–302 (2019). https://doi.org/10.1016/j.rcim.2018.12.007,https://www.sciencedirect.com/science/article/pii/S0736584517302867
3. Feng, H., et al.: A novel feature-guided trajectory generation method based on point cloud for robotic grinding of freeform welds. Int. J. Adv. Manuf. Technol. **115**(5–6), 1763–1781 (2021)
4. Kochenderfer, M.J., Wheeler, T.A.: Algorithms for Optimization. The MIT Press, Cambridge (2019)
5. Kumar, V., Todorov, E.: MuJoCo HAPTIX: a virtual reality system for hand manipulation. In: 2015 IEEE-RAS 15th International Conference on Humanoid Robots (Humanoids), pp. 657–663 (2015). https://doi.org/10.1109/HUMANOIDS.2015.7363441
6. Lončarević, Z., et al.: Specifying and optimizing robotic motion for visual quality inspection. Robot. Comput.-Integr. Manuf. **72**, 102200 (2021). https://doi.org/10.1016/j.rcim.2021.102200,https://www.sciencedirect.com/science/article/pii/S0736584521000831
7. Nelder, J.A., Mead, R.: A simplex method for function minimization. Comput. J. **7**(4), 308–313 (1965). https://doi.org/10.1093/comjnl/7.4.308
8. Peng, R., Navarro-Alarcon, D., Wu, V., Yang, W.: A point cloud-based method for automatic groove detection and trajectory generation of robotic arc welding tasks (2020). https://doi.org/10.48550/ARXIV.2004.12281,https://arxiv.org/abs/2004.12281
9. Santos, A., Ortiz de Solórzano, C., Vaquero, J.J., Peña, J.M., Malpica, N., Del Pozo, F.: Evaluation of autofocus functions in molecular cytogenetic analysis. J. Microsc. **188**(3), 264–272 (1997). https://doi.org/10.1046/j.1365-2818.1997.2630819.x,https://onlinelibrary.wiley.com/doi/abs/10.1046/j.1365-2818.1997.2630819.x
10. Yang, J., Li, S., Wang, Z., Dong, H., Wang, J., Tang, S.: Using deep learning to detect defects in manufacturing: a comprehensive survey and current challenges. Materials **13**(24) (2020). https://doi.org/10.3390/ma13245755,https://www.mdpi.com/1996-1944/13/24/5755
11. Zbiss, K., Kacem, A., Santillo, M., Mohammadi, A.: Automatic collision-free trajectory generation for collaborative robotic car-painting. IEEE Access **10**, 9950–9959 (2022). https://doi.org/10.1109/ACCESS.2022.3144631
12. Zheng, H., Cong, M., Dong, H., Liu, Y., Liu, D.: CAD-based automatic path generation and optimization for laser cladding robot in additive manufacturing. Int. J. Adv. Manuf. Technol. **92**(9–12), 3605–3614 (2017)

# A Parameter-Linear Formulation of the Optimal Path Following Problem for Robotic Manipulator

Tobias Marauli(✉), Hubert Gattringer, and Andreas Müller

Institute of Robotics, Johannes Kepler University Linz,
Altenberger Street 69, 4040 Linz, Austria
{tobias.marauli,hubert.gattringer,a.mueller}@jku.at

**Abstract.** In this paper the computational challenges of time-optimal path following are addressed. The standard approach is to minimize the travel time, which inevitably leads to singularities at zero path speed, when reformulating the optimization problem in terms of a path parameter. Thus, smooth trajectory generation while maintaining a low computational effort is quite challenging, since the singularities have to be taken into account. To this end, a different approach is presented in this paper. This approach is based on maximizing the path speed along a prescribed path. Furthermore, the approach is capable of planning smooth trajectories numerically efficient. Moreover, the discrete reformulation of the underlying problem is linear in optimization variables.

**Keywords:** Trajectory planning · Path following · Smooth trajectory generation · Optimization · Non-linear Optimization

## 1 Introduction

Short duration times are highly requested to maximize the productivity of a robotic system. Therefore, time-optimal motion is desirable. Frequently, the task for a robot is defined by a path that should be followed. Examples are welding and painting robots, CNC-machines as well as autonomous vehicles. Thus, several approaches for following paths in a (time-) optimal way are made in the literature [8, 10, 11, 14, 15]. A conventional approach is to reformulate the optimization problem, where the time is minimized, in terms of a path parameter. Solving the reformulated problem can be carried out numerically efficient [5, 19]. This approach comes with several drawbacks. First, singularities due to the reformulation in terms of the path parameter are present, which have to be considered. Consequently, the use of transcription methods e.g. collocation [2] or shooting methods [3, 16] is quite challenging, and the solution is computationally expensive. To overcome these issues, a different approach to minimize the travel time is presented in this paper. This approach is based on maximizing the path speed along the path. The latter approach is free of singularities, which copes well with transcription methods for smooth trajectory planning. Moreover, the resulting discrete optimization problem is linear in the approach of the path parameter.


T. Petrič et al. (Eds.): RAAD 2023, MMS 135, pp. 342–349, 2023.
https://doi.org/10.1007/978-3-031-32606-6_40

## 2 Optimal Path Following Problem Formulation

In this section, the parameter-linear optimal path following (OPF) problem is derived. Consider a $n$-DOF robotic manipulator with $\mathbf{q} \in \mathbb{V}^n$ denoting the vector of joint coordinates. The equations of motion (EOM)

$$\mathbf{M}(\mathbf{q})\ddot{\mathbf{q}} + \mathbf{h}(\mathbf{q}, \dot{\mathbf{q}}) = \boldsymbol{\tau} \tag{1}$$

describe the dynamic behavior of the manipulator, where $\mathbf{M}(\mathbf{q})$ is the mass matrix and $\boldsymbol{\tau}$ are the joint torques. The Coriolis, centrifugal and gravitational terms as well as the Coulomb and viscous friction are described by the vector $\mathbf{h}(\mathbf{q}, \dot{\mathbf{q}})$. If not necessary, the dependencies are omitted for simplicity and readability. Without loss of generality, a geometric joint path $\mathbf{q}(\sigma)$ is supposed to be given and parameterized in terms of a path parameter $\sigma \in [0, 1]$, with the geometric derivative defined as $()' = \partial()/\partial\sigma$. The time dependency of the path follows from the relation $\sigma(t)$ between the path parameter and the time. Furthermore, it is assumed that the path is followed only in forward direction $\dot{\sigma}(t) \geq 0$, $\forall t \in [0, t_{\mathrm{T}}]$. Moreover, the introduction of the squared path speed $z(\sigma) := \dot{\sigma}^2$ is beneficial for the optimization as discussed in [6, 19]. The introduction of latter leads to the relations

$$\dot{\sigma} = \sqrt{z}, \quad \ddot{\sigma} = \frac{z'}{2}, \quad \dddot{\sigma} = \frac{z''}{2}\sqrt{z} \tag{2}$$

for higher time derivatives of $\sigma$. Using the chain rule and upper relations the joint velocities, accelerations and jerks along a prescribed path $\mathbf{q}(\sigma)$ can be written as

$$\dot{\mathbf{q}}(\sigma) = \mathbf{q}'(\sigma)\sqrt{z} \tag{3a}$$

$$\ddot{\mathbf{q}}(\sigma) = \mathbf{q}''(\sigma)z + \frac{1}{2}\mathbf{q}'(\sigma)z' \tag{3b}$$

$$\dddot{\mathbf{q}}(\sigma) = (\mathbf{q}'''(\sigma)z + \frac{3}{2}\mathbf{q}''(\sigma)z' + \frac{1}{2}\mathbf{q}'(\sigma)z'')\sqrt{z}. \tag{3c}$$

The substitution of (3a) and (3b) in (1) yields the path projected EOM [19]

$$\mathbf{a}(\sigma)z' + \mathbf{b}(\sigma)z + \mathbf{c}(\sigma) + \mathbf{d}(\sigma)\sqrt{z} = \boldsymbol{\tau}(\sigma), \tag{4}$$

where $\mathbf{a}(\sigma)$ is related to the mass matrix and $\mathbf{b}(\sigma)$ includes the path-projected Coriolis and centrifugal terms, mixed with higher path derivative terms. The path depending gravitational and the Coulomb friction terms are described by $\mathbf{c}(\sigma)$ and $\mathbf{d}(\sigma)$ denotes the path projected viscous friction.

For planning time-optimal trajectories, the conventional approach of minimizing the terminal time

$$t_{\mathrm{T}} = \int_0^{t_{\mathrm{T}}} 1\,\mathrm{d}t = \int_0^1 \frac{1}{\sqrt{z(\sigma)}}\,\mathrm{d}\sigma, \tag{5}$$

is primary used [13–15]. Due to the division by $\sqrt{z(\sigma)}$ in above integral, singularities, where $z(\sigma) = 0$, have to be taken into account. This has to be considered when using numerical methods e.g. shooting or collocation methods for discretizing the time-optimal path following problem. Therefore, time-optimal trajectory planning with arbitrary smoothness while maintaining low computational effort, is quite challenging. For example, the numerically efficient approach presented in [6, 19] where the time-optimal path following problem is reformulated as second-order-cone problem is not capable of planning trajectories with arbitrary smoothness, due to the aforementioned singularities.

Planning time-optimal trajectories inevitably leads to increasing the speed $z$ along the path, to minimize the required terminal time (5) as addressed in [8]. An upper bound of $z$ is given by the maximum velocity curve (MVC) [11], defined by the velocity, acceleration and torque constraints. Therefore, maximizing the path speed $z$ gives rise to a different approach of reducing the terminal time $t_{\mathrm{T}}$. Since $z \geq 0$, the integral of the path speed

$$\int_0^1 z(\sigma)\,\mathrm{d}\sigma, \tag{6}$$

is a concave functional in $z(\sigma)$ [12]. Using above integral no singularities have to be taken into account, since the path speed is bounded and no division by $z$ is present. The optimization problem of latter approach can be written as

$$\begin{aligned}
\max_{z(\sigma)} \int_0^1 & z\,\mathrm{d}\sigma, \\
\text{s.t.}\quad & \underline{\dot{\mathbf{q}}} \leq \dot{\mathbf{q}}(\sigma) \leq \overline{\dot{\mathbf{q}}}, \qquad \underline{\ddot{\mathbf{q}}} \leq \ddot{\mathbf{q}}(\sigma) \leq \overline{\ddot{\mathbf{q}}}, \\
& \underline{\dddot{\mathbf{q}}} \leq \dddot{\mathbf{q}}(\sigma) \leq \overline{\dddot{\mathbf{q}}}, \quad \underline{\boldsymbol{\tau}} \leq \boldsymbol{\tau}(\sigma) \leq \overline{\boldsymbol{\tau}}, \\
& z(0) = \dot{\sigma}_0^2, \qquad\quad z(1) = \dot{\sigma}_{\mathrm{T}}^2, \\
& z(\sigma) \geq 0, \qquad\quad \text{for } \sigma \in [0, 1],
\end{aligned} \tag{7}$$

with the joint velocity, acceleration and jerk of (3) and the generalized torques obtained from (4) as constraints. The lower and upper bounds of the constraints are denoted by $\underline{()}$, $\overline{()}$. Solving the problem yields a maximized squared path speed profile $z(\sigma)^*$. Since optimization problems are solved numerically, the discrete reformulation of (7) is discussed in Sect. 3.

## 3 Parameter-Linear Reformulation

This section shows that the discrete reformulation of (7) leads to a parameter-linear optimization problem, when the joint jerk constraints (3c) and the viscous friction in (4) i.e. $\mathbf{d}(\sigma) = \mathbf{0}$ are not considered. To this end, the path parameter is discretized on $[0, 1]$, with $N+1$ grid points $\sigma_k \in [0, 1]$, for $k = 0 \ldots N$. Secondly, the function $z(\sigma)$ is parameterized in terms of a finite number of variables. To emphasize that the parameter-linear problem formulation can be expressed with an arbitrary representation of the path speed, 1) a polynomial and 2) B-spline approach are presented.

### 3.1 Piecewise-Polynomial Approach

The piecewise-polynomial approach is capable of planning optimal trajectories with controllable smoothness, by choosing the degree of the polynomial. For example, a polynomial of degree three leads to smooth joint jerk trajectories. In order to validate the presented approach in terms of the computation time and optimization result quality, the same approach as used in [18, 19], for discretizing the conventional time-optimal problem, is used. The piecewise-linear approach is given by

$$z(\sigma) = z_k + (z_{k+1} - z_k)\frac{\sigma - \sigma_k}{\sigma_{k+1} - \sigma_k} \tag{8}$$

for $\sigma \in [\sigma_k, \sigma_{k+1}]$, where the optimization variables are assigned on the grid points $z(\sigma_k) = z_k$. Similar as in [18], the function $z'(\sigma)$ is evaluated in between the grid points, namely $\sigma_{k+1/2} = (\sigma_{k+1} + \sigma_k)/2$. Thus, the joint acceleration (3b) and torque constraints (4) are evaluated at $\sigma_{k+1/2}$ accordingly. By introducing (8) in (3a) the joint velocity constraints can be reformulated as linear constraints, by carefully squaring them, similar to [19]. Substituting the piecewise-linear approach (8) in (3), (4) and (6) the discrete optimization problem is written as

$$\begin{aligned} \max_{z_k} \; & \sum_{k=0}^{N-1} \frac{1}{2}(z_k + z_{k+1})\Delta\sigma_k, \\ \text{s.t.} \; & \mathbf{q}'(\sigma_k)^2 z_k \le \bar{\dot{\mathbf{q}}}^2 \text{ for } \mathbf{q}'(\sigma_k) \ge 0, \; \mathbf{q}'(\sigma_k)^2 z_k \le \underline{\dot{\mathbf{q}}}^2 \text{ for } \mathbf{q}'(\sigma_k) < 0, \\ & \underline{\ddot{\mathbf{q}}} \le \ddot{\mathbf{q}}(\sigma_{k+1/2}) \le \bar{\ddot{\mathbf{q}}}, \; \underline{\dddot{\mathbf{q}}} \le \dddot{\mathbf{q}}(\sigma_{k+1/2}) \le \bar{\dddot{\mathbf{q}}}, \; \underline{\boldsymbol{\tau}} \le \boldsymbol{\tau}(\sigma_{k+1/2}) \le \bar{\boldsymbol{\tau}}, \\ & z_0 = \dot{\sigma}_0^2, \; z_N = \dot{\sigma}_{\mathrm{T}}^2, \; z_k \ge 0, \; \text{for } k = 0 \ldots N, \end{aligned} \tag{9}$$

with $\Delta\sigma_k = \sigma_{k+1} - \sigma_k$. For sake of brevity only, the joint velocity constraints are shown exemplary. The latter optimization problem is linear in $z_k$, without considering the joint jerk constraints and the viscous friction in (4) i.e. $\mathbf{d}(\sigma) = \mathbf{0}$. Since linear problems can be solved numerically efficient, a sequential convex programming (SCP) [5, 6] framework is used to incorporate the joint jerk constraints and the viscous friction in the torque constraints in the optimization process.

### 3.2 B-Spline Approach

A more flexible approach of representing the path speed results from the use of B-splines, since the number of control points can be varied independently from the number of grid points $N$ and the degree of the basis functions. Furthermore, B-splines are successfully used in smooth trajectory planning as discussed in [7]. Thus, the path speed represented by B-splines is written as

$$z(\sigma) = \sum_{i=0}^{n} p_i B_i^d(\sigma), \tag{10}$$

where $p_i$ are the control points, $n$ the number of control points and $B_i^d(\sigma)$ are the basis functions of degree $d$. Introducing (10) in (3), (4) and (6) the discrete optimization problem yields to

$$\begin{aligned} &\max_{p_i} \sum_{i=0}^{n} p_i \int_0^1 B_i^d(\sigma)\,\mathrm{d}\,\sigma, \\ &\text{s.t.}\;\; \underline{\dot{\mathbf{q}}} \le \dot{\mathbf{q}}(\sigma) \le \overline{\dot{\mathbf{q}}},\; \underline{\ddot{\mathbf{q}}} \le \ddot{\mathbf{q}}(\sigma_k) \le \overline{\ddot{\mathbf{q}}},\; \underline{\dddot{\mathbf{q}}} \le \dddot{\mathbf{q}}(\sigma_k) \le \overline{\dddot{\mathbf{q}}}, \\ &\quad\;\; \underline{\boldsymbol{\tau}} \le \boldsymbol{\tau}(\sigma_k) \le \overline{\boldsymbol{\tau}},\; z_0 = \dot{\sigma}_0^2,\; z_N = \dot{\sigma}_{\mathrm{T}}^2,\; z_k \ge 0,\; \text{for } k = 0 \ldots N, \end{aligned} \tag{11}$$

with a linear objective function in the control points $p_i$. The joint velocity and acceleration constraints are also linear in $p_i$. Thus, problem (11) corresponds to a linear optimization problem without considering the joint jerk constraints and the viscous friction in (4) i.e. $\mathbf{d}(\sigma) = \mathbf{0}$. To incorporate the latter, again a SCP approach is used.

## 4 Numerical Example

### 4.1 Smooth Trajectory Planning

An Intel(R) Core(TM) i5-9500 CPU @ 3.00GHz CPU running Windows 10 was used for all computations. For a 6-DOF *Comau Racer3* robotic manipulator the optimal path following problem (7) is solved in the parameter-linear form within a SCP framework for considering the joint jerk and torque constraints. The latter was implemented in MATLAB using YALMIP [9]. As solver MOSEK [1], a tailored solver for linear, quadratic, conic problems is chosen. For simplicity the upper and lower bounds of (7) are chosen symmetrically i.e. $\underline{()} = -\overline{()}$.

As example for smooth optimal trajectory planning the joint path as illustrated in Fig. 1 is chosen. The latter path is designed with smooth derivatives $\mathbf{q}'$, $\mathbf{q}''$ and $\mathbf{q}'''$ for smooth jerk trajectory generation. Thus, $z$, $z'$, $z''$ and $z'''$ have to be smooth for smooth joint jerk trajectories, according to (3). The OPF problem is solved with 1) the linear approach in $z$ (9) and 2) the B-spline ($d = 3$, $n = 61$) approach (11) with $N = 100$. Obviously the piecewise-linear approach in $z$ is not capable of planning smooth jerk trajectories, which is why it is only used as reference strategy to show the increase of the computational effort. The optimal solution $z^*$ of the two OPF problems is shown in Fig. 2. From the evolution of $z^*$ and the first derivative, it is evident that both solutions are nearly the same. The resulting second derivative is only smooth with the B-spline approach (11) and thus yields smooth joint jerk trajectories, since $\mathbf{q}(\sigma)$ is smooth up to the third derivative. The resulting terminal and computation times with 1) are $t_{\mathrm{T}} = 0.429\,\mathrm{s}$ and $t_{\mathrm{comp.}} = 0.515\,\mathrm{s}$ and 2) are $t_{\mathrm{T}} = 0.427\,\mathrm{s}$ and $t_{\mathrm{comp.}} = 0.899\,\mathrm{s}$. As the results show, the computational effort is not increasing drastically with a increase in the smoothness. Using the conventional approach comes with a huge increase of the computation time when increasing the smoothness of $z$, which is discussed in [4,8,19]. Since the path speed is maximized up to the MVC while satisfying the considered constraints, (at least) a sub time-optimal solution is achieved with the presented approach (7).

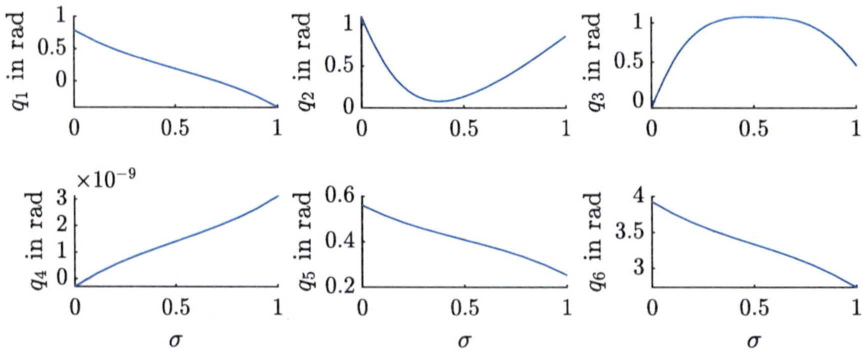


**Fig. 1.** Exemplary path in the joint space $\mathbf{q}(\sigma)$.

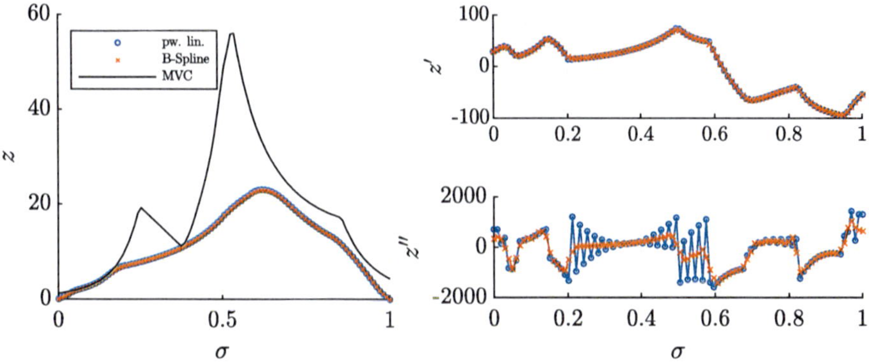


**Fig. 2.** Optimal path speed $z^*$ as well as first and second geometric derivative for following the arbitrary joint path. Blue circles: optimal solution with (9) using a piecewise-linear approach in $z$. Orange crosses: optimal solution with (11) using B-splines (Color figure online).

## 4.2 Computational Times of Proposed vs Conventional Approach

In this section the 1) conventional approach ($\min t_{\mathrm{T}}$), addressed in [18,19], is compared with the 2) presented approach (9) (using the piecewise-linear approach in $z$) in terms of computation and terminal time. The corresponding time values are depicted in Table 1, for different number of samples $N$ and four paths. The paths include A) the exemplary joint path, B) a rectangle with rounded corners (inspired by the ISO-9283 norm) as EE-path, C) a meander shaped EE-path and D) an arbitrary spatial curve. Due to the lack of space, the reader is referred to [17] for more information of the last three paths. In addition, the relative difference $p_{t_{\mathrm{T}}}$ and $p_{t_{\mathrm{comp.}}}$ compared to the 1) conventional approach are also listed in the table. Positive numbers reflect a reduction in time (e.g. computational time) and vice versa. Concluding the results in Table 1, the terminal time of all four paths is equal for the 1) conventional and 2) presented approach.

**Table 1.** Numerical values of the terminal and computation time as well as their relative differences compared to the conventional time-optimal approach. Denoting with A) the exemplary joint path Fig. 1, B) a rectangle with rounded corners as EE-path, C) a meander shaped EE-path, D) an arbitrary spatial curve, 1) the conventional approach ($\min t_T$) and 2) the presented approach (9).

| | | | A) | | B) | | C) | | D) | |
|---|---|---|---|---|---|---|---|---|---|---|
| | | | 1) | 2) | 1) | 2) | 1) | 2) | 1) | 2) |
| $N = 100$ | $t_T$ | in s | 0.429 | 0.429 | 0.858 | 0.858 | 1.936 | 1.936 | 0.892 | 0.892 |
| | $t_{comp.}$ | in s | 4.969 | 0.490 | 7.363 | 1.030 | 8.733 | 1.781 | 5.047 | 0.650 |
| | $p_{t_T}$ | in % | – | 0.000 | – | 0.000 | – | 0.000 | – | 0.000 |
| | $p_{t_{comp.}}$ | in % | – | 90.132 | – | 86.014 | – | 79.602 | – | 87.120 |
| $N = 200$ | $t_T$ | in s | 0.428 | 0.428 | 0.860 | 0.860 | 1.958 | 1.958 | 0.893 | 0.893 |
| | $t_{comp.}$ | in s | 7.019 | 0.623 | 14.089 | 1.502 | 11.663 | 2.647 | 11.248 | 1.251 |
| | $p_{t_T}$ | in % | – | 0.000 | – | 0.000 | – | 0.000 | – | 0.000 |
| | $p_{t_{comp.}}$ | in % | – | 91.117 | – | 89.34 | – | 77.303 | – | 88.877 |

Further, the resulting trajectories are also equal, which is not shown due to the lack of space. These four paths show that (7) is capable of computing time-optimal solutions. To the best authors' knowledge, there exists no proof that the two optimization problems are equal. Nevertheless, the computational effort is reduced by more than 70% when using (9), which is a huge improvement in computing (sub) time-optimal trajectories. This also holds for planning trajectories with arbitrary smoothness as indicated in Sec. 4.1.

# 5 Conclusion and Outlook

This paper introduces an alternative approach to minimizing the terminal time $t_T$ by maximizing the speed $z$ along the path. In the presented method no singularities (due to zero speed $z = 0$), have to be considered. Furthermore, it is shown that an arbitrary linear parameterization of the path speed (for smooth trajectory planning) results in a parameter-linear problem reformulation, when the joint jerk constraints and viscous friction in (4) i.e. $\mathbf{d}(\sigma) = \mathbf{0}$ are not considered. The comparison of the computation time in Table 1 between the presented and the conventional approach shows a overall reduction of more than 70%, while keeping the same terminal time and optimal solution for the viewed paths. Moreover, it is shown that the computational effort is increasing slightly in the case of smooth joint jerk trajectory planning.

The drastic decrease in computation time suggests the presented approach as good candidate for online applications. By using methods similar to those used in model predictive control, the computation time could be further decreased. Using the presented approach in an online application framework will be addressed in future work.

**Acknowledgement.** This research was funded in whole, or in part, by the Austrian Science Fund (FWF) [I 4452-N].

## References


1. ApS, M.: The MOSEK optimization toolbox for MATLAB manual. Version 9.3.21 (2022). http://docs.mosek.com/9.3/toolbox/index.html
2. Betts, J.T., Huffman, W.P.: Path-constrained trajectory optimization using sparse sequential quadratic programming. J. Guid. Control Dyn. **16**, 59–68 (1993)
3. Bock, H., Plitt, K.: A multiple shooting algorithm for direct solution of optimal control problems. IFAC Proc. Vol. **17**, 1603–1608 (1984)
4. Constantinescu, D., Croft, E.A.: Smooth and time-optimal trajectory planning for industrial manipulators along specified paths. J. Robot. Syst. **17**, 233–249 (2000)
5. Debrouwere, F., et al.: Time-optimal path following for robots with convex-concave constraints using sequential convex programming. IEEE Trans. Robot. **29**, 1485–1495 (2013)
6. Debrouwere, F., et al.: Time-optimal path following for robots with trajectory jerk constraints using sequential convex programming. In: 2013 IEEE International Conference on Robotics and Automation (2013)
7. Gasparetto, A., Zanotto, V.: A new method for smooth trajectory planning of robot manipulators. Mech. Mach. Theor. **42**, 455–471 (2007)
8. Kaserer, D., Gattringer, H., Müller, A.: Nearly optimal path following with jerk and torque rate limits using dynamic programming. IEEE Trans. Robot. **35**, 521–528 (2019)
9. Löfberg, J.: Yalmip : A toolbox for modeling and optimization in MATLAB. In: Proceedings of the CACSD Conference, Taipei, Taiwan (2004)
10. Oberherber, M., Gattringer, H., Müller, A.: Successive dynamic programming and subsequent spline optimization for smooth time optimal robot path tracking. Mech. Sci. **6**, 245–254 (2015)
11. Pfeiffer, F., Johanni, R.: A concept for manipulator trajectory planning. IEEE J. Robot. Autom. **3**, 115–123 (1987)
12. Rockafellar, R.: Integrals which are convex functionals. Pac. J. Math. **24**, 525–539 (1968)
13. Shiller, Z.: Time-energy optimal control of articulated systems with geometric path constraints. In: Proceedings of the 1994 IEEE International Conference on Robotics and Automation, vol. 4, pp. 2680–2685 (1994)
14. Shin, K., McKay, N.: Minimum-time control of robotic manipulators with geometric path constraints. IEEE Trans. Autom. Control **30**(6), 531–541 (1985)
15. Slotine, J.J., Yang, H.: Improving the efficiency of time-optimal path-following algorithms. IEEE Trans. Robot. Autom. **5**(1), 118–124 (1989)
16. Spangelo, I., Egeland, O.: Trajectory planning and collision avoidance for underwater vehicles using optimal control. IEEE J. Oceanic Eng. **19**, 502–511 (1994)
17. Tobias, M., Hubert, G., Andreas, M.: Time-optimal path following for non-redundant serial manipulators using an adaptive path-discretization. In: Robotica. Cambridge University Press (2023). [in press]
18. Verscheure, D., Demeulenaere, B., Swevers, J., De Schutter, J., Diehl, M.: Time-energy optimal path tracking for robots: a numerically efficient optimization approach. In: 2008 10th IEEE International Workshop on Advanced Motion Control, pp. 727–732 (2008)
19. Verscheure, D., Demeulenaere, B., Swevers, J., Schutter, J., Diehl, M.: Time-optimal path tracking for robots: a convex optimization approach. IEEE Trans. Autom. Control **54**, 2318–2327 (2009)

# Wind Aware Planning Using the Introduced RRT*-line Algorithm

Dimitrios Pediaditis[1,2](✉), Juan Galvis[1], and Nikos A. Aspragathos[2]

[1] Autonomous Robotics Research Center, Technology Innovation Institute, Abu Dhabi, United Arab Emirates
dimitrios.pediaditis@tii.ae
[2] University of Patras, Patras, Greece
https://www.tii.ae/

**Abstract.** This paper introduces a new approach: the rapidly exploring random tree star-line (RRT*-line) algorithm for sub-optimal path planning based on distance, time, or energy with knowledge of the wind field. The proposed strategy is predicated on the concept that humans approach a target by aiming directly at it when it is visible. The RRT*-line is an extension of RRT*, which is an asymptotically optimal path planner. The proposed algorithm is tested in cluttered environments with obstacles, considering the optimal distance path, and optimal energy or time under wind load. The new algorithm is explicated in this paper, and a comparison between RRT, RRT*, Dijkstra, and RRT*-line shows that the proposed approach provides bett er results in various complex environments for wind-aware path planning based on distance, time, or energy.

**Keywords:** Sampling-based path planning · Random Geometric graphs · RRT · Trajectory and Path Planning

## 1 Introduction

Nowadays unmanned aerial vehicles (UAVs), unmanned surface vehicles (USVs), and autonomous underwater vehicles (AUVs) are limited to following user-defined way-points in outdoor environments and are not invariably time and energy efficient under a random wind field. A noteworthy limitation in developing practical and compact, unmanned vehicles is the energy required for long-range and long-endurance operations.

Researchers are considering UAV optimal path planning in flow fields using the traveling time and energy to be the optimization criteria [1,2]. A*-DCE [3] introduces the weights of distance, energy, and time to generate paths with significantly different costs. Shangding Gu et al. [4] aiming at the problem of

This work was supported by Autonomous Robots Research Center of the Technology Innovation Institute, United Arab Emirates and the Robotics Group at the University of Patras, Greece.


T. Petrič et al. (Eds.): RAAD 2023, MMS 135, pp. 350–357, 2023.
https://doi.org/10.1007/978-3-031-32606-6_41

motion planning for USVs considering the effects of wind and current by proposing a motion planning method based on regularization- trajectory cells. Thomas Bucher et al. [5] approaches the same problem at the path-planning stage similar to this study, and the results showed that considering the wind and vehicle dynamics in the planning stage leads to a more repeatable and predictable UAV path-following behavior. Sampling-based optimal path algorithms were used by [6], who applied PRM to UAVs wind aware motion planning. In this study, the spatial wind gradients were taken into consideration for the UAV path planning.

Path planning can be performed using grid-based approaches [7], artificial potential fields [8], or sampling-based methods. Sampling-based planning methods such as RRT [9] are non-optimal planning techniques ideal for high-dimensional problems while RRT* [10] is an asymptotically optimal extension of the RRT algorithm [9]. Modifications like informed RRT* [11] and RRT*Smart [12] improve further the performance of sampling-based algorithms.

In this paper, a new path planner called the rapidly exploring random tree star-line (RRT*-line) is proposed. It emulates the way in which humans solve the path-planning problem. In this planner, we introduce an additional step to the existing RRT* algorithm aiming to the goal, which reduces the computational time. This paper is organized as follows. Section 2 describes the introduced RRT*-line path-planning approach; Sect. 3 presents the cost functions that were used for the optimal path planning; Sect. 4 presents the experimental results on distance optimal paths without taking into consideration the wind field and the wind-aware energy and time-optimal paths. Finally, Sect. 5, presents the conclusions and future work.

## 2 Rapidly-exploring Random Trees -Line (RRT*-line)

There are three main areas that robotic researchers are focusing their research on to improve the performance of sampling-based planners: intelligent sampling, fewer obstacle collision checks, and path optimization. In this work we focused on path optimization. RRT*-line is a simple modification to the existing RRT* algorithm, which can find near-optimal solutions. This section methodically describes the steps of RRT*-line (algorithm 1) and underlines the relevant modifications made to the RRT* algorithm.

**Initialization - RRT*:** RRT*-line starts by initializing the search tree with the starting point as its root. Like RRT*, it searches for the near-optimal path, to a planning problem by incrementally building a tree in the free space, consisting of a set of vertices and edges. Points are randomly generated. New vertices are added by growing the tree such that the cost of the nearby vertices is minimized. After a vertex has been connected to the nearest neighbor (see Fig. 1a), the neighbors are again examined. Neighbors are checked if being rewired to the newly added vertex will make their cost decrease. If the cost does indeed decrease, the neighbor is rewired to the newly added vertex.(see Fig. 1b)

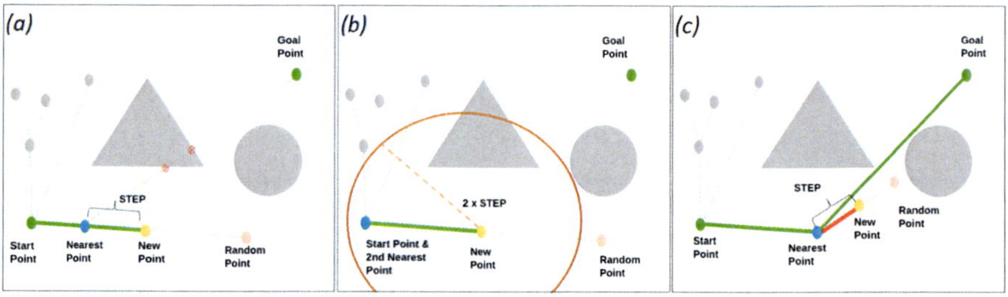


**Fig. 1.** Comparison between RRT*(red) and RRT*-line (green) tree expansion. (Color figure online)

**Algorithm 1:** RRT*-line

PathPlanning ($Point_{start,goal}, Wind - map, step, N$);
**Input** : $Point_{start}, Point_{goal}$: $X, Y$ coordinates, Wind-map: $W \times H$, *step*: length of edge, $N$:Number of iterations, *Obstacles*: Point Cloud
**Output**: *path*: List of x-y coordinates
*Optimization_state* = False;
*directly_to_goal_state* = False;
**while** *True* **do**
  Pick a random point;
  Find the nearest point;
  Find the new point (in distance step);
  Check for Near-by vertices (in distance 2 x step);
  Local collision check;
  **Aim direct towards the goal**;
  Append $Point_{new}$ to *tree*;
  **if** *Optimizationstate* **then**
    **if** *Distance*($Point_{new}, Point_{goal}$) $<$ *step* || *directly_to_goal_state* **then**
      $min\_cost = cost - of - new - point$;
      *Optimization_state* = True;
      *N_iterations* = 1;
    **end**
  **end**
  **else**
    **if** *cost_of_new_point* $<$ *min_cost* **then**
      *min_cost* = *cost_of_new_point*;
      *Optimization_state* = True;
      *N_iterations*++;
    **end**
  **end**
  **if** *N_iterations* $>$ $N$ **then**
    return *path*
  **end**
**end**

**Aim Direct Towards the Goal:** The following modification to the RRT* algorithm introduced to build the proposed Algorithm 1. RRT*-line it searches for an optimal solution by directly connecting each new nearby vertex with the goal point if there is no obstacle collision. (see Fig. 1c) Fig. 1 shows the tree expansion of the RRT*-line till the initial solution. Comparing the growth of the RRT* and the proposed approach's tree evidently the RRT* tree expansion will require more vertices to reach the goal. During the experiment, the sub-function aiming directly toward the goal seemed to speed-up the optimization process without adding high computational effort.

## 3 Wind Aware Cost Functions

This section presents, the cost functions that are used in the proposed planner to search the optimal paths based on distance, wind energy, and time by considering non-uniform wind flow constant at the time field. Due to space limitations, the derivation of the Eq. 1 and 2,assumptions on the wind field and vehicle kinematics are not presented in this paper, but it could be found in [1].

**Time Minimizing Cost:** For the cost formula of the time optimal paths in a 2D planar wind field we consider the speed of the vehicle in still (relative to the flow) $V_{still} = V_{max}$ a fixed maximum still-speed for the vehicle. Based on [1] we calculate the cost function equal $dc1 = dT$:

$$dT = \frac{|w|}{|w|^2 - V_{max}{}^2} dx_2 - \frac{\sqrt{V_{max}{}^2(dx_2{}^2 + dy_2{}^2) - |w|^2 dy_2{}^2}}{|w|^2 - V_{max}{}^2} \quad (1)$$

where $\mid w \mid$ is the speed of the wind in point $p$, $dx_2$ and $dy_2$ is the differential displacement in the local coordinate system (Wind-parallel) Fig. 2.

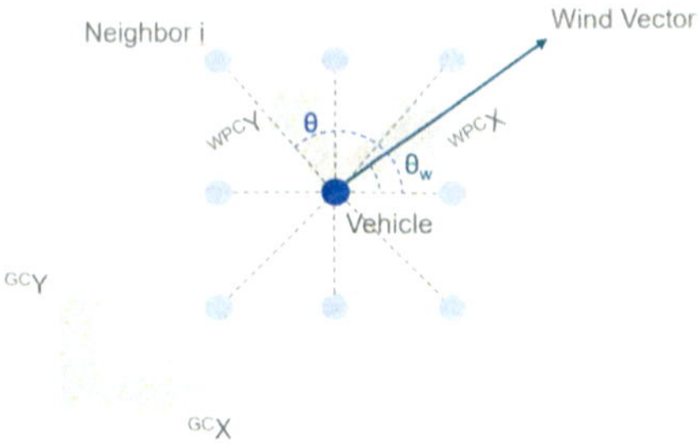


**Fig. 2.** Analysis of local and global coordinate frames

**Energy Minimizing Cost:** The second cost function represents the consumed wind energy moving along a differential displacement $[dx, dy]^T$ in time $dt$ which is equal to $dc_2 = dE = F_d * dx_{still}$ where $F_d = c_d V_{still}$ and $dx_{still} = V_{still} * dt$. Finally, the form of the cost function $dc_2 = dE$ is equal:

$$dE = \rho c_d A \sqrt{dx_2{}^2 + dy_2{}^2} \mid w \mid (\sqrt{dx_2{}^2 + dy_2{}^2} - dx_2) \quad (2)$$

where $\rho$ is the mass density of the fluid and $c_d$ is the drag coefficient.

## 4 Experimental Results

The proposed RRT*-line was tested in environments with a different number of obstacles, by considering the optimal distance with null wind. It was then compared with RRT and RRT* to show the advantages of RRT*-line. The results confirmed that the proposed approach determines paths traveled in a shorter time with fewer vertices than RRT and RRT* in most of the tested configurations with the best results in configurations with a high percentage of free space. Next, a comparison between a uniform graph Dijkstra [1] and RRT*-line for sub-optimal paths based on a given wind field was evaluated.

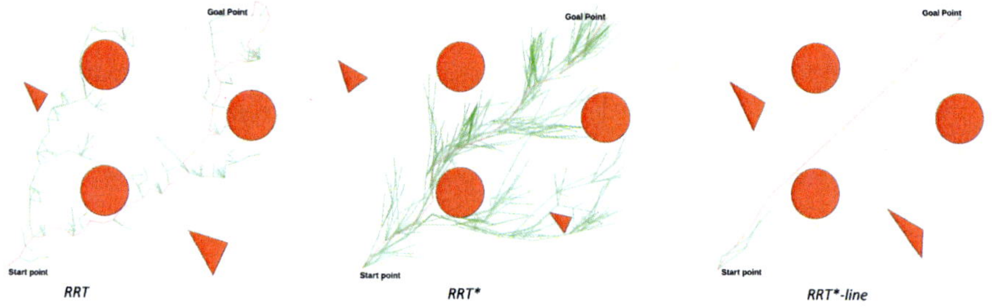


**Fig. 3.** Case A: A comparison between the resulting graphs from the initial path expansion of the algorithms RRT, RRT* and RRT*-line

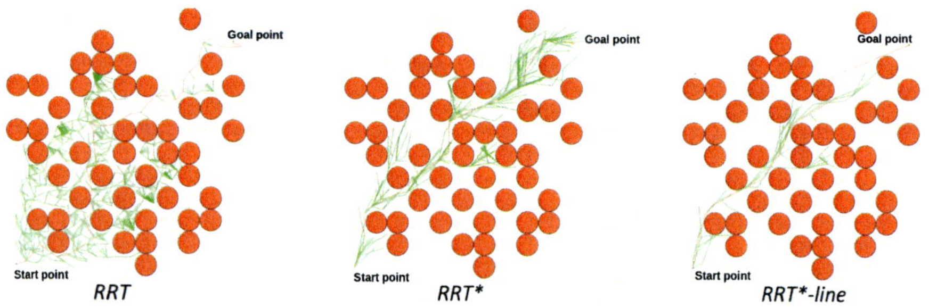


**Fig. 4.** Case B: A comparison between the resulting graphs from the initial path expansion of the algorithms RRT, RRT* and RRT*-line

**Path Planning:** Figure 3 and 4 present two cases for sub-optimal path planning with the same starting (0, 0) and goal (10, 10) points, case A (Fig. 3) and case B (Fig. 4) presents an environment cluttered with a different number of obstacles (red circles and triangles). We compared the paths from RRT*-line to RRT and RRT* based on the minimum travel distance without considering the wind. The comparison was based on three indices: the determined number of vertices, the path cost, and the processing time for the initial solution. (see Table 1) Evidently, the number of vertices of the provided path by the introduced method is much less compared to the two other planners, and the initial solution of the introduced approach is closer to the optimal path.

**Table 1.** Near-optimal path costs based on distance criteria (best, average, and worst results based on repetitive runs of the same planning problem are presented)

| | RRT | | | RRT* | | | RRT*-line | | |
|---|---|---|---|---|---|---|---|---|---|
| Case | Vertices | Time[ms] | $C_{best}$ | Vertices | Time[ms] | $C_{best}$ | Vertices | Time[ms] | $C_{best}$ |
| A | 186 | 145 | 17.2 | 122 | 84 | 14.6 | **6** | **4** | **14.2** |
| A | 149 | 106 | 18.2 | 140 | 103 | 15.3 | 12 | 7 | 14.4 |
| A | 177 | 135 | 19.2 | 148 | 108 | 15.4 | 16 | 8 | 14.9 |
| B | 204 | 169 | 19.1 | 142 | 107 | 15.9 | **49** | **44** | **15.7** |
| B | 310 | 350 | 20.4 | 259 | 270 | 17.1 | 79 | 64 | 16.6 |
| B | 303 | 278 | 23.1 | 368 | 394 | 18.8 | 106 | 78 | 17.1 |

**Wind Aware Path Planning:** A grid map was used to simulate the wind field, which was applied in a discretized area $xE[0, 10]$ and $yE[0, 10]$. In each simulated case, the execution time, and the extra needed energy due to the wind was determined, considering that the vehicle speed was 40km/h, and the average speed of wind was $10km/h$. The color map was created according to the vector (black arrow) and the starting $(0, 0)$ and the goal $(8, 8)$ points. Three cases of simulated wind fields were considered as it is shown in Fig. 5. In case (a) we simulated a tailwind in a rotated way and in case (b) a configuration of tailwind and headwind. Moreover, in cases (c) and (d) we simulated a more complex wind configuration, and we examined the results in a complex windy environment with multiple obstacles. In all cases, using RRT*-line the UAV avoided the headwind (red color) and followed the resolution optimal path based on the uniform graph. By using energy and time cost functions we calculated the energy and the time for the path of the current point to the goal point by dividing the path into equivalent int(distance/step) paths. Thus we could calculate the correct wind velocity each time.

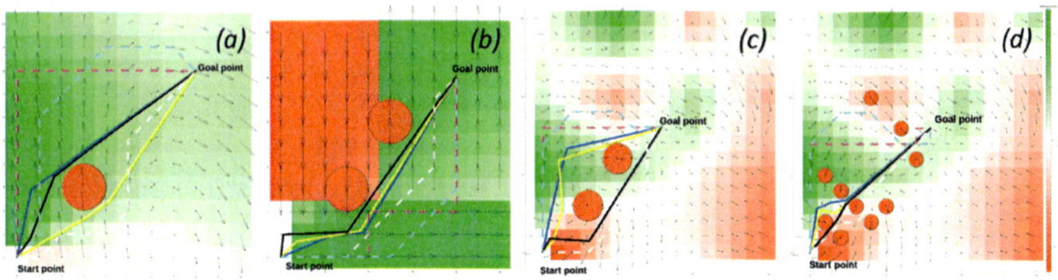


**Fig. 5.** CASE (a) Case (b) Case (c) Case (d): RRT*-line time optimal (Blue path), RRT*-line energy optimal path (yellow path), RRT*-line distance optimal path (black path), time resolution optimal magenta dashed path, energy resolution optimal cyan dashed path, distance resolution optimal white dashed path (Color figure online)

Table 2 presents the results of the near-optimal paths based on RRT*-line and the uniform graph for the 1st case (Fig. 5(a)). As we can see on the semi-optimal paths based on distance our approach gives a shorter path than the uniform path due to the resolution and the size of the obstacle. We notice the same results for the time-optimal paths. On the other hand, the energy optimal path of the uniform graph is more optimal with an execution time 3 times more than our approach, considering the small size of our graph this could indicate an even bigger execution time in a real-world scenario. This drawback can be solved by combining RRT*-line with a more advanced sampling method like Cloud RRT* [13] which samples the free space in a more efficient way, at the same time the execution time will remain less than the uniform graph approach. The algorithms were executed in MATLAB implementation running in a Intel(R) Core(TM) i9-10885H CPU @ 2.40 GHz with 32 Gb RAM.

**Table 2.** Comparison between wind aware path planning based on uniform graph and RRT*-line.

| Path | Uniform graph | RRT*-line |
|---|---|---|
| Distance Optimal | 13.071 | 11.812 |
| Time Optimal | 0.318 | 0.249 |
| Energy Optimal | 113.864 | 293.150 |
| Processing time[ms] | 390 | 84 |

## 5 Conclusion

This paper presents a new path planner called RRT*-line, which extends the RRT* by emulating the way humans approach a target. The proposed planner improves the convergence time and determines the energy and time-optimal paths, by providing a path that minimizes energy consumption caused by the known wind field or minimizes the travel time which can be critical for UAVs and USVs. The results show that RRT*-line gives better results in an environment with a wide free space compared to RRT*. The extra step seems to increase the computational cost in environments with a low percentage of free space. However, again the results seem better compared to RRT and RRT*. Moreover, the study showed that RRT*-line has less dependence on the dimension and depends mostly on the number of obstacles between the current position and the goal. Future work would be to extend this approach to a three-dimensional wind field and test it on real platforms to confirm the results shown in this paper. Moreover, the most advanced sampling-based algorithms like PRM* [14], ABIT* [15] and Informed RRT* [16] can implement this goal-focused concept to find near-optimal paths.

**Acknowledgment.** This work was supported by the Technology Innovation Institute (TII) of United Arab Emirates and the Robotics Group at the University of Patras, Greece.

## References


1. Kularatne, D., Bhattacharya, S., Hsieh, M.A.: Time and energy optimal path planning in general flows. In: Robotics: Science and Systems (2016)
2. Liu, L., Sukhatme, G.: A solution to time-varying markov decision processes. IEEE Rob. Autom. Lett. **3**, 1631–1638 (2018)
3. Song, Z., Zhang, J., Wu, D., Tian, W.: A novel path planning algorithm for ships in dynamic current environment. Available at SSRN 4331220
4. Gu, S., Zhou, C., Wen, Y., Xiao, C., Knoll, A.: Motion planning for an unmanned surface vehicle with wind and current effects. J. Marine Sci. Eng. **10**(3), 420 (2022)
5. Bucher, T., Stastny, T., Verling, S., Siegwart, R.: Robust wind-aware path optimization onboard small fixed-wing uavs. In: AIAA SCITECH 2023 Forum, p. 2640 (2023)
6. Thanellas, G., Moulianitis, V., Aspragathos, N.: A spatially wind aware quadcopter (UAV) path planning approach. IFAC-PapersOnLine **52**, 283–288 (2019)
7. Dijkstra, E.: A note on two problems in connexion with graphs. Numerische Mathematik **1**, 269–271 (1959)
8. Hwang, Y., Ahuja, N.: A potential field approach to path planning. IEEE Trans. Rob. Autom. **8**, 23–32 (1992)
9. LaValle, S.: Rapidly-exploring random trees : a new tool for path planning. In: The Annual Research Report (1998)
10. Karaman, S., Frazzoli, E.: Incremental sampling-based algorithms for optimal motion planning. ArXiv, abs/1005.0416 (2010)
11. Gammell, J., Srinivasa, S., Barfoot, T.: Informed rrt*: optimal sampling-based path planning focused via direct sampling of an admissible ellipsoidal heuristic. In: 2014 IEEE/RSJ International Conference on Intelligent Robots and Systems, pp. 2997–3004 (2014)
12. Nasir, J., et al.: Rrt*-smart: a rapid convergence implementation of rrt*. Int. J. Adv. Rob. Syst. **10** (2013)
13. Kim, D., Lee, J., Yoon, S.: Cloud rrt*: sampling cloud based rrt*. In: 2014 IEEE International Conference on Robotics and Automation (ICRA), pp. 2519–2526 (2014)
14. Karaman, S., Frazzoli, E.: Sampling-based algorithms for optimal motion planning. Int. J. Rob. Res. **30**, 846–894 (2011)
15. Strub, M.P., Gammell, J.D.: Advanced bit*(abit*): sampling-based planning with advanced graph-search techniques. In: 2020 IEEE International Conference on Robotics and Automation (ICRA), pp. 130–136. IEEE (2020)
16. Gammell, J., Srinivasa, S., Barfoot, T.: Bit*: batch informed trees for optimal sampling-based planning via dynamic programming on implicit random geometric graphs. ArXiv, abs/1405.5848 (2014)

# An Encoder-Decoder Architecture for Smooth Motion Generation

Zvezdan Lončarević[1,2(✉)], Ge Li[3], Gerhard Neumann[3], and Andrej Gams[1,2]

[1] Jožef Stefan Institute, Jamova cesta 39, 1000 Ljubljana, Slovenia
{zvezdan.loncarevic,andrej.gams}@ijs.si
[2] Jožef Stefan International Postgraduate School, Jamova 39, Ljubljana, Slovenia
[3] Karlsruhe Institute of Technology, Adenauerring 4, 76131 Karlsruhe, Germany
{ge.li,gerhard.neumann}@kit.edu
https://abr.ijs.si/, https://alr.anthropomatik.kit.edu/

**Abstract.** Trajectory generation for a dynamic task, where the outcome of the task is not ensured by simple repetition of a motion, is a complex problem. In this paper we explore a methodology for motion generation that retains the correspondence of the executed dynamic task. Throwing, which is not explored in this paper, is a very illustrative example. If we just imitate human throwing motion with a robot, the outcome of the throw with a robot will most likely not be very similar to the demonstrated one. In this paper we explore a deep encode-decode architecture, combined with ProDMP trajectory encoding in order to actively predict the behavior of the dynamic task and execute the motion such that the task is observed. Our example is based on the task of dragging a box across a surface. Guided by future work on transferability, in this paper we explore the parameters of the approach and the requirements for effective task transfer to a new domain.

**Keywords:** generalization · neural networks · movement primitives · robotic manipulation · dimensionality reduction · imitation learning

## 1 Introduction

Trajectory generation has been studied in robotics in many different setting, including in learning by demonstration (LbD) [2]. When the final position of the robot is directly linked to the success of the task, e. g., in reaching, imitation learning and proper encoding of the task might produce accurate results even for different setting – for a different starting point of reaching or even to a different reaching point. Examples where the encoding solves some of the issues are dynamic movement primitives (DMPs) [5], Gaussian mixture models (GMMs) [1], ProMPs [10], etc. For example, in the DMPs, the goal of the DMP enforces the final position of the robot [5]. Task specific generalization extend the approach for more diverse tasks, and was shown to work for the case of reaching [11].


T. Petrič et al. (Eds.): RAAD 2023, MMS 135, pp. 358–366, 2023.
https://doi.org/10.1007/978-3-031-32606-6_42

The problem is different where the final position of the robot is not directly related with the outcome of the task. Example is throwing, where the position and motion of the robot at the release of the projectile is related to the outcome of the task [8]. There, methods as DMPs by themselves do not generalize well for variations of the task. The aforementioned generalization has shown much better success [9,11]. Deep neural networks for skill modelling exploit generalization properties and were shown to perform well for such cases. However, the transferability of such methods, for example between simulated and real-world environments, requires re-training and adaptation of the skill model, or the training needs to be so broad that the domain randomization methods cover the properties of the target environment [7].

In this paper we explore a deep neural network based encoder-decoder motion architecture that generalizes well between task instances, but is also prepared in a manner that will allow for seamless transfer to a new domain. We consider a dynamic task, where the final position of the robot is not directly related to the successful outcome of the task. In this sense, the imitation learning of this approach transfer maintains the correspondence of the task. The approach is, as the name suggest, based on an encoder-decoder architecture. We tested two different architectures and combined the motion generation with a novel, ProDMP encoding methodology [6].

The applicability of the proposed approach depends on the size of the latent space, the means of sampling and the level of aggregation of data before predicting the behavior of the executed dynamic task. The transferability, on the other hand, depends on the diversity of data. In this paper we explore these aspects of motion generation that should maintain the corespondence of a dynamic task.

## 2 Experimental Setup and Definition of the Problem

Our imitation learning approach was tested on the problem of robot manipulating the box by dragging it on the table using only a rod as an end-effector. Since the task was performed in the task space of the robot, in order to avoid problems with calculation of the inverse kinematics, we used redundant Franka Panda robot with 7-DoF. The presented task is particularly hard even for human because it requires a lot of patience and corrections in order to achieve desired box position and orientation. Our setup is shown in Fig. 1 - right. The box was initialized in a random position (shown in red) and the goal for the robot was to move the box to the desired and always the same final position (shown in green).

Neural networks have a great potential in generalizing the knowledge acquired by human demonstration, thus they were the obvious choice for being the core of our imitation learning procedure. The database for training the neural network was acquired by human demonstration where the person was kinestetically guiding a simulated robot in order to push the box from the randomly initialized position to the final position.

In our work, two different architectures presented in the following section were tested on this task with two different ways of choosing the action. The action could be chosen from the distribution by only randomly sampling from

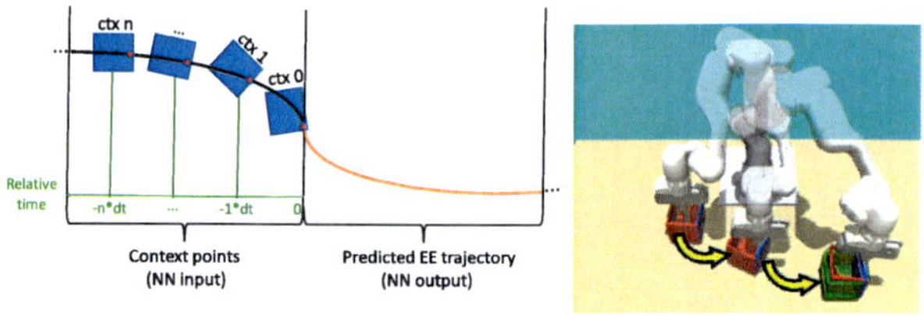


**Fig. 1.** Left: Inputs and outputs of the trajectory prediction architecture: Input to the network are several context points. Each context point contains information about box position and orientation (blue) and information about robot end-effector position (red) together with relative time in which it happened (green axis). Output of the architecture is the predicted end-effector trajectory (orange). Right: Execution of the each replanned part of trajectory on simulated robot. (Color figure online)

the distribution or by taking the mean of sampled parameterized actions. With both architectures, the input of the neural network were the current state of the box (position and orientation) and the current state of the robot (position of the end-effector). In this paper, we also introduce so-called aggregation where we do not only take the current state in consideration when choosing the further action, but also take into the consideration multiple previous robot/box states. In addition to this, experiments determining the required latent space size of the network were performed.

## 3 Learning Architecture

The goal of both architectures used in this work is to predict the trajectory based on the state of the setup. The state of the setup can be described using information about current box position in $x$ and $y$ axis and orientation about $z$ axis (denoted with $\alpha_{box}$) as well as robot end-effector position and velocity in $x$ and $y$ axis. It should be noted that $z$ position of the end-effector is held constant and its shape was cylindrical so that it's orientation doesn't have influence on the movement of the box. For this reason, these two values were left out. Together with the information about the relative time at which the system was in the given state, we obtain the necessary information for predicting the trajectory - Fig. 1 (left). This information is called context and each context point contains eight values: $\mathbf{ctx} = [x_e, y_e, \dot{x}_e, \dot{y}_e, x_{box}, y_{box}, \alpha_{box}, t]$.

As shown in depictions of the architectures in Figs. 2 and 3, several ($m$) context points are first sent to two encoders ($\mu$ and $\sigma^2$) in order to get the latent space observation ($\boldsymbol{\mu}_r$) and an uncertainty $\boldsymbol{\sigma}_r^2$ measuring how important this observation is. The following step is to obtain latent state posterior $p(\boldsymbol{z}|\mathbf{ctx})$ which is given by a factorized multivariate Gaussian distribution $\mathcal{N}(\boldsymbol{z}|\boldsymbol{\mu}_z, \boldsymbol{\sigma}_z^2)$. For this purpose, aggregator is used. As an input, it takes the set of latent observations and their uncertainties together with the prior distribution: $(\boldsymbol{\mu}_A, \boldsymbol{\sigma}_A^2)$. Prior distribution was set to be normalized Gaussian distribution with zero mean. For the both architectures, this part is the same. The difference comes from the part where the trajectory representation is sampled and decoded.

### 3.1 Architecture with PB-Decoder

The first architecture that we used inlcuded Parameter Based (PB) Decoder [12]. This decoder consists of the two decoders that take posterior distribution as an input and output mean value of the ProDMP weights and goal ($\mu_{w_g}$) as well as Cholesky decomposition of the covariance matrix of the weights and goal distribution ($\boldsymbol{L}_{w_g}$). This is further used by ProDMP architecture that is, when given initial position and velocity of the end-effector ($\boldsymbol{y}, \dot{\boldsymbol{y}}$), able to convert this values into the mean and covariance of the trajectories: $\boldsymbol{\mu}_{\tau}$, $\boldsymbol{\Sigma}_{\tau}$. For training of this network, negative log-likelihood cost function that increases the possibility of observing the trajectory $\boldsymbol{y}$, given the distribution with mean $\boldsymbol{\mu}_{\tau}$ and covariance $\boldsymbol{\Sigma}_{\tau}$ that are determined with decoder NN parameters $\boldsymbol{\theta}$, was used:

$$\mathcal{C}(\boldsymbol{\tau}, \boldsymbol{y}_b, \dot{\boldsymbol{y}}_b, \mathbf{ctx}) = -\log \mathcal{N}(\boldsymbol{y}|\boldsymbol{\mu}_{\tau}, \boldsymbol{\Sigma}_{\tau}, \boldsymbol{\theta}). \tag{1}$$

An important thing to note with this architecture is that the trajectory is sampled from the distribution of ProDMP parameters when executing the forward pass in the evaluation so that only one trajectory is decoded.

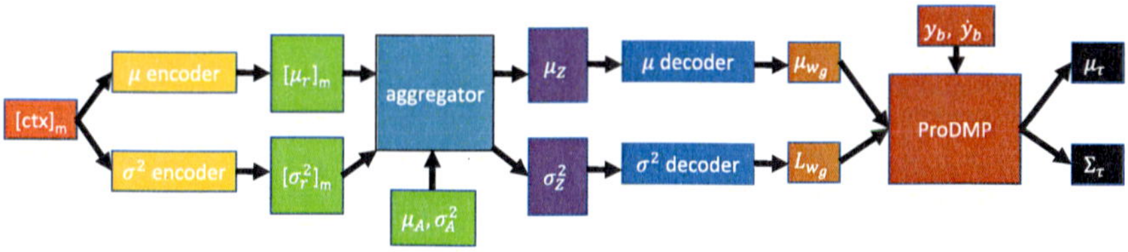


**Fig. 2.** Architecture with PB decoder

### 3.2 Architecture with MC-Decoder

The second, Monte Carlo (MC) architecture, was based on taking the action sample directly in the latent space instead of taking the sample from the space of ProDMP weights and goal. This architecture is presented in Fig. 3. It takes also posterior distribution as the starting point for the decoding process. From this distribution it takes $n$ samples $[\boldsymbol{z}]_n$ and they are input to one decoder network that has weights and goal as the output: $[\boldsymbol{w}_g]_n$. Variance of the posterior is directly forwarded to the second decoder that outputs distribution of the weights and goal - $\boldsymbol{\sigma}^2_{w_g}$. From this two values, ProDMPs are able to calculate $m$ trajectories $[\boldsymbol{\tau}]_m$ and variance $\boldsymbol{\sigma}^2_{\tau}$. Cost function used to train this network was negative marginal log-likelihood. This function increases probability of observing ground-truth trajectory given the several ($n$) samples from the distribution and variance (both determined by decoder parameters $\boldsymbol{\theta}$):

$$\mathcal{C}(\boldsymbol{\tau}, \boldsymbol{y}_b, \dot{\boldsymbol{y}}_b, \mathbf{ctx}) = -\log \mathcal{N}(\boldsymbol{y}|[\boldsymbol{\tau}]_n, \boldsymbol{\sigma}^2_{\tau}, \boldsymbol{\theta}). \tag{2}$$

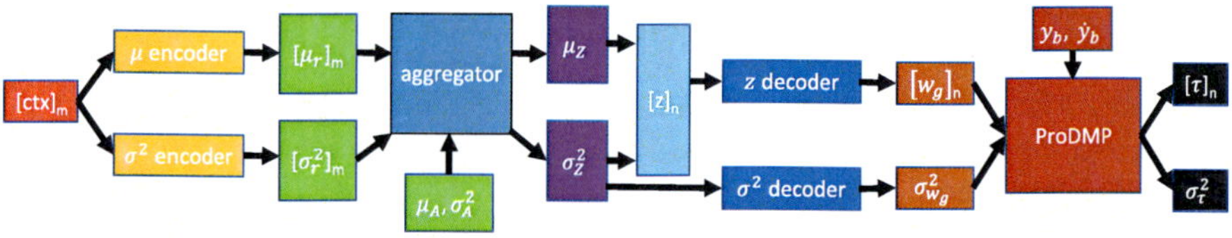


**Fig. 3.** Architecture with MC decoder

An important thing with this architecture is to note that the trajectory is sampled in the latent space distribution. Consequently when using the forward pass in the evaluation, only $z$ decoder is used and only one trajectory is predicted.

### 3.3 Probabilistic Dynamic Movement Primitives (ProDMPs)

The last part of our architecture is based on ProDMPs. This new method [6] enables for combining the benefits from the both Dynamic Movement Primitives (DMPs) [4] and Probabilistic Movement Primitives (ProMPs) [10]. Having DMP weights and goals determined by decoder network and given the initial position and velocity as the initial conditions $(\boldsymbol{y}_b, \dot{\boldsymbol{y}}_b)$, this method is able to provide trajectory distribution given as a multivariate Gaussian distribution.

In this paper we use the benefits of the method being able to make a smooth transition between the trajectory segments and presenting the predicted trajectory as a distribution. Detailed information including auxiliary math describing the ProDMPs is given in [6].

## 4 Training and Experimental Protocol

In first step of our research, the complete database of the human trajectories was collected. For this purpose, we used a simulated Panda Franka robot in MuJoCo environment. Using the computer mouse for kinesthetic guidance of the robot, we acquired the database of 300 trajectories and stored all the relevant data - context points. All the trajectories lasted 2 s.

This database was further used to train both of the architectures presented in Figs. 2 and 3. Both mean and variance encoder consisted of the input layer sizes of 8 that corresponds to the dimensionality of our context. All the encoders used in this paper consisted of 128 neurons in 3 hidden layers. Output size, determining also the latent space size was chosen to be $S \in \{8, 16\}$. Consequently, decoders had input size S. Three hidden layers consisting of 128 neurons were used here just as with the encoders, and output size was set to be 52. This was caused by us choosing for both degrees of freedom to use 25 radial basis functions in ProDMP and one goal position. Activation function used in our case was Leaky-ReLu.

We trained both architectures with (PB+agg and MC+agg) and without aggregation (PB and MC). All the training was done in 10000 epochs and (for the sake of completeness of the results) in 10 different seeds. For the cases where aggregation was used, we used 5 last context points, while in other cases we used only the last. For the both cases of using MC decoders, we used $n = 16$ Monte-Carlo samples for choosing the latent space variable. For each randomly chosen point of the trajectory, only the following 0.5 s were considered when training. When evaluating the trained architectures, we considered following cases:

- **MC + agg (mean)**: MC architecture combined with aggregation was used with 5 context points in the both training and evaluation. In the evaluation case, only mean $z$ variable was used and decoded.
- **MC (mean)**: MC architecture without aggregation was used in both training and evaluation. In the evaluation case, only mean $z$ variable was used and decoded.
- **PB + agg (mean)**: PB architecture combined with aggregation was used with 5 context points in both training and evaluation. In the evaluation case, only mean of the trajectory distribution was used.
- **PB (mean)**: PB architecture without aggregation was used in both training and evaluation. In the evaluation case, only mean of the trajectory distribution was used.
- **MC + agg (stochastic)**: MC architecture combined with aggregation was used with 5 context points in both training and evaluation. In the evaluation case, $z$ variable was sampled from the distribution and decoded.
- **MC (stochastic)**: MC architecture without aggregation was used in both training and evaluation. In the evaluation case, $z$ variable was sampled from the distribution and decoded.
- **PB + agg (stochastic)**: PB architecture combined with aggregation was used with 5 context points in both training and evaluation. In the evaluation case, trajectory was sampled from the final trajectory distribution.
- **PB (stochastic)**: PB architecture without aggregation was used in both training and evaluation. In the evaluation case, trajectory was sampled from the final trajectory distribution.

In the evaluation of the learned policy, neural network was always predicting the following 0.5 s of the trajectory based on the provided context. Total allowed length of the trajectory was 4 s, thus leading to 8 replanning steps.

## 5 Results

Figure 4 shows the average error in the box position and orientation for the latent space sizes of 8 and 16. Higher latent space sizes were also tested, but they did not bring any increase in the performance. Lower latent space sizes were excluded because they were insufficient to catch the complexity of the observation. From the graphs, it can be seen that using the mean rather than randomly sampled variable resulted in better performance. Also, it is noticeable that aggregation increases the performance when using the PB architecture while the difference when applied on MC architecture was much lower. The best performance was observed when using the MC+agg and PB+agg methods.

The performance was tested for all the experiments in 10 different seeds and with using 20 test trajectories. Histograms showing the distribution of the error in position and orientation are shown in Fig. 5.

Since stochastic methods are, due to their diversity in the data, important for the transferability of the learned policy to the situations where the parameters of the task are slightly different (changed friction or mass of the box), we additionally tested PB+agg method and MC+agg method for the error increase with respect to the increase in standard deviation of the trajectories.

Average error is given as $\sum_{s=1}^{S}\sum_{k=1}^{K}(|e|_{s,k}^{x} + |e|_{s,k}^{y} + 0.1|e|_{s,k}^{\varphi})$, where S is the number of seeds, K is the number of test trajectories for the each seed and $|e|_{s,k}^{x}$, $|e|_{s,k}^{y}$, $|e|_{s,k}^{\varphi}$ are errors in the final box position in $x$, $y$ axis given in meters and orientation about $z$ axis given in degrees, respectively. For obtaining error increase as a consequence of standard deviation, entropy of the MC policy was varied by multiplying the standard deviation $\sigma_z^2$ with coefficients $[0, 0.1, 0.5, 1.0, 2.0, 5.0, 10, 15]$ and for the case of PB multiplying Cholesky of the covariance matrix with coefficients $[0, 0.1, 0.5, 1.0, 2.0, 5.0]$. Larger coefficients consequently resulted in higher standard deviation of the resulting trajectories. Graphs show that MC+agg method offers solutions of higher diversity while maintaining the smaller error. This makes it more suitable for acquiring the policy that will be later transferred into the slightly different environment [3].

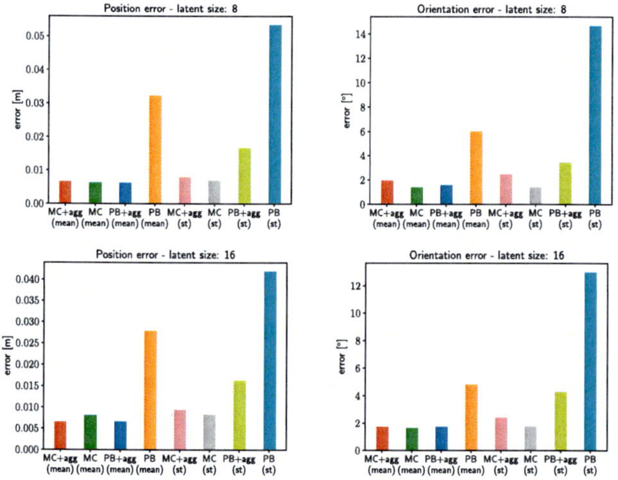


**Fig. 4.** Bar graphs showing the average error in position and orientation for all the tested cases for latent space sizes 8 and 16.

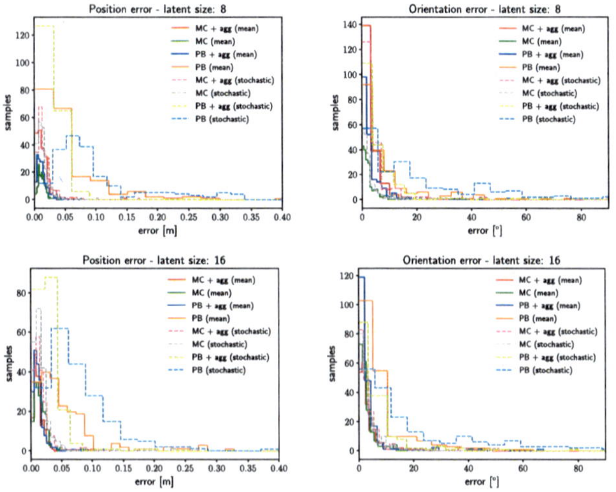


**Fig. 5.** Histogram showing the distribution of the error in position and orientation for different latent space size.

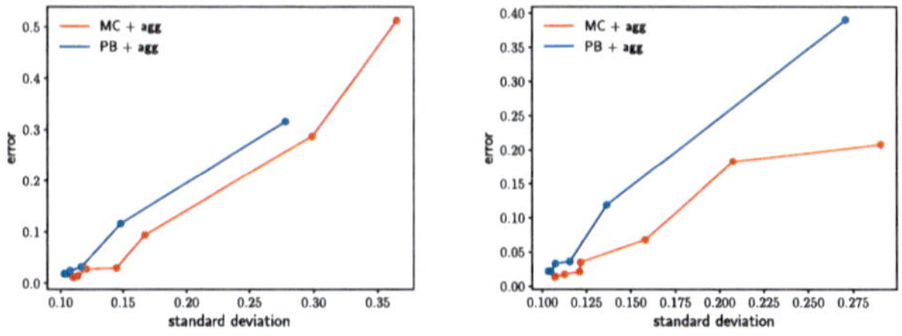


**Fig. 6.** Error with respect to the increase in standard deviation of the learned policy.

# 6 Conclusion

Presented approaches show the benefits of pairing ProDMP approach with neural network in order to obtain successful imitation learning policies for dynamic tasks. Moreover, it shows benefits of sampling the trajectory directly in the latent space of encoder-decoder network and positive influence of aggregation to the final policy. We plan to continue research in the direction of applying RL for few shot transfer learning to a new environment. In the future we will use the MC or MC+agg models since they will provide higher precision, as shown in results, while also having higher diversity of the data, as shown in Fig. 6, thus making the transfer learning easier as reported in [3].

**Acknowledgement.** This research was funded by ARRS research grant Robot Textile and Fabric Inspection and Manipulation - RTFM (J2-4457), ARRS program group Automation, Robotics, and Biocybernetics (P2-0076) and DAAD Grant 57588366.

## References


1. Calinon, S., Alizadeh, T., Caldwell, D.G.: On improving the extrapolation capability of task-parameterized movement models. In: International Conference on Intelligent Robots and Systems, pp. 610–616 (2013)
2. Dillmann, R.: Teaching and learning of robot tasks via observation of human performance. Robot. Auton. Syst. **47**(2–3), 109–116 (2004). https://doi.org/10.1016/j.robot.2004.03.005
3. Eysenbach, B., Levine, S.: Maximum entropy RL (provably) solves some robust RL problems (2021). https://doi.org/10.48550/ARXIV.2103.06257
4. Ijspeert, A., Nakanishi, J., Pastor, P., Hoffmann, H., Schaal, S.: Dynamical movement primitives: learning attractor models for motor behaviors. Neural Comput. **25**(2), 328–373 (2013)
5. Ijspeert, A.J., Nakanishi, J., Hoffmann, H., Pastor, P., Schaal, S.: Dynamical movement primitives: learning attractor models for motor behaviors. Neural Comput. **25**(2), 328–373 (2013)
6. Li, G., Jin, Z., Volpp, M., Otto, F., Lioutikov, R., Neumann, G.: ProDMPs: a unified perspective on dynamic and probabilistic movement primitives (2022). https://doi.org/10.48550/ARXIV.2210.01531
7. Lončarević, Z., Simonič, M., Ude, A., Gams, A.: Combining reinforcement learning and lazy learning for faster few-shot transfer learning. In: 21st International Conference on Humanoid Robots (Humanoids), pp. 285–290 (2022)
8. Pahič, R., Lončarević, Z., Ude, A., Nemec, B., Gams, A.: User feedback in latent space robotic skill learning. In: 18th International Conference on Humanoid Robots (Humanoids), pp. 270–276 (2018)
9. Pahič, R., Lončarević, Z., Gams, A., Ude, A.: Robot skill learning in latent space of a deep autoencoder neural network. Robot. Auton. Syst. **135**, 103690 (2021)
10. Paraschos, A., Daniel, C., Peters, J., Neumann, G.: Probabilistic movement primitives. In: Neural Information Processing Systems, pp. 2616–2624 (2013)
11. Ude, A., Gams, A., Asfour, T., Morimoto, J.: Task-specific generalization of discrete and periodic dynamic movement primitives. Trans. Robot. **26**(5), 800–815 (2010)
12. Volpp, M., Flürenbrock, F., Grossberger, L., Daniel, C., Neumann, G.: Bayesian context aggregation for neural processes. In: International Conference on Learning Representations (2021)

# Cloth Flattening with Vision-to-Motion Skill Model

Peter Nimac[1,2](✉) and Andrej Gams[1]

[1] Jožef Stefan Institute, Jamova cesta 39, SI-1000 Ljubljana, Slovenia
peter.nimac@ijs.si

[2] Jožef Stefan International Postgraduate School, Jamova cesta 39, SI-1000 Ljubljana, Slovenia

**Abstract.** The handling of textiles by robots remains a largely unexplored and underdeveloped area of robotics. This is mainly due to the complexity of the actions resulting from the properties of the textiles and the difficulty in accurately determining the state of the textiles. Due to the considerable variability in the shape and size of planar, non-rigid objects, we have addressed this challenge by using advanced deep learning methods. In this work, we demonstrate a vision-to-motion DNN (Deep Neural Network ) trained to straighten a single crumpled corner on a rectangular piece of fabric that was deformed and then flattened inside a simulated environment. The neural network was trained to identify a correct grab point at which to grab the simulated fabric, and also a correct drop point to which to move the grabbed piece of fabric. For this simplified example, our trained model was able to achieve good results with an average error of 4.4 mm in determining the grab point position and an average error of 4.2 mm in determining the drop point position. Using the predicted points, the robot performed a smoothing motion to bring the deformed fabric almost to its canonical state.

**Keywords:** Vision-to-Motion · Deep Neural Network · Cloth Flattening

## 1 Introduction

Robotic manipulation of textiles is still a largely unexplored field and has only recently begun to gain noticabletraction. As a result, the handling of clothing, fabrics and other textile elements is still done manually, except for the most basic tasks. The properties of textiles, in particular their deformability, self-collision and self-occlusion; lack of general approaches on the detection and robotically

Supported by young researcher grant (PR-11324), research grant Robot Textile and Fabric Inspection and Manipulation – RTFM (J2-4457) and program group Automation, Robotics, and Biocybernetics (P2-0076), all by the Slovenian Research Agency (ARRS).


T. Petrič et al. (Eds.): RAAD 2023, MMS 135, pp. 367–374, 2023.
https://doi.org/10.1007/978-3-031-32606-6_43

handle such materials; and the wide variety of actions involved in handling textiles, i.e. for different types of clothing, are prohibitive factors for robotising the handling of textiles. Due to these characteristics, robots need to perform complex actions to successfully handle textiles. Estimation of the state of garments has also only recently been tackled [7]. The sheer number of possible states of textile objects and the actions associated with them makes it not only difficult to manipulate a textile object, but even just to describe it. However, advances in machine learning methods have made it possible to generate end-to-end perception-action pairs for robots.

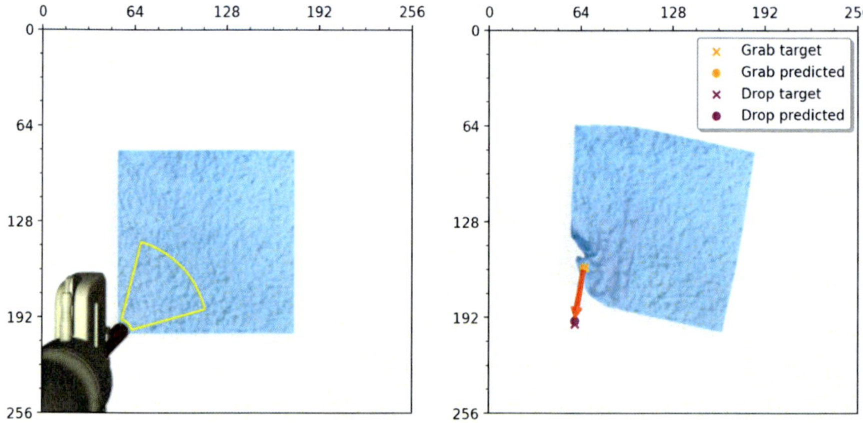


(a) (b)

**Fig. 1.** Subfigure a shows the area where the robot pushed the bottom left corner during data acquisition. Subfigure b shows predictions of grab and drop points for rectangular cloth manipulation based on a single input image. Crosses mark targeted grab (orange) and drop (purple) points, while dots mark the predicted grab and drop points (Color figure online).

In this paper, we demonstrate a method for smoothing fabrics in a simulated environment where the skill is trained with a DNN describing a vision-to-motion model, as shown in Fig. 1. It is a continuation of our work [4], where the robot uses the information from the DNN to perform the smoothing motion on a simulated piece of fabic. Similar methods were shown in [3,6,8,9]. In [3] Lee et al. recently showed that fabric handling can be learned directly in the real world by the random interaction of the robot with a piece of cloth. To this end, they used the DQL (Deep Q-Learning) approach with a self-supervised learning pipeline to learn fabric handling through autonomous data collection.

In [6] Seita et al. have proposed a DIL (Deep Imitation Learning) model-free approach to fabric smoothing that does not use reinforcement learning. Unlike reinforcement learning, imitation learning is a type of supervised learning in which the agent learns a control policy by analysing demonstrations of the policy performed by an algorithmic or human supervisor [2]. For smoothing fabrics, Seita et al. chose an approach in which the fabric is smoothed by being pulled at its corners. With this approach, the policy can be easily defined in a simulator and coded as an algorithmic supervisor.

In [8], Tsurumine et al. have demonstrated the use of DRL (Deep Reinforcement Learning ) algorithms that combine value function-based reinforcement learning with automatic feature extraction from high-dimensional observations in DNN to improve the sample efficiency and learning stability with fewer samples.

In [9] Wu et al. proposed another image-based DRL method to speed up training by training the pick and place as two separate policies instead of training both as a single set of actions. The latter approach would work for rigid objects because an entire object is moved uniformly. This is not the case for non-rigid objects, as they deform during movement, which can lead to inefficient learning.

We use the approach where the RGB image of the deformed cloth is used as input to an image-to-motion DNN and the initial and final points of the robot's movement are the output.

## 2 Dataset

The data samples were collected in MuJoCo simulated environment. The fabric was represented in the simulation as a square piece of cloth with a side of $a_s =$ 28 cm and a thickness of $a_t = 5$ mm. It was described as a grid of flexibly connected capsule nodes with a raster of $20 \times 20$ as shown in Fig. 2.

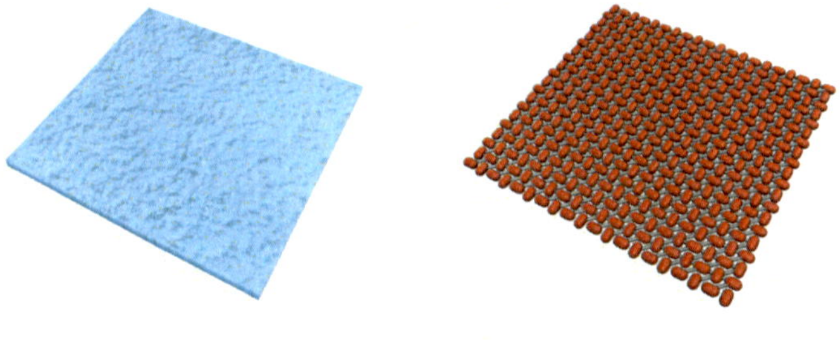

**Fig. 2.** Cloth raster representation inside the simulation.

To generate the database of deformed cloths, we simulated 2000 deformations. In each simulation, we initialised the cloth in a canonical (flat) state and used the robot to push the corner of the cloth linearly towards its centre while pressing on it with a round stick. In each run, it pushed the corner a random distance between 3 cm and 14 cm and at a random angle between 20° and 70° (Fig. 1a). At the end of each simulation cycle, we recorded an image of the deformed state of the fabric and a pair of points where the robot started and concluded the deforming motion.

Generated this way, the data pairs $D$ in the dataset have the following structure:

$$D = \{C_j,\ M_j\}_{j=1}^{P}, \tag{1}$$

where $P$ is the number of generated data pairs, constructed of images $C_j \in \mathbb{R}^{3\times H\times W}$ with height $H$ and width $W$ and corresponding grab and drop points of linear smoothing motions

$$M_j = \{\mathbf{g}_j,\ \mathbf{d}_j\}. \tag{2}$$

In above equation, vectors $\mathbf{g}_j$ and $\mathbf{d}_j$ are defined as

$$\mathbf{g}_j = \{g_{x,j},\ g_{y,j}\}, \tag{3}$$

$$\mathbf{d}_j = \{d_{x,j},\ d_{y,j}\}, \tag{4}$$

and represent $x$ and $y$ pixel coordinates for grab $\mathbf{g}_j$ and drop $\mathbf{d}_j$ locations.

## 3 DNN Architecture for Vision-to-motion Skill Encoding

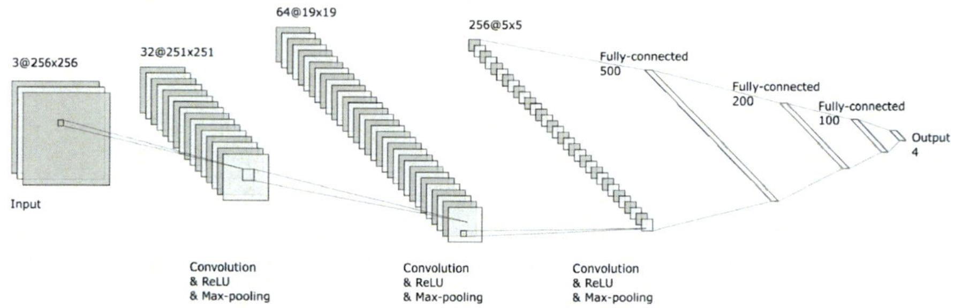


**Fig. 3.** The architecture of image-to-motion DNN that was used to generate grab and drop point pairs.

Based on the work of Pahič et al. [5], we constructed an image-to-motion network (shown in Fig. 3) used in our experiments, which is built as follows. The architecture consists of three convolutional layers and three fully-connected layers. The convolutional layers take a $3 \times 256 \times 256$ pixel RGB image as input. After convolution has been performed on the input data, it is followed by ReLU and Max Pool operations respectively on each convolution layer. After the input data is processed by the last convolution layer, it is squished into a one-dimensional vector and passes through three fully connected layers. The final output is a pair of grab and drop points in a 2D plane.

### 3.1 Performance

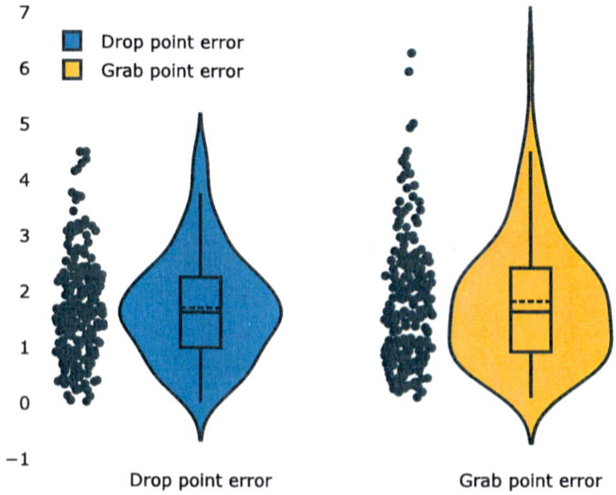


**Fig. 4.** The distribution of drop point and grab point errors is shown with a violin plot and a box-and-whisker plot. On the left-hand side of each plot, the dark dots represent the actual calculated errors for all samples that were tested. Samples that fall within the range of the boxes represent 50% of all samples in the test set.

The network was set to train over the course of maximum 1500 epochs, with the learning rate set to $1 \times 10^{-7}$ and the weight decay rate set to $1 \times 10^{-4}$ with a batch size of 10. The loss during training was calculated as the MSE (mean square error) between the predicted and target pixel coordinates. Training was stopped after 100 consecutive epochs with no decrease in validation loss. This way we ensured that the network did not overfit to the training data despite the number of trainable parameters being very high. The Adam optimiser [1] was used to update the network parameters during training.

We evaluated the model's achieved accuracy in two parts. First, we tested our model on a test dataset of 200 images, that were generated the same way as a training set. Model's accuracy, in the form of drop point error and grab point error from the test set, is shown in Fig. 4. To estimate the model's output error, we calculated the Euclidean distance between the target and predicted points. The trained model on average misses the target drop point by $\bar{d}_{err} = 1.73$ px and it misses the the target grab point by $\bar{g}_{err} = 1.85$ px. The median values of both errors are slightly lower, with $\tilde{d}_{err} = 1.65$ px and $\tilde{g}_{err} = 1.66$ px for drop point error and grab point error, respectively. The largest tested drop point error was $d_{err,\,max} = 4.52$ px and largest grab point error was $g_{err,\,max} = 6.30$ px. On the contrary, the smallest drop point and grab point errors were measured at $d_{err,\,min} = 0.04$ px and $g_{err,\,min} = 0.12$ px. In most cases, the model should miss the drop point by less than $d_{err} = 2.27$ px and grab point by less than $g_{err} = 2.45$ px. On a $256 \times 256$ image, where 1 px $\approx 2.4$ mm, this translates to an error of $d_{err} \approx 5.5$ mm and $g_{err} \approx 5.9$ mm.

(a) Canonical cloth state, no deformation.

(b) Surface area deformed.

(c) Surface area flattened.

**Fig. 5.** Cloth surface area estimation before and after flattening. On the left and right are shown two examples of cloths that have been smoothed from a deformed state (middle row) to a flattened state (bottom row).

Second, after putting the trained model into practice, we evaluated the surface area of the cloth after the robot flattened it in the simulation using the grabing and dropping points calculated by the model. We evaluate the cloth surface area by counting the black pixels of the projected surface, as shown in Fig. 5. In this case, we use the $1000 \times 1000$ resolution images for better accuracy. Table 1 shows the results of 15 flattening cycles and all surface area estimates are compared to the surface area of the cloth in its canonical state (Fig. 5a), which we have taken as our baseline. Figure 6 shows the robot performing the flattening operation in successive frames (from left to right).

**Fig. 6.** Sequence of images representing the flattening operation.

**Table 1.** Cloth's projected surface area.

| Surface area deformed | | Surface area flattened | |
|---|---|---|---|
| Pixels [px] | Baseline percentage | Pixels [px] | Baseline percentage |
| 217052 | 95.39% | 226779 | 99.67% |
| 213061 | 93.64% | 223857 | 98.39% |
| 223688 | 98.31% | 229543 | 100.88% |
| 218404 | 95.99% | 225178 | 98.97% |
| 214122 | 94.11% | 227677 | 100.06% |
| 218234 | 95.91% | 225166 | 98.96% |
| 220592 | 96.95% | 226823 | 99.69% |
| 216920 | 95.34% | 227746 | 100.09% |
| 218076 | 95.84% | 227847 | 100.14% |
| 213783 | 93.96% | 228491 | 100.42% |
| 212058 | 93.20% | 229596 | 100.91% |
| 223677 | 98.31% | 228954 | 100.63% |
| 211943 | 93.15% | 230192 | 101.17% |
| 216792 | 95.28% | 228301 | 100.34% |
| 213979 | 94.04% | 229400 | 100.82% |

## 4 Discussion

This paper describes our initial research on robotic manipulation of textiles, where an experiment using a single cloth was carried out to predict flattening motions. Unlike our previous work [4], the cloth model we used in MuJoCo was far cruder compared to the one used in PyBullet, as both simulators evaluate collisions differently. The main limiting factor here was the computational speed. With the chosen raster (20 $\times$ 20) we could only achieve a simulation speed that was about 10% of the real time. By sacrificing the simulation speed, we could increase the raster density and obtain more accurate results. On the other hand, it is possible that this level of accuracy is sufficient for sim- to-real transfer.

We were also able to achieve better accuracy in predicting grab and drop point pairs. Although this is most likely the result of a smaller deformation area, it is also possible that the images from the new set have more recognisable features that the neural network can identify. However, the reduced deformation area also reduced the number of possible deformation states. This resulted in the cloth retaining most of its initial surface area. Part of this error can be attributed to the automatic measurement procedure. Post-processing of the images representing the state of the fabric sometimes resulted in slightly more black pixels than pixels of the actual surface. This is most noticeable in cases where the surface area of the smoothed cloth was larger than the surface area of the baseline.

## 5 Conclusion

We tested the performance of our model by evaluating the surface area and general shape of the cloth after flattening. We tested our methodology in a new simulated environment and executed the motions calculated by our model. On average, the model made a computational error of 1.85 px in predicting the grabing point position and an error of 1.73 px in predicting the drop point position. This corresponds to an error of 4.4 mm and 4.2 mm, respectively, which did not affect the flattening performance. In the future, we will extend the dataset to include more deformations and apply our model to a real robot.

## References

1. Kingma, D.P., Ba, J.: Adam: A method for stochastic optimization. In: The International Conference on Learning Representations (ICLR) (2015)
2. Laskey, M., Lee, J., Fox, R., Dragan, A., Goldberg, K.: DART: noise injection for robust imitation learning, p. 14 (2017)
3. Lee, R., Ward, D., Cosgun, A., Dasagi, V., Corke, P., Leitner, J.: Learning arbitrary-goal fabric folding with one hour of real robot experience. CoRR **abs/2010.03209** (2020)
4. Nimac, P., Mavsar, M., Gams, A.: Cloth smoothing simulation with vision-to-motion skill model. Društvo Slovenska sekcija. IEEE (2022)
5. Pahič, R., Ridge, B., Gams, A., Morimoto, J., Ude, A.: Training of deep neural networks for the generation of dynamic movement primitives. Neural Netw. **127**, 121–131 (2020)
6. Seita, D., et al.: Deep imitation learning of sequential fabric smoothing policies. CoRR **abs/1910.04854** (2019)
7. Triantafyllou, D., Mariolis, I., Kargakos, A., Malassiotis, S., Aspragathos, N.A.: A geometric approach to robotic unfolding of garments. Robot. Auton. Syst. **75**, 233–243 (2016)
8. Tsurumine, Y., Cui, Y., Uchibe, E., Matsubara, T.: Deep reinforcement learning with smooth policy update: application to robotic cloth manipulation. Robot. Auton. Syst. **112**, 72–83 (2019)
9. Wu, Y., Yan, W., Kurutach, T., Pinto, L., Abbee l, P.: Learning to manipulate deformable objects without demonstrations. arXiv:1910.13439 [cs] (2020). arXiv: 1910.13439

# Motion Planning Through Model Inversion for a Gantry Crane Moving a Double Pendulum

Jason Bettega, Dario Richiedei(✉), Iacopo Tamellin, and Alberto Trevisani

Department of Management and Engineering (DTG), University of Padova, Stradella San Nicola 3, 36100 Vicenza, Italy
{jason.bettega,dario.richiedei,iacopo.tamellin,
alberto.trevisani}@unipd.it

**Abstract.** This paper proposes a method for precise motion planning of a gantry crane moving a double pendulum. The payload to be moved is composed by a pendulum and a rigid-body attached to the pendulum terminal mass. The desired output coordinates are the ones of the rigid-body tip. This choice yields a non-minimum phase system, whose internal dynamics is unstable. In this paper it is stabilized through the output redefinition technique, i.e., by assuming a fictitious output within the internal dynamics. The numerical integration of the stabilized ODEs enables to compute the motion commands for the crane platform. Numerical simulations are provided to assess the effectiveness of the proposed method which is corroborated by the low contour and tracking errors obtained while performing a prescribed planar trajectory.

**Keywords:** Motion planning · Inverse dynamics · Underactuated multibody systems · Internal dynamics · Gantry crane · Double pendulum

## 1 Introduction

Underactuated multibody systems are characterized by a number of degrees of freedom (DOFs) that is lower than the number of independent actuators used to control the system [1]. Some examples are: vibratory feeders [2], mechanisms with flexible links or passive joints [3, 4], robotic arms and grippers, underwater and unmanned vehicles, cranes [5, 6]. The latter ones are widely adopted also in industry to move suspended loads, and underactuation makes the load behave like a pendulum; hence, during the motion usually oscillations may arise if the system is not carefully controlled [7] or if its motion is not adequately planned [8]. This yields to inaccurate trajectory and path tracking during the motion as well as unprecise positioning once the motion has been completed. Planning the motion of such systems is not trivial for two reasons: first, the system is underactuated hence not all the system coordinates can be directly imposed through the actuators. Secondly, tip control usually leads to non-minimum phase, i.e., the internal dynamics is unstable [3]. In this light, this paper proposes a method to precise motion planning in underactuated, non-minimum phase systems. The proposed method relies on the partitioning of the system into actuated and unactuated coordinates. Then, the


T. Petrič et al. (Eds.): RAAD 2023, MMS 135, pp. 375–382, 2023.
https://doi.org/10.1007/978-3-031-32606-6_44

output reference is defined as a non-linear separable function and the internal dynamics is stabilized through the output redefinition technique [9]. Numerical integration is adopted to solve the non-linear ODEs describing the stabilized internal dynamics and the motion command are obtained through exact non-linear kinematic inversion. The effectiveness of the method is corroborated by its application to the model of a double-pendulum moved through a gantry crane.

## 2 Model of the System

The system is sketched in Fig. 1: a platform moves along the X and Y directions and is actuated by two independent forces. A 4 DOFs double-pendulum is attached to the platform: it consists of a rigid cable (whose constant length is $l_1$) with a terminal lumped mass $m_1$ that connects a cylindrical end-effector. The system has 6 DOFs, and the degree of underactuation is 4; hence, just 2 outputs can be imposed [1, 3, 9].

The positions of the hook (point H) and of the end-effector tip (point E) are:

$$\begin{cases} x_H = x_P + l_1 s(\vartheta_{1x}) \\ y_H = y_P + l_1 c(\vartheta_{1x}) s(\vartheta_{1y}) \end{cases} \quad \begin{cases} x_E = x_H + l_2 s(\vartheta_{2x}) \\ y_E = y_H + l_2 c(\vartheta_{2x}) s(\vartheta_{2y}) \end{cases} \tag{1}$$

where the swing angles of the pendulum are $\vartheta_{1x}$, $\vartheta_{1y}$ and the rotation angles of the rigid-body are $\vartheta_{2x}$, $\vartheta_{2y}$ (*s()* and *c()* denote the sine and cosine).

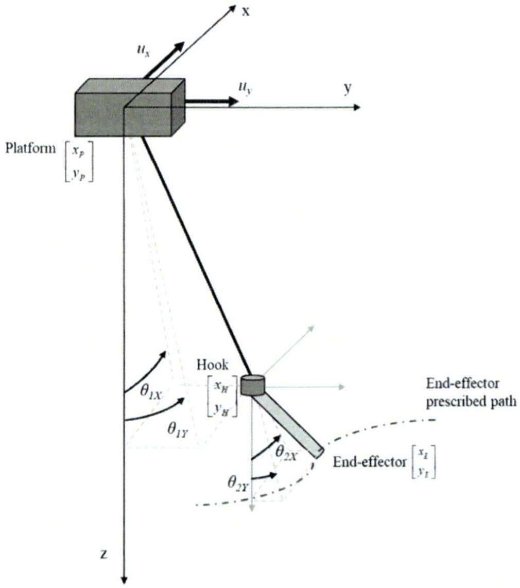


**Fig. 1.** Sketch of the system.

The non-linear dynamic model of the underactuated system under analysis is written in the following concise form, that recalls the usual one for multibody systems:

$$\mathbf{M}(\boldsymbol{q}_U) \begin{Bmatrix} \ddot{\boldsymbol{q}}_A \\ \ddot{\boldsymbol{q}}_U \end{Bmatrix} + \boldsymbol{c}(\dot{\boldsymbol{q}}_A, \boldsymbol{q}_U, \dot{\boldsymbol{q}}_U) = \boldsymbol{f}(\boldsymbol{q}_U) + \mathbf{B} \begin{Bmatrix} u_x \\ u_y \end{Bmatrix} \tag{2}$$

The unactuated coordinates are collected in $\boldsymbol{q}_U = \{\vartheta_{1x}, \vartheta_{1y}, \vartheta_{2x}, \vartheta_{2y}\}^T$; the actuated coordinates, i.e., those related to the platform, are collected in $\boldsymbol{q}_A = \{x_P, y_P\}^T$. The control forces driving the platform are $u_x$ and $u_y$ and affect the system dynamics through the force distribution matrix $\mathbf{B} = \begin{bmatrix} \mathbf{I}_2 \\ \mathbf{0}_4 \end{bmatrix}$, where $\mathbf{I}_2$ is the dimension-2 identity matrix and $\mathbf{0}_4$ is the 4x2 null matrix. The external forces due to gravity acceleration are collected in vector $\boldsymbol{f}(\boldsymbol{q}_U)$ (the meaning of the parameters is described in Table 1)

$$f(q_U) = -g\left\{0 \;\; 0 \;\; l_1(m_1+m_2)c(\vartheta_{1y})s(\vartheta_{1x}) \;\; l_1(m_1+m_2)s(\vartheta_{1y})c(\vartheta_{1x}) \;\; \tfrac{l_2}{2}m_2c(\vartheta_{2y})s(\vartheta_{2x}) \;\; \tfrac{l_2}{2}m_2s(\vartheta_{2y})c(\vartheta_{2x})\right\}^T \quad (3)$$

$\mathbf{M}(\boldsymbol{q}_U)$ is the position-dependent mass matrix [5]:

$$\mathbf{M}(\boldsymbol{q}_U) = \begin{bmatrix} M_x+m_1+m_2 & 0 & m_{13} & 0 & \frac{l_2}{2}m_2c(\vartheta_{2x}) & 0 \\ 0 & M_y+m_1+m_2 & m_{23} & m_{24} & m_{25} & m_{26} \\ m_{13} & m_{23} & l_1^2(m_1+m_2) & 0 & m_{35} & m_{36} \\ 0 & m_{24} & 0 & m_{44} & m_{45} & m_{46} \\ \frac{l_2}{2}m_2c(\vartheta_{2x}) & m_{25} & m_{35} & m_{45} & I_{yy}+\frac{l_2^2}{4}m_2 & 0 \\ 0 & m_{26} & m_{36} & m_{46} & 0 & m_{66} \end{bmatrix} \quad (4)$$

with:

$$\begin{aligned} &m_{13} = l_1(m_1+m_2)c(\vartheta_{1x}) \;\; m_{23} = -l_1(m_1+m_2)s(\vartheta_{1x})s(\vartheta_{1y}) \;\; m_{24} = l_1(m_1+m_2)c(\vartheta_{1x})c(\vartheta_{1y}) \\ &m_{25} = -\tfrac{l_2}{2}m_2s(\vartheta_{2x})s(\vartheta_{2y}) \;\; m_{26} = \tfrac{l_2}{2}m_2c(\vartheta_{2x})c(\vartheta_{2y}) \\ &m_{35} = \tfrac{l_1l_2}{2}m_2\left(c(\vartheta_{1x})c(\vartheta_{2x}) + s(\vartheta_{1x})s(\vartheta_{2x})c(\vartheta_{1y}-\vartheta_{2y})\right) \;\; m_{36} = \tfrac{l_1l_2}{2}m_2s(\vartheta_{1x})c(\vartheta_{2x})s(\vartheta_{2y}-\vartheta_{1y}) \\ &m_{44} = l_1^2(m_1+m_2)(c(\vartheta_{1x}))^2 \;\; m_{45} = \tfrac{l_1l_2}{2}m_2c(\vartheta_{1x})s(\vartheta_{2x})s(\vartheta_{1y}-\vartheta_{2y}) \\ &m_{46} = \tfrac{l_1l_2}{2}m_2c(\vartheta_{1x})c(\vartheta_{2x})s(\vartheta_{1y}+\vartheta_{2y}) \;\; m_{66} = I_{zz}(s(\vartheta_{2x}))^2 + \left(I_{xx}+\tfrac{l_2^2}{4}m_2\right)(c(\vartheta_{2x}))^2 \end{aligned} \quad (5)$$

By introducing the following variables,

$$\begin{aligned} &\bar{m}_{12} = l_1(m_1+m_2) \quad \bar{m}_1 = l_1^2(m_1+m_2) \quad \bar{m}_2 = l_1l_2m_2 \\ &\bar{\vartheta}_{1xy} = \dot{\vartheta}_{1x}\dot{\vartheta}_{1y} \quad \bar{\vartheta}_{2xy} = \dot{\vartheta}_{2x}\dot{\vartheta}_{2y} \quad v_{1xy} = \left(\dot{\vartheta}_{1x}^2 + \dot{\vartheta}_{1y}^2\right) \\ &v_{2xy} = \left(\dot{\vartheta}_{2x}^2 + \dot{\vartheta}_{2y}^2\right) \; I_5 = \left(I_{xx} - I_{zz} + \tfrac{l_2^2}{4}m_2\right) \; I_6 = \left(I_{zz} - \left(I_{xx} + \tfrac{l_2^2}{4}m_2\right)\right) \end{aligned} \quad (6)$$

the vector of the speed-dependent contributions $\boldsymbol{c}(\dot{\boldsymbol{q}}_A, \boldsymbol{q}_U, \dot{\boldsymbol{q}}_U)$ becomes:

$$c = \left\{\begin{array}{c} -\bar{m}_{12}s(\vartheta_{1x})\dot{\vartheta}_{1x}^2 - \frac{l_2}{2}m_2s(\vartheta_{2x})\dot{\vartheta}_{2x}^2 + d_x\dot{x}_t \\ -\bar{m}_{12}c(\vartheta_{1x})s(\vartheta_{1y})v_{1xy} - \frac{l_2}{2}m_2c(\vartheta_{2x})s(\vartheta_{2y})v_{2xy} - 2\bar{m}_{12}c(\vartheta_{1y})s(\vartheta_{1x})\dot{\bar{\vartheta}}_{1xy} - l_2m_2c(\vartheta_{2y})s(\vartheta_{2x})\dot{\bar{\vartheta}}_{2xy} + d_y\dot{y}_t \\ \frac{\bar{m}_1}{2}s(2\vartheta_{1x})\dot{\vartheta}_{1y}^2 - \frac{\bar{m}_2}{2}c(\vartheta_{1x})s(\vartheta_{2x})\dot{\vartheta}_{2x}^2 + \frac{\bar{m}_2}{2}c(\vartheta_{2x})s(\vartheta_{1x})c(\vartheta_{1y}-\vartheta_{2y})v_{2xy} + \bar{m}_2s(\vartheta_{1x})s(\vartheta_{2x})s(\vartheta_{1y}+\vartheta_{2y})\dot{\bar{\vartheta}}_{2xy} + d_{\vartheta_1}\dot{\vartheta}_{1x} \\ -2\bar{m}_1s(\vartheta_{1x})c(\vartheta_{1x})\dot{\bar{\vartheta}}_{1xy} + \frac{\bar{m}_2}{2}c(\vartheta_{1x})c(\vartheta_{2x})s(\vartheta_{1y}-\vartheta_{2y})v_{2xy} - \bar{m}_2c(\vartheta_{1x})s(\vartheta_{2x})c(\vartheta_{1y}+\vartheta_{2y})\dot{\bar{\vartheta}}_{2xy} + d_{\vartheta 1}\dot{\vartheta}_{1y} \\ I_5\frac{s(2\vartheta_{2x})}{2}\dot{\vartheta}_{2y}^2 + \frac{\bar{m}_2}{2}c(\vartheta_{1x})s(\vartheta_{2x})s(\vartheta_{1y}+\vartheta_{2y})v_{1xy} - \frac{\bar{m}_2}{2}s(\vartheta_{1x})c(\vartheta_{2x})\dot{\vartheta}_{1x}^2 - \bar{m}_2s(\vartheta_{1x})s(\vartheta_{2x})s(\vartheta_{2y}-\vartheta_{1y})\dot{\vartheta}_{1x}\dot{\vartheta}_{1y} + d_{\vartheta 2}\dot{\vartheta}_{2x} \\ \frac{\bar{m}_2}{2}c(\vartheta_{1x})c(\vartheta_{2x})s(\vartheta_{2y}-\vartheta_{1y})v_{1xy} - \bar{m}_2s(\vartheta_{1x})c(\vartheta_{2x})c(\vartheta_{1y}-\vartheta_{2y})\dot{\bar{\vartheta}}_{1xy} + I_6s(2\vartheta_{2x})\dot{\bar{\vartheta}}_{2xy} + d_{\vartheta 2}\dot{\vartheta}_{2y} \end{array}\right\} \quad (7)$$

## 3 Motion Planning Through Dynamic Inversion

The goal of motion planning is to make the controlled output $\boldsymbol{y}$, defined as the planar motion of point $E$ (i.e., the tip of the end-effector), track the desired reference trajectory in time, set through $\boldsymbol{y}^{\boldsymbol{ref}} = \left\{x_E^{ref}, y_E^{ref}\right\}^T$. The controlled output can be written as a non-linear separable function of the actuated and unactuated coordinates as follows:

$$\boldsymbol{y} = \boldsymbol{q}_A + \boldsymbol{g}(\boldsymbol{q}_U) \tag{8}$$

with:

$$\boldsymbol{g}(\boldsymbol{q}_U) = \left\{ \begin{array}{c} l_1 s(\vartheta_{1x}) + l_2 s(\vartheta_{2x}) \\ l_1 c(\vartheta_{1x}) s(\vartheta_{1y}) + l_2 c(\vartheta_{2x}) s(\vartheta_{2y}) \end{array} \right\} \tag{9}$$

The reference tip accelerations can therefore be written in the compact form:

$$\ddot{\boldsymbol{y}}^{\boldsymbol{ref}} = \ddot{\boldsymbol{q}}_A + \mathbf{J_g}(\boldsymbol{q}_U)\ddot{\boldsymbol{q}}_U + \boldsymbol{\gamma}_{\boldsymbol{g}}(\boldsymbol{q}_U, \dot{\boldsymbol{q}}_U) \tag{10}$$

where $\mathbf{J_g} \in \mathbb{R}^{2\times 4}$ is the Jacobian of $\boldsymbol{g}$, and $\boldsymbol{\gamma}_{\mathbf{g}} \in \mathbb{R}^2$ contains the centripetal terms:

$$\begin{aligned} \mathbf{J_g}(\boldsymbol{q}_U) &= \begin{bmatrix} l_1 c(\vartheta_{1x}) & 0 & l_2 c(\vartheta_{2x}) & 0 \\ -l_1 s(\vartheta_{1x}) s(\vartheta_{1y}) & l_1 c(\vartheta_{1x}) c(\vartheta_{1y}) & -l_2 s(\vartheta_{2x}) s(\vartheta_{2y}) & l_2 c(\vartheta_{2x}) c(\vartheta_{2y}) \end{bmatrix} \\ \boldsymbol{\gamma}_{\mathbf{g}}(\boldsymbol{q}_U, \dot{\boldsymbol{q}}_U) &= -\left\{ \begin{array}{l} l_1 s(\vartheta_{1x})\dot{\vartheta}_{1x}^2 + l_2 s(\vartheta_{2x})\dot{\vartheta}_{2x}^2 \\ l_1 c(\vartheta_{1x}) s(\vartheta_{1y}) v_{1xy} + 2 l_1 s(\vartheta_{1x}) c(\vartheta_{1y}) \bar{\dot{\vartheta}}_{1xy} + l_2 c(\vartheta_{2x}) s(\vartheta_{2y}) v_{2xy} + 2 l_2 s(\vartheta_{2x}) c(\vartheta_{2y}) \bar{\dot{\vartheta}}_{2xy} \end{array} \right\} \end{aligned} \tag{11}$$

Let us now partition the dynamic model in Eq. (2), as follows:

$$\begin{bmatrix} \mathbf{M_{AA}}(\boldsymbol{q}_U) & \mathbf{M_{AU}}(\boldsymbol{q}_U) \\ \mathbf{M}_{\mathbf{AU}}^T(\boldsymbol{q}_U) & \mathbf{M_{UU}}(\boldsymbol{q}_U) \end{bmatrix} \left\{ \begin{array}{c} \ddot{\boldsymbol{q}}_A \\ \ddot{\boldsymbol{q}}_U \end{array} \right\} + \left\{ \begin{array}{c} \boldsymbol{c}_A(\boldsymbol{q}_U, \dot{\boldsymbol{q}}_U) \\ \boldsymbol{c}_U(\boldsymbol{q}_U, \dot{\boldsymbol{q}}_U) \end{array} \right\} = \left\{ \begin{array}{c} \boldsymbol{f}_A(\boldsymbol{q}_U) \\ \boldsymbol{f}_U(\boldsymbol{q}_U) \end{array} \right\} + \begin{bmatrix} \mathbf{B_A} \\ \mathbf{0} \end{bmatrix} \boldsymbol{u} \tag{12}$$

By replacing Eq. (10) in the second row of Eq. (12) leads to the system internal dynamics:

$$\left(\mathbf{M_{UU}} - \mathbf{M}_{\mathbf{AU}}^T \mathbf{J_g}\right)\ddot{\boldsymbol{q}}_U + \left(\boldsymbol{c}_U - \mathbf{M}_{\mathbf{AU}}^T \boldsymbol{\gamma}_{\boldsymbol{g}}\right) = \boldsymbol{f}_U - \mathbf{M}_{\mathbf{AU}}^T \ddot{\boldsymbol{y}}^{\boldsymbol{ref}} \tag{13}$$

Once the trajectory of the unactuated coordinates is computed by numerically integrating the non-linear ODEs in Eq. (13), the commanded trajectory of the actuated coordinates is obtained through the exact kinematic inversion in Eqs. (8) and (10). The integration of Eq. (13) is, however, a critical issue in non-minimum phase systems, as often happens in tip control [9], due to the instability of the internal dynamics resulting from such an output choice. As a consequence, the integration of Eq. (13) diverges leading to unfeasible trajectories. In this paper the internal dynamics of the system is stabilized through the output redefinition technique. In particular, in the system under consideration, the lengths of the pendulum and of the rigid-body are respectively redefined to $\tilde{l}_1 = \gamma_1 l_1$ and $\tilde{l}_2 = \gamma_2 l_2$ with $0 \leq \gamma_1 \leq 1$ and $0 \leq \gamma_2 \leq 1$. It follows that the redefined output reference acceleration is:

$$\ddot{\tilde{\boldsymbol{y}}}^{\boldsymbol{ref}} = \ddot{\boldsymbol{q}}_A + \tilde{\mathbf{J}}_{\mathbf{g}}(\boldsymbol{q}_U)\ddot{\boldsymbol{q}}_U + \tilde{\boldsymbol{\gamma}}_{\boldsymbol{g}}(\boldsymbol{q}_U, \dot{\boldsymbol{q}}_U) \tag{14}$$

with $\tilde{\mathbf{J}}_{\mathbf{g}}$ and $\tilde{\boldsymbol{\gamma}}_g$ obtained through Eq. (11) by adopting $\tilde{l}_1$ and $\tilde{l}_2$. The internal dynamics in Eq. (13) is recast in terms of the redefined output as:

$$\left(\mathbf{M}_{\mathrm{UU}} - \mathbf{M}_{\mathrm{AU}}^T\tilde{\mathbf{J}}_{\mathbf{g}}\right)\ddot{\boldsymbol{q}}_U + \left(\boldsymbol{c}_U - \mathbf{M}_{\mathrm{AU}}^T\tilde{\boldsymbol{\gamma}}_g\right) = \boldsymbol{f}_U - \mathbf{M}_{\mathrm{AU}}^T\ddot{\tilde{\boldsymbol{y}}}^{ref} \tag{15}$$

Once $\gamma_1$, $\gamma_2$ are chosen, the numerical integration of Eq. (15) returns a feasible trajectory for the unactuated coordinates $\ddot{\boldsymbol{q}}_U^*$, $\dot{\boldsymbol{q}}_U^*$ and $\boldsymbol{q}_U^*$. Such time histories can be used together with the redefined output reference to compute the actuated coordinates reference values $\ddot{\boldsymbol{q}}_A^{ref}$, $\dot{\boldsymbol{q}}_A^{ref}$ by inverting the kinematic model as follows:

$$\begin{aligned}
\boldsymbol{q}_A^{ref} &= \tilde{\boldsymbol{y}}^{ref} - \tilde{\boldsymbol{g}}(\boldsymbol{q}_U^*) \\
\dot{\boldsymbol{q}}_A^{ref} &= \dot{\tilde{\boldsymbol{y}}}^{ref} - \tilde{\mathbf{J}}_{\mathbf{g}}(\boldsymbol{q}_U^*)\dot{\boldsymbol{q}}_U^* \\
\ddot{\boldsymbol{q}}_A^{ref} &= \ddot{\tilde{\boldsymbol{y}}}^{ref} - \tilde{\mathbf{J}}_{\mathbf{g}}(\boldsymbol{q}_U^*)\ddot{\boldsymbol{q}}_U^* - \tilde{\boldsymbol{\gamma}}_g(\boldsymbol{q}_U^*, \dot{\boldsymbol{q}}_U^*)
\end{aligned} \tag{16}$$

These values are, finally, commanded to the control system driving the platform to compute the control forces (through its feedback or feedforward control actions).

## 4 Numerical Application

The system parameters are listed in Table 1. The tip-end of the rigid-body should perform a circular path whose radius is equal to 0.2 m in 10 s by tracking a 7th degree polynomial motion law, as often used in underactuated systems. The linearization of the actual internal dynamics reveals two poles with positive real parts, therefore output redefinition is required. Figure 2 shows the stability of the internal dynamics by varying $\gamma_1$ and $\gamma_2$. A first set of parameters, hereafter denoted as "Redefinition A", is defined by setting $\gamma_1 = 0.9893$ and $\gamma_2 = 6.7\text{e} - 05$, that lead to negative real part poles.

By exploiting "Redefinition A", very accurate tracking is obtained: Fig. 3 and Fig. 4 corroborate the method effectiveness since the desired and actual load trajectories are almost overlapped, with a maximum tracking error equal to 3.2 mm, and a maximum contour error equal to 2.8 mm. These figures show how the proposed method shapes

**Table 1.** System parameters.

| Symbol and unit | Description | Value |
|---|---|---|
| $M_x, M_y$ [kg] | Platform mass along the X and Y direction | 30; 30 |
| $m_1, m_2$ [kg] | Mass of the lumped mass and of the rigid-body | 0.1; 0.049 |
| $l_1$ ,$l_2$[m] | Length of the cable and of the rigid-body | 0.75; 0.15 |
| $R$ [m] | Radius of the rigid-body | 0.010 |
| $I_{xx}$,$I_{yy}$,$I_{zz}$ [kgm$^2$] | Moments of inertia of the rigid-body | 2.8e-8; 2.8e-8; 2.5e-6 |
| $d_x$,$d_y$ [Ns/m] | Viscous friction coefficients of the platform | 0.5; 0.5 |
| $d_{\theta 1}$,$d_{\theta 2}$ [Ns] | Viscous friction coefficients of the pendulum | 1e-3; 1e-3 |

the trajectory of the platform, to compensate for the unavoidable presence of centrifugal forces. Finally, small residual oscillations are experienced. In contrast, if the set "Redefinition B" is adopted ($\gamma_1 = 0.55$, $\gamma_2 = 0.2$) the contour error increases as shown in Fig. 5, although good performances are still obtained.

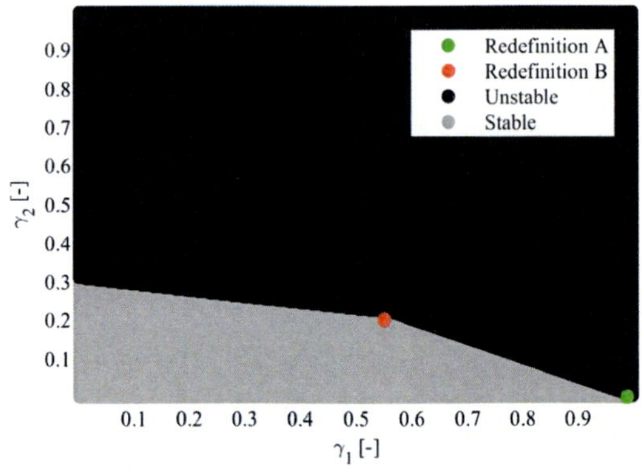


**Fig. 2.** Stability of the internal dynamics varying $\gamma_1$ and $\gamma_2$.

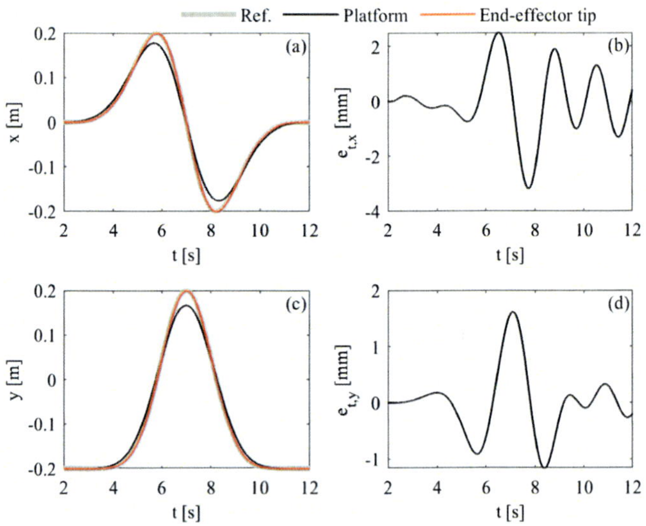


**Fig. 3.** Time histories of the platform, end-effector tip and reference displacements (a), (c) and end-effector tip tracking errors (b), (d).

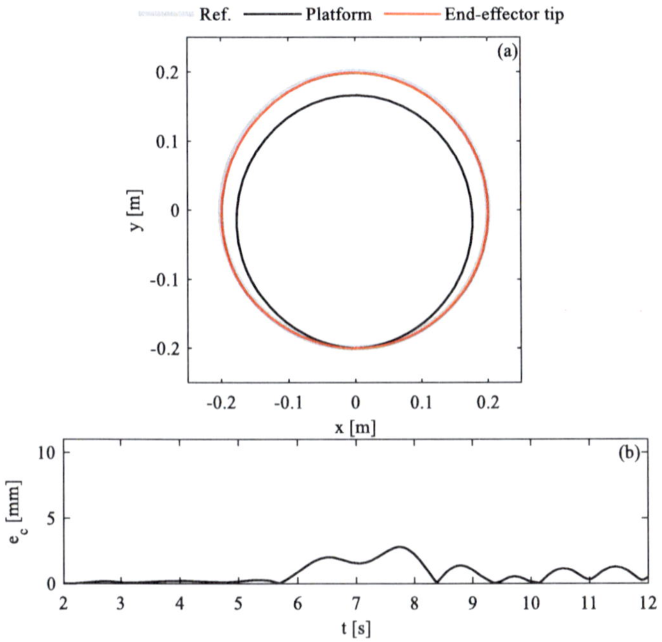


**Fig. 4.** Cartesian plane: reference, platform and end-effector tip paths (a) and end-effector tip contour error (b) with the "Redefinition A".

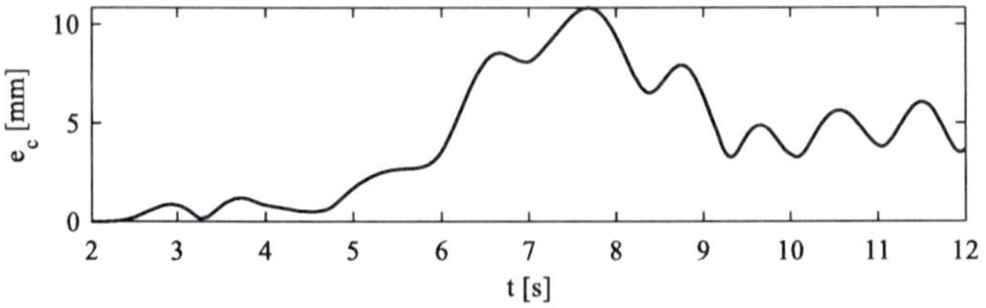


**Fig. 5.** End-effector tip contour error with the "Redefinition B".

## 5 Conclusions

This paper exploits the inversion of the dynamic model to perform precise motion planning in an underactuated, non-minimum phase system. The system consists of a gantry crane that moves a double pendulum which in turn is composed by a pendulum and a rigid-body that mimics an end-effector. The system has 6 DOFs, that are controlled by just 2 forces driving the planar, Cartesian motion of the platform. The goal is to track a prescribed trajectory for the end-effector tip. Such an output is described through a non-linear separable function in terms of the actuated and unactuated coordinates that allows exactly obtaining the internal dynamics, without any approximation at this stage. In the case under investigation, the choice of the controlled output yields unstable internal dynamics that must be stabilized to ensure feasible command. Hence, the output redefinition technique is adopted to stabilize the system internal dynamics. The effectiveness of the proposed method is assessed through the numerical application on a double-pendulum that should track a circular path. The small computational effort of the method and its simple implementation make it a good candidate for motion planning in this kind of multibody systems. More results further highlighting the method effectiveness will be provided in the conference presentation.

## 6 Fundings

This research was funded by the Italian Ministry of University and Research by the PRIN 2020 "Extending Robotic Manipulation Capabilities by Cooperative Mobile and Flexible Multi-Robot Systems (Co-MiR)" (Prot. 2020CMEFPK).

## References

1. Seifried, R.: Dynamics of Underactuated Multibody Systems. 1st edn. Springer Cham, Heidelberg New York Dordrecht London (2014)
2. Belotti, R., Richiedei, D., Tamellin, I., Trevisani, A.: Response optimization of underactuated vibration generators through dynamic structural modification and shaping of the excitation forces. The Int. J. Adva. Manuf. Technol. **112**(1–2), 505–524 (2020). https://doi.org/10.1007/s00170-020-06083-2
3. Bettega, J., Richiedei, D., Tamellin, I., Trevisani, A.: Stable inverse dynamics for feedforward control of nonminimum-phase underactuated systems. J. Mechani. Robot. 1–42 (2022)
4. Blajer, W., Kołodziejczyk, K.: A case study of inverse dynamics control of manipulators with passive joints. J. Theor. Appl. Mech. **52**(3), 793–801 (2014)
5. O'Connor, W., Habibi, H.: Gantry crane control of a double-pendulum, distributed-mass load, using mechanical wave concepts. Mechanical Sciences **4**(2), 251–261 (2013)
6. Blajer, W., Kołodziejczyk, K.: A geometric approach to solving problems of control constraints: theory and a DAE framework. Multibody Sys.Dyn. **11**(4), 343–364 (2004)
7. Richiedei, D., Trevisani, A.: Delayed-reference anti-swing control of overhead crane systems. In: 2008 10th IEEE International Workshop on Advanced Motion Control, pp. 92–97. IEEE, Trento, Italy (2008)
8. Boscariol, P., Richiedei, D.: Robust point-to-point trajectory planning for nonlinear underactuated systems: Theory and experimental assessment. Robot. Comp. Integra. Manufact. **50**, 256–265 (2018)
9. Bettega, J., Richiedei, D., Tamellin, I., Trevisani, A.: Model inversion for precise path and trajectory tracking in an underactuated, non-minimum phase, spatial overhead crane. J. Vibr. Engineering Technologies, 1–17 (2022)

# Fixed Point Iteration-Based Adaptive Control Improved with Parameter Identification

Bence Varga[1,2(✉)], József K. Tar[1,3,4], and Richárd Horváth[2]

[1] Doctoral School of Applied Informatics and Applied Mathematics, Óbuda University, Bécsi út 96/B, Budapest 1034, Hungary
tar.jozsef@nik.uni-obuda.hu

[2] Bánki Donát Faculty of Mechanical and Safety Engineering, Óbuda University, Budapest, Hungary
{varga.bence,horvath.richard}@bgk.uni-obuda.hu

[3] John von Neumann Faculty of Informatics, Óbuda University, Budapest, Hungary

[4] Antal Bejczy Center of Intelligent Robotics, Óbuda University, Budapest, Hungary

**Abstract.** This paper reports the freshest element of the chain of investigations tackling the combination of the Fixed Point Iteration-based adaptive controller with parameter identification. Though this controllers does not necessarily use the identified model, it is expected that the use of the more precise model improves its various properties. Simulation investigations are made for a cylindrical robot for replacing the Particle Swarm Optimization with simple regression-based approach. It is concluded that the use of the best identified model still makes it expedient the application the FPI-based adaptivity. The remaining imprecisions seem to be related to the not well balanced structure of the experimentally collected and analyzed data. It can be expected that this approach can be realized real-time with appropriate hardware.

**Keywords:** Adaptive control · Fixed point iteration · Linear regression

## 1 Introduction

In robotics as well as in other control application areas the model-based approaches efficiently can utilize some *ab ovo available information* contained by a dynamic model of the controlled system. This means great advantage in comparison with the model-free solutions combined with statistical approaches and iterative learning control that are very popular in our days (e.g., in robotics [5,15], in dynamic control of vehicles [2,10]), especially in the initial phase of the control in which the "model-free model" is very imprecise. It became clear from the early nineties that practically it is hopeless to develop precise enough dynamic models for robots (e.g., [1,3]). For the compensation of the effects of modeling imprecisions various adaptive techniques were developed. In the early


T. Petrič et al. (Eds.): RAAD 2023, MMS 135, pp. 383–390, 2023.
https://doi.org/10.1007/978-3-031-32606-6_45

nineties the Adaptive Inverse Dynamics Controller (AIDC) and the Slotine-Li Adaptive Robot Controller [16,18,23] used Lyapunov's direct method [12] to achieve asymptotic convergence on the basis of this *a priori* information. On the reason of numerical stability these methods learned the system parameters slowly, therefore in the initial phase of the control they were very imprecise. Beside the signal adaptive approaches as Model Reference Adaptive Control (e.g., a classical solution from the nineties is [14]) that do not learn the model parameters, an alternative solution was the Fixed Point Iteration-based (FPI) Adaptive Control [19] that besides the integrated tracking error, the tracking error itself and its first time-derivative fed back the second time-derivative of the observed motion (in the case of a 2nd order system). This approach also used an imprecise available dynamic model, learned in a very fast manner without identifying the system's parameters, and in contrast to the model-free and the classical parameter learning approaches even in the initial phase of the control provided good tracking. It was more noise sensitive as well as the Acceleration Feedback Controllers (e.g., [7]) that also feed back noisy signal. However, refinement of the available approximate model would increase the range of convergence and precision of this approach. It was successfully combined with the AIDC and the Slotine-Li Adaptive Robot Controller in 2014 [6,20]. In these approaches the original tuning method was used for the dynamical parameters but in the place of the kinematically prescribed PID-type feedback terms that of the FPI-based Adaptive Controller were written. The result was slow parameter learning but immediately precise trajectory tracking. In [8] and in [21] the Particle Swarm Optimization [9] was combined with the FPI with the conclusion that some modeling imprecisions always remained that made it expedient to maintain the FPI-based correction even when the best identified approximate model was in use in the background. In the present paper an attempt is done to replace the PSO with a simple regression-based approach utilizing the formal properties on which the AIDC approach was built up, too. Simulation investigations were made for the control of a cylindrical robot. The conclusion is the same as in the case of the PSO-based study: it is not possible to provide a well balanced teaching set for precise parameter identification, and it is expedient to use the best identified model in the background of the FPI-based adaptive control.

## 2 Proposed Parameter Identification Method

In general the equations of motion for robot manipulators are written in form $H(q)\ddot{q}+h(q,\dot{q})=Q$, where $H(q)$ is a positive definite inertia matrix, $h(q,\dot{q})$ is an array containing various difficult non linear terms (e.g., describing centrifugal, Coriolis, friction and gravitational forces) and $Q$ is the control force (in $[N]$ for prismatic- and in $[Nm]$ for rotary joints). Each array is composed of multiple dynamic and kinematic parameters which makes it inadequate for parameter estimation in this form. Instead, a similar approach is applied as in AIDC where each coefficient is redefined as a separate parameter and a linear relationship is obtained as,

$$H(q)\ddot{q}+h(q,\dot{q})=Y(q,\dot{q},\ddot{q})\Theta=Q, \tag{1}$$

in which $Y \in \mathbb{R}^{n \times r}$ known as the regressor matrix. Its structure is precisely known on kinematic basis: $n$ denotes the degree of freedom of the system, and $r$ is the number of the independent dynamic parameters. On the other hand, the $\Theta \in \mathbb{R}^r$ array contains all the independent dynamic parameters. Utilizing the linear relationship and the fact that in each $t_i$ (which is the time of taking the $i^{th}$ sample by the controller) $Q^{(i)} = Y^{(i)}\Theta^{(i)}$ a simple regression model is constructed as

$$\tilde{Y}\Theta^{(i)} = \begin{bmatrix} Y^{(i-m+1)} \\ \vdots \\ Y^{(i-1)} \\ Y^{(i)} \end{bmatrix} \begin{bmatrix} \Theta_1^{(i)} \\ \Theta_2^{(i)} \\ \vdots \\ \Theta_r^{(i)} \end{bmatrix} = \begin{bmatrix} Q^{(i)} \\ Q^{(i-1)} \\ \vdots \\ Q^{(i-m+1)} \end{bmatrix} = \tilde{Q}, \tag{2}$$

where $m$ is the number of samples used for the regression that can be arbitrarily set by the user. In (2) $\tilde{Y} \in \mathbb{R}^{k \times r}, k = nm$ and $\tilde{Q} \in \mathbb{R}^k, k = nm$ are constructed from various instances of $Y^{(i)}$ and $Q^{(i)}$ arrays and $\Theta^{(i)} \in \mathbb{R}^r$ is the array of parameters under estimation. Solving (2) for $\Theta^{(i)}$ is not very easy, as this set of linear equations has many redundant elements which yield different solutions, as there is no straight line which fits all sampled points. A small error can be expected for each point. So our task is to find the line which best fits all the points, by minimizing the error as

$$H = \sum_{l=1}^{nm} \left( \tilde{Q}_l - \sum_{s=1}^{r} \tilde{Y}_{ls}\Theta_s^{(i)} \right)^2 = \min, \tag{3}$$

which has a minimum, if for each $\Theta_u^{(i)}, u = 1, 2...r$,

$$\frac{\partial H}{\partial \Theta_u^{(i)}} = \sum_{l=1}^{nm} 2 \left( \tilde{Q}_l - \sum_{s=1}^{r} \tilde{Y}_{ls}\Theta_s^{(i)} \right) \tilde{Y}_{lu} = 0 \tag{4}$$

that can be written in matrix form as $\tilde{Y}^T\tilde{Q} - \tilde{Y}^T\tilde{Y}\Theta^{(i)} = 0$, which yields the estimated parameters as

$$\Theta^{(i)} = (\tilde{Y}^T\tilde{Y} + \mu I)^{-1}\tilde{Y}^T\tilde{Q}. \tag{5}$$

Equation (5) can be solved via the inversion of a quadratic matrix if it is not singular (rather well conditioned). In the case of near singular or singular matrices the deformation of the task applied by Levenberg and Marquardt in a similar problem can be applied in it with a small positive parameter $\mu$ [11,13]. Reliable results naturally can be expected from well conditioned matrix.

## 3 Simulation Results and Conclusion

In this paper the proposed regression based parameter identification method in combination with a fixed point iteration-based control scenario was investigated through simulation programs written in Julia language. The simulations were made for dynamic model of a cylindrical robot [17] which is shown in Fig. 1.

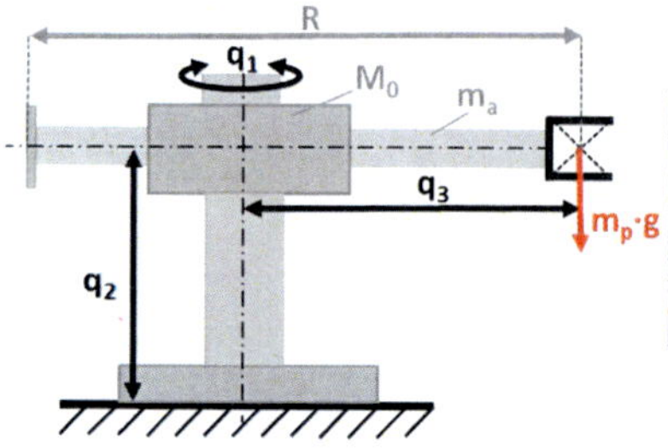


| Parameter | Exact Value | Approximate Value |
|---|---|---|
| $R$ (m) | 2 | 3 |
| $M_0$ (kg) | 20 | 35 |
| $m_a$ (kg) | 12 | 20 |
| $m_p$ (kg) | 2 | 5 |
| $g$ (m/s$^2$) | 9.81 | 10 |
| $\theta_0$ (kg m$^2$) | 50 | 75 |

**Fig. 1.** Simplified model of cylindrical robot with dynamic parameters ($M_0$ - weight of the hub, $m_a$ - weight of the horizontal arm, $R$ - length of the horizontal arm, $m_p$ - weight of the load) and kinematic parameters ($q_1, q_2, q_3$)

The dynamic equations of the cylindrical robot, are given below,

$$\begin{bmatrix} \theta_0+\theta_h & 0 & 0 \\ 0 & M_h & 0 \\ 0 & 0 & m_a+m_p \end{bmatrix} \begin{bmatrix} \ddot{q}_1 \\ \ddot{q}_2 \\ \ddot{q}_3 \end{bmatrix} + \begin{bmatrix} (m_a(2q_3-R)+2m_pq_3)\dot{q}_3\dot{q}_1 \\ M_hg \\ -\frac{1}{2}(m_a(2q_3-R)+2m_pq_3) \end{bmatrix} = \begin{bmatrix} Q_1 \\ Q_2 \\ Q_3 \end{bmatrix} \quad (6)$$

where $M_h = M_0+m_a+m_p$ is the weight of the full assembly of the horizontal arm, $\theta_0$ is the inertia of the vertical segment and $\theta_h$ is the inertia of the horizontal arm which was calculated based on the position of the horizontal segment (modeled with homogeneous weight distribution), $\theta_h = m_a\left(q_3^2 - Rq_3\right) + \frac{1}{3}R^2m_a + m_pq_3^2$. All the other dynamic parameters are given in Fig. 1. The equations were transformed, according (1), for the investigated cylindrical manipulator as,

$$Y\Theta = \begin{bmatrix} \ddot{q}_1 & q_3^2\ddot{q}_1+2q_3\dot{q}_3\dot{q}_1 & q_3\ddot{q}_1+\dot{q}_3\dot{q}_1 & 0 & 0 \\ 0 & 0 & 0 & \ddot{q}_2 & 1 \\ 0 & 0 & 0 & \ddot{q}_3-q_3\dot{q}_1^2 & -\frac{1}{2}\dot{q}_1^2 \end{bmatrix} \begin{bmatrix} \theta_0+\frac{1}{3}R^2m_a \\ m_a+m_p \\ -Rm_a \\ M_v \\ M_vq \end{bmatrix}. \quad (7)$$

To simulate modeling imprecision an *"exact"* and an *"approximate"* parameter set was introduced, shown in Fig. 1. The approximate parameters were used to calculate the $Q$ control force required to track the $q^N$ nominal trajectory and the system response (which would be measured in a real application) was calculated using the exact parameters. The first simulation results are shown in Fig. 2 where $m = 500$ samples were used for parameter estimation. The nominal trajectory was defined in joint space as $q_j^N = A_j\sin(\omega_jt)+S_j$, where k $A = [5.5\,\text{rad} \;\; 1.5\,\text{m} \;\; 0.8\,\text{m}]^T$, $\omega = [1.2 \;\; 1.6 \;\; 0.1]^T\frac{1}{s}$ and $S = [0.0\,\text{rad} \;\; 1.5\,\text{m} \;\; 0.8\,\text{m}]$. The control algorithm was implemented using PID feedback as $\text{k}\ddot{q}^{Des} = \ddot{q}^N + \Lambda^3\int_{t_0}^{t}\left(q^N(\xi)-q(\xi)\right)\mathrm{d}\xi + 3\Lambda^2(\mathrm{q^N(t)-q(t)}) + 3\Lambda(\dot{\mathrm{q}}^{\mathrm{N}}(\mathrm{t})-\dot{\mathrm{q}}(\mathrm{t}))$ ($\Lambda = 6s^{-1}$) and the adaptive deformations were made through abstract rotations [4]. Since the adaptive control is independent from the parameter estimation, at beginning the FPI controller uses the original parameter set and later it transitions to the estimated values. The applied time resolution in the simulations is $\delta t = 1e^{-3}$ s which also applies for the sampling time of the controller ($t_{sC} = 1e^{-3}$ s) and the sampling time of the parameter estimation ($t_{sP} = 1e^{-3}$ s). The simulation was made through 500000 steps.

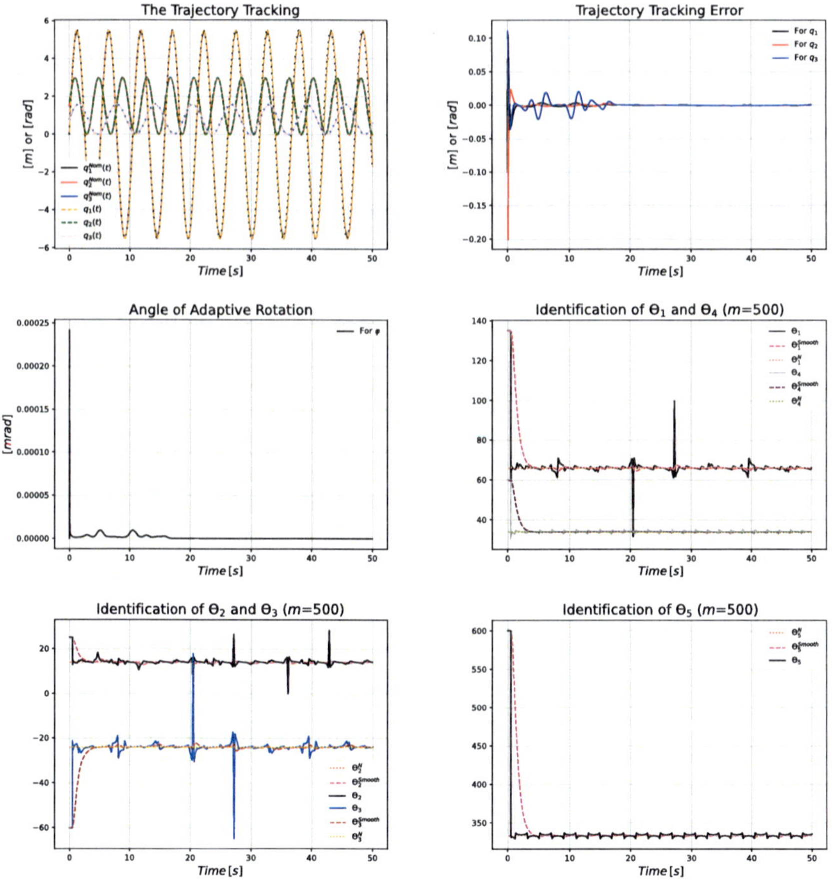


**Fig. 2.** Linear regression-based parameter identification with $m = 500$ samples

Similar observations can be made as in [8]. If the sampled data is not well balanced, which means that some of the non-zero elements of $Y(q, \dot{q}, \ddot{q})$ are approaching 0, no sufficient information can be extracted about some of the elements of $\Theta$, and the parameter identification becomes obscure. This effect can be compensated by applying a simple low pass filter as $(\Lambda_s + \frac{d}{d\,t})^2 \Theta^{Smooth} = \Lambda_s^2 \Theta$, which yields $\ddot{\Theta}^{Smooth} = \Lambda_s^2(\Theta - \Theta^{Smooth}) - 2\Lambda_s \Theta^{Smooth}$ ($\Lambda_s = 2$), from which

$\Theta^{Smooth}$ can be obtained through Euler integration. Similar idea was applied in inverse kinematics of industrial robots in [22]. This solution nicely attenuates the high peaks. Despite the small deviations which are still present in the smoothed parameters ($\Theta^{Smooth}$), the quality of the control significantly improved. The trajectory tracking error reduced almost to 0 after the controller transitioned to the estimated parameters. The precision of the parameter estimation is also well exhibited by the angle of adaptive rotation in Fig. 2 as it is related to the magnitude of adaptive deformation. Based on the figure it is clear that almost no adaptive deformation was required using the estimated parameters. Although, it is still expedient to maintain the adaptive deformation due to the deviation of the estimated parameter set.

To decrease the computational demand of the proposed solution the number of samples taken for parameter estimation can be decreased. However, in such scenario we have to take into consideration that we need data from various significantly different operating points of the investigated system to make precise linear regression. If the sampled time period is too short this criteria cannot be met, so we do not have sufficient information about the system for the parameter identification. In Fig. 3 some simulation results are displayed where the sampling time was $t_{sP} = 0.01\,\mathrm{s}$ and $m = 50$ samples were used for the parameter estimation (the time resolution for the simulation and sampling time for the controller was kept the same, $\delta t = t_{sC} = 1e^{-3}\,\mathrm{s}$). A slight increase can be observed in the deviation of the estimated parameters.

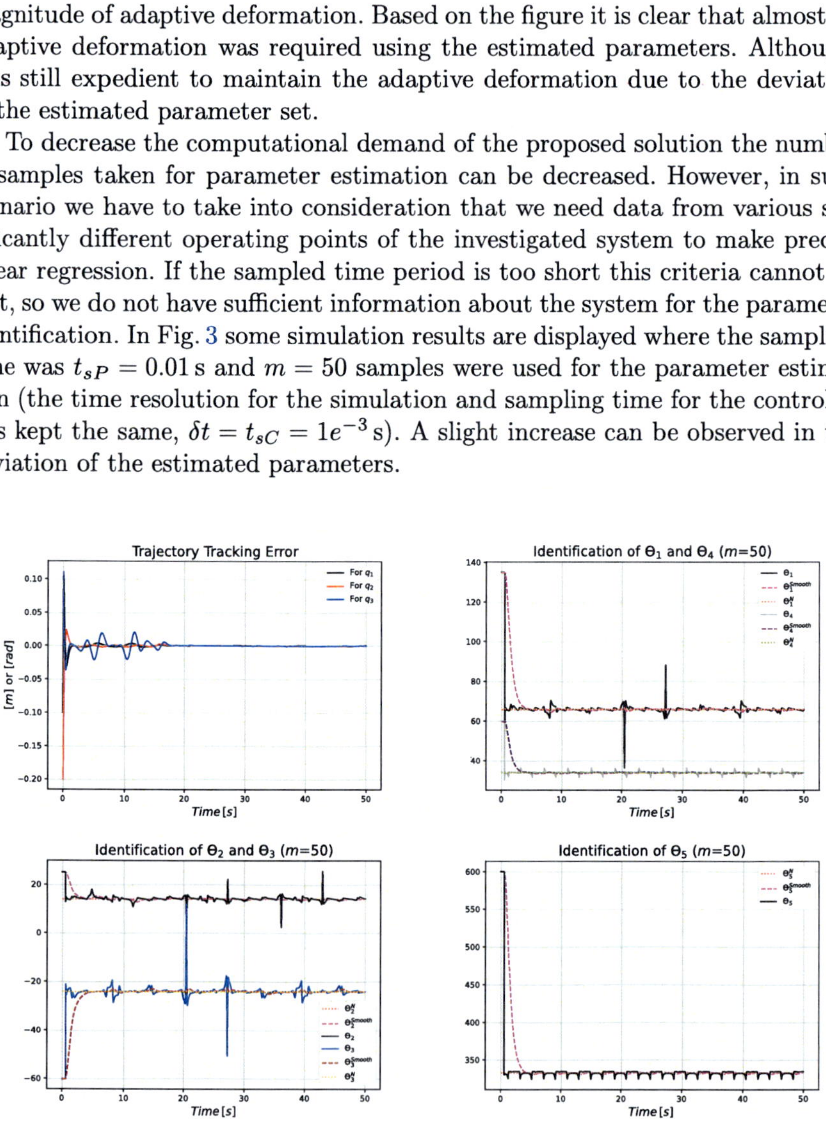


**Fig. 3.** Linear regression-based parameter identification with $m = 50$ samples

However, the increase in the estimated values are nicely handled by the adaptive controller so there is no significant change in the trajectory tacking error. On the other hand such decrease in the number of samples can spare much computational power (shown in Fig. 4) which is very important in a real application.

In conclusion the proposed Linear Regression-based parameter identification method can further improve the tracking quality in a robotic system (in other applications it is expected as well) with an FPI controller. It is also expected that this method can be appropriate for real time application as it has significantly less computational requirement compared to genetic algorithms which were tested in the past (such as Particle Swarm Optimization).

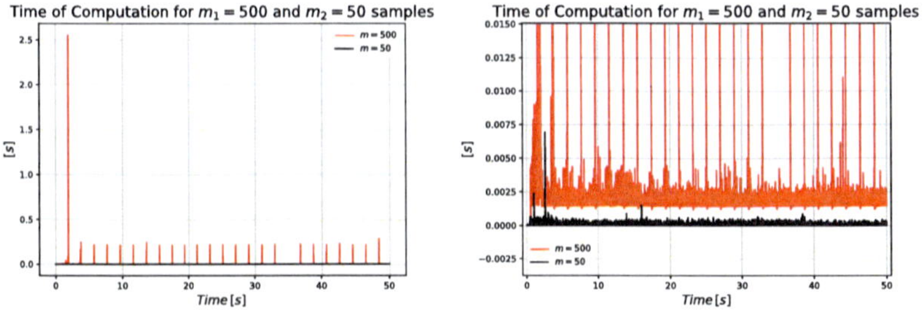


**Fig. 4.** Time of computation for different number of samples for regression (left), zoomed in to the area of interest (right) (Simulation on: HP 250 G6 laptop with Intel i3-7020U processor, under Windows10 operating system).

## References


1. Armstrong, B., Khatib, O., Burdick, J.: The explicit dynamic model and internal parameters of the PUMA 560 arm. In: Proceedings of the IEEE Conference on Robotics and Automation, pp. 510–518 (1986)
2. Bécsi, T., Szabó, A., Kővári, B., Aradi, S., Gáspár, P.: Reinforcement learning based control design for a floating piston pneumatic gearbox actuator. IEEE Access **8**, 147295–147312 (2020). https://doi.org/10.1109/ACCESS.2020.3015576
3. Corke, P., Armstrong-Helouvry, B.: A search for consensus among model parameters reported for the PUMA 560 robot. In: Proceedings of the IEEE Conference on Robotics and Automation, pp. 1608–1613 (1994)
4. Csanádi, B., Galambos, P., Tar, J., Györök, G., Serester, A.: A novel, abstract rotation-based fixed point transformation in adaptive control. In: Proceedings of the 2018 IEEE International Conference on Systems, Man, and Cybernetics (SMC2018), Miyazaki, Japan, 7–10 October 2018, pp. 2577–2582 (2018)
5. Deniša, M., Ude, A., Gams, A.: Adaptation of motor primitives to the environment through learning and statistical generalization. In: Borangiu, T. (ed.) Advances in Robot Design and Intelligent Control. AISC, vol. 371, pp. 449–457. Springer, Cham (2016). https://doi.org/10.1007/978-3-319-21290-6_45
6. Dineva, A., Várkonyi-Kóczy, A., Tar, J.: Combination of RFPT-based adaptive control and classical model identification. In: Proceedings of the IEEE 12th International Symposium on Applied Machine Intelligence and Informatics (SAMI 2014), Herl'any, Slovakia, pp. 35–40 (2014)

7. Hamandi, M., Tognon, M., Franchi, A.: Direct acceleration feedback control of quadrotor aerial vehicles. In: 2020 IEEE International Conference on Robotics and Automation (ICRA), pp. 5335–5341. IEEE (2020)
8. Issa, H., Tar, J.K.: Improvement of an adaptive robot control by particle swarm optimization-based model identification. Mathematics **10**(19) (2022). https://doi.org/10.3390/math10193609
9. Kennedy, J., Eberhart, R.: Particle swarm optimization. In: Proceedings of the 1995 IEEE International Conference on Neural Networks, vol. 4, pp. 1942–1948 (1995)
10. Kővári, B., Hegedüs, F., Bécsi, T.: Design of a reinforcement learning-based lane keeping planning agent for automated vehicles. Appl. Sci. **10**(20) (2020). https://doi.org/10.3390/app10207171
11. Levenberg, K.: A method for the solution of certain non-linear problems in least squares. Q. Appl. Math. **2**, 164–168 (1944)
12. Lyapunov, A.: Stability of Motion. Academic Press, New York (1966)
13. Marquardt, D.: An algorithm for least-squares estimation of nonlinear parameters. J. Soc. Ind. Appl. Math. **11**(2), 431–441 (1963)
14. Nguyen, C., Antrazi, S., Zhou, Z.L., Campbell, C., Jr.: Adaptive control of a Stewart platform-based manipulator. J. Robot. Syst. **10**(5), 657–687 (1993)
15. Radac, M.B., Precup, R.E., Petriu, E.: Constrained data-driven model-free ILC-based reference input tuning algorithm. Acta Polytechnica Hungarica **12**(1), 137–160 (2015)
16. Slotine, J.J.E., Li, W.: Applied Nonlinear Control. Prentice Hall International Inc., Englewood Cliffs (1991)
17. Somló, J., Lantos, B., Pham, T.C.: Advanced robot control (2002)
18. Spong, M., Ortega, R.: On adaptive inverse dynamics control of rigid robots. IEEE Trans. Autom. Control **35**(1), 92–95 (1990). https://doi.org/10.1109/9.45152
19. Tar, J., Bitó, J., Nádai, L., Tenreiro Machado, J.: Robust Fixed Point Transformations in adaptive control using local basin of attraction. Acta Polytechnica Hungarica **6**(1), 21–37 (2009)
20. Tar, J., Rudas, I., Dineva, A., Várkonyi-Kóczy, A.: Stabilization of a modified Slotine-Li adaptive robot controller by robust fixed point transformations. In: Proceedings of Recent Advances in Intelligent Control, Modelling and Simulation, Cambridge, MA, USA, pp. 35–40 (2014)
21. Varga, B., Tar, J., Horváth, R.: Tuning of dynamic model parameters for adaptive control using particle swarm optimization. In: Szakál, A. (ed.) Proceedings of the IEEE 10th Jubilee International Conference on Computational Cybernetics and Cyber-Medical Systems, ICCC 2022, Reykjavík, Iceland, 6–9 July 2022, pp. 197–202. IEEE Hungary Section, Budapest, Hungary (2022)
22. Varga, B., Issa, H., Horváth, R., Tar, J.: Sub-optimal solution of the inverse kinematic task of redundant robots without using Lagrange multipliers. Syst. Theory Control Comput. J. **1**(2), 40–48 (2021). https://doi.org/10.52846/stccj.2021.1.2.25
23. Wang, H., Xie, Y.: Adaptive inverse dynamics control of robots with uncertain kinematics and dynamics. Automatica **45**(9), 2114–2119 (2009). https://doi.org/10.1016/j.automatica.2009.05.011

# Service Robotics

# Navigation of a Mobile Platform in a Greenhouse

Jakob Gimpelj(✉), Matevž Semolič, Marko Munih, Sebastjan Šlajpah, and Matjaž Mihelj

Faculty of Electrical Engineering, University of Ljubljana, Tržaška Cesta 25, 1000 Ljubljana, Slovenia
jakob.gimpelj@gmail.com

**Abstract.** This paper details the design and development of a mobile platform specifically for use in a greenhouse, as well as the implementation of navigation capabilities to transition from a concrete path to parallel heating pipes, which also function as rails for the platform. The primary objective was to create a mobile platform that could be used for various tasks, equipped with multiple tools and robotic manipulators, and capable of aligning itself autonomously with the heating pipes. Navigation was accomplished using a depth camera to capture a point cloud, which was then compared to a reference image. The resulting transformation matrix provided the necessary rotational and translational data. Movement of the platform was achieved through simple linear and rotational motions based on odometry calculations of the platform's wheels.

**Keywords:** Mobile robot · Greenhouse · ROS · Control · Navigation · Differential drive

## 1 Introduction

The use of mobile robotic systems is rapidly increasing due to the availability of advanced technologies in their development. These systems have found applications in various sectors such as industry, laboratories, healthcare, catering, military, aerospace, and agriculture, providing many advantages. In modern agriculture, robotics plays a crucial role due to the need for increased food production, the demand for accessible technology, large arable areas, and the demanding and monotonous nature of agricultural work. Robotic manipulators, mounted on mobile platforms, can perform tasks such as pick-and-place transport, planting, pharmaceutical application, weeding, and harvesting. Furthermore, modern robots can collect and integrate data from the field, analyze parameters that affect crops and plants, and make better-informed decisions and predictions, leading to improved farming outcomes. The overall advantages of robotic technologies in agriculture, including repeatability, precision, autonomy, and the ability to handle monotonous tasks, align with the main objectives of farming, which are to increase crop quantity and quality and improve working conditions for farmers [1–3].

In this case, the goal is to design a mobile platform, which can mount sensors for analyzing tomatoes or robotic manipulators that can interact with the crops. Additionally, the platform should be able to autonomously navigate the greenhouse using a depth camera [4].


T. Petrič et al. (Eds.): RAAD 2023, MMS 135, pp. 393–400, 2023.
https://doi.org/10.1007/978-3-031-32606-6_46

## 2 Greenhouse Layout

The greenhouse platform workspace is divided into two parts as shown in Fig. 1. The first part, located at the center, is a concrete area used by the greenhouse staff and carts to transport crops and tools. The second part is the space between the rows where heating pipes are placed, and workers carry out their tasks. For this purpose, they use special carts that are designed to move on both the heat pipes and the concrete path.

The central concrete section, which is 4 m wide, serves as the main passage for staff and machinery to move between the rows. Due to its high occupancy rate, collision and obstacle detection potential in this area are increased. The central part is used to turn and align the carts with the pipes, enabling them to move on top of them.

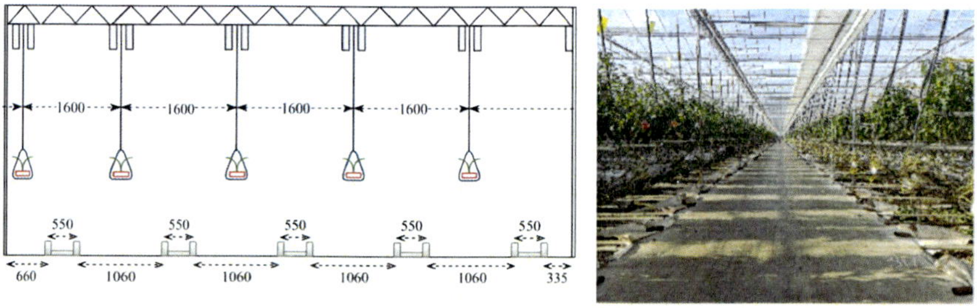


**Fig. 1.** A sketch of the greenhouse layout (units in mm) and the view along the concrete path.

The distance between adjacent rows of crops where the pipes are located is 1.6 m. Carts and platforms that move on the pipes must not exceed a maximum width of 0.8 m and a maximum height of 0.4 m, otherwise they risk damaging the crops above the pipes. The lengths of the rows in the greenhouses range from 60 m to 120 m. The ends of the heat pipes overlap with the concrete path, lying flat on it.

## 3 Design of the Mobile Platform

To ensure proper function and performance within the greenhouse, the platform frame must have suitable load-bearing capacity and shape. Due to the advantages of aluminum slotted profiles, we opted to use aluminum Bosch profiles. The frame is divided into two parts: the base section, which supports the equipment and holds the attached components, and the top plate, which serves as the attachment point for robotic manipulators and other tools.

### 3.1 The Platform Chassis

The base frame was constructed from five transverse and two longitudinal square aluminum profiles, fixed together with bolts and hammer nuts. Four vertical profiles were used to connect a top plate to the base frame. The top plate, designed to support various tools and manipulators, was assembled using three 900 mm long and 180 mm wide aluminum plates. The front and rear panels included 3D printed edges with an emergency stop button opening.

### 3.2 The Drive System

For the undercarriage of the mobile system, two separate drive wheels and two omnidirectional wheels were chosen. Industrial wheels with a maximum load capacity of 400 kg and made from polyamide were chosen for the two drive wheels. These wheels enable movement on the concrete path and have added surfaces with a smaller radius on the inner side to facilitate driving on top of the heat pipes. The wheels can rotate at the same speed in opposite directions, enabling the mobile system to rotate on its own axis. A servomotor AM8113 and a planetary gear reducer with a gear ratio of 1:10 power the drive, with the motors connected to a control module on a Beckhoff controller located on the base frame. We selected 203 mm diameter wheels from Oz Robotics [5] for the omnidirectional wheels. These wheels are made up of two aluminum plates and an intermediate spacer. Each plate contains eighteen rollers with sliding bearings, positioned at a 90° angle on the plate. This allows for an additional direction of movement, parallel to the left and right, while rotating. Each omnidirectional wheel has a load capacity of 50 kg and features an extra surface on the inner side for driving on pipes. The load capacity of the entire platform is 100 kg.

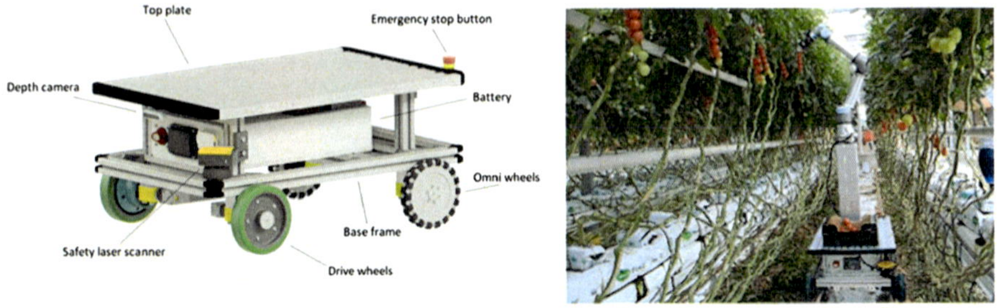


**Fig. 2.** The left image shows the final platform frame with wheels, AM8113 motors, the battery in the middle, two nanoScan3 sensors attached on opposite corners and the depth camera attached at the front. The right image shows the mobile platform in one of its intended use cases, where a robot arm is mounted on it and is used to interact with crops.

### 3.3 The Sensory System

To guarantee safety during autonomous operation, two SICK nanoScan3 safety laser scanners were installed diagonally on two opposing corners of the base frame. These scanners prevent collisions by stopping all motion when an object is detected within their safety zone. For navigation purposes an Intel RealSense D435i RGB depth camera was mounted in the middle of the front side at a height of 22 cm from the ground and tilted downwards by 11° from the horizontal. This tilt enables better image capture of low objects that are lying on the ground in front of the camera. The model of the complete mobile platform is shown above in Fig. 2.

## 4 Kinematic Model

The mobile platform has two drive wheels that form a differential drive, whereas the two omnidirectional wheels are passive and independent of the drive wheels. The differential drive is a widely used and simple drive mechanism with the main advantage of allowing the wheels to rotate independently. The mobile platform's movements are computed based on wheel odometry while considering the kinematic model shown in Fig. 3.

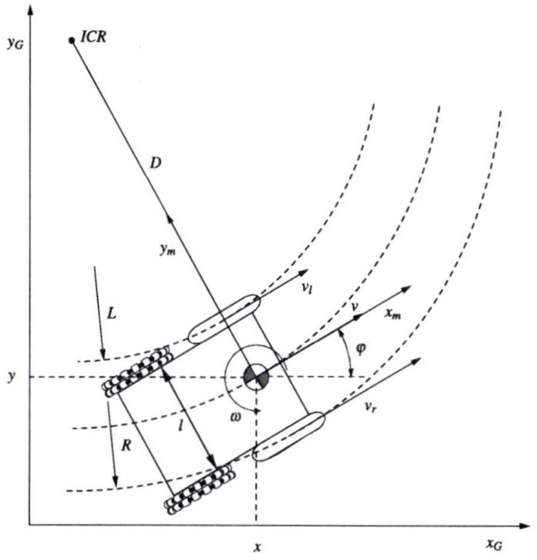


**Fig. 3.** The mobile platform with a differential drive relative to the global coordinate system.

The wheels of the differential drive rotate at angular velocities $\omega_r$ and $\omega_l$. By using the radius of the wheels $r$, we can express the circumferential velocities of the right $v_r$ and left wheels $v_l$ with Eqs. (1) and (2), respectively [6].

$$v_r = \omega_r r \tag{1}$$

$$v_l = \omega_l r \tag{2}$$

The linear $v$ and angular velocity $\omega$ of the entire mobile platform are calculated using Eq. (3) for linear velocity and Eq. (4) for angular velocity, where $l$ represents the distance between the two drive wheels.

$$v = (v_r + v_l)/2 \tag{3}$$

$$\omega = (v_r - v_l)/l \tag{4}$$

For basic localization, we have implemented odometry, which reads encoder data from the motors and determines the position and orientation of the platform at any given moment. Odometry data is given relative to the platform's starting position. This is based on the integration of the platform's kinematic model. If the circumferential and angular velocities remain constant during the sample time $T_s$, we can calculate the odometry using the Euler integration method, as shown in Eqs. (5), (6), and (7), where $x_k$, $y_k$, and $\varphi_k$ represent the position and orientation of the platform at time step $k$.

$$x_{k+1} = x_k + vT_s \cos(\varphi_k) \tag{5}$$

$$y_{k+1} = y_k + vT_s \sin(\varphi_k) \tag{6}$$

$$\varphi_{k+1} = \varphi_k + \omega T_s \tag{7}$$

## 5 Navigation System

For the mobile platform to carry out its tasks, it must be able to navigate the concrete paths in the greenhouse and autonomously transition onto and drive on the heating pipes. We will mainly focus on the transition from the path to the pipes [7].

To ensure a successful transition, a 3D point cloud obtained from the RealSense D435i depth camera was used to calculate the rotational and translational movements the platform must make to align itself with the pipes. The camera communicates with the main program, designed in the Robot Operating System (ROS), which runs on an on-board mini-PC. Its task is to capture a depth image from which it extracts a discrete set of data points, resulting in a point cloud. Each point in the point cloud contains positional and color information. The Open3D library was used for processing the point cloud in the Python programming environment.

### 5.1 Depth Image Processing

After receiving the point cloud from the camera, the program splits it into two parts: the positions ($x$, $y$, and $z$ coordinates) of the points, and the RGB values. Only the position part is used, as the color info is not relevant for this approach.

The captured point cloud represents the entire field of view of the camera, and only those points representing the target (pipe) need to be extracted. To ensure that the coordinate frame of the point cloud and the coordinate frame of the mobile platform are aligned, the point cloud must be rotated by the camera's angle of 11° around the horizontal axis (Fig. 4a).

All points that logically cannot be part of the target are removed from the point cloud. The target is defined as the end of the heat pipes laying on the path. This means removing all points lower than the camera's height and everything further than 1.8 m from the origin. This leaves the point cloud of the pipe, the path, as well as some extra points caused by noise (Fig. 4b).

A plane extraction function, which utilizes the RANSAC algorithm to find points that make up a plane, is used to remove points that represent the concrete path under the pipes [8]. All points corresponding to the plane are then removed. The remaining point cloud is further processed to remove any noise and excess points not belonging to the target object. A function checks for neighboring points within a specified radius of

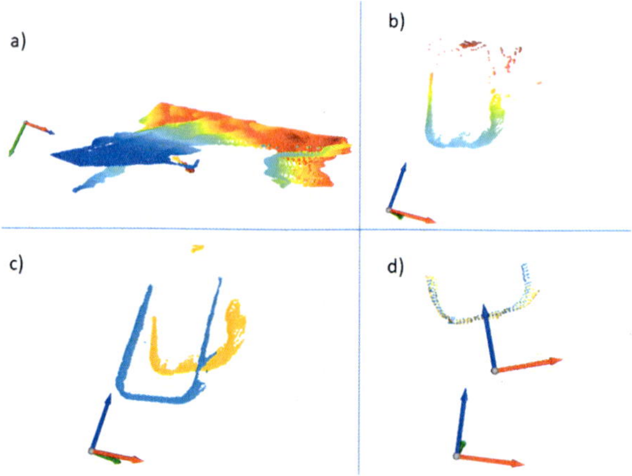


**Fig. 4.** The process of point cloud processing (a – c) and registration (d).

a randomly selected point and removes it if the number of neighboring points is below a given threshold [9]. As a result, the final point cloud contains only the desired target object (Fig. 4c).

### 5.2 Point Cloud Registration

To perform the alignment of the platform with the pipes, a RANSAC-based point cloud alignment function is utilized, which requires a reference point cloud of the target pipe. To capture the reference image, the platform is positioned in an optimal location where it can easily move straight onto the heat pipes. As all pipes have a similar geometry, only one reference point cloud is needed. Once the reference image is obtained, the alignment function can align the point cloud of the target from any arbitrary position with the reference point cloud (Fig. 4d). If the alignment error is below a certain threshold, meaning that the alignment was successful, then the function will return a homogeneous transformation matrix. If the alignment error was above the threshold, the system would retry aligning the point clouds. If more than 10 unsuccessful attempts occurred, the camera would capture another image.

Apart from the error threshold, two other important parameters are the maximum number of iterations and probabilistic confidence. As a rule, the larger these two parameters are, the more accurate the result will be, and the more time-consuming it will be. To lower the computation time, the input point clouds are downsampled.

### 5.3 Movement Planning

From the resulting transformation matrix, the translation and rotation data are extracted, and the rotation values are converted to Euler angles for each axis.

The movement of the platform is divided into three basic steps: rotation towards the target location (Fig. 5a), linear movement towards it (Fig. 5b), and rotation into the final orientation upon arrival (Fig. 5c). The Euclidean distance is used to calculate the distance to the target location. The current location and orientation of the platform are obtained from the odometry calculation. A proportional regulator is used to control the movements.

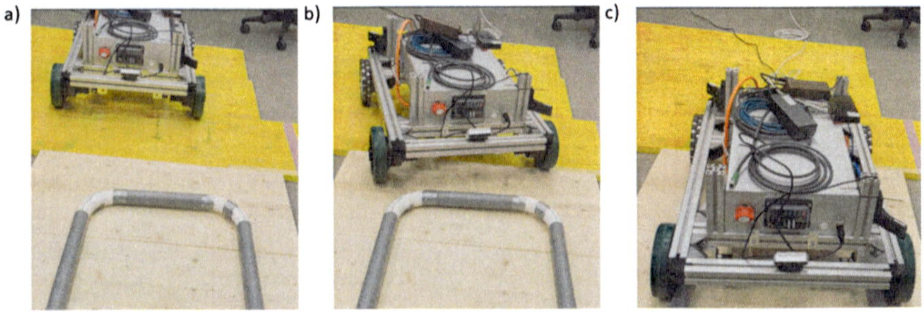

**Fig. 5.** The mobile platform in the process of alignment with the pipes.

# 6 Testing and Findings

## 6.1 Testing the Vision System

The basic image processing software was tested on a plastic mock-up of real pipes with the same dimensions. From these tests, it was found that the point-to-point alignment approach has limitations in both speed and accuracy.

The program was able to compute one full alignment at a speed of approximately 1 Hz, achieved by sufficiently downsampling the point cloud and adjusting the number of iterations parameter. However, further downsampling had a negative impact on alignment accuracy.

The position of the camera also affected the results, as the target object must be within a certain distance from the camera for the comparison to be made with sufficient accuracy and repeatability. If the camera is too close to the target, the resulting point cloud will be cut due to the limited field of view of the camera. However, if the camera is too far away, the point cloud of the object will be distorted too much.

## 6.2 Movement Testing

Testing of the alignment performance and the resulting displacement to the target point was also carried out using a mock-up tube. The computed values obtained from the transformation had an accuracy of 0.1 m. The design of the inner surfaces of the wheels additionally helps with alignment with the pipes.

The drive motors turned out to lack enough power to accurately perform smaller turns on the spot and are a top priority for future improvements. Slipping of the drive wheels caused an incorrect computation of the platform's odometry and therefore an incorrect position estimation. However, the wheels have good grip on the concrete path of the greenhouses, reducing this error.

## 7 Conclusion

In recent years, agriculture has become an industry with major investments in automation and robotics. Both in outdoor and indoor farming, the arable areas are large. Our objective to create a mobile system that can be controlled remotely and that allows autonomous driving within the greenhouse was achieved. The navigation system currently lacs a way to navigate within the broader greenhouse. Testing GNSS navigation inside the greenhouse proved to be too inaccurate due to the signal interference (greenhouse metallic elements) and blockage (plants). The navigation problem could be solved using SLAM. However, a simpler solution would be to use an additional camera mounted on the side of the platform to track visual markers placed along the main path. The RGB depth camera, which is already in use, can also be utilized for real-time localization. This would result in a more precise alignment of the platform with the heat pipes, without the need for an additional sensor.

## References


1. Autonomous Robotic Platforms for Greenhouses, https://research.qut.edu.au/qcr/Projects/autonomous-robotic-platforms-for-greenhouses/, last accessed 2022/12/14
2. Priva, Priva Octinion combine Kompano robotic activities — Priva, last accessed 22/08/28
3. Autonomous mobile platform for pest detection and control inside greenhouses, https://www.tekniker.es/en/autonomous-mobile-platform-for-pest-detection-and-control-inside-greenhouses, last accessed 2022/12/14
4. IntelRealsense for ROS, https://dev.intelrealsense.com/docs/ros-wrapper, last accessed 2022/12/14
5. O. Robotics, 203mm Double Aluminum Omni Directional Wheel Basic – 14124 – Oz Robotics, https://ozrobotics.com/shop/203mm-double-aluminum-omni-directional-wheel-basic-14124, last accessed 22/08/28
6. Klančar, Gregor, Zdešar, Andrej, Blažič, Sašo, Škrjanc, Igor.: Avtonomni mobilni sistemi. Založba FE, 2021
7. LiDAR point-cloud based 3D object detection implementation with colab, by Gopalakrishna Adusumilli, https://towardsdatascience.com/lidar-point-cloud-based-3d-object-detection-implementation-with-colab-part-1-of-2-e3999ea8fdd4, last accessed 2022/12/14
8. Ning, X., Li, F., Tian, G., Wang, Y.: An efficient outlier removal method for scattered point cloud data. PLoS ONE **13**(8), e0201280 (2018). https://doi.org/10.1371/journal.pone.0201280
9. Global registration, Open3D, http://www.open3d.org/docs/release/tutorial/pipelines/global_registration.html, last accessed 2022/12/14

# Implementation and Testing of a Shoe Polishing Process with a Collaborative Robotic System

Matteo Forlini(✉), Marianna Ciccarelli, Alessandra Papetti, Luca Carbonari, and Giacomo Palmieri

Department of Industrial Engineering and Mathematical Sciences, Polytechnic of Marche University, Ancona, Italy
m.forlini@pm.univpm.it

**Abstract.** Nowadays, automated processes in manufacturing industries are on the rise due to the need to increase productivity and product quality, but also to reduce operator cognitive-physical fatigue. This need is more felt in companies, such as footwear companies, where production is done entirely by hand and the success of the product relies totally on the skill of experienced artisans. The work presents the automation with a collaborative robot of the shoe polishing. This process is very delicate because of the high variability of leather types and the maximum quality to be achieved, as well as very tiring for the operator. An operational methodology for carrying out the polishing of a real leather shoe is proposed. Starting from the design of polishing trajectories, implementing them on UR5e, controlling the contact force of the tool, toe shoe polishing is performed, achieving a good quality standard. Experimental tests and their results are presented.

**Keywords:** Leather shoe polishing · Collaborative robotic application · Human robot collaboration

## 1 Introduction

At the moment many manufacturing companies feel the need to automate processes, in order to have better results in terms of quality and higher production, decreasing the possibility of errors [1]. The operator performs high value-added tasks that machines would not be able to do, decreasing their physical and cognitive stress. In this regard, collaborative robots [15], which are machines that can automate a production process but also work alongside the operator by interacting with it, are becoming increasingly popular in recent years [2]. This can improve the psychophysical well-being of the operator and reduce his cognitive stress [6]. For example, in the footwear industry, most production processes are carried out by hand, and only a few processes are dedicated to specialized machinery, such as cementing and cutting operations [8]; while the use of robots is generally dedicated to ordinary repetitive tasks, such as pick and place [9],


T. Petrič et al. (Eds.): RAAD 2023, MMS 135, pp. 401–408, 2023.
https://doi.org/10.1007/978-3-031-32606-6_47

glue deposition on the sole [3], and robotic grip of the sole during assembly operations [11]. The use of robots and profound industrial automation is limited in this industry by the fact that products with considerable variability and totally different processing requirements are handled, and the finished products must have high precision, quality and comfort. Only human labor can guarantee these targets, making the company competitive. Characteristic aspects of shoe companies operating in the luxury sector that stand in total opposition to the possibility of automating a process are: low productivity due to artisanal activities [10], high levels of customization and with the management of micro-lots [17], difficulty in managing and transmitting knowledge of skilled artisans [13]. A central process in the production of shoes is the polishing of them. Polishing is almost always done by hand, takes a long time of 20 min or more for shoe, and is a very delicate operation. The ability and manual dexterity of the operator is crucial to achieving a quality polish. It is a difficult process to automate because each individual shoe has different polishing behaviour since it depends on multiple factors, such as leather color, type of leather, leather processing, and so on. Fundamental parameters to be kept under control to achieve good polishing through a robot are the contact force between tool and shoe, which must always be the same and not too high. The robot must be compliant in the direction perpendicular to the surface by always applying the same force. The force control can be implemented by passive compliance control or active force control [5,7]. In passive compliance control, the polishing tool is deformed and adapts to the surface, absorbing him the deformation energy and always providing the same force to the surface. In active control, on the other hand, thanks to sensors, the force applied is measured and kept constant [14]. Today, such operation is possible thanks to the use of collaborative robots, which are characterized by high mobility, and control of the force exerted on the workpiece thanks to high-tech sensor technology. There are studies in the literature on polishing operations of objects with even complex surfaces using robots [12,16] but polishing of shoes is lacking. In [4] preliminary tests for a shoe polishing are presented. The objective of this paper is to propose an operational methodology to achieve high quality shoe polishing using a collaborative robot. This methodology starts with the definitions of polishing trajectories on a software cam and then moves on to their implementation on a collaborative robot. An active control of the polishing force was implemented, allowing for an optimal result in terms of product quality. The implementation of this methodology allows for the total automation of shoe polishing achieving good results, which is today performed by hand. No such solution exists in the literature, which could be innovative for the entire footwear sector. The whole process has been tested on a real shoe form and then the quality of the polishing has been analyzed by a chemical laboratory of a footwear company. Some of the following figures present blurred zones, which aren't essential to the correct comprehension of the work, due to company confidentiality agreement. The paper is organized as follows: Sect. 2 describes the polishing process, Sect. 3 illustrates how the process has been implemented, Sect. 4 discusses the experimental tests and the results obtained. Finally, concluding remarks and future works are summarised.

## 2 Polishing Process for Leather Shoes

The purpose of this section is to illustrate the process performed by hand by the operators and then in later sections will be automated and replicated on tests shoe. Polishing starts with visual inspection of the pair of shoes allows one to understand what color of polish is needed for that shoe, how many times the operation needs to be repeated to give uniformity of polish. A last is then inserted into the shoe to give rigidity. The first step consists of four main operations. The operator places the cloth on his fingers and picks up the polish by applying pressure with his fingers inside the container. The withdrawn excess polish is deposited by rubbing the cloth on a white tap. In such way is possible to avoid staining the shoe on first contact, which could compromise the success of the entire process. The application of polish on the shoe should always be done starting from different points depending on the shade and tone you want to give the shoe, because these are the points where the shade will be most pronounced. In the second phase, the same operations are performed, but a polish with a different consistency and viscosity is used. Step 1 and step 2 are repeated $n$ and $m$ times, respectively, where $n$ and $m$ are defined from time to time based on the goodness of the polishing process.

Starting from here, the exact same process was replicated, taking all these cautions into consideration, with a collaborative robot to complete the first polishing step, which is the most delicate and exhausting for the operator. Next, the operator must complete the polishing with the finishing step. The workstation of the robot and the polishing tool are defined in [4]. It was chosen to use the polishing tool with a 30 mm diameter backing pad with fixed rotation axis and angular velocity of 3000 rpm. In this article only the toe part of the shoe has been polished through the use of a rigid last covered with leather. Such a last is the same one that is used to manufacture the shoe and polish it.

A fundamental step is to define the trajectories for the robot to perform the polishing. The robot has to polish the same area several times but cannot do it by performing the same trajectories otherwise it would leave streak signs on the leather. In order to avoid this phenomenon, 16 different types of trajectories were designed to work the toe area, thanks to NX manufacturing software and "Multi-axis-deposition" processing, which allows creating and simulating the deposition of a polishing layer on the surface. The 3D CAD of the shoes was divided into several zones, one for the toe. We designed several paths on the toe surface starting from guide lines, for example, the edge line of the shoe sole. The different types of trajectories are shown in Fig. 1. Only the 4 most representative of the 16 are reported, the others have slightly different shapes with different start and end points.

As can be seen, the trajectories are organized so that they never have the same start and end points (indicated by the letters S and E). In addition, the first trajectories all start from specific points, to give that shade of tone described earlier. There are different paths, either zigzag or spiral. The step between two adjacent lines is 5 mm for the first path after resuming polished and the last path before resuming polish, for the others it is 8 mm. In the cycle presented in

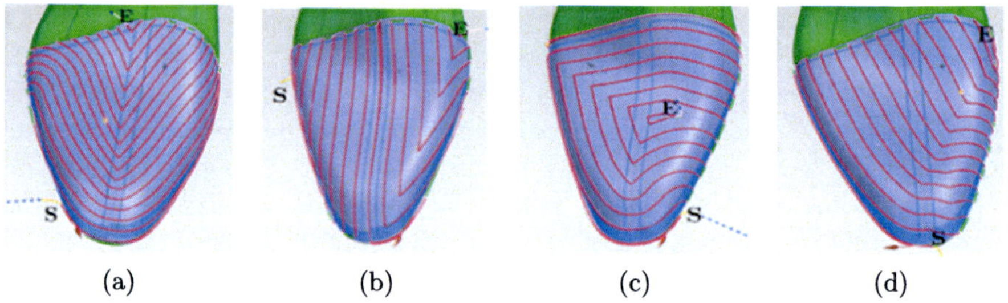


**Fig. 1.** Four different types of trajectories

the Fig. 2, the polish is taken twice: before trajectory 1 and before trajectory 9. The polish is dispensed in a precise amount from a dispenser into a container, the robot with its tool goes to the position of the container, gets in contact with the container for two seconds, and picks up the polish with the rotating backing pad. The precise amount of polish dispensed cannot be reported here for confidentiality reasons. Next this operation, the robot switches to the hemispherical surface to rub off excess polish and homogenise it. Then the polishing starts with the different trajectories.

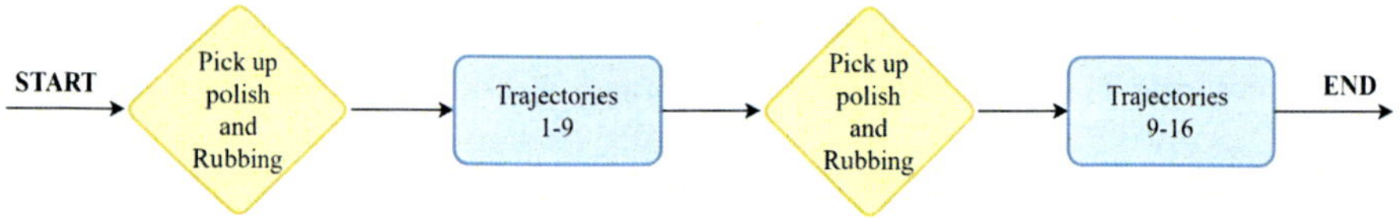


**Fig. 2.** Experimental polishing process

## 3 Implementation of the Process on a Collaborative Robot

To implement the polishing process on a collaborative robot, in this case UR5e, we had to generate the robot program from the trajectories generated by NX manufacturing. After setting the path in NX multi-axis-deposition, it is possible to export a .cls file, in which there is a list of path poses with orientation perpendicular to the shoe surface. The robotic system is replicated in ROBODK software, with the UR5e in an upside-down configuration and 3D CAD of the shoes (Fig. 3).

The .cls file of the trajectory was imported to ROBODK thanks to the "Robot Machining Project," setting the correct parameters such as tool, reference frame, preferred tool orientation and robot configuration, and so on. This made it possible to run the polishing process designed in NX with the UR5e. The compiled

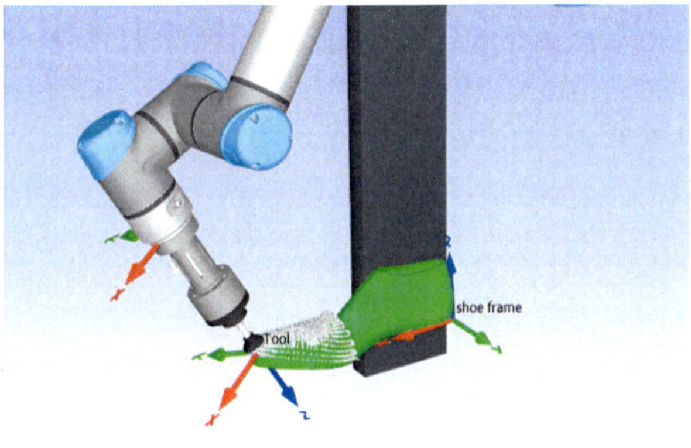


**Fig. 3.** Robodk station to obtain an executable trajectory from the robot

URscript from ROBODK Universal Robot post processor was exported and further command such as the control force setting was added. A URscript was generated for each trajectory, then implemented and executed on the real robot. All the poses of the trajectory are referred to the shoes frame.

The contact force is controlled through UR5e force torque sensor installed on the robot flange. This sensor ensures a good sensibility at the contact with the leather. To realize a control force along the direction of the tool, which changes continuously during the execution of the task, a thread operating with a frequency 100 Hz was implemented that updates the tool position from time to time and activates the force on the direction of the tool, pushing on the surface with 5 N. This value was chosen by trial and error after several tests, it allows good quality. At the end of the tool, there is a spring before the backing pad that contributes to compliance movement of the robot in the perpendicular direction of the surface and helps keep the force constant.

## 4 Experimental Tests and Results

To perform the experimental tests, the shape of the shoe was placed on a specially constructed support that exactly replicates the shape of the shoe heel. The reference system in the Fig. 4 is the reference system of the shoe. In this way, by knowing where the reference system of the support is positioned in the workspace, automatically the position of the reference system of the shoe is known, which is otherwise difficult to measure because it is in a position that is not known at the beginning. To measure the support reference position and orientation respect to the robot base, three different point are taken with the robot. These three points define the support reference system.

After this setup, knowing the position of the shoe reference, the polishing process has been started as shown in Fig. 2. The goodness of polish is given by the right compromise between three main parameters, which are the force with

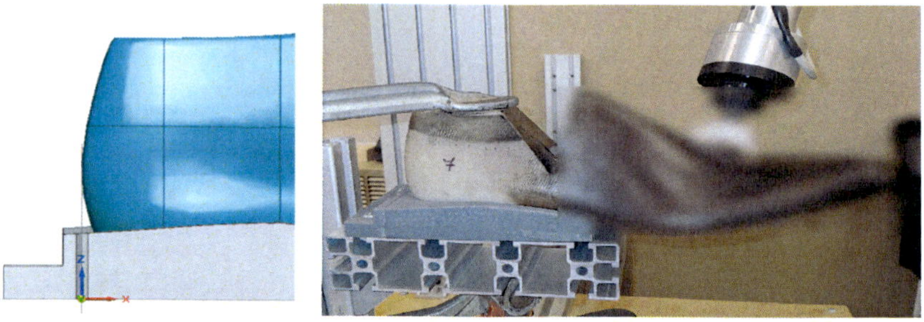

(a) *Shoe support* (b) *Shoe position during polishing*

**Fig. 4.** Last shoe with its support

which the shoe is worked, the number of repetitions of trajectories that are performed, and the speed of the movements. The first two are discussed in Sects. 2, 3 the latter can neither be too high otherwise the polish is not deposited and neither too low otherwise the cycle time is compromised. For footwear company confidentiality agreement, the speed value used cannot be given. The necessity to perform different trajectories of polishing on the toe, repeated several times with not too high speed constrains the cycle time of the operation. However, it is a fair compromise to achieve the excellent quality that is mandatory in this scenario. In Fig. 5, in the area bounded by the red rectangle, the polish on the surface of the shoe, left by the rotating backing pad, is clearly visible.

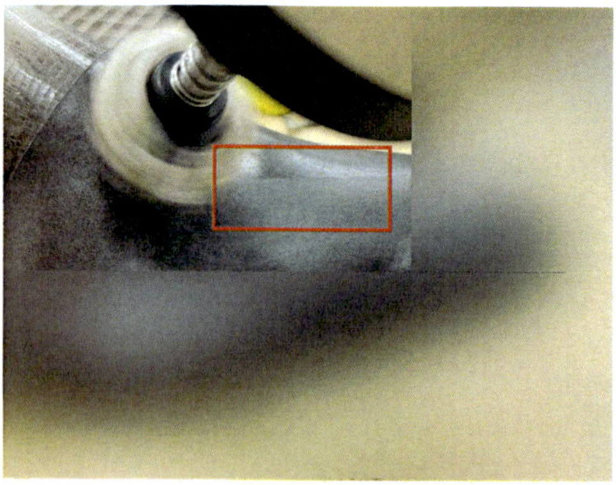

**Fig. 5.** Polishing operation

After the entire polishing process was completed, the polished toe was examined by a chemical laboratory, experienced in leather polishing, with a careful visual inspection of the shoe and other chemical tests that cannot be reported here due to confidentiality agreements. The polishing satisfies the quality standard for shoe polishing and it can be equated with an handmade polishing obtained before the second phase of finishing. At the moment, by choice of the footwear company, human and robot work in separate areas monitored by a laser scanner, but collaboration between them is also possible. Figure 6 shows the region of the polished toe at the end of the process. The difference between the polished and unpolished area is clearly visible.

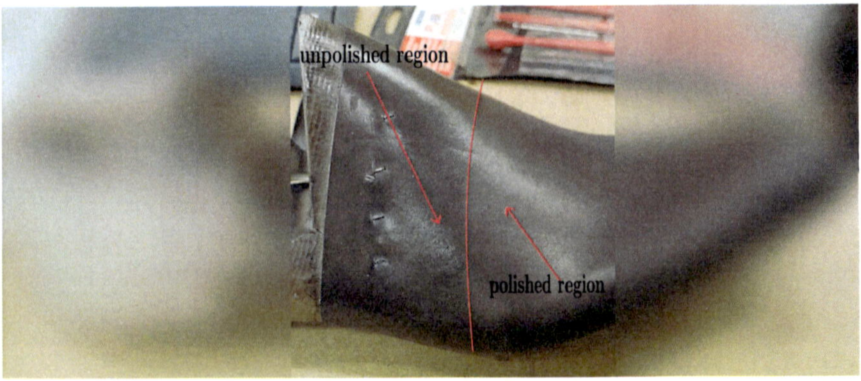


**Fig. 6.** Shoe polished after the entire process

## 5 Conclusions

This work addressed the need by a shoe company to automate a shoe polishing process, that was previously done totally by hand, in order to decrease the physical stress on the worker. An operating methodology was implemented to automate a shoe polishing process using a collaborative robot. The results obtained in terms of polishing quality are satisfactory, demonstrating the feasibility in automating this procedure. The only limitation concerns the cycle time, which is around 20 min to perform the shoe toe polishing. This time is caused by the large number of polishing passes to be made to achieve optimal polishing quality that meets the high quality standards of the handmade luxury footwear industry. The toe area is the most important area of the shoe because it is the most visible. However, this time is comparable to the time it takes the operator to polish a shoe. Thanks to this solution, the operator will no longer perform this repetitive and tiring operation, but will be able to check whether the automated polishing achieves good quality and simply finish the polishing, allowing him to spend his time on high value-added activities. As next developments, to reduce the cycle time, we will adjust the polishing speed and optimize the number of polishing. Considering the good results found on the toe, this methodology will be adopted to polish a whole shoe.

## References

1. Axmann, B., Harmoko, H.: Robotic process automation: an overview and comparison to other technology in industry 4.0. In: 2020 10th International Conference on Advanced Computer Information Technologies (ACIT), pp. 559–562. IEEE (2020)
2. Callegari, M., et al.: Tools and methods for human robot collaboration: case studies at i-Labs. Machines **10**(11), 997 (2022)
3. Castelli, K., Zaki, A.M.A., Dmytriyev, Y., Carnevale, M., Giberti, H.: A feasibility study of a robotic approach for the gluing process in the footwear industry. Robotics **10**(1), 6 (2020)
4. Chiriatti, G., et al.: Human-centered design of a collaborative robotic system for the shoe-polishing process. Machines **10**(11), 1082 (2022)
5. Dong, Y., Ren, T., Hu, K., Wu, D., Chen, K.: Contact force detection and control for robotic polishing based on joint torque sensors. Int. J. Adv. Manuf. Technol. **107**(5), 2745–2756 (2020)
6. Gualtieri, L., Rauch, E., Vidoni, R.: Emerging research fields in safety and ergonomics in industrial collaborative robotics: a systematic literature review. Robot. Comput.-Integr. Manuf. **67**, 101998 (2021)
7. Lakshminarayanan, S., Kana, S., Mohan, D.M., Manyar, O.M., Then, D., Campolo, D.: An adaptive framework for robotic polishing based on impedance control. Int. J. Adv. Manuf. Technol. **112**(1), 401–417 (2021)
8. Maurtua, I., Ibarguren, A., Tellaeche, A.: Robotic solutions for footwear industry. In: Proceedings of 2012 IEEE 17th International Conference on Emerging Technologies & Factory Automation (ETFA 2012), pp. 1–4. IEEE (2012)
9. Méndez, J.B., Perez-Vidal, C., Heras, J.V.S., Pérez-Hernández, J.J.: Robotic pick-and-place time optimization: application to footwear production. IEEE Access **8**, 209428–209440 (2020)
10. Mosca, F., La Rosa, E.: 4.0 technology within fashion and luxury production. Symphonya. Emerging Issues Manag. (2), 82–94 (2019)
11. Oliver, G., Gil, P., Gomez, J.F., Torres, F.: Towards footwear manufacturing 4.0: shoe sole robotic grasping in assembling operations. Int. J. Adv. Manuf. Technol. **114**(3), 811–827 (2021)
12. Perez-Vidal, C., Gracia, L., Sanchez-Caballero, S., Solanes, J.E., Saccon, A., Tornero, J.: Design of a polishing tool for collaborative robotics using minimum viable product approach. Int. J. Comput. Integr. Manuf. **32**(9), 848–857 (2019)
13. Rossi, M., Papetti, A., Germani, M., Marconi, M.: An augmented reality system for operator training in the footwear sector. Comput. Des. Appl. **18**, 692–703 (2020)
14. Tian, F., Lv, C., Li, Z., Liu, G.: Modeling and control of robotic automatic polishing for curved surfaces. CIRP J. Manuf. Sci. Technol. **14**, 55–64 (2016)
15. Vicentini, F.: Collaborative robotics: a survey. J. Mech. Des. **143**(4) (2021)
16. Wang, K.B., Dailami, F., Matthews, J.: Towards collaborative robotic polishing of mould and die sets. Procedia Manuf. **38**, 1499–1507 (2019)
17. Zhang, C., Chen, D., Tao, F., Liu, A.: Data driven smart customization. Procedia CIRP **81**, 564–569 (2019)

# Road Curb Detection: ADAS for a Road Sweeper Vehicle

Ivan Bilić[(✉)], Goran Popović, Tibor Bataljak Savić, Ivan Marković, and Ivan Petrović

Faculty of Electrical Engineering and Computing, Laboratory for Autonomous Systems and Mobile Robotics, University of Zagreb, Zagreb, Croatia
{ivan.bilic,goran.popovic,tibor.savic,ivan.markovic,ivan.petrovic}@fer.hr

**Abstract.** Recent development of the concept of smart cities has led to an increasing demand for advanced technological solutions that drive forward the design and capabilities of utility vehicles operating in urban environments. One possibility is to exploit recent advances in computer vision to introduce a certain level of autonomy into some of the vehicle's functionalities, e.g., an advanced driver-assistance system. Modern road sweeper vehicles are designed to possess multiple systems for maintaining the road quality and city cleanliness, such as brushes, vacuums, and great vehicle maneuverability. Introducing autonomy to these control systems lowers the burden on the human operator, thereby increasing work efficiency and overall safety, as well as making a positive impact on worker health. This paper considers a 3D curb detection system that supports autonomous road sweeping. In order to achieve this, we utilize a vision-based approach that leverages stereo depth estimation and a pretrained semantic segmentation model. In addition, we implement a simple LiDAR-based curb detection baseline. Finally, we collected our own dataset comprised of driving sequences resembling our use-case, which is used to conduct qualitative experiments.

**Keywords:** curb detection · driver assistance system · utility vehicles

## 1 Introduction

The global impacts of climate change, increasing air pollution and population in urban areas, and the development of the concept of smart cities, which represents the key strategic backbone for the development of cities in the 21st century, are making an impact on the industry of utility vehicles. The need for technology that can help reduce adverse effects (e.g., the impact of air pollution on human health), as well as increase efficiency and safety is becoming more and

This work has been supported by the European Regional Development Fund under the grant KK.01.2.1.02.0115 - Development of environmentally friendly vehicle for cleaning public surfaces with autonomous control system based on artificial intelligence (EKO-KOMVOZ).


T. Petrič et al. (Eds.): RAAD 2023, MMS 135, pp. 409–416, 2023.
https://doi.org/10.1007/978-3-031-32606-6_48

**Fig. 1.** RASCO LYNX Charge - an electric road sweeping utility vehicle.

more emphasized. Utility vehicles, such as road sweepers, can be designed to possess multiple systems for road quality maintenance and city cleanliness, such as brushes and suction mouth (Fig. 1), in order to keep the roads and streets clean and free of undesirable objects. Furthermore, brush and suction mouth control pose additional load on the human operator, forcing them to perform multiple tasks simultaneously.

Presuming that the vehicle is always on the road, a reliable curb detection system acts as an enabler for autonomous cleaning, thereby lowering the load on the human operator. Hence, introducing autonomy in brush and suction mouth control systems can have a positive impact on both the efficiency and health of the human operator, while also contributing to the overall safety of the vehicle and its environment. Existing work on road curb detection is mostly comprised of 3D methods that obtain 3D information using sensors such as light detection and ranging (LiDAR) [4,10] or rely on stereo vision only [1,7]. Furthermore, in [3] authors investigate fusing semantic information with 3D data. Finally, real-time semantic segmentation for autonomous driving applications is a well active area of research [2,9].

In this paper, we investigate the problem of implementing a real-time system for road curb detection with the aim of supporting autonomous brush and suction mouth control. To tackle this problem, we first design a LiDAR-camera sensor rig that we mounted on the utility vehicle shown in Fig. 1 and used it to collect the dataset. This processes is described in Sect. 2. Furthermore, we consider two different, simple approaches for curb detection, discussed in Sect. 3; a vision-based, and a LiDAR-based approach, respectively. Section 4 describes our calibration procedure, while the conclusion is given in Sect. 5.

## 2 Dataset Collected with the Lynx Utility Vehicle

The first task was to collect a dataset comprised of driving sequences in an urban environment, collected by driving the road sweeping utility vehicle shown in Fig. 1 throughout the streets of Zagreb. Concretely, the data was collected through seven driving sequences (~15 min each), each on a different day, during daytime and under normal weather conditions, i.e., in the absence of fog, rain, and snow. Prior to data collection process, we mounted the custom rig system on the vehicle that consisted of two color cameras and a solid-state LiDAR. The rig system was fixed on the vehicle's windshield from the inside, using two vacuum plates, as shown in Fig. 2. Images were captured using two FLIR's Blackfly S GigE color cameras. These are global shutter cameras that offer high image resolution of $2048 \times 1536$, as well as high frame rate of 35 frames per second. As for the LiDAR, we used RS-LiDAR-M1, an automotive grade solid-state LiDAR designed for mass production vehicles. This LiDAR has a long range (up to 200 m), with horizontal and vertical field-of-view of 120° and 25°, respectively. It is also characterized by a decent horizontal resolution of 0.2°, in addition to operating on a satisfying rate of 10 Hz. Finally, the calibration process, i.e., the process of obtaining the camera intrinsic and extrinsic parameters with respect to each other and the LiDAR, was carried out before and after each driving sequence, to check for considerable parameter changes.

The cameras were powered through a Power-Over-Ethernet (PoE) switch, while the LiDAR has it's own power supply. However, it was also connected to the same switch, thus cameras' and LiDAR measurements were synchronized via Generalized Precision Time Protocol (gPTP). As for the device drivers, both LiDAR and camera device drivers are publicly available and provided by the manufacturers. LiDAR drivers support Robot Operating System (ROS). However, this is not the case for FLIR cameras. Since we used ROS to implement the driver assistance software, we modified the camera drivers accordingly and wrapped them as a ROS node[1].

## 3 Road Curb Detection

The goal of this section is to describe the pipeline that we engineered in order to estimate the road curb in 3D space. To this end, we implemented two road curb detection methods. First, we describe a vision-based method for curb detection, i.e., one without exploiting LiDAR data and then we describe a LiDAR-based method which operates on the point cloud data only. All the processing components are implemented as ROS nodes, using Python and C++.

### 3.1 Vision-Based Curb Detection

By vision-based curb detection, we refer to estimating the 3D points belonging to the road curb using only color images. To this end, we developed a pipeline

[1] https://bitbucket.org/unizg-fer-lamor/stereo_camera_driver.

**Fig. 2.** A custom sensor rig system used to collect the data. Two color cameras and a solid-state LiDAR are fixed on the rig. The rig is attached to the windshield of the utility vehicle using two vacuum plates, and an extra band for safety.

that consists of stereo depth estimation, and a semantic image segmentation model, with "curb" in its class definition. Once the depth and semantic maps are obtained, a set of 3D points belonging to the curb can easily be found.

To recover the depth maps, we use Semi-Global-Matching (SGM) [5] – a stereo depth estimation algorithm that combines the concepts of global and local stereo methods for pixel-wise matching. The algorithm takes rectified stereo image pairs and outputs a disparity map. While different SGM variants have been proposed [6], we use the publicly available OpenCV's implementation. A good trade-off between accuracy and computational cost makes SGM a favorable choice for many practical applications [6].

In order to identify the curb in an RGB image, we use semantic segmentation which outputs pixel-level labels belonging to a predefined set of classes. However, a publicly available dataset that resembles our use-case containing pixel-wise or 3D curb annotations was not available. Most datasets contain "road" and "sidewalk" classes, so the curb could be approximated as a boundary between those two classes. Nonetheless, we opted for a pre-trained model that explicitly contains the curb in its class definition. To this end, we use a road segmentation model[2] developed by Intel, trained to segment each pixel into one of the four classes; road, curb, mark, and background. Unfortunately, only the pre-trained model's binary is publicly available – source code and training data are not. Nevertheless, the model is evaluated on 500 images of Mighty AI dataset, achieving 0.727 mIOU and 83.1% accuracy on the "curb" class with high computational efficiency, making it adequate for our use-case. We conducted qualitative experiments on our data to further test the model's performance, using Intel's Neural Compute Stick 2 (NCS2) for model inference.

Examples of qualitative inspection are depicted in Fig. 3. The model generalizes well on our dataset, successfully detecting the road edge for different types

[2] https://docs.openvino.ai.

(a) (b) (c)

**Fig. 3.** Examples of road curb segmentation on our custom dataset. The curb mask is overlaid on the images and shown in green. Qualitative inspection reveals satisfactory model generalization capabilities.

of curbs (Fig. 3a, Fig. 3b), and even drains (Fig. 3c), although less consistently. On the contrary, we expect degraded performance for types of curbs that are under-represented in the training data, as well as in poor visibility and lighting conditions.

In general, given a 3D point $(X, Y, Z)^T$ in the camera coordinate frame, its projection on the image plane is governed by:

$$\lambda \begin{pmatrix} u \\ v \\ 1 \end{pmatrix} = \mathbf{K} \begin{pmatrix} X \\ Y \\ Z \end{pmatrix} \tag{1}$$

where $(u, v, 1)^T$ denote the pixel coordinates, $\lambda$ is the scale factor, and $K$ represents the camera intrinsics, obtained through the camera calibration process. Thereby, the back-projection, i.e., the inverse operation is given by:

$$\begin{pmatrix} X \\ Y \\ Z \end{pmatrix} = \lambda \mathbf{K}^{-1} \begin{pmatrix} u \\ v \\ 1 \end{pmatrix}. \tag{2}$$

The curb pixel coordinates $(u, v)^T$ are provided by the semantic mask, while the depth map contains the $Z$-coordinates. Thus, given the semantic mask and (2), we obtain a set of 3D points corresponding to the curb.

### 3.2 LiDAR-Based Curb Detection

In addition to vision-based curb detection, we also employ a simple LiDAR-based curb detection method. Our implementation is inspired by [4], where a road edge localization method is proposed and considered in the context of driver-assistance systems, among others. The key idea is that a set of 3D points belonging to a curb possesses certain geometric features, based on which the curb can be differentiated from other parts of the scene. That is, LiDAR data is processed to compute a geometric feature in the form of the angular distance to the ground normal. The remainder of the chapter describes the developed implementation.

Given that our application is an urban road driving scenario, we assume that the vehicle is always on the road. Consequently, extraction of the ground plane is a must. We first fit the plane from the point cloud using Random sample consensus (RANSAC), considering only points farther than 2 m and closer than 20 m from the LiDAR. We use 300 iterations and set the distance threshold for inlier selection to 0.05 m. When RANSAC converges, parameters of the best model are used together with the corresponding inlier set.

Prior to calculating the geometric features, the point cloud is transformed in the ground plane's coordinate frame, given that it's defined in the LiDAR coordinate frame. This transformation is computed by using the ground plane parameters estimated with RANSAC. First, the magnitude of the ground plane normal is computed. The ground plane normal $\mathbf{z}$ is defined as $(a, b, c)^T$, where $a$, $b$ and $c$ are the ground plane parameters and the magnitude $\|\mathbf{z}\|$ is defined as $\|\mathbf{z}\| = \sqrt{a^2 + b^2 + c^2}$. Next, an affine transformation matrix $\mathbf{T}$ is formed as:

$$\mathbf{T} = \begin{bmatrix} \cos(\phi) + \alpha^2 \cdot (1 - \cos(\phi)) & \alpha \cdot \beta \cdot (1 - \cos(\phi)) & \beta \cdot \sin(\phi) & 0 \\ \alpha \cdot \beta \cdot (1 - \cos(\phi)) & \cos(\phi) + \beta^2 \cdot (1 - \cos(\phi)) & -\alpha \cdot \sin(\phi) & 0 \\ -\beta \cdot \sin(\phi) & \alpha \cdot \sin(\phi) & \cos(\phi) & d \\ 0 & 0 & 0 & 1 \end{bmatrix}, \quad (3)$$

where $\alpha = \frac{b}{\sqrt{a^2+b^2}}$, $\beta = \frac{-a}{\sqrt{a^2+b^2}}$, $\cos(\phi) = \frac{c}{\|\mathbf{z}\|}$ and $\sin(\phi) = \frac{\sqrt{a^2+b^2}}{\|\mathbf{z}\|}$. Then, the rotation matrix $\mathbf{R}$ and the translation vector $\mathbf{t}$ are easily extracted, given that $\mathbf{T} = [\mathbf{R}|\mathbf{t}]$. Finally, the transformed point $\mathbf{X_t}$ is obtained by taking:

$$\mathbf{X_t} = \mathbf{R}\mathbf{X} + \mathbf{t}, \quad (4)$$

and this transformation is applied to every point $\mathbf{X}$.

Next, geometric features from which the curb is determined are computed. Let $\mathbf{n}_p$ denote the normal vector at the considered 3D point. Given the ground plane normal $\mathbf{z}$, the angular distance $\theta$ is expressed as:

$$\theta = \arccos(\mathbf{n}_p \cdot \mathbf{z}), \quad (5)$$

with the direction of the normal $\mathbf{n}_p$ always oriented such that $\mathbf{n}_p \cdot \mathbf{z} > 0$, hence $\theta \in [0, \frac{\pi}{2}]$. Finally, the curb is determined using the computed angular distance $\theta$, by labeling a 3D point as a curb if $\theta$ falls in a predefined sub-interval, i.e., if $\theta \in [\frac{\pi}{8}, \frac{\pi}{4}]$ holds as proposed in [4]. The feature calculation process is done using the Point Could Library [8].

The process of normal estimation is heavily affected by the chosen search radius $R$. If set improperly, the feature calculation process degrades. When considering a 3D point, this radius defines an area from which the points are taken into account to fit a plane. Following manual inspection of the results on our data, we set $R$ to 0.3 m.

LiDAR-based curb detection is shown in Fig. 4. The two leftmost images depict an exemplar scene as a color image (Fig. 4a) and a LiDAR scan (Fig. 4b). The estimated ground plane is shown in Fig. 4c, while the curbs, extracted by

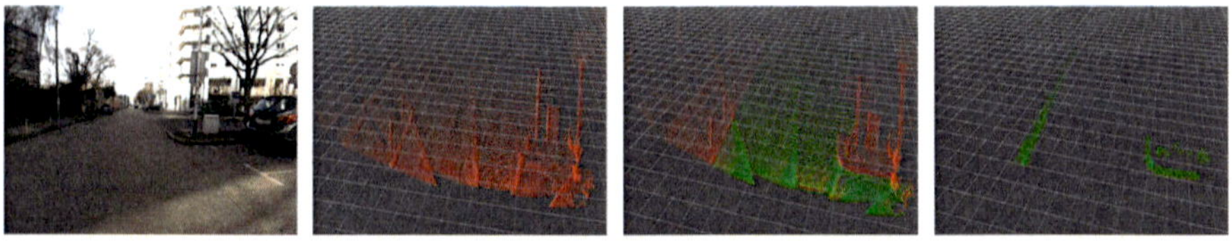

(a) (b) (c) (d)

**Fig. 4.** Visualization of LiDAR-based curb extraction process via geometric features. First two images display sensor data; (a) RGB image (not used for computation, displayed for convenience), (b) LiDAR scan, which represent the scene. The following two images show; (c) ground-plane estimation, (d) estimated curbs.

using the angular distance feature are displayed in Fig. 4d. Finally, by means of qualitative inspection, we observe that this method exhibits a higher false positive rate (Fig. 4d), in contrast to the described vision-based approach. However, we note that this method is fairly straightforward, so this comparison cannot serve as a conclusion regarding superiority of vision-based over LiDAR-based methods in general. Nonetheless, LiDAR is useful in poor lighting conditions. As future work, we plan to enhance this approach by introducing additional informative features, as well as filtering (e.g., a Kalman filter).

## 4 Camera-LiDAR Calibration Method

The major challenge of multi-sensor calibration lies in the fact that the only thing that can be surely measured in both sensor domains is the planar shape of a calibration checkerboard. The calibration process expects a set of synchronized camera images and LiDAR point clouds, where each pair contains measurements of the checkerboard. First, the parameters of the commonly used pinhole radtan camera model are determined. OpenCV functions for corner extraction and camera calibration are used to compute the intrinsic camera parameters and relative transformation from the calibration target reference frame to the camera reference frame $T_T^C$ for each frame in the measurement set. Furthermore, in each point cloud we seek to find the closest planar surface assuming it is the calibration target. We find this assumption reasonable since the calibration is performed in a controlled environment, and the ground plane can be easily filtered out by eliminating surfaces within a certain range around the ground plane normal vector.

The point cloud filtering step leaves us with points $p_L(i)$ corresponding to the checkerboard, expressed in the LiDAR reference frame. Given a valid LiDAR to camera transformation $T_L^C$, the point cloud is transformed from the LiDAR reference frame to the calibration target reference frame using the following transformation chain:

$$p_T(i) = (T_T^C)^{-1}\ T_L^C\ p_L(i). \tag{6}$$

Expressed in the target reference frame, all measured LiDAR points $p_T(i)$ should have a zero $Z$ coordinate. Extrinsic parameters estimation between the camera and the LiDAR is thus achieved by minimizing the following cost function:

$$T_L^C = \underset{T_L^C}{\arg\min} \left( \sum_i (T_T^C)^{-1} \, T_L^C \, p_L(i) \right)_Z . \tag{7}$$

## 5 Conclusion

In this paper, we have tackled the problem of estimating curbs in urban driving scenes for the purpose of controlling the brushes of a road sweeping utility vehicle. To this end, two different approaches were developed and qualitatively compared: a vision-based and LiDAR-based curb detection. Furthermore, we also described our data generation process – the camera-LiDAR rig preparation, data collection, and sensor setup calibration. As future work, we plan to investigate fusion of the vision-based and LiDAR-based approach, employ filtering on the LiDAR data, and explore learning-based approaches for curb detection in the LiDAR point cloud.

## References


1. Fernández, C., Izquierdo, R., Llorca, D.F., Sotelo, M.A.: Road curb and lanes detection for autonomous driving on urban scenarios. In: 17th International IEEE Conference on Intelligent Transportation Systems (ITSC), pp. 1964–1969 (2014)
2. Gao, R.: Rethink dilated convolution for real-time semantic segmentation. arXiv preprint arXiv:2111.09957 (2021)
3. Goga, S.E.C., Nedevschi, S.: Fusing semantic labeled camera images and 3D lidar data for the detection of urban curbs. In: 2018 IEEE 14th International Conference on Intelligent Computer Communication and Processing (ICCP), pp. 301–308. IEEE (2018)
4. Hervieu, A., Soheilian, B.: Road side detection and reconstruction using lidar sensor. In: 2013 IEEE Intelligent Vehicles Symposium (IV), pp. 1247–1252 (2013)
5. Hirschmuller, H.: Accurate and efficient stereo processing by semi-global matching and mutual information. In: 2005 IEEE Computer Society Conference on Computer Vision and Pattern Recognition (CVPR 2005), vol. 2, pp. 807–814. IEEE (2005)
6. Hirschmüller, H.: Semi-global matching-motivation, developments and applications. Photogram. Week **11**, 173–184 (2011)
7. Kellner, M., Hofmann, U., Bouzouraa, M.E., Stephan, N.: Multi-cue, model-based detection and mapping of road curb features using stereo vision. In: 2015 IEEE 18th International Conference on Intelligent Transportation Systems, pp. 1221–1228 (2015)
8. Rusu, R.B., Cousins, S.: 3D is here: Point Cloud Library (PCL). In: IEEE International Conference on Robotics and Automation (ICRA), Shanghai, China. IEEE (2011)
9. Xu, J., Xiong, Z., Bhattacharyya, S.P.: PIDNet: a real-time semantic segmentation network inspired from PID controller. arXiv preprint arXiv:2206.02066 (2022)
10. Zhang, Y., Wang, J., Wang, X., Dolan, J.M.: Road-segmentation-based curb detection method for self-driving via a 3D-lidar sensor. IEEE Trans. Intell. Transp. Syst. **19**(12), 3981–3991 (2018)

# Adaptive Robotic Levering for Recycling Tasks

Boris Kuster[1,2](✉), Mihael Simonič[1], and Aleš Ude[1]

[1] Humanoid and Cognitive Robotics Lab, Department of Automatics, Biocybernetics and Robotics, Jožef Stefan Institute, Ljubljana, Slovenia
boris.kuster@ijs.si

[2] Jožef Stefan International Postgraduate School, Ljubljana, Slovenia

**Abstract.** A common step in autonomous robotic disassembly (recycling) of electronics is levering, which allows the robot to apply greater forces when removing parts of the devices. In practical applications, the robot should be able to adapt a levering action to different device types without an operator specifically recording a trajectory for each device. A method to generalize the existing levering actions to new devices is thus needed. In this paper we present a parameterized algorithm for performing robotic levering using feedback-based control to determine contact points and a sinusoidal pattern to realize adaptive levering motion. The algorithm can deal with devices of different shapes. After the initial adaptation process, the subsequent executions of the learnt levering action can be sped up to improve performance.

**Keywords:** Robotic disassembly · Levering · Force control

## 1 Introduction

Recycling faces the problem of small batch sizes and large variety of recycled items [1,2]. In such circumstances, the effort of robot programming to perform autonomous disassembly of generic electronics is one of the main reasons for the slow deployment of robotic-based solutions.

During the disassembly of electronics and other items, various robotic skills are needed, one of which is levering. Levering is a process whereby mechanical advantage can be gained using a rigid beam (lever) and a fixed hinge (fulcrum), which allows a greater force to be exerted on the load (the levered object). Some operations where levering is needed include removing pins and nails or separating different device parts. While this is often an easy task for humans as they can rely on vision, force, and pressure sensing supported by previous experience and generalisation capabilities, robotic disassembly applications still commonly use pre-recorded trajectories obtained by learning from demonstration (LfD) [3]. These trajectories, however, cannot be applied to different electronic devices without a properly implemented adaptation process.


T. Petrič et al. (Eds.): RAAD 2023, MMS 135, pp. 417–425, 2023.
https://doi.org/10.1007/978-3-031-32606-6_49

To perform a generalized levering action, we encode it with a sinusoidal pattern. The execution is controlled by monitoring the external forces and torques acting on the end-effector. One of the benefits of the proposed framework is that adaptation can be performed even with noisy vision information, which can occur due to the poor lighting conditions and occlusions often encountered in recycling processes [4]. The proposed algorithm for adaptive levering was implemented on a collaborative Franka Emika Panda robot within a modular robotic workcell [5] and applied to disassembling heat cost allocators (electronic devices shown in Fig. 2).

The paper is organized as follows: in Sect. 2.1, the levering setup is presented. Section 2.2 describes a search algorithm for automatically detecting contact points which must be known in order to perform a levering action. In Sect. 2.3 we discuss the application of the sinusoidal pattern to encode a levering action. An algorithm for detecting the completion of the levering task based on force-torque measurements is also presented. In Sect. 2.4, the adaptation of the levering movement is explained. The experimental evaluation of our approach is presented in Sect. 3. We conclude with a critical discussion and plans to improve the proposed algorithm.

## 2 Methodology

The proposed adaptive levering procedure is based on our knowledge about the geometry of the task.

### 2.1 Levering Setup

Figure 1a shows a typical levering setup, where the lever is attached to the robot's flange. In the following the terms lever and tool are used interchangeably. In our system, the tool is integrated into qbRobotics Variable Stiffness Gripper. Subsequently, we refer to the levered object (in our experiments a printed circuit board) as part. To increase mechanical advantage, the lever is positioned against the fulcrum.

In Fig. 1b, the object coordinate system (c.s.) ($x$, $y$, $z$ axes), the robot flange c.s. ($x_0$, $y_0$, $z_0$), the tool end c.s. ($x_1$, $y_1$, $z_1$) and the fulcrum c.s. ($x_2$, $y_2$, $z_2$) are shown. The Tool Center Point (TCP) coincides with the origin of tool end c.s. The fulcrum c.s. is estimated once the lever establishes a contact with the edge of the device housing. In our work, we make use of force-torque (FT) measurements ($F_{x0}$, $F_{y0}$, $F_{z0}$, $M_{x0}$, $M_{y0}$, $M_{z0}$), which are calculated in the robot flange c.s.

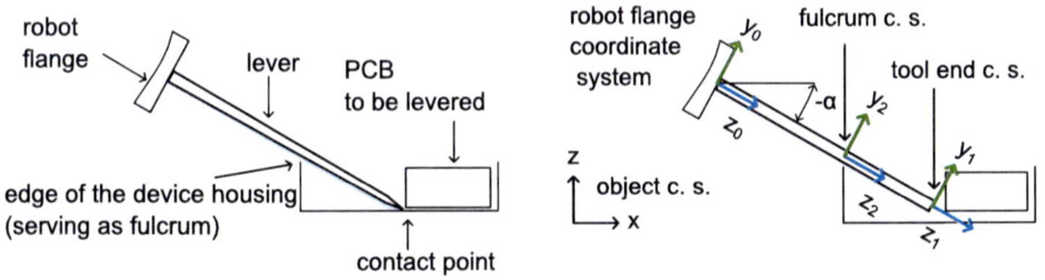


**Fig. 1.** (a) Elements of the levering process. (b) Coordinate systems used during the levering process.

To transform the forces from the flange to the object coordinate system, we define the rotational matrix between flange and object c.s. as seen in Eq. (4). The vector of forces $\mathbf{f}_{obj} = [F_x, F_y, F_z]$ in object c.s. can be estimated as follows

$$\mathbf{f}_{obj} = \mathbf{R}_{flange\_to\_obj}\, \mathbf{f}_{flange}, \tag{1}$$

$$\mathbf{R}_{flange\_to\_obj} = \mathbf{R}_{flange}^{\top} \mathbf{R}_{obj}, \tag{2}$$

where $\mathbf{R} \in \mathbb{R}^{3\times3}$ is a rotation matrix. The torques are not transformed to the object c.s. since the $M_{x,0}$ measurements are used, which are calculated in the robot flange c.s. To map the robot movements from the robot base c.s. to the object c.s., we define the rotational transformation between them as

$$\dot{\mathbf{p}}_{base} = \mathbf{R}_{base\_to\_obj}\, \dot{\mathbf{p}}_{obj}, \tag{3}$$

$$\mathbf{R}_{base\_to\_obj} = \mathbf{R}_{base}^{\top} \mathbf{R}_{obj}. \tag{4}$$

## 2.2 Fulcrum and Part Contact Point Search Algorithm

To perform a levering action, the robot must first place the lever into the gap between the object and the fulcrum. A safe initial position is in the middle of the gap. We detect the gap by using 3-D vision. Examples of devices and the gaps for levering are shown in Fig. 2, where the gaps are marked with a red line. For some devices, the gaps are quite large, while for others, the gap is very narrow and challenging to detect.

While vision is sufficiently accurate to approximately position the tool in the middle of the gap, it is not possible to place the lever at the object based on vision results only. To establish contact, the robot first moves to a fixed height above the middle of the gap and then starts moving in the negative $z$ direction until it hits the bottom of the device housing. The motion is stopped once the measured force $F_z$ at the tool center point in vertical direction exceeds a predefined threshold $F_{max}$. The force-torque estimation is performed by the Franka Emika robot control system using internal joint torque sensors. Based on the half length of the gap, we determine also the initial inclination of the lever.

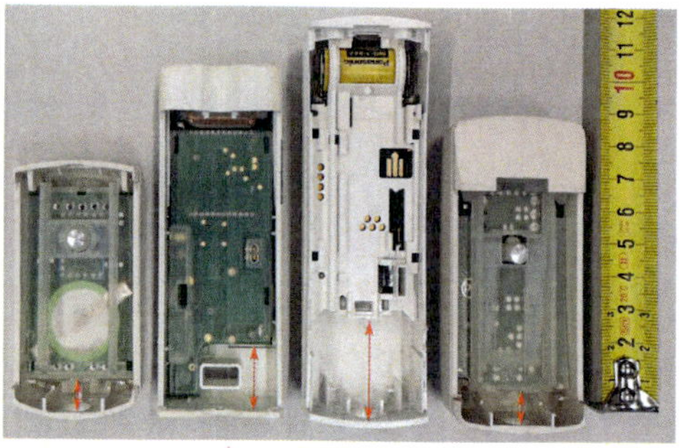

**Fig. 2.** Example electronic devices that need to be dismantled. Levering is used to remove the PCB from the housing.

A minimal threshold distance $d_{z,min}$ is set beforehand. If it is not exceeded before detecting contact, we consider that the initial position for levering (the gap) was incorrectly determined. In this case, we select another initial position with stochastic search. It is implemented by adding a small random vector $\varepsilon$ to the previous starting point

$$\begin{bmatrix} x \\ y \end{bmatrix} = \begin{bmatrix} x_0 \\ y_0 \end{bmatrix} + \begin{bmatrix} \varepsilon_x \\ \varepsilon_y \end{bmatrix}, \tag{5}$$

where the mean of the random vector is set to zero, while the standard deviation is determined as a fraction of the half gap width of a particular device type (the red line in Fig. 2).

In the next step, the robot moves along the positive $x$ axis in the object c.s. until the horizontal force $F_x$ exceeds the force threshold $F_{max}$. At this point the lever (tool) is in contact with the PCB to be levered out of the device. Next the robot determines the fulcrum position by performing a rotational motion around the $x$ axis of the tool end c.s. (positioned at the tip of the lever). The motion is stopped once the torque $M_{x,0}$ exceeds the threshold $M_{max}$, which signifies that the contact between the lever and the fulcrum has been established.

### 2.3 Adaptive Levering Algorithm

The robot can now proceed with levering out the PCB. However, it is not easy to manually program a general levering action because the required forces and amplitude of motion that need to be applied to pry out the PCB cannot be analytically determined in advance because they depend on the geometry of the device and its current state of damage. We have therefore developed an adaptation algorithm that modifies the generic levering action so that it becomes suitable for the current device that needs to be dismantled.

Humans often use periodic movements when levering, especially when they do not know the force required to dislodge an object with the lever. In doing so, they slightly increase the force on the lever in each period.

We generate a single degree-of-freedom (DOF) sinusoidal cycle with an amplitude $A$ and cycle time $t_c$, which represents the angle of the end-effector relative to the initial angle $\alpha$ at which the robot is in contact with the fulcrum and part. This results in the following trajectory around the $x$-axis of the fulcrum c.s.

$$\varphi_{x,2}(t) = g(t) + A \sin\left(\frac{2\pi t}{t_c}\right). \tag{6}$$

The other two angles around fulcrum coordinate axes are set to $\varphi_{y,2}(t) = \varphi_{z,2}(t) = 0$. The levering algorithm is parameterized with the following parameters:

- initial amplitude of the sinusoidal cycle, set to $A = 10°$ in our experiments.
- duration of the sinusoidal cycle, set to $t_c = 3\,\text{s}$.
- offset increment, set to $g(t) = 0.1At/t_c$.

When the lever is in the initial position, touching both the PCB and the fulcrum, the execution of the sinusoidal pattern begins. An example of a three period sinusoidal pattern with a constant amplitude is shown in Fig. 3.

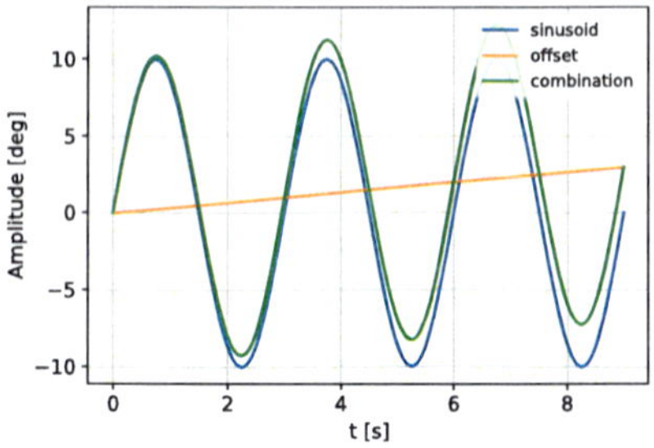


**Fig. 3.** Three periods of a sinusoid with an increasing offset $g$.

During levering, the lever must remain in contact with the part. Velocity-resolved control is used to ensure that a desired constant force on the part is maintained [6]. While the angle of the tool is determined by the sinusoidal pattern, the commanded velocity in the $x_{obj}$ direction is determined based on a desired preset contact force $F_d$, the actual current contact force $F_x(t)$ and the coefficient $k_F$. The commanded position in the $x$ direction of the object c.s. is computed by integration

$$\dot{x}(t) = k_F(F_d - F_x(t)), \tag{7}$$

$$x(t + \Delta t) = x_0 + \dot{x}(t)\Delta t. \tag{8}$$

The other two coordinates remain constant during the execution of the levering movement, i.e. $y = y_0,\ z = z_0$.

The goal of levering is to pry out the PCB out of the device housing. We therefore let the levering out trajectory (3) running until the success signal (9) reaches the threshold. Given the time $t$ at which the levering action was successful, we record the amplitude as $g(t) + A$. Thus next time we can start the levering action using this amplitude instead of starting from $A$ and gradually increasing the amplitude. Note that we cannot just start with a very large amplitude as this could cause unsafe robot movements or uncontrolled motion of the PCB being levered out. The above adaptation procedure has turned out to be sufficient for our application. If needed more advanced learning algorithms [7] could be used to optimize the learnt levering out behavior.

We can use force-torque measurements $[F_{x,0}, F_{y,0}, F_{z,0}, M_{x,0}, M_{y,0}, M_{z,0}]^\top$ at the robot's flange to detect when the levering action succeeded at prying out the PCB. To determine when the PCB (part) has been pried out, we monitor the torque $M_{x,0}$ measured at the robot's flange. During levering the torque increases, while a sharp torque drop-off is observed at the moment when the PCB is dislodged. The levering is successful if the difference between the maximum and minimum torque value within the width of the signal observation window is greater than a prespecified threshold $M_{max}$

$$\max_{t \in W}\{M_{x,0}(t)\} - \min_{t \in W}\{M_{x,0}(t)\} > M_{max}, \tag{9}$$

where $W = \{t_k - t_w, \ldots, t_k\}$, $t_k$ denotes the current sample time, and $t_w = 1.5s$ is the size of the sliding window. This condition can only be triggered while the lever angle $\alpha$ is decreasing, meaning it's applying a force to the part. The size of the sliding window is constant for all device types. The value of the threshold $M_{max}$ is constant for all devices on which we tested the algorithm, however for novel device types it might require tuning.

When recycling old electronics, devices can be in various states of damage. It can sometimes happen that the part that needs to be levered out is very loose and does not provide a large resistance force, so a drop in $M_{x,0}$ will not be detected. Therefore, the secondary condition for levering success is when the lever angle $\alpha$ becomes higher than a prespecified angle $\alpha_{max}$. In our case, shown in Fig. 1a, the contact point with the part is always lower than the fulcrum in the object $z$ axis. Thus the levering action is stopped if the lever angle is greater than the horizontal tool placement ($\alpha_{max} = 0°$).

### 2.4 Levering After Adaptation

The time required to perform an operation is particularly important in industrial robotics, where fast cycle times are required to optimize the productivity of the robotic workcell. The knowledge about the performed operations can be used to accelerate the performance. To improve the execution time when the levering operation is performed several times, we record the robot's joint trajectory performed during the initial adaptation of the generic levering action. If the device of the same type is encountered for the second time, we can achieve the necessary contacts without searching. In addition, the levering trajectory can be sped up.

To speed up the levering process, we record the initial position at the start of the levering action and the robot pose reached after establishing the contact of the tool with the levered part (in our case, the PCB) and the fulcrum (edge of the device housing), both in object c.s. The highest amplitude of the adapted levering operation is also recorded. During execution of the adapted levering operation, the robot can directly move from the initial position to the posture where the tool establishes contact with the PCB and the fulcrum. Thus we perform only one movement at a higher speed instead of two. Next the recorded levering operation defined by Eq. (6) is executed at the maximum recorded amplitude and at a higher speed, i.e. by decreasing $t_c$. The position trajectory defined by Eq. (8) is sped up in the same way. The success signal is monitored as per Sect. 2.3. When success is not detected, the adaptive levering procedure defined in Sect. 2.3 is performed. However, here the search process can start at the highest previously recorded amplitude.

## 3 Experimental Evaluation

To test the robustness of the algorithm, we tested altogether 5 exemplars of two different device types and performed the levering for each of them. Some selected devices are in good condition and require a higher levering force, while some are already worn out and require less force.

Figures 4a and 4b show the average $M_{x,0}$ torques (which we use as a feedback signal) observed during the trials of each device type in various states. Initially the robot is only touching the fulcrum, so the torque values are negative. Upon contacting the PCB, the torque values rise. It can be seen that particularly for devices of type 1, the required levering torque differs significantly depending on the particular device. A torque drop-off is observed at around normalized time $t = 0.8$ in Fig. 4a and around $t = 0.95$ in Fig. 4b, which indicates successful levering completion.

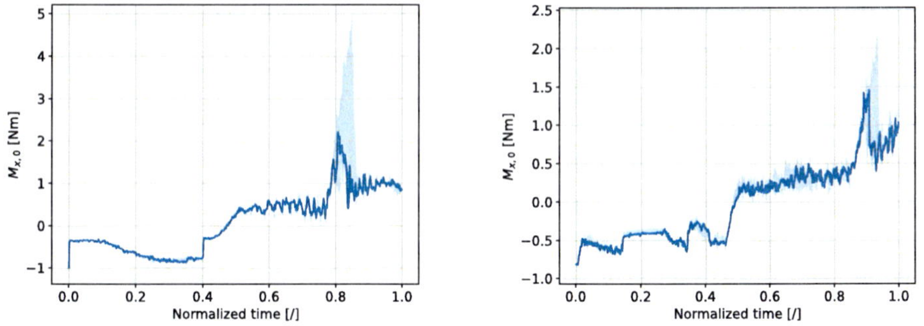


(a) Avg. torque and the standard deviation during 5 experiments - device type 1

(b) Avg. torque and the standard deviation during 5 experiments - device type 2

**Fig. 4.** The torques measured while performing the levering operations.

A comparison of the adaptation (initial) trajectory and the adapted trajectory is shown in Figs. 5a and 5b. Instead of searching for contact, the robot immediately moves to the previously learned contact point. Figure 5c shows the comparison of average execution duration, as well as the standard deviation of this duration, both for the case of the initial search and after adaptation, for each of the two device types. It can be seen that after adaptation, the levering is faster and the duration is deterministic.

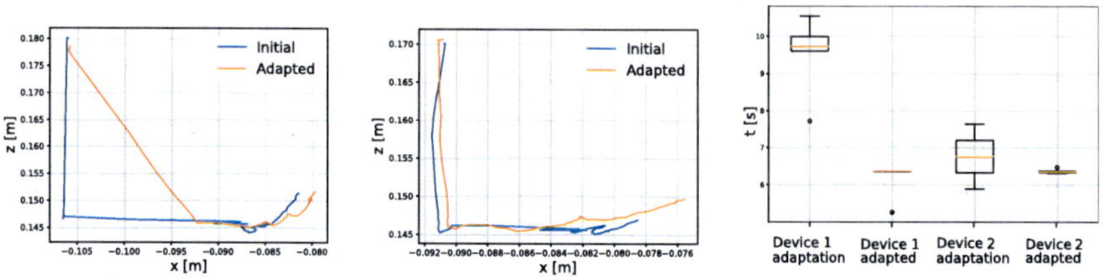


(a) Device type 1 (b) Device type 2 (c) Duration comparison

**Fig. 5.** Comparison of initial and adapted trajectories for two device types (a,b) and duration comparison of initial and adapted levering for both types (c).

## 4 Conclusion and Further Work

We presented a parameterized levering algorithm, composed of two sub-tasks, searching and levering. The search algorithm automatically detects contact points after the device pose is estimated by vision. Force control is then used to first establish contact between the tool and the device housing and then position the tool so that it establishes contact with the part to be pried out and the fulcrum. Levering is performed using a sinusoidal motion pattern and force-torque feedback. After the initial learning step, subsequent executions can be sped-up. We have demonstrated the algorithm's robustness to different device types.

With the applied robot, end-effector forces and torques acting are calculated from internal joint torque measurements, which can be noisy, particularly in or near singular joint configurations. This can be solved by using a dedicated force-torque sensor mounted on the end-effector. However, even precise force measurement cannot assure totally reliable classification of the levering process outcome. Additional modalities, such as the gripper encoder feedback signal, could be used to more reliably determine the outcome.

## References

1. Foo, G., Kara, S., Pagnucco, M.: Challenges of robotic disassembly in practice. Proc. CIRP **105**, 513–518 (2022)
2. Gašpar, T., et al.: Smart hardware integration with advanced robot programming technologies for efficient reconfiguration of robot workcells. Robot. Comput.-Integr. Manuf. **66**, 101979 (2020)
3. Simonič, M., Petrič, T., Ude, A., Nemec, B.: Analysis of methods for incremental policy refinement by kinesthetic guidance. J. Intell. Robot. Syst. **102** (2021)

4. Laili, Y., Wang, Y., Fang, Y., Pham, D.: Optimisation of Robotic Disassembly for Remanufacturing (2022)
5. Radanovič, P., Jereb, J., Kovač, I., Ude, A.: Design of a modular robotic workcell platform enabled by plug & produce connectors. In: 2021 20th International Conference on Advanced Robotics (ICAR), pp. 304–309 (2021)
6. Gams, A., Do, M., Ude, A., Asfour, T., Dillmann, R.: On-line periodic movement and force-profile learning for adaptation to new surfaces. In: 2010 10th IEEE-RAS International Conference on Humanoid Robots, pp. 560–565 (2010)
7. Kroemer, O., Niekum, S., Konidaris, G.: A review of robot learning for manipulation: challenges, representations, and algorithms. J. Mach. Learn. Res. **22**(1), 1395–1476 (2021)

# Robotic Assembly of a Wooden Architectural Design

Federico Neri[1] (✉), Roberto Cognoli[2], Giacomo Palmieri[1], and Roberto Ruggiero[2]

[1] Department of Industrial Engineering and Mathematical Sciences, Polytechnic University of Marche, Ancona, Italy
federico.neri@pm.univpm.it, g.palmieri@univpm.it
[2] Scuola di Architettura e Design, Università di Camerino, 62032 Camerino, Italy
{roberto.cognoli,roberto.ruggiero}@unicam.it

**Abstract.** This paper presents a fast and cost-effective solution for developing and implementing a collaborative robotic station to assemble wooden modules. For this purpose, the Grasshopper plug-in is proposed, which allows the management of the Rhinoceros 7 design environment to which it is connected to obtain information about the CAD of the robot station. Moreover, with this tool, it is possible to verify the movements of the robot through simulation and finally create a program that allows the control of the selected robot. The article describes the advantages of this design methodology, which allows a quick modification of the robot control in case of changes to the CAD project. Finally, the implementation of an automated station for assembling wooden elements developed using the described methodology is shown.

**Keywords:** Robotic assembly · Collaborative robotics, architectural design · Timber · Offline programming

## 1 Introduction

In recent years, there has been a growing interest in the application of industrial automation and robotics within the construction industry. This trend is driven not only by the industry's pursuit of greater efficiency but also by the need to address the increasing demand for sustainable construction practices. Many studies examine how robotic systems can enable the use of fewer materials, the employment of new materials, or the more intelligent use of traditional materials [1–3].

The use of wood construction has acquired a key role in reducing CO2 emissions [4]. The demand for appropriate solutions to achieve environmental sustainability goals, efficiency and quality in the construction industry is growing. Robotics for automating the assembly of wood structures is now a developing sector that is increasingly able to meet these needs. Many researchers are developing solutions for assembling wood elements at different scales employing both industrial and collaborative robots. Gramazio Kohler's research at the ETH studied the robotic assembly of building components such


T. Petrič et al. (Eds.): RAAD 2023, MMS 135, pp. 426–433, 2023.
https://doi.org/10.1007/978-3-031-32606-6_50

as walls and ceilings, but also pavilions, based on a layering of customised wood elements joined with adhesives, nails or non-reversible screws [5, 6]. The CREATE Lab at SDU studies methods of robotic assembly and disassembly of discrete wooden structures to promote the adoption of circular economy principles in the construction industry [7]. In professional practice, the RAP Studio built a robot cell to compose a wooden partition wall from a large number of wooden parts obtained from industrial waste material [8]. Companies such as Leko Labs [9] and Coded Façade [10] are exploiting the potential of robotic cells by utilising a fully digital design workflow, resulting in increased efficiency and innovative products.

This paper outlines an approach for the design and assembly automation of layered timber structures, utilising a small and flexible collaborative robot cell. The proposed approach incorporates advanced assembly procedures and digital design techniques to enable the creation of non-standard timber structures.

This project is part of a larger study on the role of automation and robotics in promoting circular processes in the construction industry. The study is conducted by the School of Architecture and Design, University of Camerino, the Department of Industrial Engineering and Mathematical Sciences, Polytechnic University of Marche and the company Joytek. The aim of the paper is to demonstrate how, to date, complete integration and collaboration between architectural design and robotic programming is possible and easy to apply. Furthermore, the cost-effectiveness and versatility of the implementation offer a significant opportunity to innovate the timber construction industry also in SME (Small Medium Enterprise). The study emphasises the importance of integrating a robot cell into a conventional production line, as it transforms the process and the end product.

## 2 Project Aim

The proposed computational workflow aims to facilitate the interlinking of design, fabrication, and robotic assembly by introducing an integrated approach to support the virtualisation of automation workflow and the simulation of the production process. The workflow is established using the Grasshopper for Rhino environment comprises three primary modules:

- design,
- manufacturing,
- robotic simulation and control,

This workflow creates the design digital twin and production digital twin to enable performance-driven design and assembly process simulation. A flexible robot setup is also introduced to enhance the robotic assembly process. The proposed workflow presents a clear and objective approach to interconnecting various stages of the design and production process. It provides a uniform and logical flow that facilitates a seamless module transition. In this paper the design and the robotic simulation and control for assembly will be presented.

## 3 Architectural Design

The proposed design for a timber façade cladding (Fig. 1) utilizes an algorithm based on Grasshopper software to generate variations in the design. The cladding component consists of small wooden battens that are stacked alternately in layers at a 90° angle. The design algorithm provides a range of variations, enabling a degree of customization in the final product, including modifying the angle as a parameter. This feature is beneficial for architectural and construction applications, creating unique and adaptable structures. It can optimize the building envelope to achieve specific performance criteria, such as the amount of filtered solar radiation or the level of visual permeability, based on the number of battens and the angle between them.

The façade module measures 60 × 70 × 25 cm and comprises wooden elements with a cross-section of 3 × 3 cm and a length of 21 cm, each having a series of 6 mm holes. Some special elements, half the size of the standard ones 10.5 cm, are used to close the side edges. The elements were machined with a numerically controlled milling machine from waste materials with a larger cross-section. Elements are fastened together by embedded nuts and threaded bolts, making the structure fully reversible.

The use of parametric design allows for a highly iterative design process with multiple options and a greater degree of customization. Furthermore, the integration of control and robotic simulation into the parametric design process provides an additional layer of assessment to the design process. This integration allows for early-stage assessments of construction process constraints and feasibility, facilitating the optimization of the design for efficient construction and material usage.

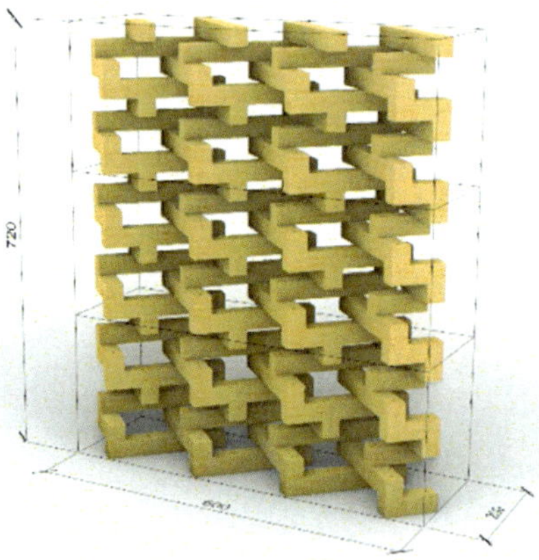

**Fig. 1.** Timber façade cladding module.

# 4 Robotic Simulation and Control

## 4.1 Robotic Cell Setup

The robotic cell consists of a UR5e collaborative robot equipped with a Robotiq Hand-E gripper with fingertips specially designed. They allow gripping the blocks or screws through grooves. In addition, there is a wooden support, which serves as a warehouse for the blocks to be picked up, and a support for the screws made of acetal resin [POM]. A wooden base has also been placed in the area where the parts are released, to fix the first layer of the structure. Next to the workbench is a holder for the electric screwdriver, which is equipped with a proximity sensor to check its presence. The robot pauses if the sensor does not detect the presence of the screwdriver. Fig. 2 shows the workstation setup. Fig. 3 shows the station simulated with Grasshopper, which replicates the described workstation.

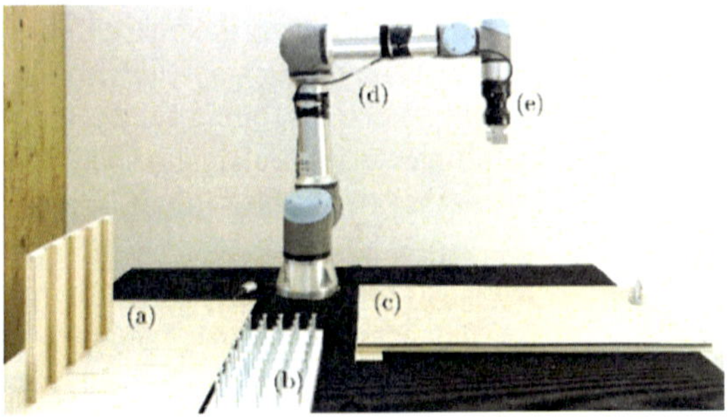


**Fig. 2.** Workstation setup: a] wooden block warehouse, b] screw holder, c] base for the assembly, d] UR5e, e] Robotiq gripper with multi-use fingertip.

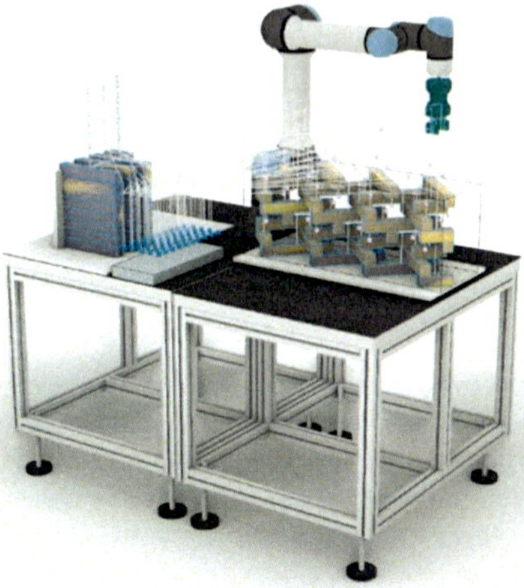

**Fig. 3.** Grasshopper setup with robot paths.

The Grasshopper environment was used to reconstruct a virtual model of the robot cell and tools, which served as the geometrical foundation for the setup's digital twin. All the elements of the robotic cell were designed in cad and machined with a CNC machine to ensure correspondence between the model for simulation and the actual model. Despite this, there may be some discrepancies which make the gripping actions of the wooden blocks inaccurate. To ensure that the real reference system corresponds to the simulated one, a position calibration of the supports fixed to the bench was performed.

## 4.2 Program Structure

In order to establish a direct connection between the design generation process HAL Robotics Framework [11] was used to translate the input geometry – which consisted of layered building blocks – into target planes and subsequently into robot commands. The system enables flexible programming of toolpaths based on *Actions*, which are fundamental robot operations such as moving to a target, changing signal states, or communicating with sensors. Actions are linked to part geometry in a parametric manner and subscribed to changes, and are combined into a *Procedure*. This provides graphical insights into how design choices affect, and are affected by, process and robot constraints, while also ensuring that trajectories remain associative even when the part design changes. This approach follows the principles of Robot Oriented Design proposed by T. Bock [12].

Two sets of toolpaths were generated for the assembly of timber blocks: (P1) pick-and-place targets, which were translated into gripper commands, for the timber elements and (P2) for the bolts. The robot path planning was performed in a parametric manner, where targets were generated based on the geometrical description and relationships between the blocks and bolts, and subsequently sorted according to a pre-planned assembly sequence. For each block, the pick-and-place toolpath involved interpolation between the center point of the block's top face, both in its picking and placing coordinates, as well as offset positions in the positive z-direction (100mm) to ensure a secure approaching trajectory. An intermediate point was also included, with a safety offset (350mm) in the positive z-direction from the current built volume. These points were referenced to the gripper TCP (Tool Centre Point), specifically the mid-point along the height of the gripper fingertips, as illustrated in Fig. 4. The path consists of the following steps. Home, approach, grip, safety position, approach, placement, safety position, for a layer of wood elements, then a path with the same steps is used for bolt placement.

Safety is essential when a human works near a robot without protective barriers; for this reason the UR5e cobot has been equipped with a collaborative gripper with force control and no sharp edges. Since the operator shares the workspace with the robot, a proximity sensor has been inserted into the screwdriver holder to stop the robot when the tool is used to fix the screws on the inserts of the wooden blocks. This avoids collisions that could injure humans. The sensor is connected to the digital inputs of the manipulator's control system: to stop the robot, the digital signal was linked to the action which pauses the program. Fig. 5 shows the hardware and software architecture used during the tests. Through the use of a computer, it was possible to send the script to the robot controller.

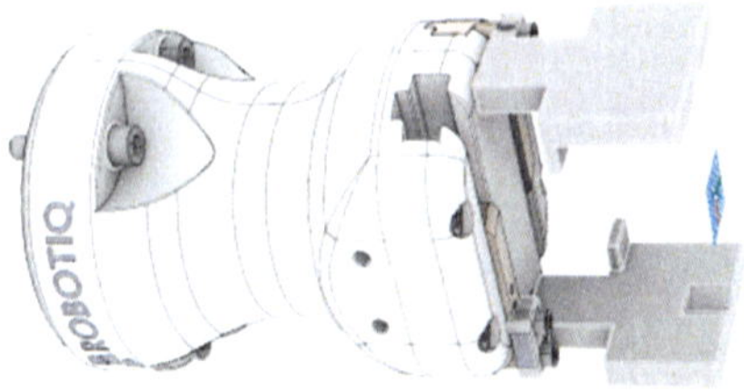


**Fig. 4.** Custom fingertips.

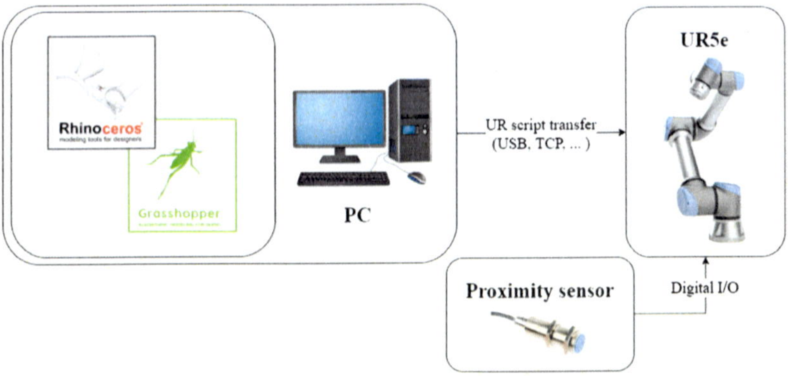


**Fig. 5.** Diagram of the communication used during the tests.

## 5 Results

During the tests, the implementation of the program on the collaborative robot allowed the wooden blocks to be assembled in the correct order and with precision. The single cluster consists of 8 layers. After positioning each layer (Fig. 6), the screws are taken from the magazine and inserted. To increase the safety of the application, the screw insertion phase is force-controlled. This way, if the robot senses high resistance, it retracts 10 cm and pauses. This solution prevents possible injury to the operator and at the same time, in the case of incorrect insertion, allows the operator to intervene. Afterwards, the robot waits for the operator to pick up the electric screwdriver to fix the blocks. Thanks to the proximity sensor, in this phase it is possible to tighten the screws with the certainty that the robot remains stationary. Once this operation is complete, the operator can insert the tool into the support allowing the robot to proceed with the next layer. The described sequence is repeated until the layers that make up the single cluster have been assembled. Once the robot has completed its activity, the operator has the task of repositioning the wooden blocks and the screws in their respective warehouses and removing the assembly from the support. In addition, the operator also has the task of connecting the individual clusters to obtain the final assembly (Fig. 7).

**Fig. 6.** Picking up and assembling the wooden blocks.

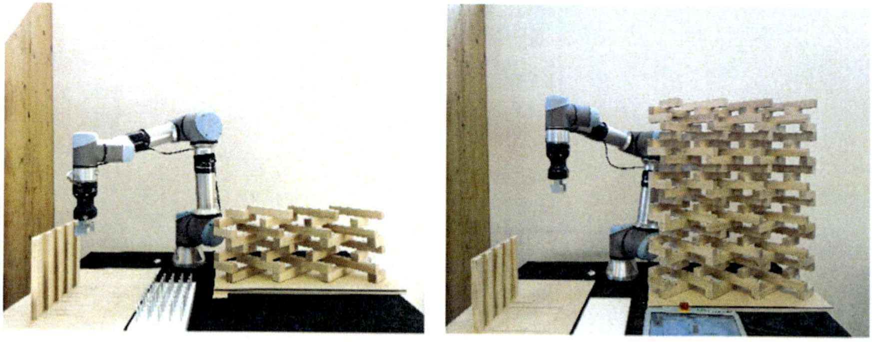

**Fig. 7.** Final assembly: a] single cluster, b] final composition with multiple clusters assembled.

## 6 Conclusions

The work described in this document examines the possibility of assembling wooden modules starting from the CAD design, which is associated with the Grasshopper plugin for planning robot trajectories. This method makes it possible to combine the mechanical design with robot software programming, optimizing the development times of the collaborative station and facilitating the management of possible errors. This approach falls under the criteria of concurrent engineering and allows a reduction in development times and associated costs and greater flexibility. Indeed, if the dimensions of the wooden blocks or the assembly structure need to be changed, it would be sufficient to modify the initial CAD to obtain a ready-to-use robot script. Despite the higher complexity of the initial design phase, this approach is advantageous and versatile. In the future, the flexibility of the application could be increased by implementing collision avoidance techniques [13] without stopping the robot.

## References


1. Søndergaard, A., et al.: Topology optimization and robotic fabrication of advanced timber space-frame structures. In: Reinhardt, D., Saunders, R., Burry, J. (eds.) Robotic Fabrication

in Architecture, Art and Design 2016, pp. 190–203. Springer International Publishing, Cham (2016). https://doi.org/10.1007/978-3-319-26378-6_14

2. Eversmann, P., Ochs, J., Heise, J., Akbar, Z., Böhm, S.: Additive timber manufacturing: a novel, wood-based filament and its additive robotic fabrication techniques for large-scale, material-efficient construction. 3D Print. Addit. Manuf. **9**(3), 161–176 (2022). https://doi.org/10.1089/3dp.2020.0356
3. Bechert, S., Sonntag, D., Aldinger, L., Knippers, J.: Integrative structural design and engineering methods for segmented timber shells – BUGA Wood Pavilion. Structures **34**, 4814–4833 (2021)
4. Arup: Rethinking Timber Buildings. Seven perspectives on the use of timber in building design and construction. Report Arup, London (2019)
5. Apolinarska, A.A., Knauss, M., Gramazio, F., Kohler, M.: The sequential roof. In: Menges, A., Schwinn, T., Krieg, O.D., Menges, A., Schwinn, T., Krieg, O.D. (eds.) Advancing Wood Architecture: A Computational Approach, pp. 45–59. Routledge, New York (2016). https://doi.org/10.4324/9781315678825-4
6. Eversmann, P., Gramazio, F., Kohler, M.: Robotic prefabrication of timber structures: towards automated large-scale spatial assembly. Constr Robot **1**, 49–60 (2017)
7. Kunic, A., Naboni, R., Kramberger, A., Schlette, C.: Design and assembly automation of the Robotic Reversible Timber Beam. Autom. Construct. **123**, 103531 (2021)
8. Circular Experience. Studio RAP. https://studiorap.nl/Circular-Experience. Last accessed 12 Feb 2022
9. Vincenzo, V., Hichri, B., Plapper, P.: Industry 4.0 – Implementation of an automated assembly line in a wooden modular house production plant: The case Leko Labs (2018)
10. Collaborative design of prefabricated façade systems. CodeFacades. https://codefacades.ch/. Last accessed 2 Dec 2022
11. Gobin, T., Andraos, S., Vriet, R., Schwartz, T.: HAL robotics framework. In: 38th International Symposium on Automation and Robotics in Construction, Dubai, UAE (2021)
12. Bock, T.: Robot-oriented design. In: 5th International Symposium on Automation and Robotics in Construction, Tokyo, Japan (1988)
13. Scoccia, C., Palmieri, G., Palpacelli, M.C., Callegari, M.: A collision avoidance strategy for redundant manipulators in dynamically variable environments: on-line perturbations of off-line generated trajectories. Machines **9**(2), 30 (2021)

# Assistance Systems for Construction Site Safety: Revealing Static Forces for Deconstruction Processes

Christoph Heuer(✉) and Sigrid Brell-Cokcan

Chair of Individualized Production (IP), RWTH Aachen University, Campus-Boulevard 30, 52074 Aachen, Germany
{heuer,brell-cokcan}@ip.rwth-aachen.de

**Abstract.** High-performance tools and construction machinery generally involve a high accident risk. Currently, there are few assistance systems on construction sites that warn employees at an early stage of potential safety risks for man and machine. Sensors can be used to detect the movement of construction machinery, opening up opportunities for the development of assistance systems. In the demolition of concrete, powerful construction machines are used that work with high process forces. In this case, the construction machine generates a constant static force by which the cutting process is stabilized. The operators have no possibility to measure and adjust the static force of the construction machine in the planning and while the process. Evaluating the horizontal static performance at an early stage can prevent subsequent damage to the machine and ensures a safe working environment. Using the example of a (semi-) automated deconstruction machine, which can be controlled (semi-) automatically via *ROS* (Robot Operating Systems), a procedure was developed which enables the personnel to determine the static process forces for a safer work environment.

**Keywords:** Assistance systems · *ROS* · Construction Site · Reachability Map · Force-Reachability Map Construction Site Safety

## 1 Introduction

A high number of serious and even fatal injuries occur on construction sites every year. In 2020 alone, there were 51 fatal accidents in Germany in the field of construction, building construction and mining [1]. Yet, the construction site is one of the least digitized industries in the world [2]. Digital assistance systems can help drivers to identify risks, whether in collision detection [3] or in detecting obstacles through a reversing warning system [4].

This paper is part of the ROBETON research project. Which is funded by the Federal Institute for Research on Building, Urban Affairs and Spatial Develop-ment (BBSR) on behalf of the Federal Ministry of Housing, Urban Development (BMWSB) with funds from the Zukunft Bau research funding program.


T. Petrič et al. (Eds.): RAAD 2023, MMS 135, pp. 434–441, 2023.
https://doi.org/10.1007/978-3-031-32606-6_51

The use of sensors to determine the joint positions of construction machines enables the development of new assistance systems [5]. Especially when cutting and removing concrete, high process forces arise that increase the challenge for the operator [6]. In previous work, a conventional demolition machine (*BROKK 170*) was converted into a (semi-) automatic deconstruction machine. The joint positions of the construction machine are captured with the help of encoders. With the framework *ROS*, a cartesian position control was implemented. The operator is provided with the necessary process force information [7].

The aim of the work is to develop a planning tool that uses encoders, a reachability map and ROS to determine the static force of a semi-automated construction machine at an early stage to support the worker in planning and operating the construction machine. This results in early detection of potentially dangerous situations in the plant cutting process.

## 2 Related Work

In previous work, a static force analysis for hydraulic machines was investigated. By means of a force equation, the resulting force at the end effector should be determined. This should be achieved using pressure sensors on the individual hydraulic cylinders, encoders for the identification of the joint stems, and the use of a pressure sensor on the end effector [5].

Reachability maps are an important component of *ROSs* navigation and planning algorithms. They are used to plan the possible movements of a robot in its environment. A reachability map represents all areas that the robot can inversely reach. The *ROS* package *reuleaux* offers the possibility to create these maps. By means of this package, the movement field of a robot can be captured. Through this, possible movements can be planned, and collisions can be avoided.

The robot is given a set of positions for inverse planning. Then it is looked which positions can be approached successfully. The data is used as a basis for further planning, such as path planning [8].

The overall goal of the work is to create a safe and efficient method for removing concrete components from concrete walls in existing buildings [6]. This will be accomplished using a combination of a concrete saw which is mounted on the mounting plate of the (semi-) automated deconstruction machine [7]. The machine, known as the *BROKK 170*, will position the concrete saw on the wall and apply a horizontal force of 17 kN to stabilize the saw and ensure a safer cut (see Fig. 1).

## 3 System Description

The following chapter three and four coupe with the aspect of the development and evaluation.

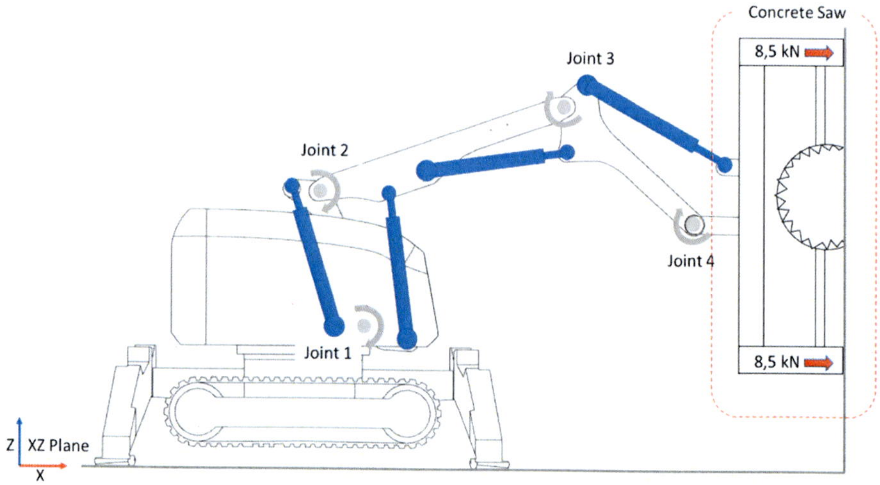


**Fig. 1.** *BROKK 170* with concrete saw, pressing the concrete saw against the wall with the required combined pressure of 17 kN. Only the arm joints (1, 2, 3, 4) are considered in the force-reachability map. The *BROKK 170* can only reach positions in the XZ-plane.

### 3.1 System Component

The *BROKK 170* is a remote-controlled demolition machine which is used for excavation and tunneling. Different end effectors can be attached to the machine to perform tasks such as drilling, demolition, cutting and trenching [10]. The machine has five degrees of freedom. It consists of an arm and a movable base. The joints are controlled by hydraulic actuators. At each joint there are motion sensors that are read and used for control. A velocity controller is implemented to control the movement of (semi-) automated deconstruction machine [7].

### 3.2 Method

In the presented work, a method for generating and calculating a force-reachability map was developed.

The basis for the present work is *reuleaux* [8], in relating to the work of Malaguti [5] on the calculation of the static force of construction machinery. The development is based on the already converted *BROKK 170* [7]. The outcome is a force-reachability map.

In the *reuleaux* library for every point a set of target positions were placed on a sphere to create the reachability map. Due to the described use-case, only one horizontal target position was checked for reachability "Eq. (1)" in the XZ-plane of the machine.

The method consists of the following steps:

1. Generating of the reachability map in *ROS*
2. Adding the static force at each point of the reachability map
3. Visualization of the results in the form of a force-reachability map

The results are evaluated on the (semi-) automated *BROKK 170*.

### 3.3 Creation of the Reachability Map

To generate the inverse reachability map based on *ROS*, a comparable approach to *reuleaux* was chosen. The inverse kinematic model of the *BROKK 170* was used to specify target positions in a set grid of 5 cm, which are then targeted by the planning algorithm. Only the arm position is considered for the reachability map. The rotational joint of the base is assumed to be static. The arm can only perform movements in 2 dimensions (XZ-plane) (see Fig. 2) which can be further extended by simply rotating the base. Furthermore, only horizontal positions in the XZ-plane "Eq. (1) defines Target Pose" are considered in the later force calculations. The later positions look as follows:

$$Target\ Pose = \{Xn, Y, Zn, 0^\circ, 0^\circ, 0^\circ\} \tag{1}$$

The target pose describes the position and orientation of an object in a 3D coordinate system, where:

1) Xn and Zn are iterated through a set of values, by a step of 5 cm between every position
2) The value of Y remains fixed at 0 and represents the position of the object along the Y axis.
3) The angles are fixed to a horizontal orientation, which means that the orientation of the object around the Xn, Y and Zn axes is always set to 0°.

The results of these movement plans are captured, and a reachability map is created that shows the reachable positions of the in the XZ-plane. In addition, the joint position is added to the map for the calculation of the force map.

### 3.4 Determining the Static Forces

In order to determine the static process forces, the resulting force was divided into a vertical and a horizontal component. The required static contact pressure of 17 kN from the concrete cutting process was used as a reference. The forces that can be achieved by the hydraulic cylinders result from the product of the system pressure of the *BROKK 170* and the corresponding contact surfaces of the cylinder. The values required for this were taken from the manual. (Compression force = 90,477 kN and tensile force 67,858 kN) (Fig. 3).

To calculate the static forces of the hydraulic cylinders, the necessary target force of the end effector is required. This can be, for example, a necessary contact pressure force or the weight force of an object to be stabilized.

If this is known, the pressure forces of the hydraulic cylinders of the individual joints required for this position and target forces can be determined by the static force and moment equilibrium. The further calculations follow the following example.

$$F\_cylinder1 = (F\_H * Z\_f1 - F\_V * Z\_f2)/(Z\_2 * cos(\alpha) + Z\_1 * sin(\alpha)) \tag{2}$$

$$B\_H = F\_H - F\_cylinder1 * cos(\alpha) \tag{3}$$

$$B\_V = F\_V + F\_cylinder1 * sin(\alpha) \tag{4}$$

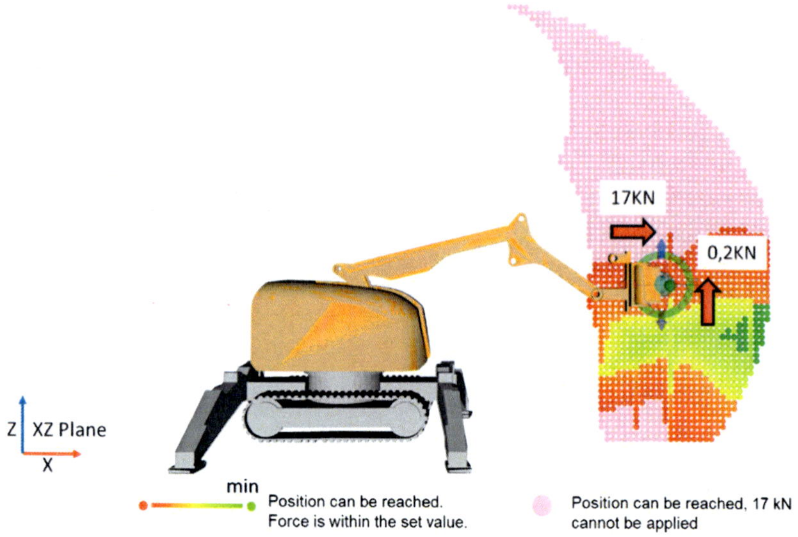


**Fig. 2.** Force Reachability map with *BROKK 170*. To calculate the color values, the most heavily loaded cylinder is extracted as a basis and then the load is applied to the color scale (red = heavily loaded, green = minimally loaded).

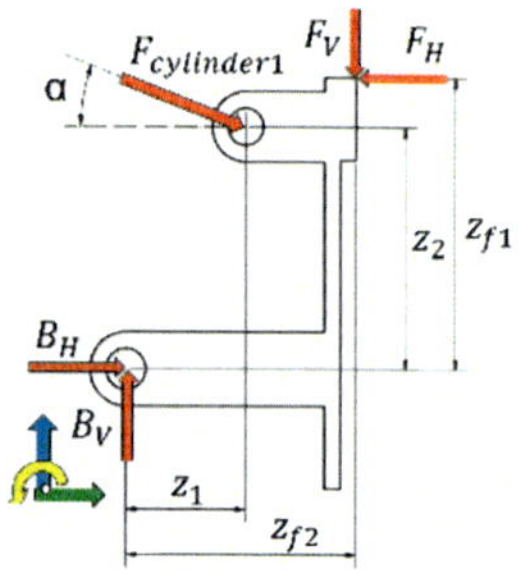


**Fig. 3.** Calculation of the static force using the example of the fourth cylinder (see Fig. 1). **F_H** horizontal nominal force, **F_V** vertical nominal force, **B_H** horizontal bearing force, B_V vertical bearing force, **Z_f1** vertical distance to the point of application of force, **Z_f2** horizontal distance to the point of application of force, **Z_1** horizontal distance to the cylinder engagement point, **Z_2** vertical distance to the cylinder engagement point, **α** angle between **F_cylinder1** and the horizontal.

This procedure successively determines all cylinder forces of the arm. The forces to be applied by the cylinders are then compared and the highest-loaded cylinder is determined. If the required force here exceeds the possible force for the cylinder, this position is not reachable for the desired process. By iterating through all geometrically achievable positions, an achievability map is created that contains all achievable positions as well as all permissible positions for the required forces.

For later color labelling of the reachable positions, a color range is used to distinguish between: 1) reachable positions, where the force value are exceeds valid value (purple) (see Fig. 4.) 2) reachable positions, where the force values are within the setpoint (range red to green).

**Fig. 4.** Four different force values were evaluated, to visualize the impact on the actual workspace.

To further investigate the working range, four additional force values were analyzed (see Fig. 4.). It becomes clear that the applied force has a decisive influence on the working range.

# 4 Experimental Evaluation

In the following, the simulation results are verified by means of experiments with the (semi-) automated deconstruction machine. In the first step, the experiments evaluate the reachability map and, building on this, the force map.

## 4.1 The Evaluation of the Reachability Map

**Fig. 5.** Performed test with the *BROKK 170*, tested with five randomly picked target positions from the reachability map.

Within this work, five different positions were evaluated to demonstrate the performance of the implemented reachability map. In this way, it was possible to check whether the position could be reached and, consequently, whether the reachability map can be used under real conditions with the real machine. For this purpose, five target positions were randomly selected from the reachability map and passed to the BROKKS 170 controller, which then attempted to approach these positions. All five positions are reachable by the *BROKK 170* (see Fig. 5).

### 4.2 Verifying the Force Values

**Fig. 6.** 1) BROKK 170 2) Direction of force 3) Load cell

To verify the force values calculated in the simulation, a load cell was mounted on the end effector of the construction (*BROKK 170*) (see Fig. 6). The horizontal force on the end effector of the construction machine was measured via the measuring device (HBM U2B 100 kN). The positions were chosen to test the force capability of the *BROKK 170*:

Position 1) resulting force is > 17 kN
Position 2) resulting force is < 17 kN

At the first position the *BROKK 170* reached a static pressure of 18,25 kN at the second position only 9,5 kN static pressure was applied.

## 5 Outlook

The work represents a first step towards further assistance systems that can support the driver of a construction machine. Overall, the work shows that with small adjustments, the results of the work can be incorporated into the planning and execution of a safer and more efficient cutting process. By adding the weight of the individual links of the machine, the accuracy of the simulation can be further increased. In later phases, additional degrees of freedom, a three-dimensional and a more universal implementation of the present assistance system are to take place.

In future work, vibration feedback could be implemented in the control unit of the construction machine. The operator would be warned via the control unit when critical threshold of the joint's values are exceeded. In this way, the created function can be extended to other construction machines. With the help of tool interfaces, tool information can be read out, which could feed into the force analysis and inform the personnel accordingly [11]. The maximum permissible force of the construction machine can also be considered to then adjust the power of the attachment to this value.

To find suitable placement locations for the remote-controlled machines, the reachability maps and force information can help the control personnel to find them faster. This can protect the machine, the workspace and save precious working time.

Reading out the joint positions provides the basis for further developments. The center of gravity and cylinder load can be recorded. When edge positions are reached, the machine control system can automatically block further movement of the axis in the direction of the danger zone to minimize the loads on the machine.

## References

1. DGUV. https://publikationen.dguv.de/widgets/pdf/download/article/4271. Accessed 30 Jan 2023
2. Abioye, S.O., et al.: Artificial intelligence in the construction industry: a review of present status, opportunities and future challenges. J. Build. Eng. **44**, 103299 (2021)
3. Bosh Rexroth. https://www.boschrexroth.com/en/de/company/press/latest-developments-for-off-highway-machinery-at-bauma-2022-10688.html. Accessed 28 Jan 2023
4. Volvo. https://www.volvoce.com/global/en/news-and-events/press-releases/2023/volvo-ce-introduces-collision-mitigation-system-for-jobsite-safety/. Accessed 28 Jan 2023
5. Malaguti, F.: Development of on-board weighting and force sensor system for construction machines: static analysis. In: Proceedings of 14th ISARC (1997)
6. Lee, H.J., Heuer, C., Brell-Cokcan, S.: Concept of a robot assisted on-site deconstruction approach for reusing concrete walls. In: Proceedings of 39th ISARC (2022)
7. Lee, H.J., Brell-Cokcan, S.: Cartesian coordinate control for teleoperated construction machines. Constr. Robot. **5**(1), 1–11 (2021). https://doi.org/10.1007/s41693-021-00055-y
8. Makhal, A., Goins, A.: Reuleaux: base placement by reachability analysis, pp. 137–142 (2018). https://doi.org/10.1109/IRC.00028
9. BROKK Website. https://www.brokk.com/product/brokk-170/. Accessed 20 Jan 2023
10. MIC 4.0. https://mic40.org. Accessed 11 Jan 2023

# A Fruit Detection Algorithm for a Plum Harvesting Robot Based on Improved YOLOv7

Jovan Šumarac(✉), Jelena Kljajić, and Aleksandar Rodić

Institute Mihailo Pupin, University of Belgrade, Belgrade, Serbia
jovan.sumarac@pupin.rs

**Abstract.** The paper presents a novel fruit detection algorithm for a plum harvesting robot. At present, the adequate recognition of plum fruits remains a particularly challenging, under-researched task. Difficulties occur due to small plum fruit sizes and dense growth, as well as numerous occlusions in their environment. A harvesting robot operating in such conditions needs to understand which fruits are reachable, in order to avoid collision and end effector damage. This makes a precise and robust visual detection system of crucial importance. Therefore, a lightweight plum detection procedure based on the improved YOLOv7 algorithm has been proposed. Firstly, the images of domestic plums (*Prunus domestica* L.) were collected in the field, and train, validation and test sets were established. Secondly, light data augmentation was performed. Next, the initial anchor box sizes used by the original YOLOv7 have been updated, based on the plum sizes in the collected dataset. Finally, an SE (Squeeze-and-Excitation) module was added to the backbone network, which helps model the channel interdependencies at almost no computational cost. The Improved-YOLOv7 model was then trained and evaluated on our dataset. The achieved Precision, Recall and mAP were 70.2%, 72.1% and 76.8%, respectively. The model has been compared with other recent models from the YOLO family, and has shown the best accuracy and generalization ability in real, complex environments. Therefore, the proposed plum detection method can provide theoretical and technical support for harvesting robots in real environments.

**Keywords:** Agriculture 4.0 · Plum Harvesting Robot · Object Detection · Yolov7 · Deep Learning Models

## 1 Introduction

In Serbia, domestic plum (*Prunus domestica* L.) is one of the most spread fruit species and it is considered a traditional species. In terms of total plum production in recent years, Serbia has consistently ranked among top five producers worldwide making plums one of the most economically significant species [1]. However, unfavorable harvesting conditions and lack of human labor led to the need to automate the picking process.

To find the solution to limited labour resources, robotic harvesters have been consistently developed over past decades. Despite this fact, they are still not in wide use


T. Petrič et al. (Eds.): RAAD 2023, MMS 135, pp. 442–450, 2023.
https://doi.org/10.1007/978-3-031-32606-6_52

in orchards. In recent years, a number of review papers has been published, examining various fruit and vegetable picking robots and end effectors for such tasks [2–4]. In [2] authors provide a detailed and comprehensive overview of intelligent fruit harvesting robots by comparing their system architectures, fruit detection methods, gripping methods and performances. They also emphasize the current challenges and future research directions in this area. A similar study [3] reviews picking robots for both fruits and vegetables, evaluating robotic arms, end effectors, vision systems, and picking environments and performance. In [4] the authors focus on end effectors as one of the most critical and challenging parts of a robotic harvesting system.

A crucial part of all robotic harvesters is the vision system, i.e. the system that detects the harvested fruits or vegetables; this has also been a subject of different recent review papers [5–7]. All studies highlight deep learning methods as the dom-inant approach in recent years. These methods also experience technical difficulties that prevent the wide use of robot harvesters in orchards. In [7] authors point out variable illumination, presence of different obstacles, occlusion filled environments and vision system fault tolerance as particular challenges.

Within the field of deep learning, different models were used for fruit detection. Among them the most represented were object detection models based on Region-CNNs [8], Single Shot Detectors (SSDs) [9], and the YOLO model family [10]. Vision system based on YOLO models were particularly present in studies [5–7], and applied in recent papers. In [11] authors used an a model based on YOLOv2 and YOLOv3 models called MangoYOLO for real time mango fruit detection and orchard load estimation. Improved YOLOv4 models were used for detecting citrus fruits in orchard environments [12] and for robust real-time pear detection on mobile applications [13]. An improved, lightweight YOLOv5 model was proposed for a picking robot to automatically recognize the graspable and ungraspable apples in an apple tree image [14]. More recently, YOLOv7 model with substantial data augmentation was used to accurately detect *Camellia oleifera* fruit in complex scenes [15].

While apples, tomatoes, strawberries, kiwi fruits and citrus fruits are consistently found among the species that are most often represented in fruit detection research, plums are found in less than 5% of papers [2, 4]. Their small size, dense growth, occlusion and obstacle filled environment makes them particularly challenging for detection. In Serbia, mostly unstructured orchards and lightly pruned trees pose another difficulty. In [16] authors used a model based on YOLOv4 for precision detection of dense *Prunus salicina* plums in orchards. A variety of state-of-the-art object detection methods were compared on *Prunus mume* plums and their growth was estimated [17]. So far, no substantial research was found on detection of *Prunus domestica* species, which is characteristic for Serbia and Europe. The goal of this paper was to collect a dataset for the detection of *Prunus domestica* plums in natural environments, and to develop an efficient detection model. At the time of this research, YOLOv7 was the latest version from the YOLO model family. Our model was based on this version, improved with data augmentation, initial anchor box estimation and a channel attention module.

## 2 Materials and Methods

### 2.1 Image Acquisition

The image dataset for this study was formed from the original images obtained in the non-arranged (unstructured) plum orchard located in the Sopot municipality (southern part of Belgrade), Serbia. The analyzed fruit belong to species of *Prunus domestica* L., specifically "Čačak's Beauty" and "Čačak's Best" varieties. The images of plum trees were obtained on a sunny day, which are typical weather conditions for the plum harvesting season, in mid-July 2022, in 3 iterations: in the late morning, in the early afternoon and in the late afternoon, to account for different illumination conditions. The majority of the plums were in the middle of maturity at the time of image acquisition.

During the image acquisition, we aimed for variety of angles and shooting distances that would simulate the real situation. The images were captured by Intel® RealSense™ Depth Camera D405. Since this camera's optimal operating range is be-tween 7cm and 50cm, our acquisition followed those guidelines. Vertically, the camera was placed between 0.5m and 2m from ground, which would be the case once the whole system is built. Since the acquisition was done in the natural environment, the images often contain overlapping fruit, both slight and heavy occlusion in form of branches and leaves, as well as diverse light intensities and angles (direct light, side light and backlight angles).

### 2.2 Dataset Preprocessing

The total of 1079 images was obtained. Of those images, 757 formed the training set, 217 formed the validation set and 105 formed the test set (the ratio of those three sets, respectively, is 70:20:10). The distribution of the images between these three sets was random. In order to prevent overfitting of the model, there were no repeating images in either of those three sets.

The objects of interest involved 3 classes: reachable plums (fruit that is close enough for the robot to pick, and there is no obstacle found in front of it; especially no obstacle in front of the top part of the fruit, where the stem is), unreachable plums (fruit that is close, but is unreachable due to at least one obstacle, in form of a branch, big leaf/leaves and/or another plum) and background plums. This third, additional class incorporates the plums that are too far in the background to be picked from the current position of the robot, but since our overall idea is to map the fruit position in the orchard in advance, this class would be used for future robot positioning and picking steps. Labeling of all the objects of interest was based on the smallest surrounding rectangle around the visible part of the fruit. In these images, there were 4465 exam-ples of reachable plums, 2255 of unreachable and 4372 of background plums. Due to the nature of our data, unreachable plums are slightly underrepresented, which was taken into account during the results analysis.

Resizing of the images had to be done – it was necessary due to potential computational heaviness (which would have prolonged the training and processing time), but on the other hand, we could not resize it too much, since the images often contain many smaller objects of interest and meaningful details. As a compromise between the two conditions, all the images were resized to $640 \times 640$.

In order to improve the quality of training and prediction accuracy, we applied the following augmentation methods: horizontal flipping of the images, image rotation by 90° (clockwise, counter-clockwise), as well as histogram equalization. As a result, we had a total of 2685 training images. Image labeling and some image augmentation was performed through a computer vision developer framework called '*Roboflow*', while additional augmentation was done in *OpenCV* library (Fig. 1).

**Fig. 1.** Examples of labeled images. Green boxes are reachable plums, purple unreachable and red are background plums. These examples showcase the variability and the complexity of the data, particularly of the various illumination conditions. Another prominent challenge is the occasional overlap between the classes that is, how to determine with certainty to which class a particular plum belongs.

### 2.3 Model Architecture

The YOLO (You Only Look Once) model family has dominated object detection landscape since the publication of the first YOLO version [10] in 2015. YOLO is a single-stage object detector, which means that it directly predicts the class label and bounding box coordinates of objects in an image, without any additional processing steps. This made YOLO generally faster and more computationally efficient than many other state-of-the-art object detectors, but with a lower accuracy. Since then YOLO has been updated iteratively, and subsequent models have improved many features including more advanced network architectures, detection of objects of different scales, anchor boxes etc. [18]. Anchor boxes are predefined bounding boxes used in object detection models and they serve as a reference for the network to predict the location and size of objects in an image. The lightweight nature and high accuracy of YOLO models have set the benchmark for the state-of-the-art methods of visual object detection.

The latest object detection model from the YOLO family at the time of this research - YOLOv7 [19] was presented in 2022. In the field of real-time object detection it outperformed most of well-known and used object detectors such as R-CNN family of models [8], and as well as other models from the YOLO family.

Figure 2 shows the architecture of YOLOv7 model. It consists of three components: input image in desired resolution, backbone network, and detection head. The back-bone network extracts feature maps from the data, while the head layer is used for prediction. The output of the backbone network is a feature map in three layers of different sizes. They then pass through the head layer, and finally the prediction result is exported through convolution and detection is made.

YOLOv7 architecture design is based on ELAN (efficient layer aggregation network). ELAN considers designing an efficient network by controlling the shortest and the longest gradient path, so that deeper networks can converge and learn effectively. YOLOv7 modifies the ELAN architecture, called E-ELAN (extended efficient layer aggregation network), which is used in two versions, ELAN1 and ELAN2 (shown at the bottom of Fig. 2). Another important feature of the YOLOv7 architecture is the SPPCSPC block, which combines the cross stage partial connection module [20] and the spatial pyramid pooling module [21].

On top of this original YOLOv7 architecture, we propose to add a Squeeze and Excitation module [22]. SE modules adaptively re-calibrate the feature maps produced by the CNNs, so that the model can focus more on the most important features for the task. The SE module consists of two parts: a squeeze and an excitation operation. The squeeze operation is used to reduce the spatial dimensions of the feature maps, typically by using global average pooling, to produce a channel-wise descriptor. The excitation operation then uses this descriptor to adaptively scale the channel-wise activations of the feature maps. This module allows the model to learn the relative importance of different feature maps. In our model, the SE module was added immediately after the SPPCSPC block, at the beginning of the head network. This is shown in the lower right corner in Fig. 2. Finally, our model used the K-means clustering method to find clusters in our labeled bounding box sizes and make a better initial guess of anchor box sizes.

# 3 Results

In this paper the proposed metrics to evaluate our model are precision (P), recall (R), and mean average precision (mAP). Precision measures the accuracy of the model in terms of the number of false positive (FP) predictions, i.e. the proportion of true positive (TP) predictions (correctly predicted positive examples) out of all positive predictions made by the model. Recall measures the completeness of the model in terms of the number of false negative (FN) predictions, i.e. the proportion of true positive predictions (correctly predicted positive examples) out of all actual positive examples.

Mean average precision (mAP) is the average of the precision values for all object categories. The average precision is computed as the area under the precision-recall curve for a particular category. mAP is a widely used metric for object detection as it takes into account both precision and recall and provides a single number that summarizes the overall performance of the object detector. The mAP@0.5 metric is a specific version

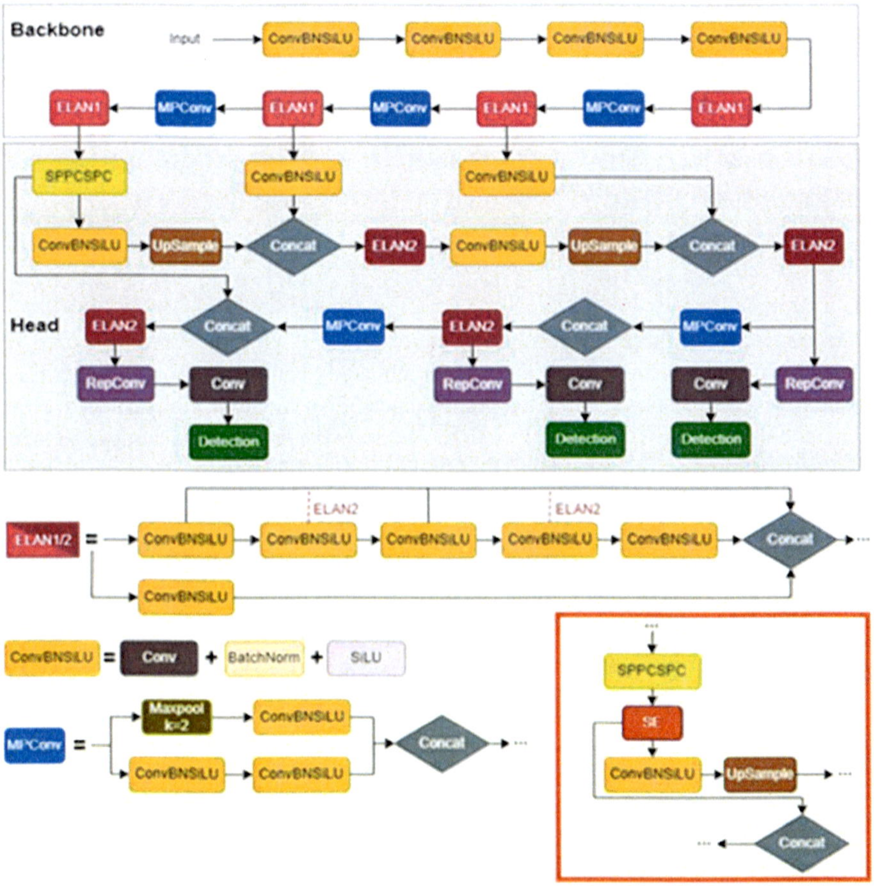


**Fig. 2.** Original YOLOv7 architecture with the added module (lower right corner in a red box).

of the mAP metric, where the Intersection over Union (IoU) threshold is set to 0.5. The IoU threshold is used to decide if a detected bounding box is considered a true or a false positive. If the IoU between the predicted and the ground truth bounding box is greater than 0.5, the detection is considered a true positive.

In this paper, the experiments were conducted on Google Collaboratory platform. The used version of Python was 3.8, Pytorch v 1.13.1, CUDA v 11.6 for the model training, and the Tesla T4 (16G) was the utilized GPU. The size of all images used for training is 640x640, batch-size is 16 and the model was trained for 30 epochs. The validation set was used to benchmark all the metrics during the training phase. The dataset and the complete code are available at request. The results of our model are shown in Table 1.

Figure 3 shows some examples of inference on the test set.

Due to the complex nature of our dataset the reached precision, recall and mAP metrics are between 70% and 80%. It is obvious that these scores are lower for background plums, which are smaller and more easily missed. The slight underrepresenta-

**Table 1.** Recognition results of plum targets using improved YOLOv7 network.

| Test Set | Number | Precision (%) | Recall (%) | mAP (%) |
|---|---|---|---|---|
| All plums | 1861 | 0.702 | 0.721 | 0.768 |
| Reachable plums | 822 | 0.888 | 0.807 | 0.919 |
| Unreachable plums | 623 | 0.602 | 0.662 | 0.705 |
| Background plums | 416 | 0.616 | 0.695 | 0.679 |

tion of unreachable plums has resulted in lower scores as well. However, the class of main interest, i.e. reachable plums, achieves significantly better scores, including a mAP of over 90%. Table 2 shows very good performance of our model in comparison to other recent models from the YOLO family. Our model beats the precision, accuracy and mAP of the best performing other model (YOLOv7) by 1.4%, 1.1% and 1.3% respectively.

**Fig. 3.** Test results using our improved YOLOv7.

**Table 2.** Performance comparison of four object detection models.

| Model | P | R | mAP50 | size |
|---|---|---|---|---|
| YOLOv5s | 0.691 | 0.679 | 0.745 | 14.4 MB |
| YOLOv6 | 0.573 | 0.470 | 0.579 | 38.74 MB |
| YOLOv7 | 0.688 | 0.710 | 0.755 | 74.8 MB |
| YOLOv7 imp | 0.702 | 0.721 | 0.768 | 74.9 MB |

## 4 Conclusion

The main contributions of this paper are: (i) The collection of a new dataset, for the *Prunus domestica* plum species. The images were collected in real environments (unstructured plum orchards with poorly pruned trees) and under different illumination conditions. ii) A new object detection model was presented based on the state-of-the-art YOLOv7 model. Our model was improved with data augmentation, initial anchor box sizes and an SE module. Iii) The fruits were classified in a way that reflects the needs of harvesting robot tasks. Plums were placed into 3 different classes based on their reachability and graspability. While this resulted in slightly poorer results, the precision, recall and mAP for the reachable plum class were 88.8%, 80.7% and 91.9%, which is relatively high and shows promise for future applications.

Future work will include collecting a larger dataset, including images of plums during cloudy days. Further improvement will be made on this model to achieve higher mAP scores across all plum classes. Future research direction is the development of a more precise detection system, able to detect the stem of the plum fruit or estimate its position, for the harvesting robot with scissors-type end effector.

**Acknowledgements.** This work is the result of research and development carried out on the project "Development of a demonstration prototype of a manipulative robot for picking edible plums - proof of concept", contract no. 623–1-22, innovation voucher no. 1216 co-financed by the Innovation Fund of the Republic of Serbia, 2022.

## References

1. Tomić, J., Štampar, F., Glišić, I., Jakopič, J.: Phytochemical assessment of plum (Prunus domesti-ca L.) cultivars selected in Serbia. Food Chem. **299**, 125113 (2019). https://doi.org/10.1016/j.foodchem.2019.125113
2. Zhou, H., Wang, X., Au, W., Kang, H., Chen, C.: Intelligent robots for fruit harvesting: recent developments and future challenges. Precis. Agric. **23**(5), 1856–1907 (2022). https://doi.org/10.1007/s11119-022-09913-3
3. Wang, Z., Xun, Y., Wang, Y., Yang, Q.: Review of smart robots for fruit and vegetable picking in agriculture. Int. J. Agric. Biol. Eng. **15**(1), 33–54 (2022). https://doi.org/10.25165/j.ijabe.20221501.7232
4. Vrochidou, E., Tsakalidou, V.N., Kalathas, I., Gkrimpizis, T., Pachidis, T., Kaburlasos, V.G.: An overview of end effectors in agricultural robotic harvesting systems. Agriculture **12**(8), 1240 (2022). https://doi.org/10.3390/agriculture12081240

5. Koirala, A., Walsh, K.B., Wang, Z., McCarthy, C.: Deep learning – method overview and review of use for fruit detection and yield estimation. Comput. Electron. Agric. **162**, 219–234 (2019). https://doi.org/10.1016/j.compag.2019.04.017
6. Tang, Y., et al.: Recognition and localization methods for vision-based fruit picking robots: a review. Front. Plant Sci. **11**, 510 (2020). https://doi.org/10.3389/fpls.2020.00510
7. Ukwuoma, C.C., Zhiguang, Q., Bin Heyat, M.B., Ali, L., Almaspoor, Z., Monday, H.N.: Recent advancements in fruit detection and classification using deep learning techniques. Math. Probl. Eng. **2022**, 1–29 (2022). https://doi.org/10.1155/2022/9210947
8. Girshick, R., Donahue, J., Darrell, T., Malik, J.: Region-based convolutional networks for accurate object detection and segmentation. IEEE Trans. Pattern Anal. Mach. Intell. **38**(1), 142–158 (2016). https://doi.org/10.1109/TPAMI.2015.2437384
9. Liu, W., et al.: SSD: single shot multibox detector. In: Leibe, Bastian, Matas, Jiri, Sebe, Nicu, Welling, Max (eds.) ECCV 2016. LNCS, vol. 9905, pp. 21–37. Springer, Cham (2016). https://doi.org/10.1007/978-3-319-46448-0_2
10. Redmon, J., Divvala, S., Girshick, R., Farhadi, A.: You only look once: unified, real-time object detection (2015). https://doi.org/10.48550/ARXIV.1506.02640
11. Koirala, A., Walsh, K.B., Wang, Z., McCarthy, C.: Deep learning for real-time fruit detection and orchard fruit load estimation: benchmarking of 'MangoYOLO.' Precision Agric. **20**(6), 1107–1135 (2019). https://doi.org/10.1007/s11119-019-09642-0
12. Chen, W., Lu, S., Liu, B., Li, G., Qian, T.: Detecting citrus in orchard environment by using improved YOLOv4. Sci. Program. **2020**, 1–13 (2020). https://doi.org/10.1155/2020/8859237
13. Parico, A.I.B., Ahamed, T.: Real time pear fruit detection and counting using YOLOv4 models and deep SORT. Sensors **21**(14), 4803 (2021). https://doi.org/10.3390/s21144803
14. Yan, B., Fan, P., Lei, X., Liu, Z., Yang, F.: A real-time apple targets detection method for picking robot based on improved YOLOv5. Remote Sens. **13**(9), 1619 (2021). https://doi.org/10.3390/rs13091619
15. Wu, D., et al.: Detection of camellia Oleifera fruit in complex scenes by using YOLOv7 and data augmentation. Appl. Sci. **12**(22), 11318 (2022). https://doi.org/10.3390/app122211318
16. Wang, L., Zhao, Y., Liu, S., Li, Y., Chen, S., Lan, Y.: Precision detection of dense plums in orchards using the improved YOLOv4 Model. Front. Plant Sci. **13**, 839269 (2022). https://doi.org/10.3389/fpls.2022.839269
17. Kim, E., et al.: CNN-based object detection and growth estimation of plum fruit (Prunus mume) using RGB and depth imaging techniques. Sci. Rep. **12**(1), 20796 (2022). https://doi.org/10.1038/s41598-022-25260-9
18. Redmon, J., Farhadi, A.: YOLOv3: an Incremental improvement (2018). https://doi.org/10.48550/ARXIV.1804.02767
19. Wang, C.Y., Bochkovskiy, A., Liao, H.Y.M.: YOLOv7: trainable bag-of-freebies sets new state-of-the-art for real-time object detectors (2022). https://doi.org/10.48550/ARXIV.2207.02696
20. Wang, C. Y., et al.: CSPNet: a new backbone that can enhance learning capability of CNN (2019). https://doi.org/10.48550/ARXIV.1911.11929
21. He, K., Zhang, X., Ren, S., Sun, J.: Spatial pyramid pooling in deep convolutional networks for visual recognition (2014). https://doi.org/10.48550/ARXIV.1406.4729
22. Hu, J., Shen, L., Albanie, S., Sun, G., Wu, E.: Squeeze -and-excitation networks (2017). https://doi.org/10.48550/ARXIV.1709.01507

# Innovative Robots Design

# Efficient Robot Hopping Mechanisms with Compliance Using Evolutionary Methods

Pramod Pal[1], Anubhab Dasgupta[2], Avinash Bhashkar[3], Shishir Kolathaya[1], and Ashitava Ghosal[1](✉)

[1] Robert Bosch Centre for Cyber-Physical Systems, Indian Institute of Science, Bangalore, India
{pramodpal,shishirk,asitava}@iisc.ac.in

[2] Department of Mechanical Engineering, Indian Institute of Technology, Kharagpur, Kharagpur, India
anubhab.dasgupta@kgpian.iitkgp.ac.in

[3] Department of Mechanical Engineering, Indian Institute of Technology, Kanpur, Kanpur, India
avinashb21@iitk.ac.in

**Abstract.** This paper presents as a study of a single leg of a quadruped robot with a compliance mechanism in its foot. The objective is to optimize the hopping behavior in terms of maximum jump height, least impact force for a minimum jump height, and power consumed in the actuators of the leg as a function of the compliance in the foot. The paper consists of numerical simulations of the leg in the MuJoCo environment and experimental results obtained in hardware. In order to find the optimum, a two-stage algorithm is used. In the first stage, an exhaustive coarse grid search of the parameter space, consisting of the compliance in the foot and controller gains at the hip and knee actuators, is done to arrive at a set of parameters. Next, an evolutionary algorithm is used to conduct a more focused search beginning from the best set of parameter values found in the first phase. Simulation results show that there is an optimum value of the compliance in the foot, which can give efficient hopping behavior with the highest jump height and lesser impact force on hitting the ground. Experimental results obtained from hardware also show an optimum spring stiffness and are in reasonable agreement with the simulation results.

**Keywords:** Quadruped · Single legged robot · Evolutionary Optimization · Design

## 1 Introduction

Compared to wheeled mobile robots, walking robots are ideal for traversing uneven terrain due to their inherent versatility and ability to adapt to the environment. A legged robot can jump or go over obstacles, negotiate soft and hard terrains, flat or sloped terrains, and change directions or motion instantaneously.


T. Petrič et al. (Eds.): RAAD 2023, MMS 135, pp. 453–460, 2023.
https://doi.org/10.1007/978-3-031-32606-6_53

One of the key features of a walking robot is jumping or hopping, and in this paper, we study the hopping behavior of a leg which is a part of a quadruped. The robotic leg in this study consists of a waist, a hip joint, a knee joint, and a foot. In nature, animals have compliant elements in their limbs in the form of the tendon, muscles, etc., which are used to minimize impact forces and store energy while walking [1]. The compliance in a legged robot is similarly expected to absorb some of the impact force, store energy during the stance phase, and release energy during the swing the phase of locomotion, thereby increasing the efficiency, durability, and speed of the robot [2].

Raibert, a pioneer in the area of hopping robots, introduced a passive prismatic joint in the hopping robotic leg developed by him – the springy link of the hopping leg was of a telescopic type. Raibert and associates went on to develop mono-ped, biped, and quadruped and analyzed various gaits [3]. Ahmadi and Buehler worked on mono-ped with compliance and showed energy efficiency [4]. Hyon and Mita modeled a simple model of a hind limb to create a robot that could run like a dog with a tensile spring as a tendon [5]. In the work by Michele Focchi, the effect of different spring stiffness is studied by plotting the torque generated by the impact in the knee joint when the leg hits the ground [6].

To the best of our knowledge, limited literature deals with the design of compliant hopping mechanisms using learning methods [7]. The most recent work involves using Bayesian techniques for obtaining an optimal set of SLIP (spring-loaded inverted pendulum) parameters [8]. However, Bayesian methods are not scalable for large samples and parameters. In this work, we explore optimization methods in the context of compliant hoppers.

## 2 Design: Simulation and Hardware

The single leg is an essential component in a walking robot and in a quadruped for its mobility and stability. Designing a leg has the following essential components, link length, actuator placement, link shape, foot design, etc. In general, a shorter link length increases stability and control, as well as reduces weight, but restricts the range of motion and produces less force. On the other hand, a longer link length enhances the range of motion and power but diminishes stability and control and adds weight. In this work, we assumed uniform link lengths of 0.17 m, which is about the same as the leg length of a medium-sized dog. A well-designed leg/foot should be capable of absorbing impact force, which prevents its components from failing. Using spring at the foot can help in mitigating the impact force and, at the same time, store and deliver the energy while performing different gaits.

### 2.1 Leg Parameters

The single leg comprises a base, thigh, shank, and foot link connected with two active revolute joints and one passive prismatic joint (see Fig. 1(a)). The thigh and shank leg lengths are denoted by $l_1$ and $l_2$, and the angles $\theta_1$ (from vertical) and $\theta_2$ (from the direction along the thigh link) denote the rotations of the links, and their ranges are shown in Table 1.

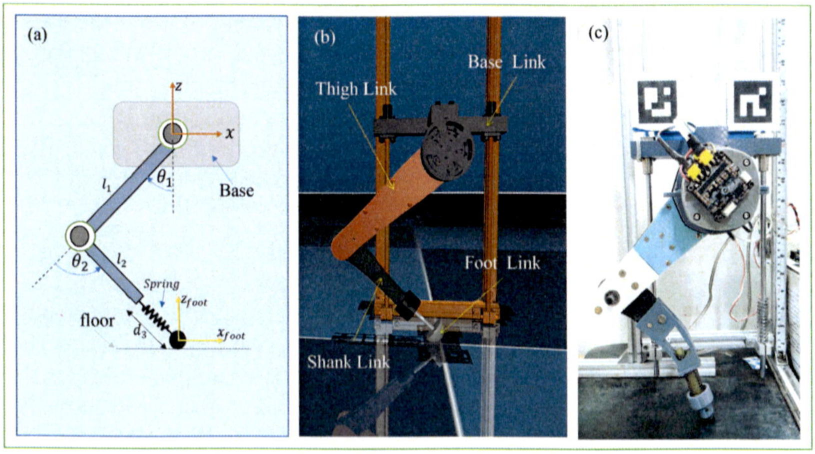


**Fig. 1.** (a) Schematic diagram of the single leg with compliant foot, (b) Single leg in MuJoCo environment, (c) Single leg hardware

**Table 1.** Motion range of the joint variables

| Joint Variable | Joint name | Type of joint | Range |
|---|---|---|---|
| $\theta_1$ | hip | Revolute | 0 to 1.6 rad |
| $\theta_2$ | knee | Revolute | −2.1 to 0 rad |
| $d_3$ | foot | Prismatic | 0 to 0.012 m |

## 2.2 Simulation Platform

The simulations are performed using the Multi-Joint dynamics with Contact (MuJoCo) [9] simulation engine. MuJoCo captures contact dynamics with state-of-the-art capabilities. It is a popular open-source program for robotic system simulation and is simple to use. The model shown in the Fig. 1(b) represents the single leg that has been designed using a CAD software (SolidWorks), and then the URDF (Unified Robotics Description Format) is created. The geometry and inertia parameters (from the CAD model) are then used for the simulation in MuJoCo. Simulation parameters, such as the step size, the number of iterations, choice of solvers are also made to govern the simulation. In the case of the leg, we have also chosen control parameters, such as actuator forces and torques, to define how the leg is to be controlled. MuJoCo also enables accurate visualization by choice of camera position and lighting conditions and environment parameters describe the attributes of the simulated environment, such as the coefficients of gravity and friction. All of these parameters work together to form a realistic depiction of the robot and its environment, enabling the precise testing and assessment of robotic systems.

We have used a few basic parameters to control the simulation of the leg – gravity allows the single leg to fall to make contact with the ground when it hits the ground plane, and we have used the proportional and derivative controller gain of a motor to simulate the behavior of a spring and damper as an added compliance mechanism to the foot of the single-legged robot. We have used a small time step of 0.001, which provides better accuracy and stability. For the solver, we have used Newton's method and a $4^{th}$-order Runge-Kutta method. The total mass of the single leg is 1.1 kg. At all joints, we have used a damping of 1.0 N s $m^{-1}$ and the torque is limited in both hip and knee joints to 7.0 N m.

### 2.3 Experimental Setup

To validate the simulation results and to obtain the effect of the compliant foot, we have used hardware as shown in Fig. 1(c). Both the hip and knee actuators were attached to the base link. The rotary motion of the knee motor is transferred to the shank link by a belt-and-pulley drive with a 1:2 gear ratio between the driver and the driven pulley. We've used a planetary gearbox with a gear ratio of 1:6 for power transfer from the motor to the links. The foot link is connected to the lower end of the shank link to the foot using a linear spring. Table 2 shows the parameters and their values, while Table 3 shows the hardware elements and their details. In the experiments, we have used different springs whose stiffness vary from 100 to 30000 N $m^{-1}$. The power consumed by the leg was obtained from the voltage and the current supplied to the motor, and the height of the hopping was measured using an ultrasonic sensor with the least count of 0.05 cm.

**Table 2.** Single leg parameters.

| Parameter | Value | Unit |
|---|---|---|
| Total mass | 1.1 | Kg |
| Thigh link length | 0.172 | m |
| Shank link length | 0.130 | m |
| Foot link length | 0.037 | m |
| Foot link maximum travel | 0.012 | m |

**Table 3.** Hardware element and their details.

| Hardware | Details |
|---|---|
| Motor | 5010 360 KV BLDC |
| Magnetic Encoder | AS5048B |
| Microcontroller Chip-set | STM32G4 |
| Driver | Moteus r4.5 |
| Communication | 5Mbps CAN-FD |
| Link materials | Polylactic acid(PLA) |

## 3 Control Strategy, Optimization Objective and Training

In this section, we describe the details of the simulation and experiments done on the single leg with compliance at the foot.

### 3.1 Control Strategy

In this work, torque is the control input in both the hardware and the simulation. The torque is computed based on feedback from measured values, its first derivative, and its corresponding desired or reference values. We denote the hip, knee, and foot positions by the symbols $x_h$, $x_k$ and $x_f$, respectively and their first derivatives by $\dot{x_h}, \dot{x_k}, \dot{x_f}$. The torque is computed as

$$\tau_h = k_{p_h} e_h + k_{dh} \dot{e_h}; \tau_k = k_{p_k} e_k + k_{dk} \dot{e_k}; \tau_f = k_{p_f} e_f \tag{1}$$

where, $e_h = \hat{x}_h - x_h$ is the error with $\hat{x}_h$ denoting the reference or desired value of the hip angle. A similar correspondence exists for $e_k$ and $e_f$.

**Inverse Kinematics for Reference Trajectory Generation:** To make the leg hop in one place, a linear, cyclic trajectory is used to approximate the foot end-point trajectory in the $X - Z$ plane. This is given by

$$x = 0; z = -0.174 + 0.026 * \sin(\phi) \tag{2}$$

where the angle $\phi$ is used to divide the trajectory into 200 points. The three joint variables $\theta_1$, $\theta_2$ and $d_3$ for a given $(x, z)$ in a leg are obtained using sequential least squares programming (SLSQP) algorithm [10]. The quantity $||\mathbf{q}||_2{}^2$ is used as the optimization objective function, where $\mathbf{q}$ denotes the vector of the hip angle, knee angle, and the change in length divided by the original length of the spring-damper system, to obtain the inverse kinematics solution of the compliant legged system. Once the reference trajectory is generated, respective torque values are calculated, and the trajectory of the single leg can be observed.

### 3.2 Data-Driven Evolutionary Optimization for Training Parameters

The motion of the leg is a function of five controller gains – proportional and derivative gains at the hip and knee motors and the stiffness of the spring at the foot. The jump height and the impact force as the leg hits the ground is a function of these 5 parameters, and symbolically they can be expressed as:

$$h = f_1(\mathcal{K}); \mathcal{F} = f_2(\mathcal{K}) \tag{3}$$

where $h$ denotes the hopping height, $\mathcal{F}$ denotes the highest impact force experienced by the foot over a cycle of the simulation and $\mathcal{K}$ denotes the vector of the controller gain values and the stiffness of the foot.

The dynamics of the foot impacting the ground is fairly complex and the task of finding the functional representation above is not tractable. In our work, we have used the MuJoCo environment to simulate the motion of the leg as a function of the 5 parameters. We use a two-step approach to obtain the optimum values of the 5 parameters. We first perform a "grid-search", where a sufficiently large range of the parameters is chosen, and an exhaustive search is performed where the program iterates through small step sizes in the space and records the value of the objective function at those nodal points. Following that comes the second step of our algorithm, where a more refined search is done using the Covariance Matrix Adaptation Evolutionary Strategy (CMA-ES) [11] only around a selected nodal point to find the best objective, which minimizes the impact force or maximizes the height. Hence, we solve two optimization problems:

$$\max_{\mathcal{K}} \quad Ce^{-(H-h)}$$

and

$$\min_{\mathcal{K}} \quad \mathcal{F}$$

Both are subject to system dynamics and in the second, a penalty term is added if the height falls below a particular value. This is done to ensure that while still minimizing the impact force, the jump height does not become poor. In the above optimization problems $H$ is an assumed constant upper bound on the jump height and $C$ is a constant.

### 3.3 Training Parameters

For the grid search, we used a range of about 500 to 1300 N m rad$^{-1}$ for the proportional controller gains and small derivative gains ranging from 0.01 to 1.5 N m s rad$^{-1}$ for the hip and knee motors. The spring stiffness for the foot is varied between 500 to 5000 N m$^{-1}$. For the objective function in CMA-ES, we took 4000 steps and calculated the objective as the maximum of objective values over an entire cycle of simulation. The motivation for choosing the pointwise maximum is well known in optimization literature for it preserves the nature of the problem.

## 4 Results and Discussion

In this section, we present the numerical simulation and the experimental results.

### 4.1 Simulation Results and Analysis

The single-leg hopping is simulated in MuJoCo. Figure 2(a) shows the convergence of the optimization process for a typical value of the spring stiffness. The maximum height with number of training steps converge to about 0.08 m. Figure 2(b) shows the impact force with number of simulation steps – the impact force converges to a minimum of about 70 N. Figure 2(c) shows the jump height with increasing number of simulations – after the initial transient, the periodic behavior is independent of the initial starting position of the leg, which demonstrates the stability and robustness of the controller with respect to initial joint states. The average jump height is computed from this figure for different values of spring stiffness and Fig. 2(d) shows the variation of jump height with spring stiffness. Figure 2(d) clearly shows that there exists an optimal spring stiffness to obtain maximum jump height jump for other parameters being fixed.

### 4.2 Experimental Results and Analysis

The experiments were done on a hard surface. Figure 3(a) shows the percentage of hopping height to the maximum range vs the spring stiffness. The result for the fixed foot was obtained when the spring was removed, and the foot was rigidly connected to the shank. As shown in Fig. 3(a), the near-optimal hopping

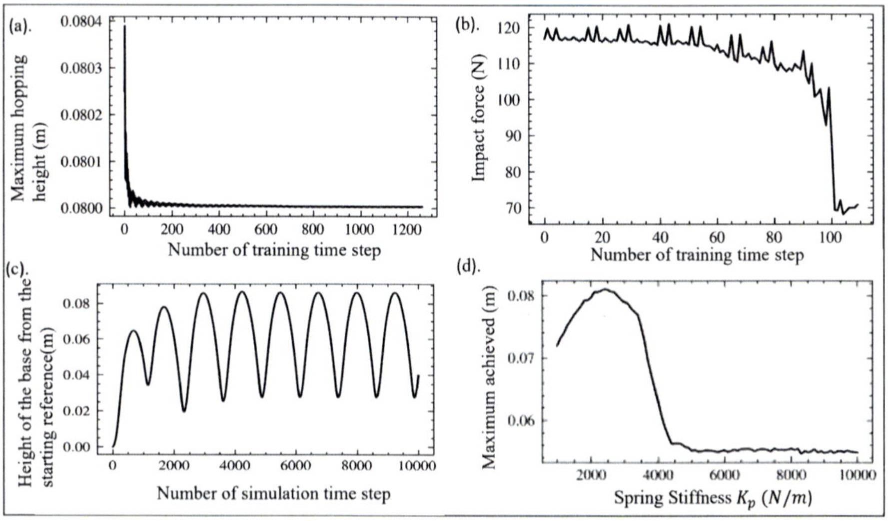


**Fig. 2.** Average hip and knee motor combined power versus foot stiffness

height range is obtained for spring with approximate stiffness 2800 N m $^{-1}$. The hopping height decreases on both sides of this value - this variation of the hopping height range with spring (feet) stiffness and the existence of an optimum value of the spring stiffness is consistent with the numerical results shown in Fig. 2. It can also be seen from Fig. 3(b) that the nature of the power consumed plot shows a minimum value for spring stiffness of 2800 N m$^{-1}$. This shows with optimal stiffness, we achieve the near-optimal hopping height along with low power consumption. The experimental results are in reasonable agreement with the simulation results obtained using MuJoCo.

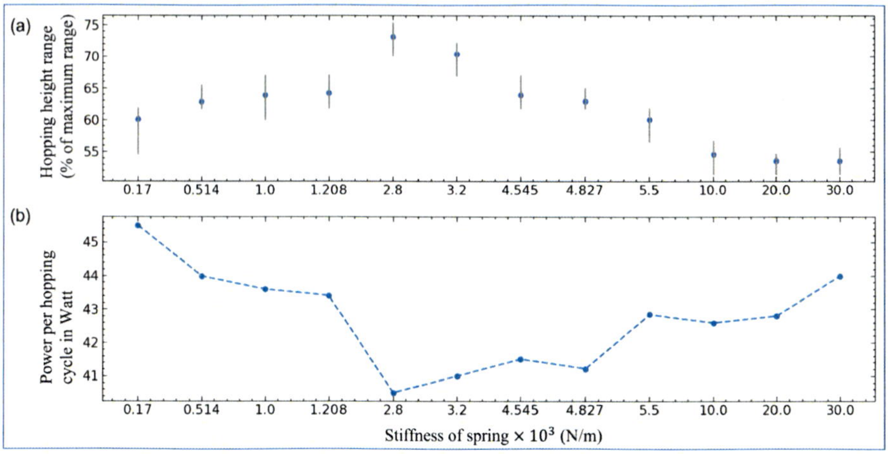


**Fig. 3.** (a) Plot of Hopping height range (%of Maximum range) to spring stiffness; (b) Plot of Power consumed per hopping cycle to different spring stiffness

## 5 Conclusion

This paper presents a study of the effect of compliance in the foot of a single legged hopping robot. The single leg contains two motors and the goal is to obtain an optimal set of parameters for maximum vertical jump and minimum power consumption using evolutionary strategies. A model of the single legged robot is analysed using MuJoCo and a physical hardware with similar mass and geometrical parameters is fabricated. The main results of the work are a) there exists an optimum range of the stiffness and control gains which results in the highest vertical jump, b) the power required to jump is lower for this stiffness, and c) the experimental results obtained from the hardware are consistent with the simulation results. This work is being extended to training using the actual hardware using model free reinforcement learning techniques. We also plan to extend this work to a quadruped where stiffness will be added to each of the four legs and investigate the effect of the compliance on the feet towards cost of transport for the quadruped.

## References

1. Hyon, S.H., Emura, T., Mita, T.: Dynamics-based control of a one-legged hopping robot. Proc. Inst. Mech. Eng. Part I: J. Syst. Control Eng. **217**(2), 83–98 (2003)
2. Degrave, J., Caluwaerts, K., Dambre, J., Wyffels, F.: Developing an embodied gait on a compliant quadrupedal robot. In: 2015 IEEE/RSJ International Conference on Intelligent Robots and Systems (IROS), pp. 4486–4491. IEEE (2015)
3. Raibert, M., Chepponis, M., Brown, H.B.: Running on four legs as though they were one. IEEE J. Robot. Autom. **2**(2), 70–82 (1986)
4. Ahmadi, M., Buehler, M.: The ARL monopod II running robot: control and energetics. In: Proceedings 1999 IEEE International Conference on Robotics and Automation (Cat. No. 99CH36288C), vol. 3, pp. 1689-1694. IEEE (1999)
5. Hyon, S.H., Mita, T.: Development of a biologically inspired hopping robot-" Kenken". In: Proceedings 2002 IEEE International Conference on Robotics and Automation (Cat. No. 02CH37292), vol. 4, pp. 3984–3991. IEEE (2002)
6. Focchi, M.: Strategies to improve the impedance control performance of a quadruped robot. Istituto Italiano di Tecnologia, Genoa (2013)
7. Ghosh, R., Pratihar, D.K., Gupta, S.: Towards the optimal design of an uncemented acetabular component using genetic algorithms. Eng. Optim. **47**(12), 1587–601 (2015)
8. Saar, K.A., Giardina, F., Iida, F.: Model-free design optimization of a hopping robot and its comparison with a human designer. IEEE Robot. Autom. Lett. **3**(2), 1245–51 (2018)
9. Todorov, E., Erez, T., Tassa, Y.: MuJoCo: a physics engine for model-based control. In: 2012 IEEE/RSJ International Conference on Intelligent Robots and Systems, pp. 5026–5033 (2012)
10. Kraft, D.: A software package for sequential quadratic programming. Forschungsbericht- Deutsche Forschungs- und Versuchsanstalt fur Luft- und Raumfahrt (1988)
11. Hansen, N., Müller, S.D., Koumoutsakos, P.: Reducing the time complexity of the derandomized evolution strategy with covariance matrix adaptation (CMA-ES). Evol. Comput. **11**(1), 1–8 (2003)

# The Effect of Deformation on Robot Shape - Changing Link

Jakub Mlotek(✉), Jiří Suder, Michal Vocetka, Zdenko Bobovský, and Václav Krys

Department of Robotic, Faculty of Mechanical Engineering, VSB - Technical University of Ostrava, 17. Listopadu 2172/15, 708 00 Ostrava-Poruba, Czech Republic
jakub.mlotek@vsb.cz

**Abstract.** The adaptability of the robot can be increased by a shape-changing link of the robotic arm that allows to achieve a suitable kinematic structure for a given task. The principle described in this paper is the use of low melting point material (LMPM) in the design of the robotic link and focuses on the behavior of this shape-changing link during a controlled deformation. The effect of the equivalent stress during the controlled deformation of the outer shell of the shape-changing link on the LMPM core is also investigated. In the introduction, the type of deformation to which the link can be deformed is determined. After the definition of the outer boundary curve along which the cell deforms, a finite element analysis (FEA) is performed for selected points on this curve. The analysis defines the location of the equivalent stress concentration and its magnitude. Real measurements define the resulting forces and torques at the end point of the link. Subsequently, the real effect of the equivalent stress on the core of the link is observed. The result is the distribution of equivalent stress in the structure of the link. The influence of the equivalent stress distribution and its magnitude at defined strain is so pronounced for LMPM that shape instabilities occur in a short period of time. The use of a given LMPM is not appropriate under the given conditions. In the conclusion chapter, possible solutions are proposed to limit the effect of equivalent stress on the LMPM core.

**Keywords:** Robot shape – changing link · system adaptability · Low Melting Point Material (LMPM) · equivalent stress · FEA · robotic manipulator

## 1 Introduction

Today, standard manipulators with straight and fixed connecting links that define a fixed working envelope are very common. But increasingly, new principles are emerging to change the structure of the manipulator by changing the shape of the connecting link. These principles could then be used to achieve the designed shape of the manipulator based on the trajectory in the collision environment [1]. Thus, it is possible to have a flexible system capable of adapting to dynamic conditions in the working environment. The known principle is that a cable-controlled deformation, extended by the system, increases the rigidity of the internal segmental structure by means of very accurate


T. Petrič et al. (Eds.): RAAD 2023, MMS 135, pp. 461–469, 2023.
https://doi.org/10.1007/978-3-031-32606-6_54

lamellas [2]. The lamellas are placed around the entire perimeter and length of the segmented structure. The lamellas are enclosed in a latex membrane. When a vacuum is created in the membrane, the lamellas are compressed, increasing friction, and thus the stiffness of the system. The advantage is the rapid change in the stiffness and thus the possibility of rapid shape change. The disadvantage is the possible loss of vacuum in case of membrane damage. In these cases, there is an immediate loss of stiffness of the whole system.

Therefore, a method that uses the energy supplied to the system only to change the shape of the structure seems to be a more appropriate solution. This eliminates the problem of losing stiffness when the vacuum is broken. These solutions use LMPM enclosed in a flexible material. When the LMPM is heated, the structure loses its stiffness and deformation is enabled. This principle was used for the reconfigurable legs of a walking robot [3]. The use of LMPM is proposed for the use of the shape - changing link of the manipulator in the work [4]. The advantage is the expected stability of deformation until heating. The disadvantage is the low strength of LMPM.

For applications using LMPM, alloys with bismuth or indium admixture dominate. These materials include Field alloy (FA) [5] which is non-toxic and has a melting point of 60 °C.

In our previous research [6], we looked at the effect of link deformation on the shape of the manipulator work surface. We explored where this deformation would find application. We build on this work and investigate an experimental link with an LMPM. Due to the low-strength material content, we focus on the area of the possible deformation and the effect of the deformation on the internal structure of the link. Therefore, we performed experiments with and without LMPM. We investigate how the deformation of the shell affects the LMPM core and applies internal stresses on it.

## 2 Material and Methods

The chapter describes an experimental shape-changing link. Furthermore, the chapter gives the parameters and describes the experimental methods performed with and without the LMPM core.

### 2.1 Description of the Shape - Changing Link Without LMPM Core and Deformation Region Boundary Curves

To investigate the effect of deformation on the LMPM core, a test shape - changing link was designed and realized. The link is based on the design of [4]. The composition of the link is shown (Fig. 1.).

For further work, we identified a deformation region. This is the region into which the end point of the link can be deformed at given boundary conditions. The deformation region was defined from the parameters listed in Table 1, namely the length for bending and minimum bending radius.

This region was defined only for bending in 2D. We used [4] to identify the edge curves of the deformation region. In our work, we have identified one curve as the involute of a circle and specified the second curve as a cardioid-type cycloid (Fig. 2(a)).

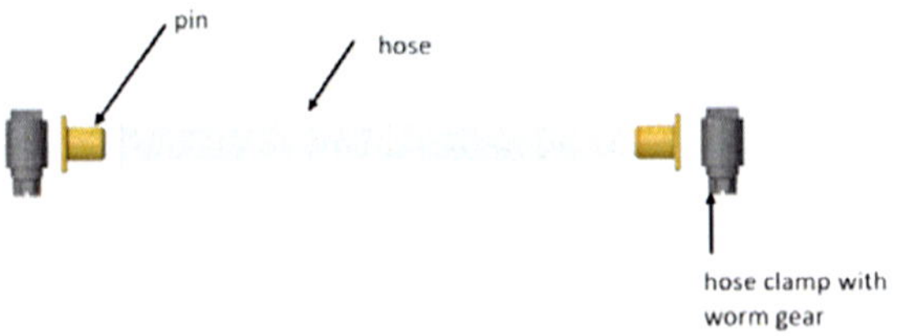


**Fig. 1.** Shape - changing link design

**Table 1.** Hose parameters [6]

| Material | Temperature resistance [°C] | Inner diameter [mm] | Outer diameter [mm] | Minimum bending radius [mm] | Length [mm] | Length for bending [mm] |
|---|---|---|---|---|---|---|
| Silicone Rubber | −60 ÷ 180 | 9.5 | 16.6 | 55 | 132 | 106 |

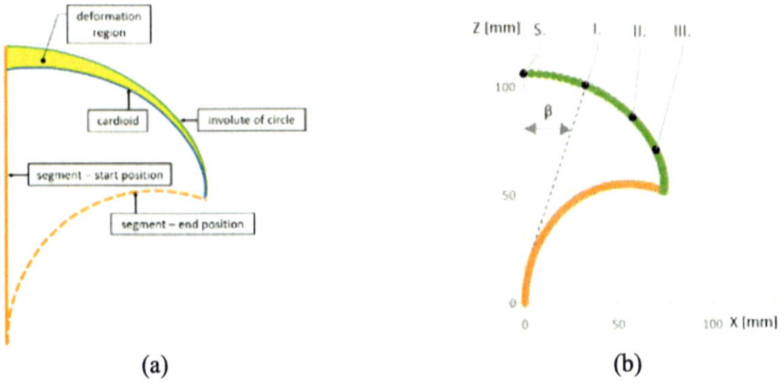


**Fig. 2.** (a) deformation area and boundary curves (b) three measuring points

To measure the equivalent stress generation and its effect on the LMPM core we chose to deform the link after involute of a circle. This is for two reasons. First, it is the maximum reach of the link during deformation. Second, for each point on the curve, there is only one shape deformation link to reach the point.

Three measurement points were defined on the selected curve in (Fig. 2(b)), to which the link was deformed. Table 2 shows the X, Z coordinates and the rotation β of the measurement points. This coordinate system was chosen because of the location of the link in space and to respect the coordinate system in the real measurements. The first two points are evenly distributed. Point III being determined as the furthest point on the curve that can be reached in the measuring stand.

**Table 2.** Points coordinates and angle of rotation

| Position | X [mm] | Z [mm] | β [°] |
|---|---|---|---|
| S | 0 | 106 | 0 |
| I | 32.83 | 100.44 | 19.88 |
| II | 58.05 | 85.64 | 41.97 |
| III | 70.16 | 70.5 | 61.84 |

## 2.2 FEA of the Shape - Changing Link Without LMPM

The FEA was performed for the described link and the defined endpoints; see Sect. 2.1. The analysis is static with a non-linear hyperelastic setting. In the FEA, only the position of the end point (Table 2.) to be reached by the link was determined. The mesh model consists of 41798 tetrahedral elements and 9185 nodes shown on (Fig. 3).

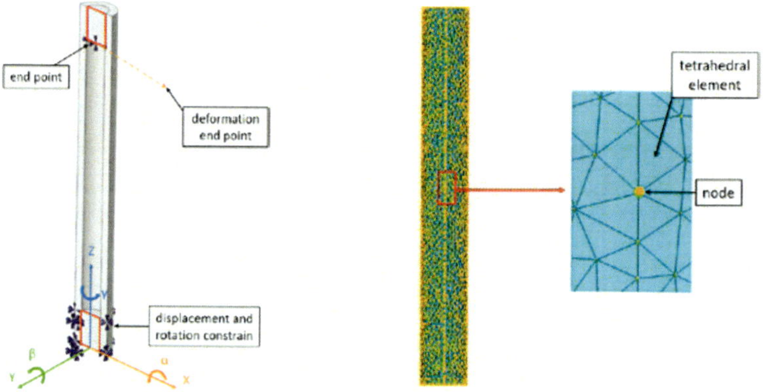


**Fig. 3.** Description of the simulation model

The industrial hose material is defined as silicone rubber, which is described by the Moon-Rivlin method [8].

## 2.3 Force and Torque Measurement in Shape - Changing Link Without LMPM

The measurement of forces and torques occurring at the end point of the deformed link was measured using available force-torque system [9]. The coordinate system of the force and torque sensor is shown in (Fig. 4.). One end of the link was attached to the stand and the other end to the robot end-effector with a force-torque sensor. The end-effector was then moved to the measuring points with the required orientation.

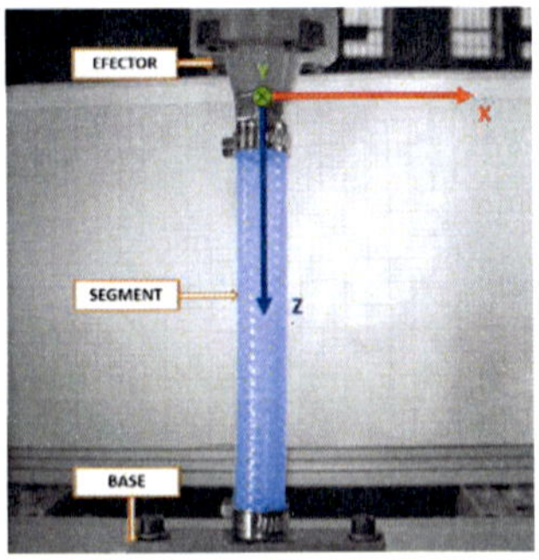


**Fig. 4.** The attached link in the workplace - measured structure

### 2.4 Measuring the Efeect of Equivalent Stress on Shape - Changing Link with LMPM

To show the effect of the equivalent stress on the LMPM core, a real link was created with the LMPM Fields alloy [5]. This link was subsequently deformed to position III, see. Table 2. The link was then clamped on a stand and the effect of deformation/equivalent stress on the LMPM core was observed. The test was stopped after 72 h due to link shape stabilization.

## 3 Results

The chapter presents the results from the FEA, the real measurements with the force-torque sensor, and the results from the measurement of the equivalent stress on the LMPM core.

### 3.1 FEA Results

In the results of the FEA, only the deformed part of the link and the upper pin are shown for the purpose of appropriate representation of stresses in the structure. (Fig. 5.) shows the resulting strain shape, area, and magnitude of equivalent stress from the FEA. For position I (Fig. 5(a)), the magnitude of the maximum equivalent stress is 0.27 MPa, for position II (Fig. 5(b)) 0.44 Mpa and for position III (Fig. 5(c)) 0.45 Mpa.

### 3.2 Results of Force and Torque Measurements

The results of the real measurements are shown in Table 3. as the average of 20 measurements The spread of the selected values is shown in (Fig. 7) (Fig. 6).

Figure 7. Shows the resulting shape of the deformation. The deformation is coloured blue.

### 3.3 Results of the Effect of Equivalent Stress on the LMPM Core

The results of the change in position over time of the highlighted endpoint E on the link are shown graphically in (Fig. 8). The coordinates of point E are then given in Table 4.

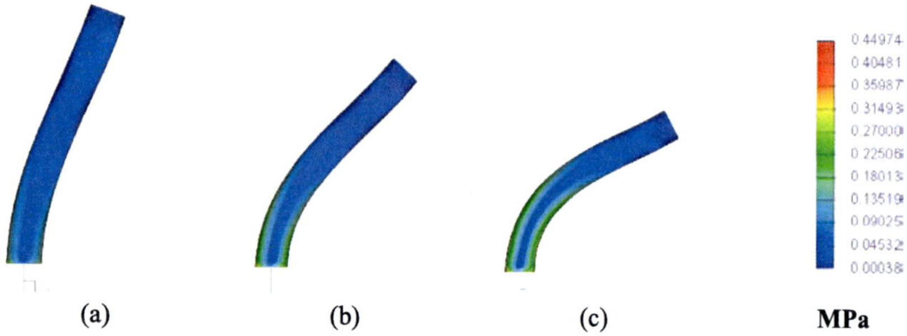


**Fig. 5.** FEA – shape, equivalent stress in silicone rubber for (a) position I, (b) position II. (c) position III.

**Table 3.** Force and bending torques real measurements

| Position | Fx [N] | Fy [N] | Fz [N] | Mx [Nm] | My [Nm] | Mz [Nm] |
|---|---|---|---|---|---|---|
| I | 1.29 | 0.17 | 10.64 | 0.07 | 0.14 | 0 |
| II | 2.45 | 0.13 | 5.75 | 0.06 | 0.24 | 0 |
| III | 2.31 | 0.08 | 0.36 | 0.06 | 0.36 | 0 |

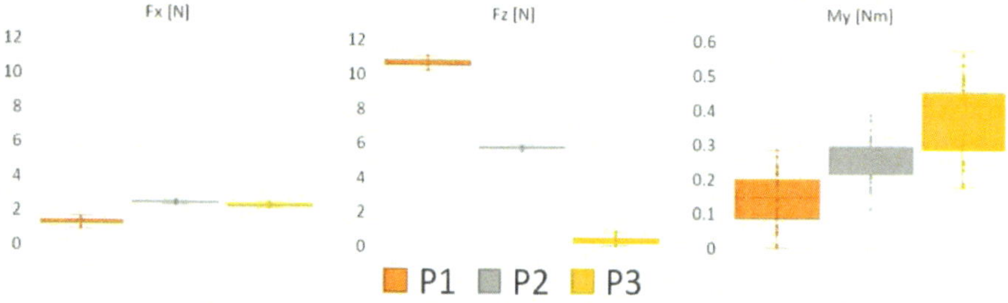


**Fig. 6.** Force and bending torques results

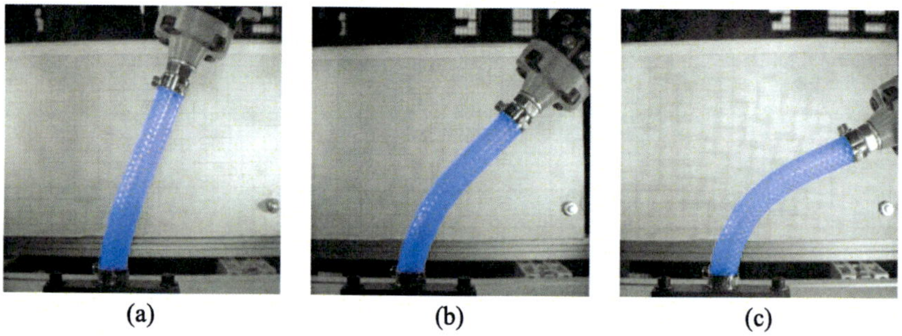


**Fig. 7.** Real deformation shapes (a) position I., (b) position II., (c) position III.

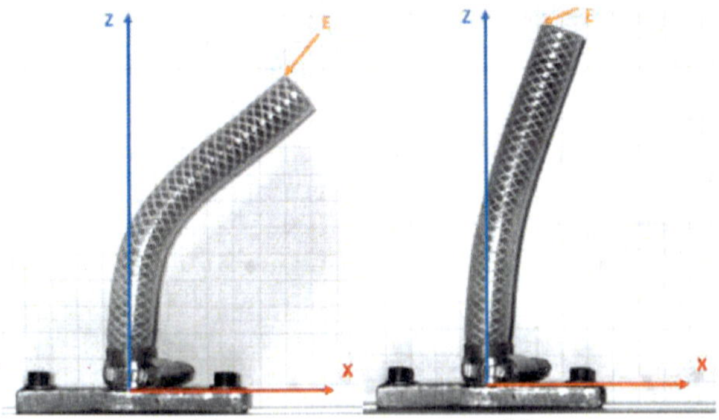


**Fig. 8.** Effect of equivalent stress on LMPM core (a) after 1 h, (b) after 72 h

**Table 4.** Coordinates of the endpoint position in the X, Z axes

| Time [h] | X [mm] | Z [mm] |
|---|---|---|
| 1 | 59 | 118 |
| 72 | 22 | 137 |

## 4 Conclusions

This paper presents the effect of equivalent stress on the core line from LMPM during its controlled shape deformation. Before the actual deformation of the link, a deformation region is defined. The deformation region defines the region where we can deform the link in a controlled deformation while maintaining the boundary conditions. The size of the region varies with the boundary conditions, but the boundary curves are always the same. For this reason, the involute of circle boundary curve is used to investigate equivalent stress generation. By using FEA, we have localized the occurrence of equivalent stress and its magnitude. The maximum magnitude of equivalent stress is 0.45 MPa for position III. The equivalent stress concentrates in the shape deformed part of the link. Real measurements then show that during deformation, the force distribution between the components significantly changes. The force on the X-axis increases from 1.29 to 2.31 N as the deformation increases. In contrast, the force on the Z-axis decreases from 10.64 to 0.36 N. The force distribution shows that the link shell has the natural property of returning to its original undeformed state and thus affects the LMPM core. The stresses and forces from the FEA and real measurements are then presented on a real link with an LMPM core. When the action of the equivalent stress is so pronounced that from the original position (Fig. 8. (a)) with coordinates (59, 118), the endpoint has shifted to coordinates (22, 137) after 72 h (Fig. 8. (b)) due to the action of the equivalent stress. Here it can be seen that the expected deformation stability mentioned in the introduction is not possible for this material. There is a significant change in shape in a relatively short time interval, and thus a change in the entire kinematic structure of the robot.

From this we can conclude that this structure with LMPM is not suitable for robotic application. Therefore, it is necessary to modify the LMPM core, increase its stiffness or use another LMPM. Another option is to use a different bending method and bend the link into a regular arc (Fig. 9.). This will reduce the magnitude of the equivalent stress and its distribution.

**Fig. 9.** Regular arc FEA – shape, equivalent stress in silicone rubber

**Acknowledgements.** This article has been elaborated with the support of the Research Centre of Advanced Mechatronic Systems project, reg. no. CZ.02.1.01/0.0/0.0/16_019/0000867 in the frame of the Operational Program Research, Development and Education. This article has been also supported by specific research project SP2023/060 and financed by the state budget of the Czech Republic.

## References

1. Huczala, D., Kot, T., Pfurner, M., Krys, V., Bobovský, Z.: Multirepresentations and multiconstraints approach to the numerical synthesis of serial kinematic structures of manipulators. IEEE Access **10**, 68937–68951 (2022). https://doi.org/10.1109/ACCESS.2022.3186098
2. CLARK, Angus B. Design and Workspace Characterisation of Malleable Robots. https://www.researchgate.net/publication/344244537_Design_and_Workspace_Characterisation_of_Malleable_Robots
3. Allen, E.A., Swensen, J.P.: Directional stiffness control through geometric patterning and localized heating of field's metal lattice embedded in silicone. Actuators **7**, 80 (2018). https://doi.org/10.3390/act7040080
4. Brandstötter, M., Gallina, P., Seriani, S., Hofbaur, M.: Task-dependent structural modifications on reconfigurable general serial manipulators. In: Aspragathos, N. A., Koustoumpardis, P. N., Moulianitis, V. C. (eds.) RAAD 2018. MMS, vol. 67, pp. 316–324. Springer, Cham (2019). https://doi.org/10.1007/978-3-030-00232-9_33
5. Field's Metal Low Melting Alloy. http://www.matweb.com/search/DataSheet.aspx?MatGUID=539fb73c246f49b3975622eacc0ec956. Accessed 12 Apr 2021
6. Mlotek, J., Bobovský, Z., Suder, J., Oščádal, P., Vocetka, M., Krys, V.: Shape-changing manipulator possibilities and the effect of the deformable segment on the size of the working area. In: Müller, A., Brandstötter, M. (eds.) Advances in Service and Industrial Robotics. RAAD 2022. Mechanisms and Machine Science, vol 120. Springer, Cham (2022). https://doi.org/10.1007/978-3-031-04870-8_32
7. Vena® TECHNOSIL. In: Hydraquip. https://hydraquip.co.uk/wp content/uploads/techsheets/Hydraquip_Technical_Data_Sheet_TS60179.pdf

8. Overview of materials for silicone rubber. MatWeb (2023). https://www.matweb.com/search/DataSheet.aspx?MatGUID=cbe7a469897a47eda563816c86a73520
9. Product manual Integrated Force Control. SHANGHAI, CHINA: ABB Engineering (Shanghai) Ltd. Robotics and Motion (2018)

# Prototyping of Robot Harvester for Plum Farmyards – Mechanical Design

Ilija Stevanović, Uroš Ilić, and Aleksandar Rodić(✉)

Institute Mihajlo Pupin, Belgrade, Serbia
aleksandar.rodic@pupin.rs

**Abstract.** This paper cover a prototyping process of a composite 6-DOF robot harvester with telescopic robot arm for plum farmyards. It is a fully autonomous robotic system, which purpose is to replace human labor inside the field. Hence the mechanism is designed to replicate human motion in harvesting process. The composite robot consists of 3DOF Cartesian robot together with added a 3DOF telescopic robot arm with a customized gripper. Robot is designed in the form of a telescope that enables the extraction and retracting motion that has a specially designed, task-oriented end-effector. The telescopic robot arm is placed on the linear guide rail Cartesian robot. The first three joints are linear and intended to reach position. The last three joints are cylindrical and define orientation of the gripper. This robot configuration enables, initially, swift positioning in the fruit neighborhood, and then fine adjustment of the gripper orientation for the final process of cutting off the fruit and harvesting. The fruit harvester was designed in such a way that the plum is carried by robot arm to storing crates and put inside a specified spot.

**Keywords:** Agricultural Robotics · 6DOFs Composite Mechanism · Cartesian Robot · Telescopic Robot Arm · Spherical Wrist

## 1 Introduction

Plum production is very promising because a quality plum is a highly demanded product on the world food market. The quality of Balkan region products belongs to the world's top quality due to the soil and climatic conditions. They are especially favorable for the plum cultivation. Given these predispositions, why is Serbia not using its maximum capacities in plum production?

The first reason is the increasingly elderly population in our country (averaging over 40 years). Negative demographic trends fueled by young people emigration. Physically capable people leave villages to the city in search of a better existence. A lack labor force in the countryside is certainly one of the biggest problems.

Another problem relates to the insecurity of product placement. Disorganized market and huge market share of corporate food chain significantly affects people's motivation to work in agriculture and in this case plum cultivation.


T. Petrič et al. (Eds.): RAAD 2023, MMS 135, pp. 470–477, 2023.
https://doi.org/10.1007/978-3-031-32606-6_55

The third reason is the lack of appropriate machinery and automation systems that would simplify the process of growing and harvesting plums. That problem can be solved by introducing highly automated and robotized processes in agriculture. The goal is to create a robot that would enable people to organize harvesting fruits without high costs for the already lacking labor force in the countryside.

### 1.1 Technical Requirements

Before starting the design of the robot system, it is needed to define a set of technical requirements that determine latter technical solutions. When they are accomplished, they should enable the technology of harvesting according to standards of correctness and quality of consumption. The technical requirements for a picking robot are reduced to the following terms determined by the tree arrangement plan on the plantation.

- **Vertical reach** of the robot arm should to be in the range of 1 to 6 m since the trees in the plantation are pruned in a way that their height should not exceed the maximum height of five to six meters. A plum tree can grow even higher in height, but in plantations that's not a practice because it affects both fertility and physical reach of human worker.
- **Horizontal reach** of the robot (arm reach left to right) is given not to be larger than the span of human hands, on average 1.5 m. The robot is placed on tractor trailer that moves between rows of plum trees. The tractor successively moves by 1–2 m along the row of trees and in this way enables the robot to harvest the entire crown of the tree in width (in the range of 4–5 m).
- **Depth reach** should be 2.5 to 3 m, bearing in mind that's the radius of the tree crown in its widest part. A robotic hand should have the ability to penetrate into the depth (to the center) of the canopy and grasp (gather) the fruits that grow on the branches inside the tree canopy.
- **Harvesting speed** is determined by the estimated cycle time picking one fruit by hand. This cycle should be as short as possible to achieve maximum efficiency of the machine. The process of picking plums by hand is quite slow. One picker picks around 30 kg per day (8 h of effective work). If we consider that the weight of one plum is on average 20–60 g (we will adopt 40 g for calculation purposes). There are 25 in 1 kg pieces of plums. A worker harvests 30 kg in 8 h or 750 pieces of fruit. This means that the human hand-picking cycle is as short as 35 s. Our goal for the robotic arm is to have harvesting speed of 4–6 pieces/minutes or that the picking cycle should be 10–15 s or 3 to 4 times faster than worker's. Thus, a manipulative picking robot would theoretically harvest 100–120 kg of fruit in 8 h of work compared to a person who only gains 30 kg on average. It should be also emphasized that the robot can work 24 h a day without stopping with adequate light provided.
- **Fruit picking technology** must be very carefully planned. Edible plums (consumable plums) are harvested nowadays exclusively by hand so they could meet the strict quality standards we are stated in the previous paragraph. The plum fruit must be grasped by its petiole, taking care not to wipe off its bluish powder while grasping. Preserved powder is an indicator that the fruit was minimally physically manipulated and its freshness is guaranteed because it's practically sent directly from the branch

to the storage. This slows down and complicates the process of fruit picking, resulting in expensive production and therefore it is justified to introduce robots in these technological tasks.

## 2 State of the Art

Even though an interest in agricultural robotics is growing rapidly there are no universal technical solutions for fruit harvesting [1]. All robot solutions of fruit harvesters known to date for picking and collecting fruit can be categorized into three basic groups: (i) shakers with collectors [14] (for olives, almonds, blueberries and raspberries and hanging fruits), (ii) fruit vacuum collectors [4–7] (for apples, pears, quinces, oranges, kiwis, etc.), and (iii) pickers (graspers) piece by piece using specialized grippers [8–14] (for strawberries, plums, apricots, peaches, cherries, etc.). There are several start-up companies and Institutes that develop picking and harvesting systems. Most promising one is Spanish E-Series [2]. This system is specialized in strawberry harvesting [3], but the mechanism is completely different than our proposed in this paper. Several other companies introduced various harvesting robotic arms, but their grasping method is different (vacuum pumps [4]-[7], various shakers [15], pickers [8–14], etc.). These types of robots are different both in harvesting operation technology and mechanism itself. The mechanism presented in this paper is based on some existing ideas but upgraded by innovative telescopic arm with specialized gripper for gentle and careful grasping of sensitive fruits.

## 3 Concept Design

This system imitates the basic principles of manual operation picking plums that does it from a mobile platform for fruit picking. Mechanical part of this robotic system consists of several functional modules:

1. Mobile platforms for transporting (moving) harvesting robot arms around the plum field. The mobile platform also serves as a carrier for the robotic system and
2. Linear (Gantry) robot for moving the telescopic robot arm in two coordinate direction - vertical (up-down) and horizontal (left- right), both within the workspace;
3. Telescopic robotic arm with a specially designed gripper for cutting and picking fruits;
4. Vertical transport system (freight lift) for delivery/removal crate/box for packing harvested fruits.

The robot is mounted on a mobile transport platform, where multiple gantry robots can be mounted. On the horizontal the linear joint (for left-right movement) of the gantry robot, so-called telescopic robotic arm is mounted. It can be extended and retracted according to the nominal depth of workspace (x-direction). When the telescopic arm is retracted (the end parking position), it has a length slightly greater than 1.0 m. In extracted position, it can reach 3 m. That is enough for the robot hand to reach center of the tree crown. At its end, the telescopic arm carries a two-fingered segmented gripper and which serves to accept the thin, green stem (petiole) of the fruit plums. The lower segment of the grip resembles a small basket that opens like a basket shell or flower and serves to accept the picked plum so that it does not fall out. The gripper opens and closes with a complex mechanism. Caught fruit is then carried to the place of storage, which is usually a plastic crate or cardboard box.

## 4 Cartesian Robot – 3D Space Positioner

Cartesian coordinate robot serves for rough positioning of the robot arm relative to the chosen target, in this case the target is the plum spotted with a depth camera. Cartesian robot represents a linear moving system along three coordinate axes - vertical (for movement up and down), lateral (for left-right movement) and longitudinal (for back and forth movement). Cartesian coordinate robot has one electric motor on each axis that, with the support of a timing belt, enables the transmission of power and achieves linear movement in the required direction (Fig. 1).

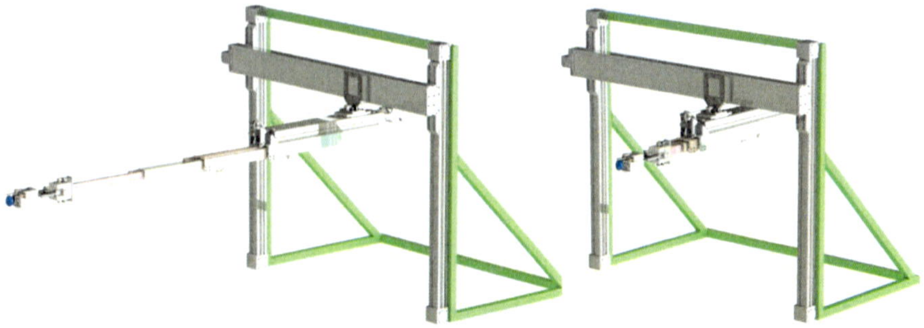

**Fig. 1.** 6DOFs robot harvester with Cartesian coordinate robot for basic positioning of the telescopic arm with a customized gripper – extended position (left) and retracted position (right)

## 5 Telescopic Arm – Approach to the Target Point

Taking into account the complexity of plum harvesting technology, as well as the size and shape of the trees, the most favorable configuration of the robot arm for that purpose is a telescopic robotic arm. Telescopic arm (Fig. 2), with its possibility of pulling in (Fig. 3) and, can be extended and retracted ±500 mm. Plum fruits that are visible by global cameras (stereovision) can be reached with a telescopic arm.

Considering that the robotic arm should enter the "depth" of the tree's crown and at the same time passes through a tangle of branches, it must be sufficiently robust and possess elements of elasticity in its structure so it would not be mechanically damaged during harvesting operation. The arm has three segments which can be moved linearly relative to each other (Fig. 2, middle). Segments are made of standard aluminum squared tubes so that their cross-sectional dimensions are different and can be inserted into each other.

Previously mentioned telescopic arm consists of one linear motor and three dependently actuated segments. These segments act as a serial pulley mechanism where each segment is attached to timing belt that's wrapped around previous pulley. In other words, former pulleys actuate latter ones. In this way a velocity addition is obtained where each segment has incremented velocity from previous'. If we take that the nominal motion of linear motor is equal to $v_0$, last segment will have a velocity of $4 \times v_0$. . Since each segment starts and stops at the same time, e.g. time cycle is same for each segment, the tip of last segment (end-effector) will move four time as much as linear motor. In this way, a single actuator enhanced with a complex motion mechanism covers four times larger distance than original one.

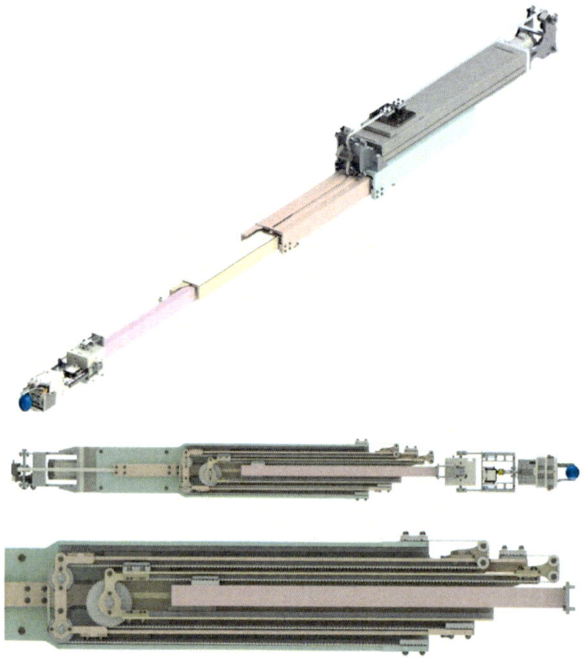

**Fig. 2.** Telescopic robot arm with gripper (top); View to the arm with the extending/retracting mechanism (middle); Principle of actuation (bottom).

## 6 Gentle Gripper

The gripper of the plum harvesting robot is a customized end-effector. It was developed in a way that it resembles a spherical joint, just like a human wrist. It is mounted on the top of the telescopic arm (Fig. 3).

To achieve maximum precision and exploitation of the workspace, this end-effector has three independently actuated rotary joints. The assembly of the gripper (Fig. 4) consists of three functional parts:

- Gripper with a camera, shown in bottom left corner in Fig. 4

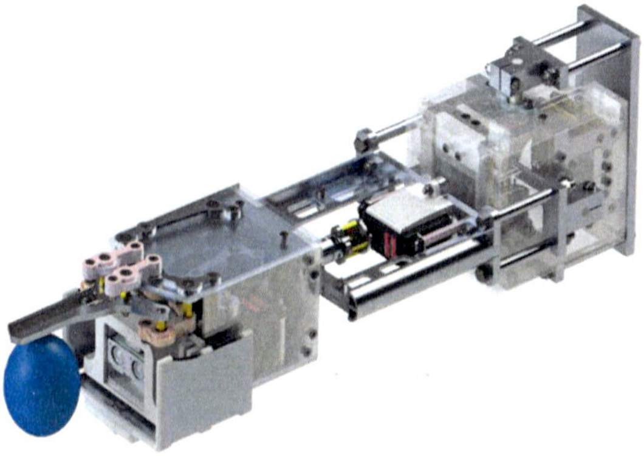

**Fig. 3.** Specially designed gripper mounted to the end of the telescopic robot arm

- Rotation along longitudinal axis shown in the center of the image
- Spherical joint for pitch and yaw motion, upper right corner (Fig. 3)

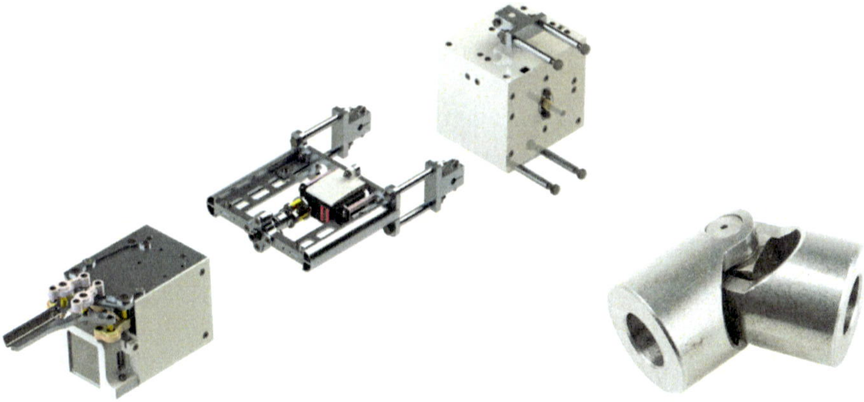

**Fig. 4.** Subassemblies of the end effector that correspond for, not only raw, pitch and yaw movement, but fruit grasping as well (left); A standard universal joint that can be found in carriage axles and shaft connection (right)

Spherical joint is based on the principle of a universal joint (Fig. 4). In this case the actuation comes inside of connecting part (in the middle) and connected parts are the ones being moved. In the Fig. 5 a spherical joint is depicted. On the left side a close up view of the joint with plastic housing mounted is presented. On the right side, the mechanism itself is shown. Actuating mechanism is powered by small RC-servo motors and timing belts. Additionally a belt tensioner is introduced to increase robot's performances, e.g. its precision and accuracy.

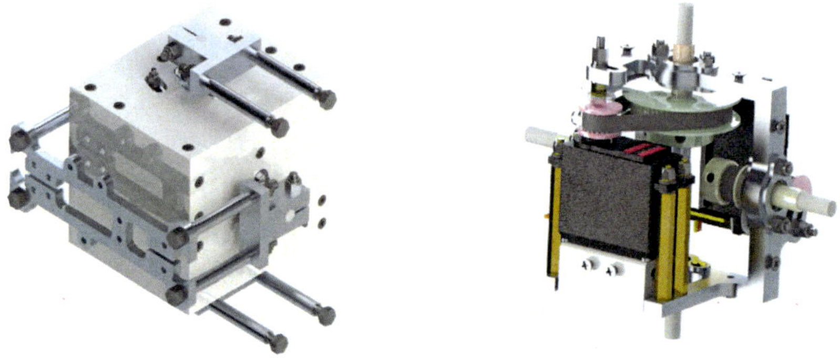

Fig. 5. Model of spherical joint with and without plastic housing.

## 7 Operation Principle

A tractor pulls the moving platform, which has several robotic arms on the left and right side. Each robot arm replaces one worker with multiple times better efficiency and speed of harvesting. Tractor stops at every 1–2 m to allow the robots to harvest that part of the field. When the platform stops, global cameras scan the entire sector of the workspace. Based on newly-created digital maps of the sector, the robot controller computes trajectories for each arm. If there is a small angular deviation alongside longitudinal axis of the telescopic arm, which carries a gripper at its end, a spherical wrist performs fine adjustment around raw, pitch and yaw axes.

The hand continues its predefined path to the target and at a distance of just 20–30 mm the gripper jaw opens. The gripper carefully grasps the plum by its peduncle. With the plum grasped, the telescope reaches its end retracted position in the bottom end position of the gantry robot carrier (the lowest position). The gripper jaws open and the plum is carefully placed in the crate. The process is repeated every 10–15 s.

## 8 Conclusion

This paper shows the entire concept of mechanical design of the manipulative robot for plum harvesting. The system consists of a supporting steel frame to be fixed to the tractor trailer, Cartesian coordinate robot (linear positioning system) with a telescopic robotic arm and end effector for cutting and grasping. Given solution is a prototype based on technical solutions and previously performed feasibility studies. All parts of the robot, except gantry robot system (which is imported), are original technical solutions subjected to the patent application.

At this phase of research, the system is pulled by a tractor since at the moment battery drives are under consideration and the autonomy of operation should be set to 8 h of continual operation on the flat terrain.

**Acknowledgments.** This work is the result of research and development carried out on the project "Development of a demonstration prototype of a manipulative robot for picking edible plums - proof of concept", contract no. 623–1-22, innovation voucher no. 1216 co-financed by the Innovation Fund of the Republic of Serbia, 2022.

## References


1. www.theenglishappleman.com/journal_2013-03-22-The-English-Apple-Man- and Under-40s-at-Skiernierwice.asp. Accessed 24 Jan 2023
2. https://www.agrobot.com. Accessed 24 Jan 2023
3. https://www.agrobot.com/e-series. Accessed 24 Jan 2023
4. https://www.nag.co.za/2019/03/26/how-do-you-like-them-robot-harvested-apples/. Accessed 24 Jan 2023
5. https://www.researchgate.net/figure/The-ARCO-robot-developed-in-Belgium-for-apple-fruit -picking-the-white-hose-end_fig3_267809044. Accessed 24 Jan 2023
6. https://www.newshub.co.nz/home/rural/2019/03/watch-world-s-first-commercial-robotic-apple-harvest-underway.html. Accessed 24 Jan 2023
7. https://www.wisfarmer.com/story/news/2021/02/25/israeli-ag-company-introduces-artificial-pollination-technology/4364725001/. Accessed 24 Jan 2023
8. https://www.naro.go.jp/english/topics/laboratory/fruit/138191.html. Accessed 24 Jan 2023
9. https://www.capitalpress.com/ag_sectors/orchards_nuts_vines/israeli-company-works-on-robotic-apple-picker/article_e2f4f29d-fdc8-52cf-9fe5-4dda8667fbdc.html. Accessed 24 Jan 2023
10. https://www.growingproduce.com/fruits/orchard-automation-is-on-the-horizon/. Accessed 24 Jan 2023
11. https://www.therobotreport.com/abundant-robotics-shuts-down-fruit-harvesting-business/, last accessed 2023/01/24
12. https://www.slow-journalism.com/delayed-gratification-magazine/easy-pickings-how-robot-farm-hands-could-revolutionise-agriculture. Accessed 24 Jan 2023
13. https://siamagazin.com/the-root-ai-virgo-tomato-harvesting-robot/. Accessed 24 Jan 2023
14. https://www.sciencelearn.org.nz/images/2566-kiwifruit-harvester-robot. Accessed 24 Jan 2023
15. https://www.agriexpo.online/prod/sicma/product-179641-151471.html. Accessed 24 Jan 2023

# Comparison of Some Mechanical Amplifiers for Micro-robotic Devices

Jaroslav Hricko(✉) and Stefan Havlik

Institute of Informatics, Slovak Academy of Sciences, Banská Bystrica, Slovakia
{hricko,havlik}@savbb.sk

**Abstract.** Mechanical amplifiers are passive compliant mechanisms for increasing motions of actuators, mainly piezoelectric, frequently used in micro-robotic devices. This paper brings a short overview of principal mechanical structures and gives some recommendations that should consider in the design phase. The performance characteristics of some selected mechanical structures of amplifiers are simulated and compared.

**Keywords:** Compliant mechanisms · piezo actuator · mechanical amplifiers · micro-robotics · elastic materials

## 1 Introduction

The micro-robotic devices are usually constructed as compliant elastic structures powered by linear motion actuators on the input port of mechanisms. There are several actuators suitable for this purpose [1]: thermal, SMA (shape memory alloys), piezoelectric and magnetic. For cases of the quasi-static operations of such robotic devices, all mentioned types of actuators are applicable. However, with demands for higher working frequencies, the SMA and thermal actuators achieve their applicability is very limited. Therefore, when a higher dynamic response of the system is required, using piezoelectric transducers (PZT) or piezo-stack actuators (PSA) are the best solutions. The PSAs provide very high positioning accuracy together with working frequencies up to several kilohertz (kHz). They can give forces up to several kilonewtons (kN), but their main drawback is the small range of output motion that must be mechanically increased. The available motion of such an actuator represents the ratio of the PSA length, which is about 0.1% [2]. Then, to get the desired range of motion an additional motion amplification mechanism is needed.

Another possible solution is using piezoelectric motors. They work on the principle of repeating a series of small periodic steps and provide long-range motion strokes (up to several millimetres) with nanoscale positioning accuracy. The motion consists of sequential steps of a quasi-static motor generated by the ultrasonic frequency. Theoretically, quasi-static motors can achieve an unlimited range of motion by an accumulation of micro-step motions. The main drawback is the low output force [3]. Further research [4, 5] orients to piezo-stack actuators connected with a compliant mechanism that uses friction for moving on the linear guidance. They are denoted as stick-slip actuators. An


T. Petrič et al. (Eds.): RAAD 2023, MMS 135, pp. 478–485, 2023.
https://doi.org/10.1007/978-3-031-32606-6_56

example is the piezoelectric linear actuator that imitates skateboarding movements with a resolution of 45 nm, load capacity of 2 N and speed of approximately 10 mm/s [4]. The advantage of such actuators lies in the adjustable size of the step. For instance, the step 102 μm with resolution12 nm was achieved [5]. One of the drawbacks of such types of actuators is serrated output motion that can be unwanted in some applications.

The motion amplifiers are usually the flexure-based mechanisms designed for a given amplification ratio with keeping the positioning accuracy. Anyways, the insertion of a compliant mechanical part results in lower working frequencies. From the structural point of view, the most frequently used are two basic configurations: the lever type and the triangular type or bridge mechanism [2]. The two-lever amplifier is studied in [8, 9]. Well known are bridge/rhombus type amplifiers and Scott-Russell linkages [6, 7]. The standard value of the amplification ratio in both basic types of amplifiers is about 5.

## 2 Related Research

### 2.1 Structures of Mechanical Amplifiers

There are numerous applications where the principal requirement is to achieve higher values of motion amplification under desired frequency range. For this purpose, hybrid or multistage displacement amplification mechanisms are considered. Such devices include a combination of both basic structures of amplifiers, mutually connected. The output stroke of the first part is then the input to the second part of the mechanism. Frequently, it is the combination of a lever and a bridge mechanism. The combination of differential lever and half-bridge mechanisms [10] provides a higher amplification ratio of 6.51, compared to the bridge amplifier of equal dimensions; but the natural frequency is increased from 95.5 Hz up to 252.6 Hz. For amplifiers based on a combination of the lever and bridge mechanisms the influence of dimensions (compliant hinges and arms) on the resonant frequency is analyzed [11]. When combining level and bridge mechanisms into one compact structure a large amplification ratio (up to 48) can be achieved [12]. Further designs combine multiple lever mechanisms [13] and the nested rhombus multi-layer mechanisms [14].

Although the combination of only two common types of amplifiers is most widely used, some recent designs consider multilevel configurations. To magnify the workspace of the micro-robotic platform the combination of the Scott-Russell and two lever mechanisms is applied [15]. The design of a three-stage displacement amplifier that combines the bridge, lever, and half-bridge mechanisms presents [16]. Here, the high amplification ratio and the relatively high own (resonant) frequency was achieved.

This work aims to point out some characteristics that should be taken into account in designing and solving mechanical motion amplifiers in connection with PSA actuators (PZT). This work focuses on comparing the characteristics of selected hybrid/multistage displacement amplifiers. All compared mechanisms will be adapted for equal input displacement produced by available PSA and with a similar amplification ratio. The work aims to select (one) suitable mechanical amplifier that will apply to our other devices (positioning stages [9, 19], gripers [20] and deformable bodies of sensors [21]) that work on the principle of elastic deflections.

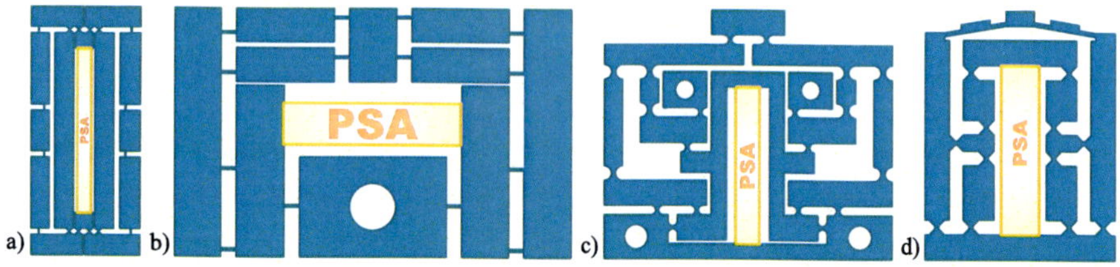


**Fig. 1.** Selected displacement amplification mechanisms a) lever-bridge displacement amplifier [12], b) amplifier with differential levers and half-bridge mechanism [10], c) symmetrical differential lever mechanism [13], d) three-stage displacement mechanism [16]

## 2.2 Characteristics and Performance Specifications

Any mechanical amplifier/reducer transforms the values of displacements **d**, or forces $f$, on the input side of mechanisms into their lower/higher levels on the output side (see Fig. 2). Let us briefly discuss some design parameters that characterize the performance of such mechanically compliant mechanisms.

The basic parameter describing transformation is the amplification/reduction ratio: as for the incremental displacement $r_d$, and the force $r_f$

$$r_d = \frac{\Delta d_{out}}{\Delta d_{in}} \text{ and } r_f = \frac{f_{out}}{f_{in}} \text{ where obviously } r_d = \frac{1}{r_f} \quad (1)$$

This ratio represents the actual sensitivity of transformation for the small displacements. When considering relatively small values input motions in cases of piezo actuators (PSA), the displacement amplification ratio is practically stable in the whole range of PSA motion. Then the available force on the output side corresponding to the actual displacement of the actuator $d_{in}$ is

$$f_{out*} = f_{out} - s.d_{in} \quad (2)$$

where $s$ - is the stiffness of the mechanism in sense of the motion.

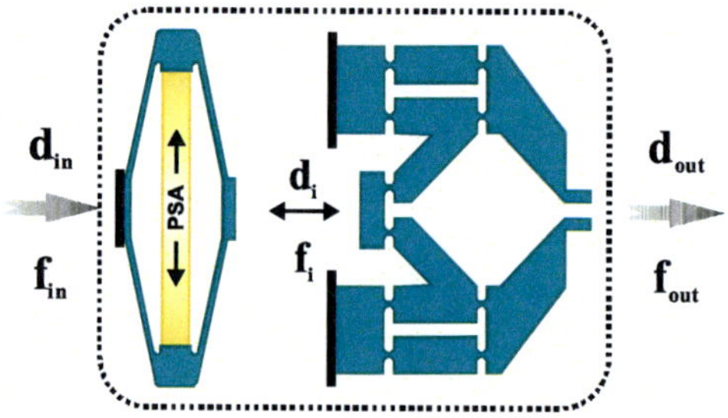


**Fig. 2.** Scheme of the compliant micro-device

The general assumption for building compliant structures for this purpose is the linearity of stress-strain characteristics in all elastic segments (joints, arms); i.e. the validity of Hooke's low within the whole range of motion/force. Then, the displacement on the input side results in the change (±) of internal stress energy inside the structure

$$\Delta E = \frac{1}{2}s.\Delta d_{in}^2 = \sum\nolimits_{(1 \div n)} \frac{1}{2}s_i.\Delta\delta_i^2 \tag{3}$$

where $\delta_i$ - are elastic deflections/strains of particular segments.

One principal requirement in using any additional mechanisms is that losses of driving force and energy should be minimal. This fact naturally leads to the design of a relatively complex structure with an equal level of elastic stress in all segments. This concept results in different thicknesses of joints/arms. Such a mechanism is suited for devices working in quasi-static operations.

The contradictory requirement is for amplification mechanisms working in dynamic modes where motion accuracy within a given operating frequency is needed. This mechanism represents the mass-spring model with only the structural/zero damping. For the purpose of a short analysis [17], we specify.

The relative amplitude error

$$\xi(\lambda) = \frac{S - S_0}{S_0} = S_\lambda - 1 \tag{4}$$

Here; S - is the total frequency response, $S_o$ - is the response corresponding to static mode and $S_\lambda$ – represents frequency dependent response.

When denoting ω - the operation frequency, $\omega_0$ - the own/natural frequency of free mechanisms oscillations and $\lambda = \omega/\omega_0$, the relative amplitude error is

$$\xi(\lambda) \div \lambda^2 \tag{5}$$

Then, at the stage of the first design, the requirement for the amplifier dynamics can be formulated as follows: Let $\omega$ be the supposed maximal frequency of the input displacement oscillations. To get the expected accuracy within the whole frequency range *0* - $\omega$, the minimal free oscillations frequency of mechanisms should be

$$\omega_0 \geq \frac{\omega}{\sqrt{\xi}} \tag{6}$$

## 2.3 Materials and Design Consideration

The basic indication is the input stroke made by the PSA. For example, this parameter of a real PSA is 41 μm (53 μm) for the input voltage range 0 V–150 V (−30 V–150 V). Dimensions are 5 × 5 × 36 $mm^3$ for whole piezo ceramics, and 7.3 × 7.3 × 41 $mm^3$ with ball ends, coated epoxy and wires. The stiffness is 24 N/μm, and the blocking force is 1000 N. The PSA is equipped with a strain-gauge bridge that enables control in closed-loop and minimizes the hysteresis of the PSA.

A material of the proposed mechanism will be considered spring steel 34CrNiNo6 with Young's modulus E = 210 GPa and Yield strength (allowable stress) 900 MPa. Such material is stiff enough and enables simple replacement (without mechanism dimensions changing) by softer materials like aluminium or thermoplastic that are used in additive manufacturing (3D print).

It is proposed the displacement amplification ratio will be around 20, which means the maximum output stroke of the mechanism is between 800 μm and 1 mm. Considering the dimensions of PSA, the minimal thickness of the whole mechanism should be 8 mm. The width and height of the mechanisms will be one of the evaluation criteria, where the minimal surface of the mechanism will have a high impact on a final decision.

To achieve high variability of the used flexure hinges, and satisfaction of requirements to the minimal losses of driving force and energy, the elliptical notch hinges are considered. These types of notches enable the creation of a right circular or corned-filleted flexural hinge that offers minimal parasitic movements of the centre of rotation or high flexural capacity, respectively. The characteristic dimension of the flexure hinge is its thickness, which is usually the boundary constraint of manufacturing technology. In our case is considered with 0.2 mm.

## 3 Comparison of Selected Structures

Four different types of hybrid displacement amplifiers are considered.

- Lever-bridge amplifier developed by Dong et al. [12] is shown in Fig. 1a)
- Amplifier with differential levers and half-bridge mechanisms developed by Ma et al. [10] shown in Fig. 1b)
- Differential lever amplifier inspired by Dang et al. [13] is shown in Fig. 1c)
- Three-stage displacement mechanism developed by Ling et al. [16] see Fig. 1d)

Adjustments were made to the proposed mechanisms to achieve the desired parameters. The kinematic analysis was made and characteristic dimensions of individual links (lever, bridges) were calculated. This provided the maximum angles of individual joints used in particular compliant hinges. To achieve minimal losses of a driving force has used Eq. (7), which expresses the dependence between admissible elastic strain $\varepsilon_{adm} = \sigma_{yield}/E$ and maximal angular deflection of flexure hinge $\varphi_{\max}$ [18]

$$|\varphi_{max}| = \frac{\varepsilon_{adm}}{6k_{M1}}\left(\frac{\beta_l}{\beta_h}\right)^{k_{M2}} + \frac{\varepsilon_{adm}}{12(1-k_{crit})k_{F1}}\left(\frac{\beta_l}{\beta_h}\right)^{k_{F2}} \tag{7}$$

where $\beta_l$, and $\beta_h$ are characteristic dimensions of flexure hinge, and $k_{M1}$, $k_{M2}$, $k_{crit}$, $k_{F1}$, and $k_{F2}$ are constants expressed by Linss et al. [18]. Flexural analysis of mechanisms was calculated by FEM, where particular dimensions were adjusted. An example of an adjusted lever-bridge amplifier is shown in Fig. 3. Considering symmetry only four hinges were analysed. For instance; the hinge on the lever mechanism (Detail A) is a right circular hinge with a radius of 0.4 mm and a thickness of 0.5 mm. The joints of the bridge have thicknesses of 0.4 mm and 0.5 mm (Detail B and C), but they have different lengths of 1.8 mm and 3.0 mm. The results of all proposed mechanisms are summarised in Table 1.

From the results in Table 1, can be concluded, that all proposed mechanisms are suitable as displacement amplification devices. They achieved desired displacement amplification ratio and sufficient resonant frequencies for work in dynamic operation modes

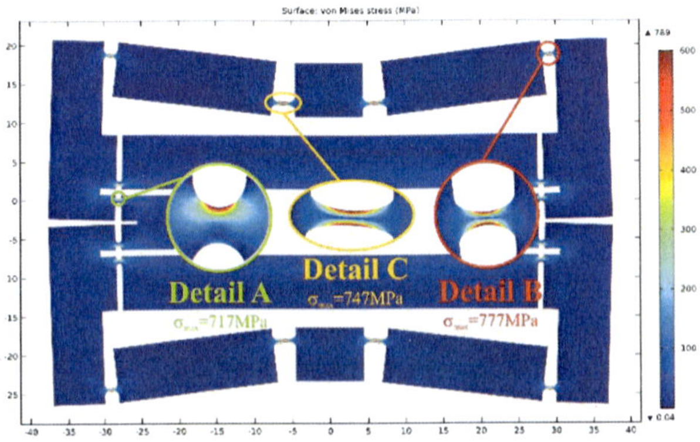


**Fig. 3.** Stress distribution in lever-bridge amplifier (displacement is 5× gained)

**Table 1.** Characteristic parameters of analysed displacement amplification mechanisms.

| Mechanism | Dimensions [mm] | Surface [mm$^2$] | Amplification ratio | Natural frequency |
|---|---|---|---|---|
| Lever-bridge amplifier | 75 × 46.6 | 3495 | 20.94 | 152.7 Hz |
| Diff. Levers and half-bridge | 67 × 30 | 2010 | 19.64 | 561.61Hz |
| Differential lever amplifier | 71.25 × 69.1 | 4923.375 | 19.00 | 338.9 Hz |
| Three-stage displacement mech | 65.5 × 56.9 | 3726.95 | 20.79 | 235.26 Hz |

too. In general, all analysed devices are satisfactory, and to evaluate the most suitable for our purposes. It is necessary to focus on specific requirements and possibilities that are connected with our laboratory and test bed. Considering that desired stroke, stiffness and dynamical properties were achieved, only dimensions, the capability of preload of PSA and other small requirements can be taken into account in device selection. The differential levers and half-bridge mechanism and three-stage displacement mechanism seem as suitable amplifiers from the studied devices. The first one achieves the smallest dimensions and highest dynamical properties, and the second enables simple adding preload screw, to achieve better dynamical properties of the used PSA.

## 4 Conclusion

Any amplification mechanism is designed for a tool and desired specifications (range of motion, force, accuracy) on the input port with respect available actuator with given parameters (max. Range of motion/stroke, accuracy, force). The common design parameter for these devices describing transmission is the motion amplification ratio. The desired amplification can be reached by choosing the suitable structure of mechanisms and their combination. Such a relatively simple approach is for cases when the motions of mechanisms are slow/quasi-static and minimal loss of energy is required. As soon as the device should work in dynamic operation modes there are further criteria, that the mechanisms must satisfy. As it is shown, the final accuracy of the positioning device is related to the resonant/own frequency of the mechanisms. To get the desired accuracy, it must have an adequate resonant frequency, which corresponds to the total stiffness of the device. So, therefore are opposite conditions, the design must satisfy. Then, a compromise solution must be adopted.

The paper brings evaluation of four different structures of multi-stage displacement amplifiers that were taken to compare their characteristics from references. For this purpose, the unique amplification ratio (20) was stated. As shown in comparisons of calculated characteristics in Table 1, there are all convenient structures for the given amplification ratio and higher operating frequency.

**Acknowledgement.** The Slovak Scientific Grand Agency VEGA under contract No. 2/0135/23 – "Intelligent sensor systems and data processing" supported this work. It was supported too by projects APVV-20-0042 – "Microelectromechanical sensors with radio frequency data transmission" and APVV-14-0076.

## References

1. Bučinskas, V., Subačiūtė-Žemaitienė, J., Dzedzickis, A., Morkvėnaitė-Vilkončienė, I.: Robotic micromanipulation: a) actuators and their application. Robot. Syst. Appl. **1**(1), 2–23 (2021). https://doi.org/10.21595/rsa.2021.22071
2. Chen, F., Zhang, Q., Gao, Y., Dong, W.: A review on the flexure-based displacement amplification mechanisms. IEEE Access **8**, 205919–205937 (2020). https://doi.org/10.1109/ACCESS.2020.3037827
3. Peng, Y., Peng, Y., Gu, X., Wang, J., Yu, H.: A review of long range piezoelectric motors using frequency leveraged method. Sens. Actuat. A: Phys. **235**, 240–255 (2015). https://doi.org/10.1016/j.sna.2015.10.015. ISSN 0924-4247
4. Wang, K., Li, X., Sun, W., Yang, Z., Liang, T., Huang, H.: A novel piezoelectric linear actuator designed by imitating skateboarding movement. Smart Mater. Struct. **29**(11) (2020). https://doi.org/10.1088/1361-665X/abb357
5. Wang, F., Zhao, X., Huo, Z., Shi, B., Tian, Y., Zhang, D.: A novel large stepping-stroke actuator based on the bridge-type mechanism with asymmetric stiffness. Mech. Syst. Signal Process. **179**, 109317 (2022). https://doi.org/10.1016/j.ymssp.2022.109317
6. Ling, M., et al.: Enhanced mathematical modeling of the displacement amplification ratio for piezoelectric compliant mechanisms. Smart Mater. Struct. **25**, 075022 (2016). ISSN 1361-665X

7. Zheng, H., et al.: Simulation and experiment of a diamond-type micro-displacement amplifier driven by piezoelectric actuator. J. Eng. **2020**(5), 141–147 (2020). https://doi.org/10.1049/joe.2019.0854
8. Chen, G., Ma, Y., Li, J.: A tensural displacement amplifier employing elliptic-arc flexure hinges. Sens. Actuat. A: Phys. **247**, 307–315 (2016). ISSN 0924-4247
9. Hricko, J., Havlík, Š: Compliant mechanisms for motion/force amplifiers for robotics. In: Berns, K., Görges, D. (eds.) RAAD 2019. AISC, vol. 980, pp. 26–33. Springer, Cham (2020). https://doi.org/10.1007/978-3-030-19648-6_4. ISSN 2194-5357
10. Ma, Y., Feng, Z., Cong, M., Sun, Y.: The design, modeling and simulation analysis of a novel positioning stage with an amplifier. In: Duan, B., Umeda, K., Kim, Cw. (eds.) Proceedings of the Eighth Asia International Symposium on Mechatronics. Lecture Notes in Electrical Engineering, vol. 885, pp. 2075–2086. Springer, Singapore (2022). https://doi.org/10.1007/978-981-19-1309-9_193
11. Ling, M., Yuan, L., Luo, Z., Huang, T., Zhang, X.: Enhancing dynamic bandwidth of amplified piezoelectric actuators by a hybrid lever and bridge-type compliant mechanism. Actuators **11**, 134 (2022). https://doi.org/10.3390/act11050134
12. Dong, W., et al.: Development and analysis of a bridge-lever-type displacement amplifier based on hybrid flexure hinges. Precis. Eng. **54**, pp. 171–181 (2018). https://doi.org/10.1016/j.precisioneng.2018.04.017
13. Fan, W., Jin, H., Fu, Y., Lin, Y.: A type of symmetrical differential lever displacement amplification mechanism. Mech. Ind. **22**, 5 (2021). https://doi.org/10.1051/meca/2021003
14. Ueda, J., Secord, T.W., Asada, H.H.: Large effective-strain piezoelectric actuators using nested cellular architecture with exponential strain amplification mechanisms. IEEE/ASME Trans. Mechatron. **15**(5), 770–782 (2010). https://doi.org/10.1109/TMECH.2009.2034973
15. Dang, M. P., Le, H.G., Chau, N.L., Dao, T.-P.: An optimized design of New XYθ mobile positioning microrobotic platform for polishing robot application using artificial neural network and teaching-learning based optimization. Complexity **2022**, 1–20 (2022). https://doi.org/10.1155/2022/2132005. Article ID 2132005
16. Ling, M., Zhang, C., Chen, L.: Optimized design of a compact multi-stage displacement amplification mechanism with enhanced efficiency. Precis. Eng. **77**, 77–89 (2022). https://doi.org/10.1016/j.precisioneng.2022.05.012
17. Havlik, S., Hricko, J.: Some quality measures in designing compliant mechanisms for robotic devices. In: Zeghloul, S., Laribi, M.A., Sandoval Arevalo, J.S. (eds.) RAAD 2020. MMS, vol. 84, pp. 438–447. Springer, Cham (2020). https://doi.org/10.1007/978-3-030-48989-2_47. ISSN 2211-0984
18. Linß, S., Schorr, P., Zentner, L.: General design equations for the rotational stiffness, maximal angular deflection and rotational precision of various notch flexure hinges. Mech. Sci. **8**(1), 29–49 (2017). https://doi.org/10.5194/ms-8-29-2017
19. Hricko, J., Havlik, S.: Design of the 2 D.o.F compliant positioning device based on the straight-line watt's mechanisms. In: Zeghloul, S., Laribi, M.A., Arsicault, M. (eds.) MEDER 2021. MMS, vol. 103, pp. 247–255. Springer, Cham (2021). https://doi.org/10.1007/978-3-030-75271-2_26. ISBN 978-3-030-75270-5, ISSN 2211-0984
20. Hricko, J., Havlik, S.: Design of compact compliant devices – mathematical models vs. experiments. Am. J. Mech. Eng. **3**(6), 201–206 (2015). https://doi.org/10.12691/ajme-3-6-9
21. Hricko, J., Havlik, S.: Flexural body for a wireless force/displacement sensor. In: Zentner, L., Corves, B., Jensen, B., Lovasz, EC. (eds.) Microactuators and Micromechanisms. Mechanisms and Machine Science, vol. 45, pp. 59–66. Springer, Cham (2017). https://doi.org/10.1007/978-3-319-45387-3_6. ISBN 978-3-319-45386-6, ISSN 2211-0984

# Robotics and Reconfigurable Mechanisms in Architecture Towards the Synthesis and Control of Reconfigurable Buildings

Eftychios G. Christoforou[1(✉)], Marios C. Phocas[2], Andreas Müller[3], Maria Matheou[4], and Loukas Georgiou[1]

[1] Department of Mechanical and Manufacturing Engineering, University of Cyprus, Nicosia, Cyprus
e.christoforou@ucy.ac.cy
[2] Department of Architecture, University of Cyprus, Nicosia, Cyprus
[3] Institute of Robotics, Johannes Kepler University, Linz, Austria
[4] Institute of Lightweight Structures and Conceptual Design, University of Stuttgart, Stuttgart, Germany

**Abstract.** The synthesis and control of a closed–chain reconfigurable mechanism is considered, which finds direct applications in reconfigurable architecture. This term refers to building structures whose shape can be adjusted towards certain objectives including energy efficiency and improved occupants' comfort. The structural concept for the building is presented together with a corresponding control methodology, which allows for energy–efficient reconfigurations as well as flexibility in motion planning. The basic concepts are presented and then exemplified through simulation and experimental studies.

**Keywords:** Reconfigurable Architecture · Reconfigurable Mechanisms · Mechanism Synthesis · Motion Planning and Control

## 1 Introduction

Architectural design has been primarily associated to fixed civil engineering structures. This continues to be the prevailing architectural design philosophy today. However, the use of mechanisms in architecture has attracted attention as part of kinetic facades, adjustable shading elements but also in relevance to foldable, erectable and transportable structures. Reconfigurable buildings provide unique advantages including shape adjustments in response to changing use requirements, improved energy efficiency (bioclimatic architecture), more comfortable conditions for the occupants, optimal performance of a photovoltaic roof, and reduced structural loads from aerodynamic wind forces. Building motions may also be used to shake snow off the roof.

Previously proposed implementations of reconfiguragle structures involved mechanisms, tensegrity structures, and the use of scissor elements [1]. It is noteworthy that inspiration for building reconfigurability was often drawn from artistic origami creations [2,3], which has also attracted attention from a robotics


T. Petrič et al. (Eds.): RAAD 2023, MMS 135, pp. 486–495, 2023.
https://doi.org/10.1007/978-3-031-32606-6_57

perspective [4,5]. Reconfigurable robotics itself is an established research area. Relevant topics include modularity and reconfigurability [6–8], expandability [9], and deployability [5]. Research has also focused on the self–X capabilities of robotic systems (self–reconfiguration, self–assembly, self–adaptation, etc.) [10], which are also relevant from the reconfigurable buildings perspective.

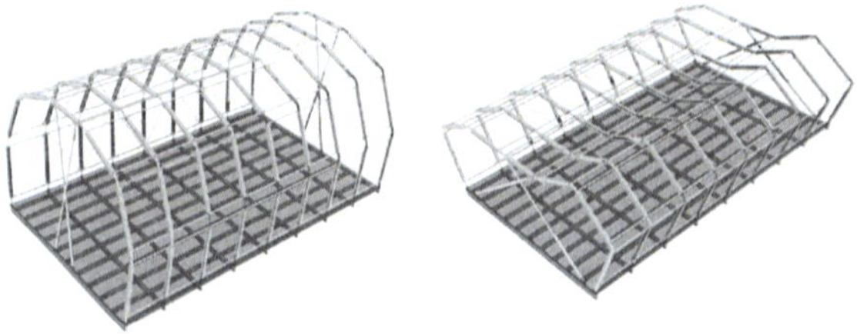

**Fig. 1.** The building structural concept. Skeleton structure at two different configurations.

In previous work by the authors [11], two structural and reconfiguration concepts were proposed, namely the *effective four–bar* and the *effective crank–slider* methods. These were further examined in subsequent works, which also included cable–actuated versions [12,13]. The integration of photovoltaic elements on a reconfigurable building was examined in [1], towards *heliotropic buildings* implementations. In the present work a different base mechanism is proposed for the reconfigurable structure and a control approach is shown to apply, as presented in Sect. 2. The motion planning and control approach is further examined in Sect. 3 through simulation and experimental studies, while the last section presents the conclusions.

## 2 Reconfigurable Building Structure and Control

### 2.1 Building Structure

The proposed reconfigurable building idea refers to a lightweight structure consisting of a parallel arrangement of planar mechanisms. Interconnections between them form the skeleton structure of the building. Depending on the use requirements different elements can be installed over it to define the building envelope, including shading elements and photovoltaic panels. An example of the architectural concept is depicted in Fig. 1. The building is shown at two different configurations, while no mechanical and actuation system details were included. Note that the two particular building shapes correspond to the simulation and experimental study presented in Sect. 3. Coordinated motion of the individual planar mechanisms accordingly adjusts the overall building shape. The basic planar mechanism consists of a slider block with two chains of serially–connected links on either side which are also pinned to the ground at fixed positions, as shown in Fig. 2a. The slider is restricted to move in the vertical direction while actuation adjusts its position. All individual rotational joints are equipped with brakes (e.g., electromagnetic or hydraulic), which are utilized by the reconfiguration control method as explained below. This planar mechanism is the basic structural and control element of the building and for this reason the rest of the paper focuses on this mechanism and its reconfiguration.

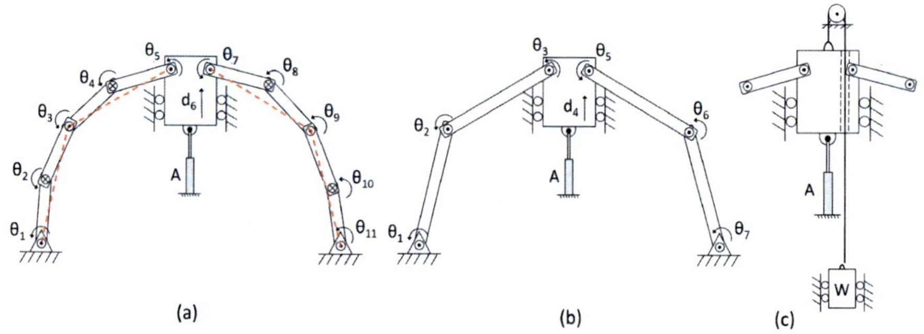


**Fig. 2.** The basic planar mechanism. (a) Basic structural and kinematics element of the building, (b) Corresponding coupled effective crank–slider mechanism (c) Counterbalancing system concept (⊗: locked joint, ⊙: unlocked joint, △: pivoted–to–the–ground joint, – – –: effective link).

## 2.2 Control Concept

The basic structural mechanism described above is a multibody system with many degrees–of–freedom (DOF). Control using conventional methods can be particularly complex and involve a significant number of actuators. This would also increase the structure's own weight leading to energy inefficient operation. To overcome these issues an alternative control approach is proposed which effectively employs the joint brakes.

By selectively applying the brakes so that only three joints on either side of the slider remain unlocked, the mechanism becomes a 1–DOF system and in principle one motion actuator suffices for position control. Note that application of the brakes between consecutive links results to a corresponding *effective link*, i.e., a continuous and stiff member. This is demonstrated in Fig. 2b, where the resulting mechanism practically reduces to two basic crank–slider mechanisms which share a common slider. The slider provides the kinematic coupling between the two parts. This system will be referred to as a *coupled effective crank–slider* (coupled ECS) mechanism. This allows for a conceptually simple multistep reconfiguration method to be implemented: Each intermediate reconfiguration step involves the selective release of brakes to define a coupled ECS followed by the use of the motion actuator to adjust one of its joint angles to its target value. This procedure is repeated until all joint angles of the mechanism are finally adjusted to their desired value.

For the coupled ECS mechanism (Fig. 2b) the number of DOF can be calculated from Grübler's formula. Using the notation of [14], the specific system comprises $L = 6$ members (including the links, the slider and the ground), $J_1 = 7$ full joints and $J_2 = 0$ half joints. For a planar system the formula for the number of DOF: $M = 3(L - 1) - 2J_1 - J_2$, yields $M = 1$, i.e., the mechanism is a 1–DOF system. As part of this work the slider is restricted to move in the vertical direction. One alternative for the basic mechanism of the building involves horizontal motion for the slider, as considered in [15]. The vertical slider motion case examined here is particularly interesting and provides distinct design options in an architectural context. Towards reducing the control effort a static counterbalancing system can be implemented to cancel out the gravity forces, as with the idea represented in Fig. 2c.

### 2.3 Motion Planning Considerations

In principle, for the coupled ECS implementation on a system with $m$ physical links (excluding the ground and the slider), a complete reconfiguration of the mechanism will require at least $(m - 3)$ steps. The final step will involve the simultaneous adjustment of the remaining six joint angles along with the position of the slider itself. Obviously, a reconfiguration between a given initial and final position can be realized using alternative control sequences. Feasible sequences are required to consider the motion actuator and joint brakes' capacities, actuation or joint travel limits, as well as collisions with other structures.

Motion planning is also concerned with the avoidance of any singular configurations of the mechanism. Kinematics simulations can be used as part of motion analysis. In general, singularities are known to affect the motion of mechanisms and their control [16,17]. For a basic crank–slider mechanism the fully–extended or fully–retracted positions are singular and the slider may not move any further. Also, any force applied to the slider side cannot turn the crank and when slightly disturbed, the crank is possible to move in either direction. Note that in the case of a coupled crank–slider mechanism, the above type of singularity may occur in one sub–chain or simultaneously in both sides. This is exemplified in Fig. 3 which shows different possible combinations. Given the set of feasible sequences an optimal one can be selected on the basis of defined objectives, as for example minimum reconfiguration times, control effort and required energy. The application of a systematic optimization procedure as part of motion planning is beyond the scope of the present work.

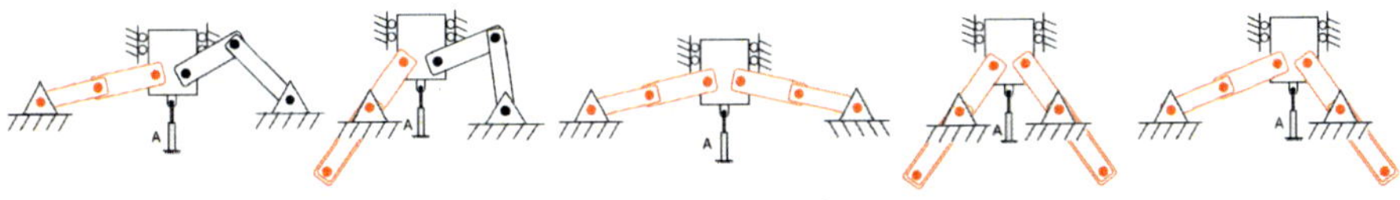

**Fig. 3.** Singular configurations of the coupled crank–slider mechanism.

## 3 Simulation and Experimental Studies

The proposed reconfiguration approach was investigated in simulations prior to experiments concerning a small–scale model shown in Fig. 4.

### 3.1 The Experimental Setup

The system has a sliding block moving in the vertical direction and is actuated by a stepper motor via a linear ball–screw mechanism. The linear rail is attached to a metal supporting structure. Between the slider and the ground are pin–joined two chains of serially–connected links. Electromagnetic brakes are installed on every rotational joint. The brakes are normally–locked and they are released upon application of the control voltage. Their fail–safe characteristic is desirable so that the mechanism does not collapse when the actuators are not active or in the case of a power failure. Optical encoders provide joint position feedback information to the control system, which manages the operation of the brakes and the motion of the slider to implement the control sequences. Each subchain comprises three links and it is pinned to the ground at fixed points. The link lengths are 19.5 cm, the distance between the ground supports is 70.5 cm and the axis of the slider joint crosses the ground at midspan.

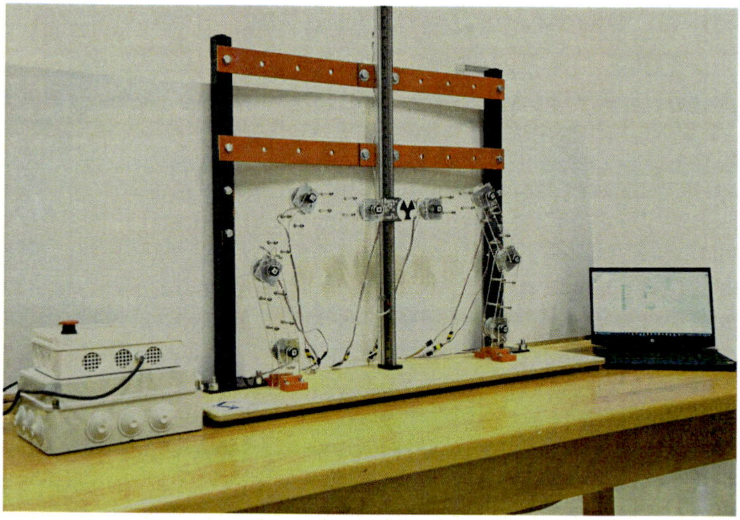

**Fig. 4.** The experimental setup.

### 3.2 Reconfiguration Control Sequences

For demonstration purposes a specific initial (symmetric) and a target (non-symmetric) position were chosen for the reconfiguration of the mechanism, as shown schematically in Fig. 5. Two feasible control sequences were selected for the demonstration, as described in Fig. 5 in the form of scheduling tables for *Sequence I* and *Sequence II*. Such tables provide an efficient way of defining the sequences prior to translating them into actual control commands.

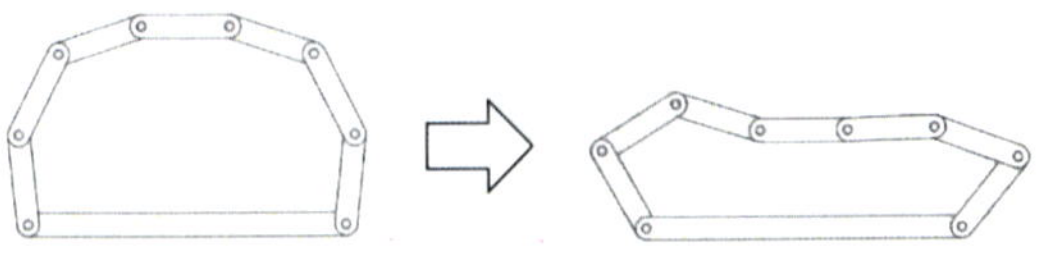

| | Joint 1 | Joint 2 | Joint 3 | Joint 4 | Joint 5 | Joint 6 | Joint 7 | Joint 8 | Joint 9 |
|---|---|---|---|---|---|---|---|---|---|
| Step 1 | △ | ⊙ | ⊙ | ⊗ | ▭ | ⊗ | ⊙ | ⊙ | △ |
| Step 2 | △ | ⊗ | ⊙ | ⊙ | ▭ | ⊗ | ⊙ | ⊙ | △ |
| Step 3 | △ | ⊗ | ⊙ | ⊙ | ▭ | ⊙ | ⊗ | ⊙ | △ |

| | Joint 1 | Joint 2 | Joint 3 | Joint 4 | Joint 5 | Joint 6 | Joint 7 | Joint 8 | Joint 9 |
|---|---|---|---|---|---|---|---|---|---|
| Step 1 | △ | ⊙ | ⊗ | ⊙ | ▭ | ⊗ | ⊙ | ⊙ | △ |
| Step 2 | △ | ⊗ | ⊙ | ⊙ | ▭ | ⊗ | ⊙ | ⊙ | △ |
| Step 3 | △ | ⊗ | ⊙ | ⊙ | ▭ | ⊙ | ⊙ | ⊗ | △ |

**Fig. 5.** Initial and final position for the reconfiguration (top), scheduling tables for Sequence I (middle) and Sequence II (bottom). (⊗: locked joint, ⊙: unlocked joint, △: ground joint. Symbols in red represent the currently adjusted joint(s)).

### 3.3 Simulation and Experimental Study

The simulation and experimental reconfiguration studies were performed and the intermediate steps are shown in Fig. 6 and Fig. 7, for Sequence I and Sequence II, respectively. The direct correspondence between simulation and experimental results can be confirmed upon examining the postures of the mechanism at the individual reconfiguration steps. The recorded motion trajectories for the rotational joints are shown in Fig. 8. At each reconfiguration step one angle is adjusted and from then on it remains locked until the overall reconfiguration is complete, which is the underlying idea of the method. At the final step, the remaining six angles are simultaneously adjusted together with the slider. A fixed linear speed for the slider was used (4 cm/s) and the resulting reconfiguration times are collected in Table 1. In terms of speed of reconfiguration, Sequence II provides a better option. Clearly, the overall reconfiguration time can provide a criterion for the generation of optimal reconfiguration sequences.

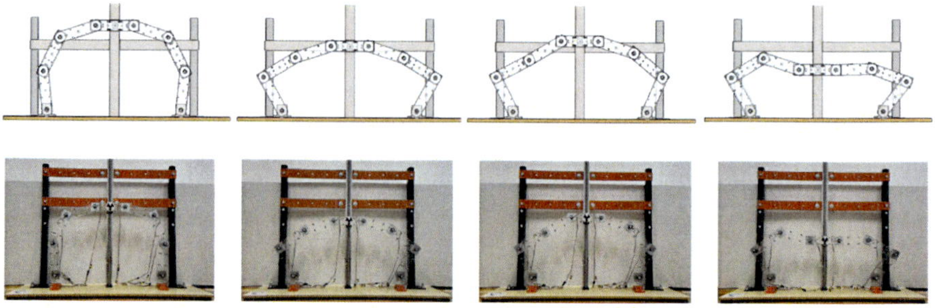

**Fig. 6.** Simulation and corresponding experiment reconfiguration steps for Seq. I.

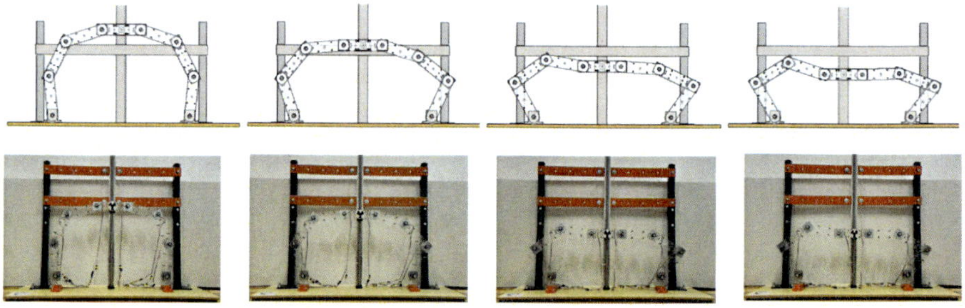

**Fig. 7.** Simulation and corresponding experiment reconfiguration steps for Seq. II.

**Table 1.** Reconfiguration times (sec).

| Sequence | Step 1 | Step 2 | Step 3 | Total |
|---|---|---|---|---|
| Sequence I | 2.5 | 0.8 | 3.71 | 7.01 |
| Sequence II | 1.7 | 3.40 | 0.31 | 5.41 |

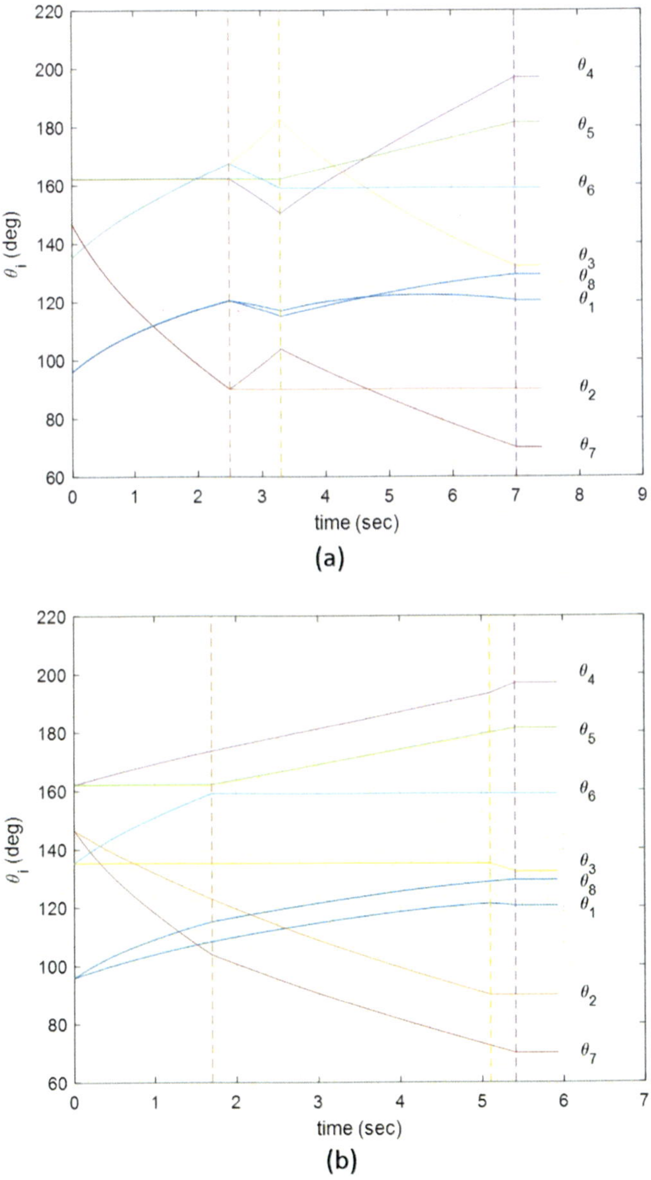


**Fig. 8.** Simulated time histories of rotational joint positions during reconfiguration. (a) Sequence I, (b) Sequence II. The vertical zones correspond to the individual reconfiguration steps.

## 4 Conclusions

The application of robotics in architecture allows for the realization of reconfigurable buildings and creates new possibilities for innovation in design. At the core of these realizations is mechanisms synthesis and control. Towards that direction a structural and control concept has been proposed and its implementation was demonstrated. Simulations were used in order to examine the selected reconfiguration sequences, before performing laboratory–scale experiments. As part of future work, the automatic generation of optimal motion sequences will be examined, while considering limitations including singularities. Extensions towards "adaptive architecture" will also be explored, which refers to automatic shape reconfigurations in response to external conditions including the sun motion.

## References

1. Phocas, M.C., Christoforou, E.G., Theokli, C., Petrou, K.: Reconfigurable linkage structures and photovoltaics integration. Build. Eng. **43** (2021)
2. Li, S., Fang, H., Sadeghi, S., Bhovad, P., Wang, K.W.: Architected origami materials: how folding creates sophisticated mechanical properties. Adv. Mater. **31**(5), 1805282 (2019)
3. Chen, Y., Peng, R., You, Z.: Origami of thick panels. Science **349**(6246), 396–400 (2015)
4. Arun, S.B., Anveeth, B.H., Majumder, A.: Advancements in origami inspired robots, a review. In: IEEE 2nd International Conference on Intelligent Computing, Instrumentation and Control Technologies, Kannur, Kerala, India, pp. 1293–1297 (2019)
5. Sedal, A., Memar, A.H., Liu, T., Menguc, Y., Corson, N.: Design of deployable soft robots through plastic deformation of Kirigami structures. IEEE Robot. Autom. Lett. **5**(2), 2272–2279 (2020)
6. Brunete, A., Ranganath, A., Segovia, S., Perez de Frutos, J., Hernando, M., Gambao, E.: Current trends in reconfigurable modular robots design. Int. J. Adv. Robot. Syst. **14**(3), 1–21 (2017)
7. Qiao, G., Song, G., Wang, W., Zhang, Y., Wang, Y.: Design and implementation of a modular self-reconfigurable robot. Int. J. Adv. Robot. Syst. **11**(3), 1–12 (2014)
8. Feczko, J., Manka, M., Krol, P., Giergiel, M., Uhl, T., Pietrzyk, A.: Review of the modular self reconfigurable robotic systems. In: IEEE 10th International Workshop on Robot Motion and Control, Poznan, Poland, pp. 182–187 (2015)
9. Hedayati, H., Suzuki, R., Rees, W., Leithinger, D., Szafir, D.: Designing expandable-structure robots for human-robot interaction. Front. Robot. AI **9**(719639), 1–22 (2022)
10. Liu, J., Zhang, X., Hao, G.: Survey on research and development of reconfigurable modular robots. Adv. Mech. Eng. **8**(8), 1–21 (2016)
11. Christoforou, E.G., Müller, A., Phocas, M.C., Matheou, M., Arnos, S.: Design and control concept for reconfigurable architecture. J. Mech. Des. **137**(4), 042302-1–042302-8 (2015)
12. Phocas, M.C., Christoforou, E.G., Matheou, M.: Design and motion planning of a reconfigurable hybrid structure. Eng. Struct. **101**(29), 376–85 (2015)
13. Christoforou, E.G., Phocas, M.C., Matheou, M., Müller, A.: Experimental implementation of the 'effective 4-bar method' on a reconfigurable articulated structure. Structures **20** (2019)
14. Norton, R.: Design of Machinery. McGraw-Hill, New York (2008)

15. Christoforou, E.G., Georgiou, L., Phocas, M.C., Louca, L.S., Müller, A.: A robotics perspective on architecture: Modelling and control of reconfigurable buildings. In: Müller, A., Brandstötter, M. (eds.) RAAD 2022. Mechanisms and Machine Science, vol. 120, pp. 290–297. Springer, CHam (2022). https://doi.org/10.1007/978-3-031-04870-8_34
16. Park, F.C., Kim, J.W.: Singularity analysis of closed kinematic chains. Mech. Des. **121**(1), 32–38 (1999)
17. Zlatanov, D., Fenton, R.G., Benhabib, B.: Identification and classification of the singular configurations of mechanisms. Mech. Mach. Theory **36**(6), 743–760 (1998)

# Author Index





T. Petrič et al. (Eds.): RAAD 2023, MMS 135, pp. 497–499, 2023.
https://doi.org/10.1007/978-3-031-32606-6